D0850112

Safety and Health Management Planning

James P. Kohn
Theodore S. Ferry

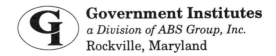

Government Institutes
a Division of ABS Group, Inc.
Rockville, Maryland

Government Institutes, a division of ABS Group Inc.
4 Research Place, Rockville, Maryland 20850, USA.

Copyright © 1999 by Government Institutes. All rights reserved.

03   02   01   00   99   5   4   3   2   1

No part of this work may be reproduced or transmitted in any form or by any means, electronic or mechanical, including photocopying, recording, or the use of any information storage and retrieval system, without permission in writing from the publisher. All requests for permission to reproduce material from this work should be directed to Government Institutes, 4 Research Place, Suite 200, Rockville, Maryland 20850, USA.

The reader should not rely on this publication to address specific questions that apply to a particular set of facts. The author and publisher make no representation or warranty, express or implied, as to the completeness, correctness, or utility of the information in this publication. In addition, the author and publisher assume no liability of any kind whatsoever resulting from the use of or reliance upon the contents of this book.

Library of Congress Cataloging-in-Publication Data

Safety and health management planning for the 21st century / [edited] by James P. Kohn,
   Theodore S. Ferry.
      p.   cm.
   Includes bibliographical references and index.
   ISBN   0-86587-634-7
      1. Industrial safety—Management. 2. Industrial safety—Planning.
   3. Industrial hygiene—Management. 4. Industrial hygiene—Planning.
   I. Kohn, James P.   II. Ferry, Theodore S.
   T55.S197 1998
   658.3'82—dc21   98-44129
                                                       CIP

"Workers' Compensation and Loss Control" by JoAnn Sullivan. Copyright ©1998 by JoAnn Sullivan. Reprinted by permission of author. "Multinational Organizations" by Kathy A. Seabrook. Copyright ©1998 by Kathy A. Seabrook. Reprinted by permission of author. "Safety Cultures and the Behavior-Based Model" by Thomas R. Krause, John H. Hidley, and Stanley J. Hodson. Copyright ©1998 by Behavioral Science Technology, Inc. Reprinted by permission of author. "Safety in Design: A Simple Perspective for the Designer of Machinery" by Robert N. Andres. Copyright ©1998 by Robert N. Andres. Reprinted by permission of author. "Safety in the Project Design Process" by Steven F. Kane. Copyright ©1998 by Steven F. Kane. Reprinted by permission of author. "System Safety Engineering" by Henry Walters. Copyright ©1998 by Behavioral Science Technology, Inc. Reprinted by permission of author.

Printed in the United States of America

Dedicated to
Ted Ferry
for his many important contributions
to the field of safety management

Summary Table of Contents

Table of Contents

About the Authors

James P. Kohn, CSP, CIH, CPE, is a professor of Industrial Technology at East Carolina University. He was previously director of the Occupational Safety Management program at Indiana State University. Dr. Kohn holds a doctorate in education from West Virginia University. He has both academic and industrial experience, including having worked as a safety, health, and environmental consultant in the telecommunications industry and as a corporate supervisor of safety and health training in the electrical utility industry.

Dr. Kohn has written six books on the topics of safety and industrial hygiene. He has also published numerous articles covering ergonomics, safety, industrial hygiene, and training. In addition, Dr. Kohn has presented numerous papers at professional conferences and has conducted training workshops both in industry and the public sector.

Dr. Kohn is currently serving as a regional vice president and member of the board of directors of the American Society of Safety Engineers. He is also a member of the American Industrial Hygiene Association and the American Conference of Governmental Industrial Hygienists. He currently serves as a member of the board of directors of the Eastern Carolina Safety and Health School.

Robert N. Andres, CSP, CPE, CMfgE, is Vice President of Oshex Associates, Inc., of Baldwinsville, New York, a company specializing in machine and facility safety, which he founded in 1974. He is also a member of the American Society of Safety Engineers, the Association of Facility Engineering, and the Society of Manufacturing Engineers. He is also Board Certified by the American College of Forensic Examiners, serves on the ANSI B11.8 and B11.20 subcommittees and chairs the ANSI B11.TR3 Risk Assessment and Reduction subcommittee. He was named ASSE Engineering Division Safety Professional of the Year in 1996.

Arthur F. Billington II, CSP, is a Senior Risk Control Consultant for Sumitomo Marine Management. His specialization is the advancement and incorporation of safety management practices in Japanese owned and operated firms in North America. Previously he was a Vice President of CHUBB Services Corporation. He currently is a Director of the National Safety Management Society and a Professional member of the American Society of Safety Engineers.

Don Bloswick, Ph.D., CPE, P.E., is an Associate Professor of Mechanical Engineering at the University of Utah in Salt Lake City, Utah. He is a member of the American Society of Mechanical Engineers, the Human Factors and Ergonomics Society, and the American Society of Safety Engineers, as well as a contributor to *Sourcebook of Occupational Rehabilitation*, published by Plenum Press, *Ergonomic Process Management*, published by CRC Press, *Health Care and Rehabilitation Ergonomics*, published by Butterworth-Heinemann, among other publications.

Jerry L. Burke, Ph.D., is President of the Burk Group, a performance and development consulting organization with more than thirty years of experience with organizations across the United States. His background and experience includes organization assessment, training for development, and consulting with responsibility for the design and implementation of strategies in organizations including health care, service, petroleum, metals, engineering, utilities, and aerospace. He has been directly involved with systems-oriented safety and health development efforts that span the executive, managerial, and individual contributor levels. Publications include contributions to three books of readings, journals, and magazine articles in support of safe performance development.

Sharon Lynn Campbell, CSP, is President of SLC Communications in Fairless Mills, Pennsylvania, and was previously an Environmental Safety and Health Coordinator at Columbia University. She is a member of the National Safety Council, the American Congress of Governmental Industrial Hygienists, and has published articles in safety and health publications.

J. Brett Carruthers, CSP, is Senior Risk Management Consultant for Wright Risk Management Company, Inc., in Niagara Falls, New York. Previously he was President of the BC Group and Safety Manager of the Occidental Chemical Corporation. He is also a member of the American Society of Safety Engineers and the National Safety Council.

Jack E. Daugherty, CIH, CHMM, P.E., is Owner and Principal of AEGIS, an environmental, health and safety compliance engineering and management service in Jackson, Mississippi. Previously, he was the Environmental and Safety Engineer for Vickers, Incorporated, a manufacturing company also in Jackson. He is a professional member of the American Society of Safety Engineers, the American Industrial Hygiene Association, and a chapter president of the American Academy of Certified Hazardous Materials Managers. He is the author of *Industrial Environmental Management* and *Industrial Safety Management*, both published by Government Institutes, Inc., and his articles have appeared in several trade journals including *Occupational Hazards* and *Industrial Safety and Hygiene News*.

Donald J. Eckenfelder, CSP, P.E., is the principal consultant with Profit Protection Consultants, a firm offering management and loss prevention consulting services in Glen Falls, New York. He was Corporate Director of Loss Prevention for Chesebrough-Ponds for 13 years and has been President of the American Society of Safety Engineers and the American Society of Safety Research. He is a Fellow of the American Society of Safety Engineers. He has taught graduate courses at several institutions and is the author of *Values-Driven Safety*, published by Government Institutes, Inc.

Mark Hansen, CSP, CPE, P.E., is Manager Safety Services for Petrochemicals, a subsidiary of ARCO, located in Houston, Texas. Mr. Hansen has over sixteen years experience in the areas of safety, health, environment, and ergonomics and is currently the Vice President of Finance on the Board of Directors of the American Society of Safety Engineers. He is a widely published author with over 100 publications in various magazines and journals. He is currently the author of the Computer Technology column of ASSE's journal *Professional Safety*. He is ASSE's 1992-1993 recipient of the Edgar Monsanto-Queeny Safety Professional of the Year Award and the 1991-1992 recipient of the Charles V. Culbertson Outstanding Volunteer Service Award.

John H. Hidely, MD, is co-founder of Behavioral Science Technology, Inc. He is member of the American Society of Safety Engineers. His publications include "Critical success factors for behavior-based safety—Avoiding common pitfalls and achieving real gains" in *Professional Safety* (July 1998).

Stanley J. Hodson is managing editor and trainer with Behavioral Science Technology, Inc. He is a member of the American Society of Safety Engineers. His publications include: "Behavior-based training for safety & health" in *Occupational Hazards* (September 1998) and "Behavior-based safety applied for environmental concerns at Celanose Clear Lake Plant" in *The Synergist* (July 1998).

Steven F. Kane, CSP, P.E., is a Systems Safety Engineer for the Relativistic Heavy Iron Collider (RHIC) Project for Brookhaven National Laboratory in Upton, New York. He is a member of the American Society of Safety Engineers, the American Welding Society, the Society of Automotive Engineers, the American Society for Materials International, and the National Association of Professional Accident Reconstruction Specialists. He has published numerous times for the International System Safety Conference, the American Society of Safety Engineers, and the American Welding Society. Mr. Kane was selected the 1995 Edgar Monsato Queeny National Safety Professional of the Year by the American Society of Safety Engineers.

Thomas R. Krause, Ph.D., is Chief Executive Officer of Behavioral Science Technology, Inc., an industrial safety consulting firm he co-founded in 1980 in Ojai, California. He is member of the American Society of Safety Engineers, the National Safety Council, and of the American Psychological Association. He is the author of two books: *Employee-Driven Systems for Safe Behavior* and *The Behavior-Based Safety Process*, both published by Van Nostrand Reinhold, and he has published more than 30 articles on the subject of behavior-based safety and is a regular speaker at the National Safety Council and ASSE conferences.

Richard W. Lack, P.E., CSP, CHCM, CPP, operates Safety and Protection Services Consulting in Cheyenne, Wyoming. Born in London, England, Mr. Lack was in the Royal Air Force before beginning a long career in the mining and chemical industries in Jamaica and the United States. He was then a Corporate Risk Control Engineer for Castle and Cooke in California, and later the Safety Officer at the San Francisco International Airport. Mr. Lack is a member of the American Society of Safety Engineers, the American Society for Industrial Security, the National Safety Council, the Institution of Occupational Safety and Health (UK), the International Institute of Risk and Safety Management (UK), and the International Commission on Occupational Health, among others. He is a regular speaker, author of numerous papers, and contributor to *Essentials of Safety and Health Management*, published by CRC Press, and *Security Management*, published by the National Safety Council.

John C. Myre has been Publisher of *Safety Times,* a product of Safety Times, Inc., of Chesterfield, Missouri, for the past six years. Formerly he was Director–Risk Management at Southwestern Bell, a company with which he was affiliated for 34 years. He is an Associate in Risk Management (ARM) and has published articles in *Professional Safety* and *Business Insurance.*

O. Dan Nwaelele, MSc, CSP, has been President of Northwest ESH, Inc., for the past seven years. His is also Safety Engineer of King County, Washington, and is a member of the American Association of Safety Engineers and the American Institute of Chemical Engineers. He is author of Health and Safety Risk Management and Your Company's Safety and Health Manual, both published by Government Institutes, Inc.

F. David Pierce, MSPH, CIH, CSP, CIH, is Vice-President of Leadership Solution Consultants, Inc., in Salt Lake City, Utah. He is also Vice-President of Alliance for Training, Inc., and a member of the American Association of Safety Engineers and the American Industrial Hygiene Association. He has published numerous books with Government Institutes, Inc., including *Total Quality for Safety and Health Professionals*, *Shifting Safety and Health Paradigms*, *Managing Change for Safety and Health Professionals: A 6 Step Process*, and *Project Management for Environmental Safety and Health Professionals.*

Kathy Seabrook, CSP, RSP, is President and Principal Consultant for Global Solutions, Inc., which specializes in global health and safety management. Previously she was UK Manager for Chubb Insurance Company of Europe, and has served as National and Senior Consultant and National Training and Marketing Coordinator for Chubb & Son, Inc., World Head Quarters in New Jersey. She is a member of the American Society of Safety Engineers, the National Safety Council, and the Institution of Occupational Safety and Health (UK), and was the American Society of Safety Engineers' 1998 International Division Safety Professional of the Year and the 1994 President's Award winner for the American Society of Safety Engineers.

Richard Sesek, MPH, CSP, is a Research Associate with the University of Utah, Department of Mechanical Engineering, in Salt Lake City, Utah. He has also been a Safety Engineer/Ergonomist with the Michelin Tire Corporation and is a member of the American Society of Safety Engineers, the Human Factors and Ergonomics Society, and the American Society of Mechanical Engineers.

Robert D. Soule, CSP, CIH, P.E., is chairman, graduate coordinator, and professor of occupational health in the Safety Science Department at Indiana University of Pennsylvania. He is a member of the American Industrial Hygiene Association, the American Conference of Governmental Industrial Hygienists, and the Industrial Hygiene Roundtable, and a member of the editorial board of *Occupational Hazards* magazine.

Paul G. Specht, Ph.D., CSP, is a Professor of Occupational Safety and Hygiene at Millersville University in Lancaster County, Pennsylvania. He is a professional member of the American Society of Safety Engineers and the American Conference of Governmental Industrial Hygienists. In addition to over twenty years experience in safety education, he served as Vice President of Allied Safety and Health Associates for five years. Dr. Specht has written extensively on accreditation of safety and industrial hygiene curricula. He has also published articles on product safety and legal aspects of safety in *Manufacturing Forum* and *Professional Safety*.

Jeffrey O. Stull, MSChE, MSEngMgt, is President of International Personnel Protection, Inc., of Austin, Texas, a consulting company in the area of personal protective equipment. He has also been past President of TRI/Environmental and Program Manager of Texas Research Institute. Mr. Stull is a member of the American Industrial Hygiene Association, the American

Society for Testing and Materials, the National Fire Protection Association, and the International Standards Organization, as well as author of *PPE Made Easy*, published by Government Institutes, Inc., in 1998.

JoAnn M. Sullivan, CSP, is Vice President of Risk Control Consulting, J&H Marsh & McLennan, Inc., in Phoenix, Arizona. She was Vice President and Manager of Technical Services for American Risk Services for 7 years and is a member of the American Society of Safety Engineers and the Society for Human Resource Management.

Ron Teichman, MD, MPH, FACP, FACOEM, is Medical Director and Administrator for Worksite Partners/CareAlliance Health Services, a charitable healthcare delivery service based in Charleston, South Carolina. Currently he is also President of Teichman Occupational Health Associates, Inc, and was a former Regional Medical Director at General Motors Corporation. He is a fellow of the American College of Occupational and Environmental Medicine and the American College of Physicians, a member of the American Society of Safety Engineers and the American Medical Association, and an affiliate faculty member of the Columbia University School of Public Health and the Medical University of South Carolina. Dr. Teichman serves as an editor and has published numerous articles in the *Journal of Occupational Medicine* and other public health and occupational medicine periodicals.

Anthony Veltri is an associate professor of environment, safety and health studies at Oregon State University. Dr. Veltri's research is focused on strategy formulation and organizational structuring of environmental, safety and health functions within business and industry. His current research is aimed at profiling the cost burdens and profitability potential of environment, safety and health activities. Dr. Veltri has consulted for semiconductor manufacturing companies, aerospace companies, electric/gas utilities, petroleum companies, steel manufacturers, coal mining operations, and the Environmental Protection Agency. He is a graduate of West Virginia University.

Henry A. Walters, Ph.D., is Senior Consultant at Behavioral Science Technology in Ojai, California. He has also been an officer in the U.S. Army and assistant professor of Occupational Safety and Health at Murray State and Indiana State Universities, from 1988 to 1996. He is a member of the American Society of Safety Engineers and the National Safety Council. He is also author of *Statistical Tools of Safety Management*, published by Van Nostrand Reinhold in 1994.

Acknowledgments

The author wishes to thank all of his professional colleagues who contributed to the development and editing of this manuscript. This project would not have been possible were it not for the hard work and materials provided by the contributing authors. Special thanks goes to Wendy Peoples and Leslie Ridings for editing and organizing this manuscript. Special thanks also goes to Mr. Russ Bahorsky and Mr. Alex Padro who helped in the organization, contact, and coordination of this project.

Several East Carolina University graduate students helped research information for my contributions to this book. Occupational Safety graduate students who contributed to this project included: Eddie Allen, Chris Austin, James Braswell, Lee Caulder, Steve Crooks, Ken Jordan, Jennifer Lewis, Manny Lourenco, Kathryn Pacha, and Mark VanDam.

I would also like to thank my wife, Carrie Kohn, M.A., for her assistance, patience, and support during the development of this manuscript.

Safety and Health Management Planning

Chapter 1

Introduction to Workplace Safety and Health Management Planning

James P. Kohn

INTRODUCTION

Safety management is like any other management function in an organization. To be effective, safety management must meet its mission of preserving the people, property, and environmental resources within the organization. It accomplishes this mission by establishing goals and objectives, and by implementing activities necessary to achieve the mission.

And like other management systems in the organization, safety management uses personnel and resources to aid in successfully achieving the established goals and objectives of the organization. To ensure that it will be successful at achieving the safety mission, the safety professionals responsible for this assignment must be effective at developing and implementing safety, health, and environmental plans.

The purpose of this chapter is to present the components of workplace safety and health management planning. This chapter serves as an introduction to subsequent chapters, which will examine components of occupational safety and health management in greater depth. The addendum, written by Mark Friend and found at the end of this chapter, presents a history of management theory that focuses on the concepts associated with planning and their application to safety management.

CASE STUDY:
THE NEED FOR A MANAGEMENT PROGRAM

Elizabeth had been with a North Carolina company for less than three months and was still "learning the ropes" of her new position as safety manager for a two hundred-employee manufacturing operation. This was her first occupational safety and health position following graduation from a Masters of Science graduate-degree program in occupational safety management. As far as safety was concerned at this facility, she was *it*.

She was fortunate to have the support of the owner and the president of the company as well as other key members of the management team. The human resources manager was one of her key supporters. He had a background in workers' compensation, but knew little else about safety. He was aware that Occupational Safety and Health Administration (OSHA) regulations existed and that the company was experiencing a considerable number of OSHA-recordable injuries.

Elizabeth's first three months had been spent "learning the business" of the company. She became familiar with the employees and the production processes. She also reviewed the OSHA 200 logs for several previous years and tried to determine where the company was experiencing problems. She also looked over all of the "safety files," attempting to understand what the company had accomplished from a safety and health perspective prior to her arrival.

Several things became very clear to her. Both the employees and most of the managers recognized the importance of a good safety and health program. Management grasped that workers' compensation costs were affecting the company's ability to make a profit. The employees did not understand the financial issues associated with health and safety, but they expressed concern that some of the work they did was dangerous.

Elizabeth was relieved to discover that labor and management had a positive relationship. She knew that poor labor-management relations made establishing good safety programs extremely difficult, if not impossible. However, the employees still felt that safety and health issues were often ignored when production concerns were raised.

It soon became apparent to Elizabeth that there was very little in the way of written safety programs, policies, goals, or objectives. She also realized that the company was in trouble with its workers' compensation insurance carrier. Having an experience modification factor of more than 1.5 indicated that trouble was brewing. She noticed that little task or safety training had been conducted. In addition, the OSHA 200 log suggested that women and some of the Hispanic workers were involved in an excessive number of accidents.

There was so much that Elizabeth recognized had to be accomplished that she felt overwhelmed. Elizabeth decided to start with the basics. She had already completed a thorough analysis of where the program presently stood. Her next task was to put together a plan and try to get the organization moving toward accomplishing that plan.

She set up a meeting with the company's president and the human resources director to share her thoughts about the strengths and weaknesses of the company from both a safety and a general management perspective. She talked with them about what steps should be taken to move forward. The president liked her approach, but he was concerned that other managers might not buy into the program. He turned to his associate and said, "Why don't you work with Elizabeth and the shift supervisors and develop a blueprint for kicking off this strategic planning project. I want to know how our company can reduce costs and improve its productivity and service to our customers. I want our company to be the best company in our industry, and I want everyone that works here to be proud to be a member of the team." When Elizabeth walked out of that meeting, she knew her real work in occupational safety and health had just begun.

MANAGEMENT PRINCIPLES AND THEORIES

Built to Last, a book by James Collins and Jerry Porras, compared eighteen world-class companies with less successful companies doing the same type of business to determine what caused differences among these companies. By comparing highly successful companies with less successful ones, the authors attempted to identify the reason for the profound success of certain companies. They found that highly successful companies had goals that extended far beyond simply making money or meeting production quotas. Goals often were built on the company's foundation of core values, which were considered extremely important. Whether these goals took the form of social missions or major organizational challenges, the authors pointed to the need for goals to exceed profits. Specific strategies for achieving the goals were found to be equally important for extremely successful companies. Management striving for socially sound and ideal goals seemed to be the formula for organizational success. How does a company pursue this formidable task?

According to the *Merriam-Webster Dictionary* (1998), the word *management* has been in use since 1598. It means "(1) the act or art of managing: the conducting or supervising of something (as a business), or (2) the judicious use of means to accomplish an end." Of course, management also requires the identification of organizational resources required for manufacturing products or providing services. Management, from a business perspective, is "the process by which people, technology, job tasks, and other resources are combined and coordinated so as to effectively achieve organizational objectives" (Ivancevich, Donnelly, and Gibson 1980). The fundamental principle of management is clear, based on the definitions provided by all of these references. Management is the process used by an organization to accomplish successfully its goals through the efficient use and effective coordination of its resources.

Although theories about organizational structure and management have existed for over five thousand years, modern management theory is commonly believed to have evolved from the start of the industrial revolution in eighteenth-century England. Charles Babbage, one of the first to study management scientifically, wrote extensively on the concept of division of labor. Individuals such as Frederick Taylor and Frank Gilbreth examined issues of efficiency and performing a job "the best way." Today, individuals such as Charles Deming and his quality management disciples are addressing process issues that affect organizational productivity and efficiency (Walton 1986).

Interestingly, students of management have observed four philosophical shifts in management theory since Babbage, Taylor, and Gilbreth first studied how people worked (Rimer 1993). The early theories studied *management by doing*. The doing phase of management was reflected by the original craftsmen who worked together with the individuals they managed. The second shift of management theory is *management by direction*. The directing phase saw the evolution of a hierarchical management structure. The hierarchical organizations provided a structure through which managers were promoted to higher levels in the organization and therefore had more responsibility, including directing larger numbers of individuals. This was modeled after the military style of workforce management. The third phase of management theory is *management by results*. This theory includes the *management by objectives* approach, in which superiors and subordinates negotiate specific work goals that help an organization achieve its goals and objectives. The fourth and recent generation of management theory is *management by methods*. Management by methods applies concepts that emphasize how organizations operate as a system. This approach studies the effects on system processes and not necessarily on the quantity of product and service itself (Rimer 1993).

Safety and health management principles and theories have mirrored the changes observed in many of their sister disciplines—especially management. An understanding of management theory and how it has evolved over the past several centuries can provide the foundation that safety and health managers must build on if they are going to be effective. Safety and health professionals often report to an individual with a BBA or MBA. Often, plant managers or corporate executive officers are individuals who are oriented in business, accounting, or finance. In other words, they are individuals trained in management theory and practice.

To be successful, safety and health professionals must be competent in health and safety theory and practice and in safety management knowledge and skills. Their competency should include the technical and managerial skills associated with the occupational safety and health profession. Perhaps even more important, however, safety and health professionals must effectively communicate to their superiors in terms that business-oriented individuals are familiar with and that are used on a daily basis. This is the challenge that confronts health and safety professionals. To determine how best to implement an effective safety program requires answers to numerous questions including

- How do I establish a safety and health management system?
- How can I effectively conserve organizational resources by establishing a quality occupational safety, health, and environment program?
- How do I obtain top management support?
- How do I get employee buy-in for my programs?
- How do I set up the process so that roles and responsibilities are clearly delineated and organizational effectiveness is clearly documented?
- How do I keep the program on track and ensure that it performs according to objectives?

Answers to these questions are at the core of the challenge for safety and health professionals. When establishing a safety and health management system, professionals must turn to the founders of management theory. By applying the universal elements of management already developed, safety professionals can establish an effective program in their own organizations. Elements of management, which were first identified by Henri Fayol more than one hundred years ago, can be effectively used today. Fayol's position was that no matter what type of organization was in question, his principles served as guidelines for effective management. One of his most important principles was the concept of management process. Management process examines the highly important responsibilities decisionmaking, and controlling. These are some of the critical components of "solid management."

IMPORTANCE OF PLANNING

For today's safety and health professionals, the starting point to an effective program must be the first part of Fayol's management process: planning. The term *planning* has been defined as, "the act or process of making or carrying out plans— specifically: the establishment of goals, policies, and procedures" (*Merriam-Webster Dictionary* 1998). From a management perspective, planning is a process that requires an assessment of the organization and its resources in the environment in which it functions, and leads to the establishment of appropriate goals and objectives. It is a method of moving the organization into the future. Drucker (1954) said, "The only thing we know about the future is that it's going to be different." Change is a constant factor in today's society and organizations. Solid management and effective safety and health management require objective and accurate assessments of an organization's characteristics and its ability to plan for an uncertain future.

Is safety and health management planning different from planning by other functions in the organization? The answer is an emphatic "No!" The process itself is exactly the same no matter which function in the organization is being discussed. As a matter of fact, to ensure safety and health function success and support in an organization, safety professionals must ensure that function goals help achieve organizational goals. Safety and health planning is the vehicle required to achieve those organizational goals.

What is safety and health planning? It is the process of determining how safety, health, and environmental goals can be met. This process involves evaluating the company and its political, legal, financial, and market environments. It also entails establishing objectives and determining exactly what will be done to accomplish those objectives. This means someone must determine and evaluate alternative courses of action and select a course that is feasible and proactive in the process. A timetable is established and available personnel and resources are applied to the process.

"To-Do-List" Planning

Carroll (1993) identified three types of management planning: "to-do-list," tactical or operational, and strategic. "To-do-list" planning refers to the short-term, narrowly focused planning that may take place on a personal level. This type of planning might include listing specific tasks such as hazard analysis projects, periodic injury frequency and severity reports, safety staff meetings, or staff professional development goals that must be accomplished on a daily, weekly, or monthly basis. Many individuals are familiar with the benefits of establishing things-to-do lists. This activity becomes an effective time-management tool that ensures the professional will complete all responsibilities in a timely manner. The importance of proactive performance and goal setting have been emphasized in numerous publications that have examined strategies for increasing the likelihood of personal management (Covey 1989), personnel management (Daniels 1989; Locke and Latham 1984), and safety management success (Petersen 1989).

The to-do-list classification of planning permits managers to prioritize the order in which tasks must be completed, determine how long a task will take, establish deadlines for task completion, and determine whether delegation of specific components of the task will be required. This strategy helps to get things done on both an individual and organizational level.

Operational Planning

Operational planning, sometimes referred to as administrative or tactical planning, permits the accomplishment of midrange time-related tasks. This type of planning can be performed for tasks and activities that must be accomplished within the next month to the next year. As Carroll (1993) points out, "Whether it be driven by an organization's budget, a personal budget or a functional area of responsibility, operational planning focuses on getting the work accomplished effectively between now and some limited time period." Typically, operational planning is a response to strategic planning, which is the process that attempts to specify long-term goals. Operational planning is the administrative management process that focuses on short-term problems. Objectives established as a result of operational plans are very specific for the unit or units of interest. Results under operational planning tend to be quantifiable, and there are minimal levels of risk associated with these plans.

For example, operational planning for safety and health professionals could include establishing a process to implement successfully a corporate-wide personal protective equipment (PPE) program that complies with Subpart I of the Occupational Safety and Health Act (29 CFR 1910.132). The components of an effective PPE program would be identified and the personnel and resources necessary to implement this program would be established over a midrange time period. In this example, safety and health professionals may wish to use the things-to-do planning approach—listing the specific activities that must be completed. They could then delegate a hazard and risk assessment of the facility to one individual, delegate the drafting of the written program to another individual, contact safety equipment suppliers to provide a comprehensive list of appropriate equipment to another individual, and delegate the development and piloting of the training to another staff member. In this example, safety and health professionals would be functioning as managers by planning, organizing or coordinating, leading, and controlling the project. Decisionmaking could be delegated to the lowest level appropriate in the process. It is also important to point out that this operational plan could and should be tied into the long- term strategic plan of the organization. To continue this example, PPE programs could be an interim step in the process of redesigning work operations to eliminate eventually the need for PPE and increase productivity. The strategic plan of the organization should provide an ideal goal to strive for with the operational plans assisting the organization in moving toward those goals.

Strategic Planning

But what exactly is strategic planning? Strategic planning is the third and final type of organizational planning. Strategic planning involves the development of a vision for the entire organization within its environment. It includes all the activities that lead to the definition of goals and to the determination of appropriate objectives to achieve those goals. The benefits of strategic planning are numerous:

- Forces managers to think ahead and to avoid being reactive and simply responding to crisis situations
- Leads to the development of performance standards that relate directly to operational plans
- Forces management to articulate clear objectives
- Helps prepare the organization for sudden changes

There are four conditions in the environment that can impact organizations, and the strategic plan must address each of these areas: economic; technological; cultural and social; and political, legal, and regulatory. These are the same conditions that can impact the safety, health, and environmental function as well. By establishing strategic plans to try to address these conditions, the organization is also establishing long-range plans to deal with the uncertainty of its future.

As was previously mentioned, Drucker (1954) recognized the ever-changing future, and he emphasized the importance of strategic planning to prepare organizations for those changes. Strategic planning addresses long-term survival. Its objectives are typically broad and wide-ranging for the entire organization. There is no one correct strategy—typically numerous options may exist. The problem with this area of planning is that it is less quantifiable than operations and administrative management planning.

Frequently a SWOT analysis is used for strategic planning. This method of planning examines an organization's **S**trengths and **W**eaknesses, as well as the environmental condition's **O**pportunities and **T**hreats. The strategic planning process involves establishing organizational missions; developing organizational goals; conducting situational analysis; developing organizational strategies; completing strategic plans for the entire organization, division, or unit; and developing administrative plans, objectives, and activities. Figure 1-1 describes the relationship of three types of planning to the planning process.

The starting point for operational safety and health management planning is the identification of the key elements of the process. In 1989 OSHA developed safety and health management program guidelines. These guidelines are comprehensive and should serve as the elements professionals use to initiate operational planning activities.

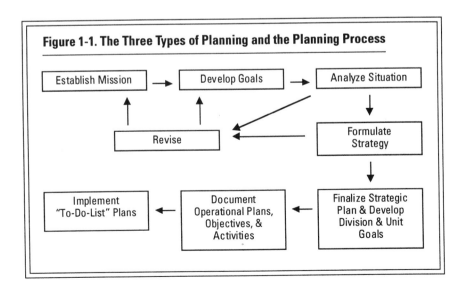

Figure 1-1. The Three Types of Planning and the Planning Process

Establish Mission → Develop Goals → Analyze Situation

Revise

Formulate Strategy

Implement "To-Do-List" Plans ← Document Operational Plans, Objectives, & Activities ← Finalize Strategic Plan & Develop Division & Unit Goals

KEY ELEMENTS OF A SAFETY AND HEALTH MANAGEMENT PROGRAM

The safety and health management program guidelines recommend that safety professionals address safety and health hazards in the industrial environment by including, in the management program, four major program elements: management commitment and employee involvement, work-site analysis, hazard prevention and control, and training and education. According to the safety and health management program guidelines:

1) Management commitment and employee involvement are complementary. Management commitment provides the motivating force and the resources for organizing and controlling activities within an organization. In an effective program, management regards workers' safety and health as a fundamental value of the organization and applies its commitment to safety and health protection with as much vigor as to other organizational purposes. Employee involvement provides the means through which workers develop and/or express their own commitment to safety and health protection, for themselves and for their fellow workers.

2) Worksite analysis involves a variety of worksite examinations, to identify not only existing hazards but also conditions and operations in which changes might occur to create hazards. Unawareness of a hazard that stems from failure to examine the work site is a sure sign that safety and health policies and/or practices are ineffective. Effective management actively analyzes the work and worksite, to anticipate and prevent harmful occurrences.

3) Hazard prevention and controls are triggered by a determination that a hazard or potential hazard exists. Where feasible, hazards are prevented by effective design of the job site or job. Where it is not feasible to eliminate them, they are controlled to prevent unsafe and unhealthful exposure. Elimination or controls is accomplished in a timely manner, once a hazard or potential hazard is recognized.

4) Safety and health training addresses the safety and health responsibilities of all personnel concerned with the site, whether salaried or hourly. If is often most effective when incorporated into other training about performance requirements and job practices. Its complexity depends on the size and complexity of the worksite, and the nature of the hazards and potential hazards at the site. (DOL 1989)

SAFETY AND HEALTH MANAGEMENT STRATEGIC PLANNING

At the highest level of the planning process, safety and health professional must establish strategic plans that interface and help achieve the organization's strategic plan. The purpose of this book is to provide readers with information that covers all the elements required for establishing these strategic plans.

In the Chapter 1 addendum, this book begins examining the planning process by giving a historical review of management theory and planning. This review is followed by an examination of the wide variety of professionals from a broad range of fields who play a role in the protection of worker health and safety as well as the conservation of organizational resources in the occupational environment. This book then studies management's role in an effective safety and health program. The safety performance of an organization depends on the resolve and attitude management has toward loss prevention. Methods for improving management attitude toward safety are discussed in Chapter 3. The use of management tools such as safety committees and audit activities is examined in later chapters.

Workers compensation is a unique system of laws and insurance designed to provide medical treatment and a portion of income to workers as a result of job-related injuries or illnesses. This book presents the reader with a detailed examination of workers' compensation and the broad field of responsibilities associated with risk management.

Since employees are an organization's most valuable resource, it is important to understand the people that make up that organization. This book studies those employees in light of the changes that have taken place in the American workforce. The composition of the American workforce has changed dramatically over the past twenty years, and those trends and the likely composition of the American workforce as it enters the twenty-first century are reviewed. In addition to the employees that make up that workforce, issues associated with staffing and developing safety and health professionals are examined. Establishing a capable loss prevention/loss control function with the appropriate structure to have an optimal impact within the organization is also discussed. And as topics of interests, behavioral and cultural issues associated with employees and organizational structure are also examined.

CONCLUSION

Not only is the American workforce changing for companies across the country, but how companies do business and the marketplace where they do business is changing as well. No longer can companies look at other companies across town as their sole source of competition. Today, companies must face competitors at home and abroad. As a consequence, many companies are merging with companies overseas or are expanding manufacturing operations into other countries. The global marketplace is discussed in depth in Chapter 7. Strategies for establishing safety and health programs in global organizations are also presented, as well as the identification of pitfalls and solutions to numerous international health and safety issues. In addition, total quality management and safety, benchmarking, designing in safety, budgeting, and cost-benefit analysis are just some of the organizational management issues that are clarified in other chapters. These are skill areas in which health and safety professionals must become more competent if they are going to communicate effectively with organization management.

Of course, this book does not ignore the basic components of occupational safety and health such as ergonomics, industrial hygiene, training, PPE, fire prevention management, systems safety, medical services, materials handling, environmental planning, computers and safety, and numerous other topics. Experts in each of these areas present vital information for professionals to use in the safety and health management planning process. Off-the-job safety and health issues are also examined.

Readers will notice, as they peruse the table of contents, an emphasis on safety in the design process. Chapters 12 and 16 examine this issue from a manufacturing and construction perspective to present to the reader useful strategies for ensuring that safety and health is integrated into all aspects of the operation of the organization. In addition, Chapter 24 examines the various tools that should be used when thoroughly evaluating hazards in the occupational environment.

As mentioned previously, the addendum at the conclusion of this chapter presents a historical account of the evolution of safety and health management planning activities. From the beginning developments of planning as a management process to its applications to off-the-job injury and illness prevention, the intent of this book is to provide readers with an extensive examination of safety and health management planning. It is the hope of this author, and all of the contributing authors, that readers will find this to be one of the most important resources on their professional library bookshelf.

The History of Safety and Health Management Planning

Mark Friend

GENERAL HISTORY OF MANAGEMENT

The history of safety management is, in essence, the history of management. Good safety is good management and good management is good safety. Many of the modern safety management principles that are touted today as coming from some of the leaders in safety are little more than rehashed ideas that have been floating around management circles for decades. Much of the thinking in management can be applied directly to safety and safety management with little or no adaptation. This is evident when reading some of the popular safety material available in books and periodicals. Comprehending modern safety management practices is dependent on an understanding of the evolution of the field of safety and the field of management as they led toward the combined field of safety management.

Following the Civil War, the country was undergoing an industrial revolution that began moving the United States from an agrarian-based economy to a manufacturing-based economy. Two significant events largely contributed to the movement. Eli Whitney contributed the concept of interchangeable parts when he introduced this idea to the army for the manufacture of weapons. Henry Ford took Whitney's idea a step further by introducing mass production via the assembly line. The early days of mass production were marked by factories with machinery that created a maze of hazards often difficult for the average worker to negotiate. As a result, workers (some who may have been mere children) were maimed and killed on a regular basis. Employers were able to fend off lawsuits successfully with defenses upheld again and again by the courts. In essence, workers were disposable: if one worker was injured or killed, another was brought in as a replacement. In the early 1900s, this began to change as a result of a number of events. The ability of Americans to access news and to transmit information was exploding.

With the development of first the print and then the broadcast media, news of tragedies and disasters rapidly reached a large percentage of the population. Relative to safety in the workplace, there was ample bad news. In 1907, in Monongah, West Virginia, 362 coal miners were killed in a horrific explosion that rocked the small town and the families and friends of the miners. That same year, the Russell Sage Foundation sponsored a twelve-month study in Allegheny County, Pennsylvania, which found that there were 526 fatalities there during that period of time. Survivors and friends bore the costs. Injured employees were simply replaced and were paid no benefits. In 1911, the Triangle Shirtwaist Factory fire in New York City killed 145 workers. A shocked public learned that deaths were directly caused by unsafe conditions. Locked doors had trapped workers in the burning building.

Public and legislative pressure began to mount. In 1911, Wisconsin passed the first successful workers' compensation laws and other states began to follow suit. With successful workers' compensation regulations being funded by employers and with public pressure mounting, management began to understand that operating safely had certain financial advantages. Managing well began to mean managing safely. Of course, the question was, how does a company manage well? The answer was elusive.

FAYOL'S PRINCIPLES

As mentioned previously, Henri Fayol (1841–1925) did some of the early, important work in management thought. Fayol, a Frenchman, worked as a mining engineer but later became interested in management and administrative issues. The elements of management that he outlined are often modified, but they serve as the basis for a systematic study of management and what managers do. These elements include planning, organizing, commanding, coordinating, and controlling. The general principles, which he outlined in 1916, are considered "rules of thumb" for administration in many modern organizations. These principles include the following:

1) **Division of work.** Work should be divided among individuals and groups to ensure that effort and attention are focused on special portions of the task work specialization as the best way to use the human resources of the organization.

2) **Authority.** Concepts of authority and responsibility are closely related. Authority is the right to give orders and the power to exact obedience. Responsibility involves being accountable and, therefore, is naturally associated with authority. When one assumes authority, one also assumes responsibility.

3) **Discipline.** A successful organization requires the common effort of workers. Penalties, however, should be applied only judiciously to encourage this common effort.

4) **Unity of command.** Workers should receive orders from only one manager.

5) **Unity of direction.** The entire organization should be moving toward a common objective, in a common direction.

6) **Subordination of individual interests to the general interests**. The interests of one person should not have priority over the interests of the organization as a whole.

7) **Remuneration.** Many variables, such as cost of living, supply of qualified personnel, general business conditions, and success of the business, should be considered in determining the rate of pay a worker will receive.

8) **Centralization.** Centralization is lowering the importance of the subordinate role. Decentralization is increasing the same importance. The degree to which centralization or decentralization should be adopted depends on the specific organization in which the manager is working.

9) **Scalar chain.** Managers in hierarchies are actually part of a chainlike authority scale. Each manager, from the first-line supervisor to the president, possesses certain amounts of authority: the president possesses the most authority; the first-line supervisor possesses the least authority: the existence of this chain implies that lower-level managers should always keep upper-level managers informed of their work activities. The existence of and adherence to this scalar chain are necessary if organizations are to be successful.

10) **Order.** For the sake of efficiency and to keep coordination problems to a minimum, all materials and persons that are related to a specific kind of work should be assigned to the same general location in the organization.

11) **Equity.** All employees should be treated as equally as possible.

12) **Stability of tenure of personnel.** Retaining productive employees should always be a top priority of managers. Recruitment and

selection costs, as well as increased reject rates, are usually associated with hiring new workers.

13) **Initiative.** Management should take steps to encourage worker initiative. Initiative can be defined as new or additional work activity undertaken through self-direction.

14) **Esprit de corps.** Management should encourage harmony and general good will among employees.

Although these classical principles of administration were designed to apply to management in general, they also have specific application to safety management. In reviewing Fayol's ideas, the only one that is arguable from a safety management standpoint is the first one, but only out of context. Fayol was not talking or even thinking about ergonomics in those days. Peruse the list and think about how it applies to safety management. The principles are nearly indispensable to any manager. They are not, however, carved in stone, but are only to be used as guidelines. Fayol knew that when he wrote them, and they have stood the test of nearly a century of use.

TAYLOR'S SCIENTIFIC MANAGEMENT

Frederick Taylor (1856–1915), now known as the "Father of Scientific Management," made astounding contributions to individual companies in his time, and much of his work is still being validated today. One of his more famous experiments, performed at the Bethlehem Steel Company, found Taylor modifying shovels to best suit the worker. To construct the "science of shoveling," he observed and experimented to answer the following questions:

- Will a worker do more work per day with a shovelful of 5, 10, 15, 20, 30, or 40 lbs.?
- What kinds of shovels work best with which materials?
- How quickly can a shovel be pushed into a pile of materials and pulled out properly loaded?
- How much time is required to swing a shovel backward and throw the load a given horizontal distance accompanied by a given height?

The obvious application of Taylor's work to the modern field of ergonomics hardly needs mentioning, but consider that modern ergonomics extends the work that Taylor and his contemporaries began.

Taylor believed that many of the management practices of his time were irrelevant. They were considered a waste of time because they were based on the assumption that an adversarial relationship was necessary between management and labor. Taylor pointed out that this relationship could be one of mutual benefit if his practices and principles were closely followed. Typical propositions of scientific management included

- The two hands should begin and complete their motions simultaneously.
- Smooth, continuous motions of the hands are preferable to zigzag or straight-line motions involving sudden and sharp changes in direction.
- Proper illumination increases productivity.
- There should be a definite and fixed place for all the tools and materials.

According to Taylor, worker output should be carefully measured and workers should be fairly remunerated based on that output. If a worker did exceptional work (and he would if he followed Taylor's principles), he should be generously rewarded. A good example of this concept occurred at Bethlehem Steel with a worker sometimes referred to as "the renowned Schmidt." Taylor observed that a pig-iron handler could load 92-lb pigs into railroad cars at the rate of 12.5 tons per man per day. The pig would be carried up an inclined plank and dropped into the railroad car. Taylor stated that "the science of handling pig iron is so great and amounts to so much that it is impossible for the man who is best suited to this type of work to understand the principles of this science." After a time-and-motion study, it was determined that each man could load 47 tons of pig iron per day. Schmidt, who weighed 130 lb., was chosen to prove the point. After receiving instructions on motions, pace, and rest, Schmidt indeed was able to load more than 106,000 lbs. of pig iron per day. His wages were raised from $1.15 to $1.85 per day. Other workers soon asked to participate in the new work methods and incentive pay scale. Taylor was a genius in his time when it came to work methods and increasing productivity.

Contemporaries of Taylor were doing similar work and productivity was soaring. Unfortunately, this approach came to be known as "speed work" and workers resisted changing. The psychology used by Taylor was based on the assumption that both managers and workers would gain by increases in money earned. However, Taylor failed to consider the possibility that after he left, wages could be lowered to their previous levels and workers would have to continue their higher rates of production.

The essence of Taylor's approach was to select carefully a worker for the job, instruct the worker in the best way to perform it, and carefully monitor the environment of the workplace to ensure maximum productivity. Taylor's work was a forerunner to many of the techniques used in safety today. Consider job-safety analysis, ergonomics, even industrial hygiene (lighting, rest fatigue, and so on).

With the obvious improvements that Taylor was able to make to the workplace, it is no wonder that during the peak of his popularity, employers reaped bonanzas from his work and similar work from his contemporaries. Taylor's work was widely embraced—and rightfully so. His ideas, and those of the scientific management movement, were a vast improvement over the vacuum that had previously existed in management thought, and they were widely embraced at the time.

Scientific Method Tested

A group from Harvard University tested some scientific management concepts at the Hawthorne Plant of General Electric. Researchers were initially perplexed by an experiment done with illumination. Management removed a group of women from the assemblers and placed them in a room by themselves for study. The women were informed of and consulted on impending changes, and then management began testing the effect of illumination on output. Illumination levels were increased to observe the effect that the changes had on production, and, as expected, when illumination levels were increased production increased. This was no surprise, but when illumination levels were decreased, production levels continued to increase. The Harvard researchers then experimented with a series of environmental changes to test their effects on productivity. They introduced breaks, light refreshments, and varied work routines. Each change was accompanied by increases in output. They finally returned to the original working conditions and found that production again increased. It seemed that the researchers could do no wrong. No matter what variables were introduced, production went up.

Following a series of interviews, the researchers finally concluded that a team spirit had developed when the women were singled out and consulted. The women felt special and formed close ties with one another (Mayo 1945).

This series of experiments and the ones that followed had a profound effect on the field of management and, of course, on safety management. The scientific management school of thought began to fade as the new human relations school emerged. Whereas, in the past, the emphasis had been primarily on the task, it had suddenly shifted to the worker. Both schools of thought incorporated the philosophy that certain variables could be manipulated and that higher productivity would result. Neither considered the possibility that a complex personality might not be easily influenced to produce more by a relatively simple formula of either optimized working conditions or social situations.

Management philosophers were looking for simple formulae to solve complex problems. Even as the field of management was developing, so was safety. The first safety book appeared in the late 1920s, and in 1931, H.W. Heinrich published *Industrial Accident Prevention,* which came to be accepted as a standard reference for decades to come. Following Heinrich's breakthrough work, much of the writing that was done in the field of safety management was little more than a takeoff on traditional management thinking or elementary psychology.

Like management theorists in general, safety management philosophers eventually figured out that a worker's motivation to work safely is often different from that of his coworker. There is no one formula for management that fits every occasion or every worker. Effective safety management can occur only when management has a thorough understanding of the workers and their needs. Of course, this is impossible on a company-wide basis for a large corporation. It is usually impractical on a plant-wide basis and may be difficult in some departments. Modern management theories and new approaches to old problems have made the job of the safety manager more complex but, similarly, more solvable. Subsequent chapters aid the safety professional in applying modern management principles to problems faced in day-to-day operations.

REFERENCES

Carroll, A. B. (1993). "Three Types of Management Planning: Making Organizations Work" *Management Quarterly* 34(1):32–36.

Certo, S. M. (1980). *Principles of Modern Management.* Dubuque, IA: Wm. C. Brown Company Publishers.

Collins, J. and J. Porras (1994). *Built to Last: Successful Habits of Visionary Companies.* New York: Harper Business.

Covey, S.R. (1989). *The 7 Habits of Highly Effective People.* New York: Simon & Schuster.

Daniels, A. C. (1989). *Performance Management: Improving Quality Productivity through Positive Reinforcement, 3rd ed.* Tucker, GA: Performance Management Publications.

Department of Labor, Occupational Safety and Health Administration (1989). "Safety Health Management Guidelines." *Federal Register* 54(18):3908–3916.

Drucker, P. F. (1954). *The Practice of Management.* New York: Harper & Row.

Heinrich, H.W., D. Petersen, and N. Roos (1980). *Industrial Accident Prevention.* New York: McGraw-Hill Book Company, p. 23.

Ivancevich, J. M., J. H. Donnelly, and J. L. Gibson (1980). *Managing for Performance.* Dallas, TX: Business Publications.

Locke, E. A., and G. P. Latham (1984). *Goal-Setting—A Motivational Technique that Works.* Englewood Cliffs, NJ: Prentice-Hall.

Mayo, E. (1945). *The Social Problems of an Industrial Civilization.* Boston: Division of Research, Graduate School of Business Administration, Harvard University, pp. 68–86.

Merriam-Webster Dictionary (1998). Springfield, MA: Merriam-Webster, Inc., Publishers.

Petersen, D. (1989). *Safe Behavior Reinforcement.* Goshen, NY: Aloray, Inc.

Rimer, E. (1993). "Organization Theory and Frederick Taylor" *Public Administration Review* 53(3):270–272.

Taylor, F. W. (1947). *Scientific Management.* New York: Harper & Brothers.

Walton, M. (1986). *Deming Management at Work.* New York: Putnam.

Chapter 2

The Safety Professions

James P. Kohn

INTRODUCTION

The occupational health and safety profession is composed of a wide variety of professionals from a broad range of fields. Worker health, worker safety, insurance and loss control, workers' compensation, occupational medicine, and industrial engineering are just some of the many professions that play a role in the protection of worker health and safety, as well as the conservation of organizational resources in the occupational environment.

How does a safety professional plan to utilize all of these players to formulate an effective program? The purpose of this chapter is to expose the reader to the various professions that impact the safety profession. It examines some of the occupations involved in the safety and health field by job titles. In addition, requirements for certification as safety professionals and industrial hygienists are presented. Finally, the academic requirements for accreditation are discussed to point out the knowledge and skills considered by practitioners in the field to be the minimum requirements necessary to enter the safety and industrial hygiene professions.

CASE STUDY: SAFETY MANAGEMENT IS A PREVENTIVE MEASURE

A small manufacturing company was experiencing repetitive motion and lifting problems. These problems had reached the point where the workers' compensation insurance carrier threatened to drop the company if they failed to take action. With this incentive, the company's human resources director was given permission by the plant manager to contact a consulting firm for assistance.

A four-member team arrived several days later to conduct a "walk-around" inspection. The team was composed of an occupational physician, an industrial engineer, an industrial hygienist/ergonomist, and a safety professional. The inspection of the facility lasted only a few hours, but it quickly became immediately apparent that the intensive manual material handling and repetitive motions required to perform many of the tasks were contributing to the ergonomic problems experienced at this facility.

The consultant group, working with the occupational nurse at the plant, quickly prepared a plan. The safety professional recommended that the various accident report files be analyzed to determine trends and identify the tasks with the highest frequency of reported incidents of ergonomic-related workers' compensation claims. The industrial hygienist/ergonomist suggested that a formal evaluation of each of those tasks be performed. The specific types of analysis would depend on the tasks being performed. At the very least, she suggested that a time-and-motion study be conducted along with anthropometric and biomechanical measures taken to identify the relationship between worker characteristics and job requirements as they presently existed.

The industrial engineer was given the responsibility of examining the workstations and the flow of work to determine whether modifications could be made to eliminate problem tasks. Finally, the occupational physician, working with the plant nurse, was to take on the injured employee screening process and develop a medical treatment strategy and return-to-work policy that would minimize lost time from work.

With the assistance of the consulting team, the human resources director presented the plan to the insurance carrier. The carrier, satisfied with the plan, extended the workers' compensation insurance coverage for another year. The carrier monitored progress while the plan was implemented.

MANAGEMENT KNOWLEDGE AND TECHNICAL SKILL

It is clear from this case study that many professions play a key role in preventing injuries, illnesses, and organizational losses. Part of the difficulty of getting management support, especially in small organizations, is that safety and health is believed to be an area of responsibility that anyone can handle. All it requires is "common sense."

Since the passage of the Williams-Steiger Act of 1970, commonly known as the Occupational Safety and Health Act, the field of occupational safety and health has become increasingly broad and technical. Perhaps at one time the individual preparing for retirement could be the "safety man," filling first aid kits and patting workers on the back. Today, technical and managerial skills are starting to be recognized by companies interested in protecting their bottom line; safety and health management programs protect the profitability of the company. The safety and health program goes beyond costs for earplugs and examines issues such as total health care costs for the company.

The degree of management knowledge and technical skill required to be a safety and/or health professional is reflected by the requirements to become a Certified Safety Professional (CSP) or a Certified Industrial Hygienist (CIH). Certification of safety and health professionals and accreditation of academic programs that formally prepare individuals to enter the safety- and health-related fields have very specific and extensive areas of competency that must be met. The profession has become quite technical.

OCCUPATIONAL SAFETY AND HEALTH BY JOB TITLES

There are many titles given to individuals who perform occupational safety and health activities. The following are just a few of those titles and their basic roles and responsibilities.

Safety Professional

Safety professionals are individuals who, by virtue of their specialized knowledge and skill and/or educational accomplishments, have achieved professional status in the safety field. They may also have earned the status of CSP from the Board of Certified Safety Professionals.

Safety Engineer

Safety engineers are individuals who, through education, licensing, and/ or experience, devote most or all their employment time to the application of scientific principles and methods for the control and modification of the workplace and other environments to achieve optimum protection for both people and property.

Safety Manager

Safety managers are responsible for establishing and maintaining the safety organization and its activities in an enterprise. Typically, safety managers administer the safety program and manage subordinates, including the fire prevention coordinator, industrial hygienist, safety specialists, and security personnel.

Industrial Hygienist

Although basically trained in engineering, physics, chemistry, or biology, industrial hygienists have acquired, by study and experience, a knowledge of the effects on health of chemical and physical agents under various levels of exposure. Industrial hygienists are involved in the monitoring and analytical methods required to detect the extent of exposure, and the engineering and other methods used for hazard control. They may also have earned the status of CIH from the American Board of Industrial Hygiene.

Occupational Nurse

Occupational health nursing is the specialty practice that provides for and delivers health care services to workers and worker populations. The practice is autonomous and focuses on the promotion, protection, and restoration of workers' health within the context of a safe and healthy work environment. Occupational nurses may also have earned the status of Certified Occupational Health Nurse from the American Board for Occupational Health Nurses.

Occupational Physician

An occupational physician is any physician who is engaged in occupational and environmental medicine. Medical specialization provides health care services for workers. The key to the occupational physician's role in occupational safety and health program management is to help prevent injuries through education, such as implementation of wellness programs, and to help promote or educate employees to notify medical staff if safety and health injuries occur. If and when injuries or illnesses do occur, the occupational physician implements the least intrusive conservative treatment strategy to employees and allows them to return to their job in a timely manner. Physicians engaged in occupational and environmental medicine on a full- or part-time basis can qualify for membership in the American College of Occupational and Environmental Medicine.

Risk Manager

The risk manager in an organization is typically responsible for insurance programs and other activities that minimize losses resulting from fire, accidents, and other natural and man-made events. According to the Risk and Insurance Management Society, risk management is a professional discipline that protects physical, financial, and human resources. The risk manager is often responsible for protecting a company's financial and physical assets; purchasing insurance or other risk-transfer products; managing an employee benefits program; administering a self-funded property, casualty, or employee benefits insurance program; or purchasing risk control services from independent suppliers

Fire Protection Engineer

A fire protection engineer (FPE), by education, training, and experience, is familiar with the nature and characteristics of fire and the associated products of combustion; understands how fires originate, spread within and outside of buildings/structures, and can be detected, controlled, and/or extinguished; and is able to anticipate the behavior of materials, structures, machines, apparatus, and processes as related to the protection of life and property from fire. Fire protection engineering is the application of science and engineering principles to protect people and their environment from destructive fires. The responsibilities of FPEs (Society of Fire Protection Engineers 1997) include the following:

- Analysis of fire hazards
- Mitigation of fire damage by the proper design, construction, arrangement, and use of buildings, materials, structures, industrial processes, and transportation systems
- Design, installation, and maintenance of fire detection and suppression and communication systems
- Post-fire investigation and analysis

Security Professional

Security professionals are practitioners who have demonstrated a professional level of competence in the security field. Security professionals are individuals in an organization that implement "measures that are designed to safeguard personnel, to prevent unauthorized access to equipment, facilities, materials, and documents, and to safeguard them against espionage, sabotage, theft and fraud" (U.S. Army 1979). They may also have earned the status of Certified Protection Professional from the American Society for Industrial Security.

Although security responsibilities may vary depending on the characteristics of the organization in which the professional is employed, seven core areas of security professional responsibility have been identified. These include the responsibilities to

> prevent the theft or loss of any of the employer's assets through the commission of a crime, prevent the loss of assets as a result of a preventable fire or a human-made disaster, prevent unauthorized persons from gaining access to the premises, protect all persons lawfully on the premises against criminal or other forms of activity that could result in injury or death, investigate questionable activities that might result or have resulted in the loss of assets, identify those persons responsible for the theft or other loss of assets, maintain liaison with local law enforcement agencies.
>
> — Burstein 1994

Health Physicist

The health physicist in an organization is typically responsible for the protection of humans and the environment from radiation. The health physicist is an individual trained in ionizing radiation physics, its associated health hazards, the means to control exposures to this physical hazard, an in establishing procedures for work in radiation areas. Health physics combines the elements of physics, biology, chemistry, statistics, and electronic instrumentation to provide information that can be used to protect individuals from the effects of radiation. Health physicists may also have earned the status of Certified Health Physicist from the American Board of Health Physics.

Ergonomist

Ergonomists are individuals who are trained in health, behavioral, and technological sciences and who are competent to apply those fields to the industrial environment. Ergonomists possess a recognized degree or professional credentials in ergonomics or a closely allied field (such as human factors engineering), and have demonstrated, through knowledge and experience, the ability to identify and recommend effective means of correction for ergonomic hazards in the workplace. Ergonomists also have earned the status of Certified Professional Ergonomist by the Board of Certification on Professional Ergonomics.

Industrial Engineer

"The Industrial Engineer analyzes and specifies integrated components of people, machines, and facilities to create efficient and effective systems that produce goods and services beneficial to mankind" (http://www.ie.ncsu.edu/generalInfo/whatIsIE.html). "Industrial engineering is concerned with the design, improvement and installation of integrated systems of people, material, information, equipment and energy. It draws upon specialized knowledge and skills in the mathematical, physical and social sciences, together with the principles and methods of engineering analysis and design to specify, predict and evaluate the results to be obtained from such systems." (http://www.iienet.org/)

Environmental Professional

Environmental professionals design and implement pollution abatement technologies to minimize the impact of human activity on the environment. Responsibilities for environmental professionals (*Environmental Engineering* 1997) include the following:

- Designing wastewater treatment facilities at an operation
- Monitoring discharges for regulatory compliance; operating systems for reducing and utilizing solid waste
- Designing hazardous waste storage or disposal sites
- Monitoring air pollution and operating control equipment
- Predicting movement of contaminants in air, water, and soil
- Devising contaminated site remediation schemes
- Developing pollution control technologies for different industries
- Responding to environmental emergencies such as oil spills

Industrial Psychologist

Industrial psychologists are the individuals in an organization concerned with promoting human welfare through the various applications of psychology. Examples of applications include selection and placement of employees, organizational development, personnel research, design and optimization of work environments, career development, consumer research and product evaluation, and other areas affecting individual performance in or interaction with organizations (Society for Industrial and Organizational Psychology 1997).

ROLES OF SAFETY AND HEALTH PROFESSIONALS

Safety Professionals

In 1996, the American Society of Safety Engineers (ASSE) published a document titled "Scope and Functions of the Professional Safety Position."

This document presented the roles of the safety professional. The major areas of roles for safety professionals relating to the protection of people, property, and the environment are as follows:

A. Anticipate, identify, and evaluate hazardous conditions and practices
  1. Developing methods for
     a. anticipating and predicting hazards from experience, historical data and other information sources.
     b. identifying and recognizing hazards in existing or future systems, equipment, products, software, facilities, processes, operations and procedures during their expected life.
     c. evaluating and assessing the probability and severity of loss events and accidents which may result from actual or potential hazards.
  2. Applying these methods and conducting hazard analyses and interpreting results.
  3. Reviewing, with the assistance of specialists where needed, entire systems, processes, and operations for failure modes, causes and effects of the entire system, process or operation and any subsystems or components due to
     a. system, sub-system, or component failures.
     b. human error.
     c. incomplete or faulty decision making, judgments or administrative actions.
     d. weaknesses in proposed or existing policies, directives, objectives or practices.
  4. Reviewing, compiling, analyzing and interpreting data from accident and loss event reports, and other sources regarding injuries, illnesses, property damage, environmental effects or public impacts to
     a. identify causes, trends and relationship.
     b. ensure completeness, accuracy and validity of required information.
     c. evaluate the effectiveness of classification schemes and data collection methods.
     d. initiate investigations.
  5. Providing advice and counsel about compliance with safety, health and environmental laws, codes, regulations and standards.
  6. Conducting research studies of existing or potential safety and health problems and issues.
  7. Determining the need for surveys and appraisals that help identify conditions or practices affecting safety and health, including those which require the services of specialists, such as physicians, health

physicists, industrial hygienists, fire protection engineers, design and process engineers, ergonomists, risk managers, environmental professionals, psychologists and others.

8. Assessing environments, tasks and other elements to ensure that physiological and psychological capabilities, capacities and limits of humans are not exceeded.

B. Develop hazard control designs, methods, procedures and programs

1. Formulating and prescribing engineering or administrative controls, preferably before exposures, accidents, and loss events occur to
   a. eliminate hazards and causes of exposures, accidents and loss events.
   b. reduce the probability or severity of injuries, illnesses, losses or environmental damage from potential exposures, accidents, and loss events when hazards cannot be eliminated.

2. Developing methods that integrate safety performance into the goals, operations and productivity of organizations and their management and into systems, processes and operations or their components.

3. Developing safety, health and environmental policies, procedures, codes and standards for integration into operational policies of organizations, unit operations, purchasing and contracting.

4. Consulting with and advising individuals and participating on teams
   a. engaged in planning, design, development and installation or implementation of systems or programs involving hazard controls.
   b. engaged in planning, design, development, fabrication, testing, packaging and distribution of products or services regarding safety requirements and application of safety principles which will maximize product safety.

5. Advising and assisting human resources specialists when applying hazard analysis results or dealing with the capabilities and limitations of personnel.

6. Staying current with technological developments, laws, regulations, standards, codes, products, methods and practices related to hazard controls.

C. Implement, administer and advise others on hazard controls and hazard control programs

1. Preparing reports that communicate valid and comprehensive recommendations for hazard controls that are based on analysis and interpretation of accident, exposure, loss event and other data.

2. Using written and graphic materials, presentations and other communication media to recommend hazard controls and hazard control policies, procedures and programs to decision-making personnel.

3. Directing or assisting in planning and developing educational and training materials or courses. Conducting or assisting with courses related to designs, policies, procedures and programs involving hazard recognition and control.

4. Advising others about hazards, hazard controls, relative risk and related safety matters when they are communicating with the media, community and public.

5. Managing and implementing hazard controls and hazard control programs that are within the duties of the individual's profession safety position.

D. Measure, audit and evaluate the effectiveness of hazard controls and hazard control programs

1. Establishing and implementing techniques, which involve risk analysis, cost, cost-benefit analysis, work sampling, loss rate and similar methodologies, for periodic and systematic evaluation of hazard control and hazard control program effectiveness.

2. Developing methods to evaluate the costs and effectiveness of hazard controls and programs and measure the contribution of components of systems, organizations, processes and operations toward the overall effectiveness.

3. Providing results of evaluation assessments, including recommended adjustments and changes to hazard controls or hazard control programs, to individuals or organizations responsible for their management and implementation.

4. Directing, developing, or helping to develop management accountability and audit programs which assess safety performance of entire systems, organizations, processes and operations or their components and involve both deterrents and incentives.

Industrial Hygienists

The functions of an industrial hygienist are as follows:

- To direct the industrial hygiene program
- To examine the work and external environments by studying work operations and processes and by measuring the magnitude of the exposure
- To interpret the results of the examination in terms of its impact on workers and the organization and present this information to management

- To determine the need for or effectiveness of control measures
- To prepare rules and regulations

SAFETY AND HEALTH RESPONSIBILITIES IN THE ORGANIZATION

In reviewing the general job titles associated with the occupational safety and health profession, one must keep in mind that the conservation of human resources is a continuing process that involves all levels of management and every employee. The ultimate goal of the occupational safety and health professional is appropriately stated in the preamble to the Occupational Safety and Health Act of 1970: "to assure safe and healthful working conditions" for "every working man and woman in the Nation." Safety and health professionals in the organization are the "in-house" consultants who assist the organization in achieving this mission. However, each and every employee in the organization has safety and health responsibilities.

Management's Responsibility

In this context, it is management's responsibility to establish overall policies and guidelines to meet this goal of healthful and safe working conditions for every U.S. work, and it is the responsibility of the safety and health manager to see that this goal is met.

A safety and health department plays an integral part in a business's overall safety and health program. It is responsible for

- Conducting an effective safety program by coordinating educational, engineering, and enforcement activities according to guidelines established by management
- Providing educational materials
- Assisting supervisors in teaching safety and health rules and procedures
- Conducting surveys of potentially hazardous areas to ensure that proper practices and procedures are followed
- Recommending changes to keep pace with technological advances.

Supervisor's Responsibility

Similarly, it is each supervisor's responsibility to maintain safe working conditions within his or her department or group and to implement the safety program directly. Program objectives should include

- Maintaining a work environment that ensures the maximum safety for employees
- Instructing each new employee in the safe performance of his or her job
- Ensuring that meticulous housekeeping practices are developed and utilized at all times
- Furnishing all employees with proper personal protective equipment (PPE) and enforcing its use
- Informing the safety department of any operation or condition that appears to present a hazard to employees.

Employees' Responsibilities

Next, we come to the employee or staff member. Every individual is responsible for contributing to the success of an environmental safety and health program. Each person's responsibility includes

- Observing all safety rules and making maximum use of prescribed PPE and clothing
- Following the practices and procedures established to conserve safety and health
- Notifying the supervisor immediately when certain conditions or practices are discovered that may cause personal injury or property damage
- Developing and practicing good habits of personal hygiene and housekeeping.

One of the first considerations in a properly oriented occupational safety and health program is creating an awareness of safety and health in each employee through an effective educational and training effort.

CERTIFICATION AND SAFETY/HEALTH PROFESSIONALS IN THE ORGANIZATION

Once staff members have been hired for the key positions in a safety and health function of the organization, keeping staff up-to-date on the cutting edge of technology should be another consideration for the safety and health manager. A professional development plan should include promotion and recognition of those individuals who pursue and maintain professional certification.

Certification confirms the demonstration of basic levels of competency as reflected by the successful completion of professional examinations. The primary benefit of completing the requirements of certification is that the successful professional can demonstrate meeting the criteria necessary to obtain the credential. These criteria are based on a statistically valid and reliable standard that has been established by the profession. Two certifications are examined next in greater detail: the CSP and the CIH.

Certified Safety Professional

The status of CSP is awarded by the Board of Certified Safety Professionals (BCSP) headquartered in Savoy, Illinois. The following are the requirements to be eligible for earning this certification.

General Requirements. To be eligible for the designation of Certified Safety Professional, an applicant must be of good character and reputation, submit an application on the proper forms, satisfy the educational and experience requirements, complete the examination requirements, and pay the required fees. The Board does not discriminate among applicants as to age, sex, race, religion, national origin, handicap, or marital status.

Certified Safety Professional. An individual who utilizes the expertise derived from a knowledge of the various sciences and professional experience, to create or develop procedures, processes, standards, specifications and systems to achieve an optimal control or reduction of the hazards and exposures which are detrimental to people and/or property by the utilization of analysis, synthesis, investigation, evaluation, research, planning, design and consultation and who has met all of the requirements for certification established by the Board of Certified Safety Professionals.

Associate Safety Professional. An individual who has successfully completed the Safety Fundamentals Examination but who has not successfully completed one of the specialty examinations.

Academic Requirements. Baccalaureate degree in safety accredited by the American Society of Safety Engineers or the Accreditation Board for Engineering and Technology, or an acceptable combination of other education and professional safety experience.

Experience Requirements. A minimum of four years of professional safety experience as defined in section 2 of the BCSP Procedures for Certification is required in addition to the academic requirement. Professional safety experience used to meet academic requirement may not be used to meet the experience requirement.

An earned graduate degree from an accredited institution may be accepted in lieu of a portion of the required professional safety experience. The specific credit allowed will depend upon the degree and the major area of study. (BCSP 1997)

Certified Industrial Hygienist

The American Board of Industrial Hygiene (ABIH), a not-for-profit corporation, was organized to improve the practice and educational standards of the profession of industrial hygiene. The ABIH is an organization responsible for certifying industrial hygiene professionals. There are two designations of certification based on the educational background, experience, and the testing status of the practitioner. The advanced industrial hygiene practitioner is recognized by the CIH designation, whereas the practitioner with basic competencies is recognized by the designation of Industrial Hygienist in Training (IHIT).

Certified Industrial Hygienist (CIH). This certification recognizes special education, long experience, and proven professional ability in the Comprehensive Practice or Chemical Practice of industrial hygiene.

Industrial Hygienist in Training (IHIT). This designation recognizes special education in and knowledge of the basic principles of industrial hygiene.

Minimum Eligibility Requirements for Admission to Examination:

- Good moral character and high ethical and professional standing.
- Graduation from a college or university acceptable to the Board with a bachelors degree in industrial hygiene, chemistry, physics, chemical, mechanical, or sanitary engineering or biology.
- Other educational backgrounds will be evaluated by the Board on an individual basis.
- One year of full-time employment in the professional practice of industrial hygiene acceptable to the Board and subsequent to the completion of an acceptable bachelors degree is required to be eligible for the Core Examination.
- Five years of full-time employment in the professional practice of industrial hygiene acceptable to the Board and subsequent to the completion of an acceptable bachelors degree is required to be eligible for the Comprehensive Practice or Chemical Practice Examination.

After satisfying the Board's bachelors degree requirements, an applicant may request credit for pre-bachelors degree experience. An applicant must be in the full-time practice of industrial hygiene at the time the application is submitted. (ABIH 1997)

ACADEMIC ACCREDITATION OF OCCUPATIONAL SAFETY AND INDUSTRIAL HYGIENE PROGRAMS

When seeking to hire entry-level candidates for their staff, safety and health professionals are often faced with a major decision. How can they hire someone who possesses the basic knowledge and skills necessary to get the job done without demanding excessive time in training. One option for increasing the likelihood of selecting competent and potentially successful candidates for entry-level positions is for safety managers to select graduates from academic programs that meet the demands and needs of the profession. How do safety and health professionals determine what these demands and needs are? One solution is to select graduates from accredited programs or programs with curriculum that can meet the criteria for accreditation.

The ASSE is a professional safety society that has been associated with academic accreditation for more than twenty-five years. In its literature, the ASSE states that the program of accreditation of academic safety programs "is designed to benefit the safety profession, the academic community, and the general public, in the following ways:

1) The *profession* will benefit through progressive improvements and refinements of degree programs designed to prepare people to enter safety positions, brought about by careful self-study and peer review.

2) The *academic community* will benefit from the opportunity to obtain unbiased, professional peer reviews of its programs and make improvements or adjustments based on those reviews.

3) The *general public* will benefit in a number of ways. Potential employers of safety students will know that certain programs have been carefully examined with respect to their objectives and their abilities to meet those objectives. Counselors helping students prepare for college will have the same information, and students will have greater information on which to base their college selection."

The Accreditation Board for Engineering and Technology (ABET), through its Related Accreditation Commission, is now the organization responsible for accrediting safety and industrial hygiene programs. According to ABET's constitution its purpose is to

- Organize and carry out a comprehensive program of accreditation of pertinent curricula leading to degrees, and assist academic institutions in planning their educational programs.
- Promote the intellectual development of those interested in engineering and engineering-related professions, and provide technical assistance to agencies having engineering-related regulatory authority applicable to accreditation.

ABET's vision and role in accreditation is that it is

. . . responsible for establishing standards, procedures, and an environment that will encourage the highest quality for engineering, engineering technology, and engineering-related education through accreditation so that each graduate possesses the skills necessary for lifelong learning and productive contribution to society, the economy, employers, and the profession.

Based on these goals and objectives, safety and health professionals will, in the long run, save time and money by selecting candidates from quality academic programs. Quality academic programs include those programs that are accredited, as well as those programs that possess the curriculum, staff, and facilities necessary to qualify for accreditation.

CONCLUSION

The goal of this chapter was to point out that the occupational health and safety profession is composed of a wide variety of professionals from a broad range of fields. The review of the numerous safety and health job titles indicates this diversity. Large organizations will have the luxury of staffing safety and health functions with personnel possessing several of these backgrounds. The small organization will, in all likelihood, have fewer individuals with safety and health responsibilities who will have to wear several of these "hats." The best safety and health function in an organization will ultimately depend on the needs and hazards present.

Certification of safety and health professionals was also discussed in this chapter, along with a brief presentation on the concept of academic accreditation. As occupational safety and health professions become increasingly technical, practitioners in these professions will be required to demonstrate the appropriate competencies. For entry-level individuals to be appropriately prepared for safety and health positions in the future, they will have to have strong foundations in the fundamentals. These fundamentals will best be obtained through educational preparation in accredited academic programs. Even after completing these programs, an individual's demonstration of competency will likely be required to manage an organization's safety and health function. This competency and continued professional development is best demonstrated by obtaining and maintaining certification credentials.

REFERENCES

Accreditation Board for Engineering and Technology (1998). *Criteria for Accrediting Engineering-Related Programs.* Baltimore, MD: Accreditation Board for Engineering and Technology, http://www.abet.org/accreditation.htm.

American Board of Industrial Hygiene (1997). www.midtown.net/~hcg/pro_abih.htm

American Society of Safety Engineers (1996). *Scope and Functions of the Professional Safety Position.* [Brochure]. Des Plaines, IL: American Society of Safety Engineers.

Board of Certified Safety Professionals (1997). www.midtown.net/~hcg/pro_bcsp.htm

Burstein, H. (1994). *Introduction to Security.* Englewood Cliffs, NJ: Prentice-Hall.

Confer, R. G., and T. R. Confer (1994). *Occupational Health and Safety: Terms, Definitions, and Abbreviations.* Boca Raton, FL: CRC Lewis Publishers.

Environmental Engineering (1997). www.mgl.ca/%7Edlinton/enveng.html

Institute of Industrial Engineers. http://iinet.org.

North Carolina State University, Department of Industrial Engineering (1998). Questions and Answers about Industrial Engineering. Raleigh, NC. http://www.ie.ncsu.edu/generalinfo/whatIsIE.html.

Occupational Safety and Health Act of 1970, Preamble.

Society for Industrial and Organizational Psychology (1997). Division 14 of the American Psychology Association Organizational Affiliate of the American Psychology Society. http://www.siop.org/TIP/SIOP/brochure.html

Society of Fire Protection Engineers (1997). Bethesda, MD: Society of Fire Protection Engineers. http://www.sfpe.org.

United States Department of Labor (1991). *All about OSHA.* Washington, DC: U.S. Government Printing Office.

U.S. Army (1979). *"FM 1930-Physical Security."* Washington, DC: U.S. Army.

Chapter 3

Safety and Corporate Management

Donald J. Eckenfelder

INTRODUCTION

Ultimately, the safety performance of an organization depends on the resolve of the management—the attitude management has toward loss prevention. This attitude is a natural result of the culture of the organization fostered by its leaders. The culture is a reflection of those leaders' beliefs and values.

Hence, the first issue that must be addressed by someone in corporate safety management is influencing what the leaders know and believe about loss prevention. This task depends on many factors including the industry involved, the recent loss history and regulatory experience, and the resident safety process.

To be effective, the safety leader or facilitator must possess a number of characteristics or be able to augment them using supplemental means. The most important attribute—for which there is probably no substitute—is credibility. Next come communication skills, organizational ability, business savvy, and technical competence. Technical competence can easily be complemented by subordinates, colleagues, or consultants if the safety leader finds himself or herself past the threshold necessary to understand and apply the know-how of others, or if he or she simply does not have the time to gain a complete understanding of this attribute.

CASE STUDIES: BENCHMARKING

First, I am going to reflect on one specific experience and then cover a generic subject that is timely in the sense that it incorporates a current management fad that will probably sustain itself for a decade or more—benchmarking.

Case 1

My CEO at Chesebrough-Pond, Ralph E. Ward, was a uniquely capable executive in many ways. He took a modest-sized health and beauty company and—through acquisitions such as Bass shoe, Stauffer chemicals, Prince tennis racquets, Ragu spaghetti sauce, Health-Tex children's clothing, Polymer Corporation, and Prince Matchabelli—built a successful and complex industrial giant. He was not a college graduate and had started with the company as a clerk. He had an almost mythical image in the organization and was viewed as somewhat enigmatic, even by many officers of the company.

The petroleum jelly processing plant where Ralph had begun his career at in Perth Amboy, New Jersey, had worked for more than a million man-hours without a lost-time accident. The company was having a presentation and luncheon to which both he and I were invited. With numerous locations all over the world, he often politely declined to attend and delegated such responsibilities to others. This luncheon was being held where he had started with the company, and many of the workers were old friends. He accepted the invitation.

I found myself seated across from him at lunch. He knew who I was, but up to that point we had had no personal contact. Our eyes met and he obviously felt the need to say something. He asked, "How are we doing on OSHA compliance?" In everyone's life there are defining moments. I sensed that I might be in the middle of one. Should I tell him what was safe—what I think he expected to hear—or should I tell it like it was? Being a confirmed risk taker, I opted for baring my soul. I responded that I wasn't really sure since I didn't spend much time on it.

He became more alert and riveted his eyes on me and suggested, "thought that was what we hired you for." I knew that there was no easy way out at that point and that I needed to be at my best if I wanted to keep my job and continue to support my young family. My life quickly flashed before me, and I delivered my short speech, which, on reflection, I believe I had prepared in quiet meditation many times as I contemplated the opportunity to actually address our CEO. The window of opportunity that I had hoped for was opening. I needed to jump through.

I said something like, "I was operating under the impression that I was hired to support the primary objectives of the corporation, which I consider to be your concern. I understand those concerns to include providing a good return on investment for the stockholders by paying solid and increasing dividends on the stock they own. My role in helping you do that, as I see it, is to be vigilant in recognizing areas where we are incurring unnecessary losses or where loss potential exists. I then communicate that to appropriate operations managers and provide suggested solutions and preventive measures, and counsel and assist them with the solutions that we devise together. The OSHA is a small part of that, but quite frankly, more of a diversion than a solution, so I don't spend much time on it. We have attended to the ten most frequently cited items and I don't think you or the corporation will ever be embarrassed by an inspection."

He looked down a little bit, then looked up. I started to have a good feeling that I can still remember and some of my anxiety left. He smiled slightly and said, "I like that, keep doing it."

I am inclined to believe that almost every CEO would respond to that kind of answer in the same way; how could they take issue with that orientation on the part of a safety leader? Companies that are in business to pay stockholder dividends invariably consider profits to hold a very high priority and desire that everything else they do supports their profit goals. Only the most enlightened companies view concerns for employees and other stakeholders to be the *best* path to building company value and profits.

Case 2

A Fortune 100 company employed me to benchmark world-class safety processes. It portrayed itself as "best in class" and wished to be one of the best in the world. I found that other companies in its class didn't see this company as best in class and had good reason for their skepticism. Closer examination revealed a denial mentality and so much decentralization that change was not likely. My messages were quickly buried and any hope of significant improvement was, at best, way off in the future.

Benchmarking can be useful, if change is a realistic expectation. If not, it can waste lots of time and money and disillusion people who might have done better without the unfulfilled promise of positive change. In safety, as in business, the formula for success is not that complicated. What is complicated is finding the ingredients and someone to mix them properly.

IMPORTANCE OF MANAGEMENT SUPPORT

Anyone who has spent any time (more than ten years) in the safety business knows that all that really matters is the culture, or simplified and in

DuPont jargon, "top management support." Without it, the accident cycle and mediocre-to-poor performance is inevitable. Once the culture gap has been breached, everything else becomes easy. Before the culture has been oriented toward accident prevention, almost nothing seems to work very well. Every safety initiative is like pushing water uphill with a sponge— you never seem to get it done and even when there appears to be progress, it is short-lived, hard to perceive, and generally unsatisfying.

Over the last decade, the importance of the social aspects of business and safety improvement have been written and talked about more than ever before. It behooves safety professionals, particularly those with primarily management responsibilities, to recognize this and other trends and adapt to them. The adaptation should come in the form of increased awareness of empowerment, self-directed work groups, broad-based quality initiatives, and other management trends. Better yet, get ahead of the curve with cutting-edge thinking. Lock in on the fundamentals; then constantly explore new horizons.

METHODOLOGIES FOR THE SAFETY PROFESSIONAL

A Firm Grounding in the Fundamentals

At the foundation of every discipline is the ability to understand and practice the fundamentals of the profession. For safety professionals at every level, this means a mastery of a wide range of skills or at least the ability to know one's shortcomings, where to find the answers, and how to apply them. Here are the skills that solid safety professionals must have or to which they must have ready access:

I. Interaction

1. Communications—oral
2. Communications—verbal
3. Employee development—resource planning and needs
4. Employee development - training and measurement
5. Influence and persuasion
6. Leadership
7. Meeting skills

II. Administration

8. Computer skills
9. Economics of safety
10. Insurance

11. Organization and recordkeeping
12. Planning and objective setting
13. Problem identification and solving
14. Public affairs
15. Safety procedures
16. Security management
17. Time management

III. Technical

18. Regulatory expertise
19. Process equipment
20. Ergonomics
21. Industrial hygiene
22. Occupational health
23. Preventive maintenance
24. Process safety management
25. Property protection systems
26. Emergency response

Interaction. Loss prevention effectiveness is heavily dependent on the facilitators' ability to enlist others in their quest. They will first need excellent communication skills, both written and oral. Then they need to recruit "apostles" who will not only practice the techniques required to produce a loss-resistant environment but will recruit others to help and support all the efforts. This calls for strategic planning and education abilities as well as certain specific training for skills development. Credibility, the capacity to influence and persuade the application of leadership, and effective meetings will need to be among the attributes of an effective corporate safety leader. These characteristics are deliberately placed above administrative and technical abilities because without them administrative and technical know-how, no matter how highly developed, will fall short of what is needed to lead an organization to excellence in loss prevention.

Administration. Computer literacy and the ability to use software for everything from data analysis and education to communication is fundamental to every manager. The corporate safety leader is no exception. Safety always needs to be sold. The main selling point is that it is just good business. The safety leader must understand everything from balance sheets and return on investment (ROI) requirements to workers' compensation costs, as well as the need to be conversant in the techniques of risk shifting and avoidance. All managers are expected to have organization and problem-solving skills—no corporate safety professional will survive long, much less prosper, without them. Security has become a part of loss prevention. Hence, the corporate safety leader needs to be conversant in the fundamentals. Safety often is a concern of those outside the organization, the domain of the public affairs department. The effective safety professional understands the fundamentals of effectively dealing with the outside community.

Technical. Many areas require technical expertise. The breadth and depth of understanding needed is dependent on the nature of the business and the support systems and persons available to the safety leader.

Knowing What Management Expects

Next, the safety leader must know what management expects. Expectations are often industry dependent. Some would say that the chemical industry is almost paranoid about safety and health. Many of the labor-intensive, low physical hazard exposure industries, such as the textile industry, pay little attention to safety, with the exception of a few of the most enlightened companies in jurisdictions where soft-tissue injuries and resulting workers' compensation costs are a significant factor. Demographics often are a factor as is individual company culture. The culture in which management demands the best in loss prevention services is the culture that would be labeled "visionary" or world class. Good loss prevention performance goes hand in hand with business prowess.

In the last ten to fifteen years, many books have described the attributes of the best companies. Here are some of the characteristics:

- Culture is at the core.
- Writings must be supported by and, better yet, preceded by behaviors.
- Persistence is vital.
- The key ingredient is people.
- Other important ingredients are courage, creativity, opportunism, spontaneity, and consistency—in a word, *balance.*
- *Proper* motivators are used.
- There is deep caring about important issues; there is little caring about minutia.
- "Honest mistakes" are tolerated and even encouraged; deviant behavior is never tolerated.
- The power of people is reverenced.
- Priorities start with employees and end with shareholders.

- High expectations are pervasive; *everybody* loves it.
- Institutional and personal values are well known and a source of pride and feelings of well being.
- Diversity is honored, reverenced, and applied; it is never discouraged, depreciated, or denigrated.
- Work and home are blended, both being strengthened rather than drained by the other.
- Emphasis is on process and not short-term results.
- Rewards are based on group accomplishments instead of individual achievement.
- Strong cultures are laden with key words and phrases.
- Solutions to almost all problems are sociological, not technical.
- We must think globally and act locally; each of us must spend more time involved in *real* education.

These characteristics or attributes must be understood and respected by those charged with leading the loss prevention process in excellent companies. In organizations that may not have this profile, the approach to loss prevention must be tailored to whatever their profile is.

Another facet to addressing what management wishes is more generic and relates more to effective staff work than specific approaches. This subject was effectively covered by Mike Hostage in a 1977 article in *Professional Safety*. At the time, Mike was President of Marriott Restaurant Operations. He outlined what a line manager expects from his safety professional (see Figure 3-1). The desires are just as relevant today as they were then.

1. Maintain technical competence in your field. Participation in professional activities in *all* ways and at *all* levels is fundamental to excellence in *all* fields, including loss prevention. In this age of change, streamlining, expanded responsibilities, and reduced resources, broadening and extending the ability to contribute is essential. Then, the salient points must be artfully conveyed to leaders to avoid "firefighting."

2. Be aware of your own technical limitations. Don't bluff on anything. It is easier for your leader to deal with your lack of knowledge and the need for time to research than with misrepresentation, which destroys trust and credibility. Misrepresentation will damage the foundation of the relationship, causing it to never be the same again. Maintain liaisons or lines of communication with top people in related fields so that you always have adequate resources.

Figure 3-1. Mike Hostage's Desired Attributes for Corporate Safety Leaders

1. Maintain technical competence in your field.
2. Be aware of your own technical limitations.
3. Develop good communications skills.
4. Be a willing resource.
5. Maintain your professional relationships.
6. Be skillful in your advisory roll.
7. Be loyal—loyalty is always admired.
8. Keep key people informed.
9. Make completed staff work a must.
10. Be the eyes and ears of your leaders.
11. Be a beacon in the darkness, a social lighthouse.

3. Develop good communications skills. The loss prevention expert interacts with almost every sector of an organization (or should). This interaction is done in writing and verbally. Today, computer literacy is fundamental to your effectiveness. Selling skills are essential—the ability to transmit your ideas and persuade others to accept them determines your success rate.

4. Be a willing resource. If you are viewed as annoying, difficult to deal with, a credit grabber, or a threat in one way or another, you will be isolated and ignored. If you appear to be pleased to help and enjoy the accomplishments of those you serve, line managers and others will line up to get your help. If clients/customers feel you will make them and their bosses look good, they will clamor for your support.

5. Maintain your professional relationships. In spite of all the formal meetings and all the correspondence, casual personal relationships are what progress invariably turns on. Job descriptions and organization charts seem to be needed in all organizations, but what really counts is how people relate to each other in spite of those artifacts, not because of them. If you understand these informal relationships and can cultivate them for accomplishment, you're an asset. Fundamentally, this will occur if those around you like to have you around—you make them feel good, secure. It is that simple.

6. Be skillful in your advisory roll. Here's where arcane language and posturing can be deadly. You must take the time to write the one-page memo. You need to sense how much a person wants something and what that person needs to know. Meet the need. If their perception is blurred or inadequate, that's where the selling comes in. You must be willing to lose skirmishes to win wars. Sense when someone has heard enough. I've found that watching the eyes is a good clue. When they start to wander or become glazed, you're through.

7. Be loyal—loyalty is always admired. I've seen people who are in high places and remain there for a long time who don't seem to have any other attribute than loyalty. Be charitable of the faults of your leaders and colleagues. Take the blame when subordinates struggle or fail. Pass on credit for successes. Be quick to compliment and slow to criticize. Compliment in public and criticize in private. Only write a critical memo or letter as a last resort—and then rip it up. Don't hesitate to place favorable impressions in writing, even laminated.

8. Keep key people informed. The caveat here is to have a heart. Recognize how much input people have to deal with, especially those who are in high positions. Empathy and a keen sense of what is really important is critical here. Avoid the temptation to enhance people's impression of you by telling them *all* you know. Get to know your audience and adopt *their* priorities. Your priorities are of personal interest only if you can influence others to share them. Recognize that that doesn't happen very often.

9. Make completed staff work a must. A reputation for never finishing things or accomplishing closure is stigmatizing. Keep a list and don't check off items until completed. Know the level of excellence expected by those you support and exceed it at all times; if you can't, get out. Remember, *what* you do is important. *How* you do it is far more important. Neither is as critical as *when* you do it. The old saying "Timing is everything in life" may not always be accurate, but you can't be too off base by believing in it.

10. Be the eyes and ears of your leaders. Your leaders normally don't have anywhere near the exposure you do. See the whole business and, without gossiping, share information that may be of concern to your leader. Staff support people should see themselves as extensions of the five senses of the people they serve: their customers, clients, and supervisors. When Ralph Ward, my CEO at Chesebrough-Ponds, asked me how a company president was doing, I was flattered. He listened carefully to my observations and asked good questions.

11. Be a beacon in the darkness, a social lighthouse. Everyone needs to be reminded at times where the boundaries are. Long-term considerations need to be weighed against expediency. Return on investment is always a consideration. The concerns of individuals can weaken an organization or strengthen it, depending on how they are handled. Few people in a company are better positioned to shed light on these issues than the person overseeing the conservation and optimizing of the physical and human resources of that organization. Being the chaplain can be very useful and enabling if you stay in touch with business realities.

If a safety professional can acquire and apply these attributes, I don't see how they can fail.

Staying Well Informed

The safety leader interacts with more different parts of an organization on a more intimate basis than almost any other person in an organization. This is both a benefit and a challenge: it is a benefit because it encourages the diversity that is enriching, challenging, and growth promoting. It is a challenge because it requires the effective and credible professional to at times be all things to all people. Safety leaders must be multilingual when it comes to management jargon; at times even more so than the CEO or COO.

Acquiring this attribute requires intensive reading of pertinent literature, continuous and varied education, and an inquiring mind fed by constant questioning by associates in the various disciplines in the organization.

Defining and Selling Benefits

Everyone is always looking for benefits. Management is no different. Management's need to see clearly the benefits of a world-class loss prevention process before committing the resources required to achieve it. Those benefits must be sold continuously. The most effective sales pitches are given by insiders—colleagues in management. But the safety professional needs to seed and stimulate the process. Once the process begins, it will in many ways be self-sustaining.

The primary purpose of any business is to make money. Certainly, profit protection can and should be used to sell safety in addition to many other "right reasons." It can be bolstered by the mountain of bonus benefits.

Profit Protection. What would you say if I claimed that loss prevention efforts represent the best way for a significant number of companies to increase their profits 10 percent or more? Although some of you would probably agree, most of you would be hard pressed to demonstrate it in a way that a chief financial officer would understand, believe, and want to go out and tell others about. Other readers are probably thinking "no way." I believe I can demonstrate the savings potential and a lot more.

My demonstration is based on the most credible aspects of the Risk and Insurance Management Society (RIMS) publication *Cost-of-Risk Survey.* There is more than a little irony in the fact that I can use RIMS's own survey to suggest not only that its name is a misnomer but that it is on the wrong track and, worse yet, has derailed most of corporate America's thinking. I'll get to that as a corollary to the main point I intend to demonstrate.

The *Cost-of-Risk Survey* is a published survey of well over 400 companies in 24 industry groups averaging about 10,000 employees. It provides a treasury of facts concerning insurance and loss costs for a representative sample of business in the United States. To lay the groundwork for my proof, I'll use a few of the publication's most defensible and easy-to-understand facts.

The cost of risk is defined as the sum of :

- Net insurance premiums
- Non-reimbursed losses (self-insured, self retained)
- Risk control and loss prevention expenses
- Administrative costs

Over the years, cost of risk has consistently averaged close to 0.50 percent of gross revenues for the companies surveyed. It has ranged from a low of 0.07 percent for the insurance industry to 2.02 percent for the "machinery" industry. For illustration, I'll run through a hypothetical situation, and in doing so use conservative numbers in all cases so that my point cannot be discredited.

For my example, I will use a $1 billion company that has an industry average cost of risk of 0.50 percent, or $5 million. The average U.S. company makes between 3 and 4 percent net profits on sales. By using the upper end of the scale, a more modest result is produced. Hence, a profit of $40 million is assumed, which can be used to pay stockholder dividends, to buy back stock, or to finance any other purpose the management desires.

My primary assumption in this model/example is that the variance between the best and worst performers will range from one-half the average to double the average cost of risk. I need to defend that up front because it is central to the example and the conclusions that I will reach. I am certain that if you research that assumption, you will find it to be *very* conservative. I have consulted for companies whose cost of risk was several times the national average for their class of industry. I know placing the worst at 50 percent above average is very conservative. If "world-class" companies in safety have incidence rates of one-fifth to one-tenth the national average (and incidence rates should correlate at least roughly with the cost of risk), the one-half bottom end is also very conservative. This means is that the upper end of the cost of risk for our sample company would be $7.5 million and the bottom—realistically achievable—would be $2.5 million (see Figure 3-2). This produces a $5 million best-to-worst spread ($7.5 million–$2.5 million = $5 million). If we assume a 40 percent effective tax rate, which also is very conservative, the difference between the best and worst companies' profits, based on the loss prevention efficiency/results, is $3 million. That's 7.5 percent of our model company's $40 million in profits.

Figure 3-2. Cost-of-Risk Range for Sampling Company

Minimum	Average	Maximum
0.25% ($2.5M)	0.50% ($5M)	0.75% ($7.5M)

Using the most conservative numbers, the average company profits are impacted between 5 and 10 percent, based on whether the company is incompetent or excellent at safety. This calculation does not take into consideration any side, intangible, or indirect benefits; only dollars are considered— that is, stockholder dividends, real profits, or earned money that can be retained or lost through careless management.

Every member of management in every organization should be aware of facts like these. But, the $3 million of extra profits for our typical company is just the tip of the economic iceberg when safety is done well.

A good rule of thumb for buying companies is to remember that they cost about what their annual revenues are. However, this rule could be easily called into question because assets, ROI, profitability, growth potential, value of trademarks, and many other factors determine the cost/value of an enterprise. Probably the most important factor, though, is an enterprise's ability to generate profits.

To simplify our discussion, consider a typical company earning about 4 percent profits after tax. At the 40 percent effective tax rate, this translates to total profits of 6.67 per cent. For every $100 million in sales, you can cover $6.67 million of excess losses. So for every $1,000 of preventable loss, it would cost an organization almost $15,000 of capital to buy the profits necessary to cover the losses. (This looks like a tactic that safety people have used frequently to justify what we do by showing how the cost of a product or number of sales must be raised to cover losses. I believe my global corporate analysis is more compelling, more like looking at the world through a wide-angle lens instead of a narrow-width field telescope.)

The alternative of loss prevention *is* very persuasive. I borrowed the slogan on my stationary, "Can you afford not to do it?" from my good friend Chuck Culbertson, who worked for Bill Marriott. It is always a good question to ask yourself. In fact, I have used it effectively even in marriage counseling. People can always see the benefits of jettisoning a partner. However, when they carefully consider the alternative, they often see their present situation in a new light.

The point could be raised that I have considered extremes and that most companies are somewhere in the middle. This may be true, but all companies certainly have the potential to be at one of the extremes, and at a given moment most are heading for one extreme or the other. Where would you like to be headed?

Other Important Benefits of World-Class Safety

Quality Parallels Can and Should Be Explored. Interest in quality and the quality process continues to grow. Many of the wise sayings of the quality gurus have been used by safety professionals for many years. I see the quality movement as a reinventing of safety.

Several years ago, I started to work for a client who pointed out that his company was on a quality binge, and I needed to be sensitive to that. He mentioned that most of the company managers had been encouraged to read Philip Crosby's book, *Quality Is Free.* I bought the book and read it. I was amazed. It contained no new information, and if I replaced the word *quality* with the word *safety* wherever it appeared, with very few exceptions it worked just fine. This book could easily become a book on safety, title *Safety Is Free*. Actually, safety is more than free; it pays dividends.

In the back of Crosby's book, there is a lively and provocative section called "Guidelines for Browsers." Let's look at a handful of his snippets of wisdom and read them two ways: first using quality and then using safety.

1) Management has to get right in there and be active when it comes to quality/safety.
2) Traditional quality control/safety programs are negative and narrow.
3) The fifth erroneous assumption is that quality/safety originates in the quality/safety department.
4) Quality/Safety improvement has no chance unless the individuals are ready to recognize that improvement is necessary.
5) If quality/safety isn't ingrained in the organization, it will never happen.

Get the point? These are just five of at least a hundred statements that could be displayed. It's a little eye opening, isn't it? Initially, I thought that quality and safety were first cousins. I now think they are brothers, perhaps twins. Many additional statements in the book don't use the words safety or quality, but would be quickly recognized and sworn to by a quality or safety professional. Here are five of them:

1) Why spend all this time finding and fixing and fighting when you could prevent the incident in the first place?
2) If you can't produce a dead dragon each week, your license may be revoked.
3) Changing mindsets is the hardest of management jobs. It is also where the money and opportunity lie.
4) Attitudes are really what it is all about.
5) People really like to be measured when the measurement is fair and open.

Where do these comparisons lead us in the discussion of safety management paralleling quality management and perhaps many other support functions, such as purchasing, human resources, and public affairs? I have taught seminars to safety professionals trying to get attendees to find the answer. I taught seminar with Barry Jessee, who had worked with Edwards Deming at Albany International. Barry is a fine presenter, a likable person, very intelligent, and well versed in the basics of quality theory. Nevertheless, I have to say we failed. If we come to understand how beliefs and values can predict outcomes, can we bridge these gaps? I think so. In one seminar, at PetroChem in Trinidad, we even had attendees do maturity grids for safety and quality and examine the comparative maturity of the two processes within their organization. They completed the activity but didn't understand our point. Our workshops were unable to lead people to be creative and envision ways that the parallels could be used to produce synergy.

Why didn't the seminars work? I am not certain, but I believe the answer is connected to not being able to think outside the box. We are so ingrained in the Humpty Dumpty school of management and so wed to our own discipline that we can't bridge to other disciplines.

How can we bridge the gap? We need to broaden our understanding of how beliefs and values predict outcomes. If values that produce an accident-free environment can be internalized and practiced, what does that do for the quality process in an organization? Or, look at it another way. What effect will keen knowledge and acceptance of quality concepts have on the safety process? We know the answers but just can't seem to apply them.

Has anyone done better than Barry and I to marry safety and quality? Perhaps, but I haven't seen it. Most books on the subject, such as F. David Pierce's *Total Quality for Safety and Health Professionals,* explain the quality movement, review some of the commonalties between the two processes, and encourage safety professionals to learn more about quality. It's good advice and the benefits of common approaches should constantly be explored.

Safety Can Drive Training. At both Merck and Chesebrough-Ponds, the justification for the first training managers, Ken McCullough at Merck and Mel Kuntz at Chesebrough-Ponds, was largely based on safety training needs. Those were and are fine companies. I suspect the training function started and evolved in most organizations in a similar fashion.

Safety training is varied and sometimes complex. It ranges from simple topics such as wearing personal protective equipment to the complexities of manufacturing hazardous chemicals without unwanted incident. Gene Bloomwell oversaw the first chemical operator training school at the Merck headquarters in Rahway, New Jersey. Safety was treated separately and woven into virtually every part of the indoctrination. Safety drove the effort. We were convinced that if we did safety well, everything else would fall into place. In retrospect, I think we were right. I'm not sure the management saw it that way, and I'm sure we didn't do as good a job of selling the role of safety in enhancing quality and manufacturing as we could and probably should have.

Safety Auditing Can Lead to Other Audit Processes

Next to the measurement of values, which I have discussed in detail in my book *Values-Driven Safety,* auditing has the greatest potential, although largely unrealized, to enrich the safety process. Auditing obviously must include behaviors as well as an inventory of physical items. Few safety audits do that, so they appear to be glorified safety inspections and, unfortunately, rarely produce much more positive change than the old safety inspections. They have a slightly longer tail but still disappear with only a hint of permanent positive change.

When corporate America thinks of audits, they mostly think of their internal auditors or auditors from one of the big accounting firms who sign the back of their annual report and tell the stockholders that all is well in the money counting department. Due to some bad publicity resulting from an environmental miscue in the middle of the twentieth century, the directors of Allied Chemical (now Allied-Signal) decided that finances were not the only thing they wanted audited. The result was a partnership with Arthur D. Little that produced some of the first environmental audits. The early environmental audits have grown into environmental *safety and health* audits in many organizations including Allied-Signal.

Interestingly, the most complex part of these audits is the safety aspects because this component contains the earliest recognition of the human element and its importance in the total equation. When IBM decided to utilize its internal auditors to participate in its safety and health audits, the company gave them the straightforward regulatory areas to measure. A team of safety and health professionals, with a sprinkling of line managers, measured the more complex areas. I find it interesting that the people who keep track of that which is most important—the money—can't be trusted to do a good job on any but the simplest of safety items. What should this be telling management about the complexity of loss prevention and the potential it has to serve as a leading edge for positive culture change. The human factor is always the most complex element and therefore affords the greatest challenge to measure. If we can become good at measuring people, the rest is easy. Safety and health can lead the way to better measurement in general and more sophisticated methods to predict outcomes.

Good Safety Is Almost Always Associated with Labor Harmony. At Chesebrough-Ponds, when we got deeply involved in the etiology of soft-tissue injuries, we initially found many confusing data. One of the facts that blurred our vision was inconsistent reports of injuries in different plants in the same state doing similar work. Reported injuries differed by several orders of magnitude. The only significant variable was the labor atmosphere. I don't want to trivialize the complexity of our analysis, but it came down to whether employees were willing to "play hurt." Everyone had the same afflictions. In operations that had good labor relations, injuries were rarely reported. In factories where the employees didn't like the management or how they were treated, the labor-management friction served as the catalyst that triggered the reporting of injuries. Bad labor relations served as a flocculent for reporting injuries and optimizing their negative effects.

When we imposed corrective measures, labor relations improved. And so did the losses and costs. The employees' perception of their afflictions changed, they developed better work habits, and they felt better about management and about themselves. In other words, everything improved.

I don't know which drives which, but labor harmony and an accident-free working environment are like conjoined twins: you don't often see one without the other. I'm convinced that the value-driven approach advocated in the present book for safety performance enhancement will cultivate and improve labor-management relationships. Every good labor relations professional or safety professional knows what I've said is correct. Why shouldn't safety in the interest of labor harmony be sold to everyone in an organization? I'm sure that stories like mine exist in virtually every organization. Why keep them a secret?

Benefits of a Loss-Resistant Environment

The great beauty of the safety movement is that everyone is in favor of it. The great tragedy is that only a handful of people are passionate about it. I think the difference between mediocrity and excellence is passion. The lack of passion leads to a lack of vision. "Where there is no vision, the people perish" (Proverbs 29:18). Well, I don't think the people will perish without adequate safety vision, but I'm sure they will have more injuries.

How many companies have off-the-job safety programs that equal their on-the-job programs? A comprehensive survey conducted by a huge company to benchmark world-class safety companies concluded that one of the characteristics needed for world class safety was a vigorous off-the-job safety emphasis process. Someday in the future, wellness and employee assistance programs will be correlated with and intimately related to loss prevention. Managing safety by values will rapidly uncover this truth.

The vision of safety as industrial accident prevention is the view through a narrow-width field telescope. We need to use the wide-angle lens. We need to see that the father with upper-extremity, soft-tissue injury can't play ball with his child; that the woman suffering from occupational lung disease can no longer hike as she once loved to do; that the once-avid golfer now struggles on the golf course since damaging his or her vision at work.

A Great Corporate Safety Culture Is the Best Insurance against a Corporate Catastrophe. Catastrophe prevention doesn't sell well in some industries, but Union Carbide (remember Bophal, India) and most of the chemical industry can be motivated by it. I'd even say that the Chemical Manufacturers Association may be a little paranoid about spills, explosions, and toxic gas emissions. Have you seen its great new television ads? They reflect the attitude of its members. I think that attitude is good for all of us.

When I used to call the presidents of the shoe and apparel companies at Chesebrough-Ponds to make an appointment, they would set the time for two weeks later and then usually would cancel. When I called Roy Sambrook, Vice Chairman for Stauffer Chemical Company, his answer was always the same: "You can come over now." He was a savvy chemical company executive. He knew that the quickest way for him to lose his job was to be involved in a serious incident that was widely reported in the media. Not allowing that to happen was his first priority. His behavior is typical of the chemical industry. Safety, particularly process safety, sells easily in that industry, and apprehension about a catastrophe is a built-in driver.

Although chemical company catastrophes are easier to envision, other industries are not immune. What is the cost when a food or cosmetic product is adulterated and must be recalled and removed from shelves? How about an automobile recall? These examples represent catastrophes and result from problems with corporate values.

Selling the Benefits Most Effectively

Obviously, the best way to promote safety is to get everyone to help you do it. How do you get others to assist you routinely? I don't have an exact answer, but I know it happens in companies that achieve excellence in safety. I also know that the voice of the safety professional gets lost in the crowd, with a few exceptions. The trick is to facilitate but stay in the background. The day I started at Chesebrough-Ponds, Ted Mullins, my boss and Director of Human Resources, asked what accomplishment I would like to be known for when I completed my work there. I said that I'd like the company to have the world's most accident-free environment and for no one to know who the safety director was. There isn't much that I said or did twenty years ago that I am comfortable with, but I am with that answer.

An example of how I tried to achieve this goal comes to mind. Chesebrough-Ponds had a corporate safety committee made up of the vice presidents of manufacturing of all the business units. It was the only time they all got together, as they were in different businesses. Apparently, the only common thread was safety. The manufacturing VP's thought the meetings were to set policy. But I knew differently. The meetings provided a rare opportunity for me to train them. I soon realized two things: 1) I would probably never convince all of them of anything; and 2) they didn't exactly see me as a font of knowledge on any subject, including safety.

I decided to let them train each other. I would pick an individual receptive to a given subject and help him or her experience success and enrichment in that area, and then have that person tell the group about his or her experiences and accomplishments. It worked like magic. The presenter felt good. The audience listened. And then everyone wanted *one of those* too (whatever it was I was selling, that is, a hearing conservation effort or better new employee training). We propagated job safety analysis, hearing conservation, ergonomics and a host of other programs throughout the company using this technique.

Perhaps a success story would best summarize the benefits of doing a good job selling the benefits of world-class safety.

CASE STUDY:
BEYOND COMPLIANCE

Chesebrough-Ponds had a hospital products division. One aspect of that business was making the now mostly obsolete mercury-in-glass thermometers. The company sold them under many different names, such as Ballo and Fachney, as a result of numerous acquisitions. The thermometers and a few related products were made in Watertown, New York.

One day, Mike Griffis, who was our industrial hygienist, came into my office and showed me a translation of a Russian article (which was several years old) on the toxicity of mercury. The study suggested that mercury could pass through the placenta and hence *could* be responsible for *some* birth defects, although there was little or no supporting literature. I brought our discovery to the attention of my leaders and they asked me what I wanted to do. I presented the information with a balanced perspective but, for one of the few times in my career, did not want to make a firm recommendation or take a rigid stand. I suggested we get the appropriate decisionmakers in the company together, present all sides of the case, and allow the leaders to make the decision.

My managers agreed and set up a meeting in our boardroom with several officers, including the leader of my company sector, Bob Bennett, a vice president who reported to the CEO, Ralph Ward. Others involved were the Hospital Products company president, John White, his leader, a group VP, some other officers (in a company that had about fifteen officers), and several other interested parties.

As background, we were following virtually all the requirements in the mercury standards and achieving generally acceptable results. Our exposure levels were below the *threshold limit values* (TLVs), but in a few areas above the *action levels* for short periods of time. We did routine sampling, urine testing, and some blood testing. For those unfamiliar with mercury, it is unlike lead in that it does not accumulate in the body, so high urine tests are not all bad. High levels of mercury in the urine do show a probable recent high exposure (the bad news) but also demonstrate that the body is doing a good job of shedding the mercury (the good news). Invariably, we found that when people had high levels of mercury due to personal hygiene problems. A little mercury under your fingernails or in your shirt or blouse can go a long way. When we forced employees to shower and change clothes at work, the problem was resolved.

You see, the plant was relatively new, but part of it had been moved from an old mill building within the last decade. Some of the employees had even worked in that old building. Some of the furnishings and workstations had been moved from the old facility. Mercury tends to collect in cracks and crevasses and evaporates slowly. A clean, modern building with new surfaces is easy to clean, an old mill building is almost impossible. Modern furniture and tables with hard, smooth surfaces are ideal; old wooden workstations are difficult to keep clean. The modern building had adequate but not exceptional ventilation, but it still had some old workstations.

We suggested that the company was in compliance with acceptable mercury levels but, at times, approached levels that would concern some people. We suggested three possible approaches in response to the concerned but fragmentary and essentially unsupported literature on mercury. In fact, we even did some crude epidemiology work and found no reason to be concerned with birth defects.

Our first approach was to do nothing but monitor the situation. The second was to make some token efforts to get the biggest bang for the buck by replacing some furniture at workstations and enhancing our clean-up procedures and providing better education and motivation for personal hygiene.

The third option was the do-it-all option. In addition to what would be done in the second option, we would provide double lockers and more showers; wash work clothes daily; install enhanced and independent ventilation systems by segmenting the plant, adding heating, and reducing recirculation and, finally, replace some floors with better surfaces. These were obviously the big-ticket items and would cost well over a million dollars but would almost guarantee cutting exposures in half.

At the end of the several-hour meeting, Bob Bennett asked all but the officers to excuse themselves, and as I walked out, he said he would call me when a decision had been reached. As further background, the hospital products business was not a big money maker for Chesebrough-Ponds. In particular, the thermometer business was already getting stiff competition from offshore manufacturers and could see electronic temperature taking clearly on the horizon. This division had a hard time getting capital for anything.

In less than an hour, Bob called me and asked me to come down to his office. I prepared myself for a letdown. Being a man who usually got right to the point, he said, "We have decided to follow all the recommendations you made." I was pretty sure he meant the practical, cost-effective second option as we had only offered the third as an illustration of just how far we could go. I inferred that that was my interpretation of what he had said, and he clarified what I had apparently misunderstood: more than $1 million would be spent to substantially update our facility.

Just a few days later, I was walking down Greenwich Avenue to the train station. I was headed into New York City for some meetings and was absorbed with thoughts concerning the day ahead. A car stopped and someone offered me a ride. It was John White, whom I hadn't seen since the meeting. As I entered his car, I was a little apprehensive. I would have preferred not to have talked to him as I knew the money to be spent on updating the plant would probably come out of marketing or production enhancements, which his company badly needed. I apologized for raising an issue that had resulted in complicating his job. He responded that he always felt that what you didn't know could hurt you—I had done my job and he appreciated that. John was not only an effective executive but a fine gentleman.

How does this case study sum up my message on selling the many benefits of safety?

1) If you are credible and sensitive to the full range of business concerns, healthy and sound organizations with moral leadership will respond very positively. They will place a high value on environmental safety and health expertise.

2) Costs will always be a concern, but doing what is right will often supersede costs.

The decision was not an isolated event in the history of the company or my relationship with the decisionmakers. They viewed me and the persons who worked with me as part of their solutions and not as unnecessary overhead. That was the background for this scenario. Without it, a different kind of decision could easily have been reached. We had made numerous deposits, so when the time came for a withdrawal, we had a balance to draw against.

The old saying "You can't soar with eagles if you work with turkeys" comes to mind here. As a consultant, I have worked for some turkeys and I was of little use to them. They wasted their money on me; I wasted my time on them. But at Chesebrough-Ponds, I soared with many eagles. The mangers made me look that good.

At Chesebrough-Ponds, I constantly sold benefits and, consequently, never lost an employee during cutbacks and always fared better during budget-paring exercises than my colleagues in other departments. I invested meaningful time educating management about the savings that a sound safety process yields, which came to color the thinking of all the members of management.

CONCLUSION

Here's a quick checklist to determine whether corporate safety management is working:

1) Do safety professionals have full access to facilities and all necessary claims and operations information?

2) Do safety positions have stature comparable to other staff support functions?

3) Does the top safety position have ready access to top executives?

4) Are all the benefits of loss prevention well understood and appreciated?

5) Do all employees assume responsibility for their own safety?

If the answer to all these questions is "yes," the process is probably working well. If you have "no" answers, here are a few things you may want to do:

1) Assess your "stature" against the basic skills suggested in this chapter and make improvements where needed.

2) Evaluate your performance against Mike Hostage's desired attributes and improve where necessary.

3) Orchestrate a program to "sell the benefits" of effective loss prevention.

4) Measure your safety culture using a maturity grid (similar to the one offered in *Values-Driven Safety*) and author a process to redesign your culture as necessary.

Once corporate management has been "sold" on safety and its culture places a high value on loss prevention, everything else will be easy. If management doesn't believe in the virtues of safety, everything done in the name of safety will be difficult to achieve and hard to sustain.

The primary mission of the corporate safety manager is to facilitate a corporate culture change that will predict a loss-resistant environment. To accomplish this mission, the manager will need to be credible and be able to define the desired culture and then move the organization toward that culture with the support of management. The book *Values-Driven Safety* provides a sample blueprint to redesign a corporate safety culture. Changing culture is not easy, but it is worth the effort. Do you know the culture profile of your organization as it relates to safety? If not, can you proceed?

REFERENCES

Cosby, P. (1979). *Quality Is Free.* New York: McGraw-Hill Book Company.

Eckenfelder, D. (1996). *Values-Driven Safety.* Rockville, MD: Government Institutes, Inc.

Hostage, G.M. (November 1997). "The Line Manager and His Safety Professional--How to Prevent Accidents." *Professional Safety.*

Pierce, F.D. (1995). *Total Quality for Safety and Health Professionals*. Rockville, MD: Government Institutes, Inc.

Tillinghast, a division of Towers, Perrin, Forster & Crosby Risk and Insurance Management Society, Inc. (1985, 1986). *Cost-of-Risk-Survey*. New York: Tillinghast.

Chapter 4

Workers' Compensation and Loss Control

JoAnn Sullivan

INTRODUCTION

Workers' compensation is a unique system of laws and insurance designed to provide medical treatment and a portion of income to workers as a result of job-related injuries or illnesses. Workers' compensation insurance is federally mandated and each of the fifty states has established specifications for insurance administration and prompt payments of benefits through legislation and case law.

We often think of the workers' compensation system as benefiting only the employee and not the employer. The latter may not be so obvious, even to managers who are extremely familiar with workers' compensation. As far as they are concerned, workers' compensation laws require them (or their insurance carriers) to pay for the benefits that are due the injured workers. However, under the workers' compensation system, liability actions against the employer are drastically reduced. The worker can collect without the employer or employee having to undertake costly legal action. This chapter examines the evolution of workers' compensation law and provides an overview of workers' compensation insurance. It also discusses opportunities for employers and safety professionals to implement strategies to reduce workers' compensation costs through loss control.

THE EVOLUTION OF WORKERS' COMPENSATION LAW AND INSURANCE

Common Law and Liability

Workers' compensation evolved as a result of the dramatic increase in accidents and injuries occurring in factories and large-scale construction

Copyright ©1998 by JoAnn Sullivan. Reprinted by permission of author.

projects during the Industrial Revolution. As European countries and the United States evolved from agricultural and craft economies into fully industrialized nations, it became evident that one significant price for this industrialization was a greater number of accidents, often resulting in serious injury or fatality to workers. As industrial technology and machine processes grew more complex and hazardous, and labor intensified, the number of severe job-related accidents increased.

Prior to the development of workers' compensation laws, an injured worker's only recourse was to initiate a tort liability action against his or her employer, based on the employer's alleged negligence as the cause of the injury. Our law system is deeply rooted in the principle of negligence, which holds that a person has a legal responsibility to exercise reasonable or due care in dealing with others. This principle asserts that a prudent employer has the responsibility to provide a reasonably safe working environment, with suitable warnings of hazards and reasonable safeguards on machinery and processes. If an employer fails to exercise due care, and this results in injury or damage to a worker, the injured worker can sue to recover the loss.

However, to win in court, the worker had to prove that his or her employer was negligent and that there was a causal connection between this negligence and the injury. The employer, in turn, had several common law defenses available, posing significant obstacles to a successful action by an employee. For example, the employer could claim "contributory negligence" on the part of the employee. If an employee contributed to the accident through his or her own failure to exercise due care, the employee could be barred from collecting any judgment.

Other defenses available to the employer were developed from the English law case of Priestly V. Fowler, in 1837. This case established the doctrines of "assumption of risk" and "common employment," which were significant defenses for the employer.

Assumption of risk means that an employee who voluntarily accepts employment knows or should have known about the risks of the job. Thus, by accepting the employment, the employee also accepts the risks. Assumption of risk also emphasizes that the employee is being paid sufficiently to compensate for the risks of the job.

The doctrine of common employment means that the best judges of an employee's competence to perform an assigned task in a safe manner are the employee's fellow workers, because they are in a far better position to know the basics as well as the hazards of the job. Furthermore, these fellow employees, should they discover unsafe actions on the part of others, have a responsibility to notify the employer. Their failure to notify the employer is the key point. If the employer did not know about the unsafe actions, how could he be held responsible for them or their consequences?

Besides these defenses, the employer had the advantage of superior economic power, which provided him or her with far better legal counsel. These factors, as well as a court system generally favorable to the employer, made it evident that injured employees would frequently be unsuccessful in tort actions. Even when an injured employee was successful, the wheels of justice moved so slowly that it was extremely difficult to obtain the award promptly. The injured worker was faced with the prospect of having no income, while medical bills and ordinary living expenses continued for the worker and his or her family.

It soon became evident that some accidents, however unfortunate, could not be attributed to any one person, but were the result of a new, complex technology and hazardous working conditions. Existing laws simply did not apply to these new conditions. Also, injured and maimed employees became an increasing burden to society, because many of them could no longer obtain gainful employment. Inevitably, society began to demand that something be done to aid the victims of industrialization, both for humane reasons and to keep them off charity and welfare rolls.

Early Developments

The concept of workers' compensation originated in Germany during the late 1830s. In 1838, Otto Von Bismark passed the Prussian Employer's Liability Law, which held railways liable to both the passengers and the employees in the event of train wrecks. In effect, this law established the worker's right to compensation. Later, the Unified Prussian States passed a Sickness Act, to which was later added an Accident Insurance Law. Under this law, an injured employee was given assistance without regard to fault or liability. The employee was required to help pay for the cost of this insurance, but it is important to note that the employer paid most of the cost. These actions established a far-reaching precedent—that of providing compensation without a determination of fault. Moreover, it laid the groundwork that ultimately required the employer to pay the cost of these benefits.

Numerous workers' compensation laws were passed in industrialized Europe during the latter part of the nineteenth century. These laws usually required an employer to provide prompt and reasonable payments for work-related injuries, regardless of fault. In return, the employee had to give up the right to sue the employer. The cost of workers' compensation, it was reasoned, should be passed on through the chain of commerce as part of the cost of production.

Workers' Compensation in the United States

The development of workers' compensation in the United States lagged behind that of industrialized Europe. However, our own experience with industrialization was taking its toll, and the problem soon made itself felt. Severe work accidents in textile mills, mines, railroads, and other new industries created a public outcry that, inevitably, moved many states to take action. This action first took the form of employer liability laws, which were passed during the period from 1900 to 1910. These laws sought to hold an employer liable for work-related accidents; most of them, however, proved either unsatisfactory or unconstitutional. In 1910 the State of New York did pass a workers' compensation law, but it too was later held unconstitutional.

In 1911, however, several states passed the first lasting workers' compensation laws. These laws, like their European counterparts, were based on the no-fault concept of compensation for injuries, without the need to establish fault or liability. By 1913 over one-quarter of the states had workers' compensation laws to protect employees, and by 1948 all states had them.

The foundation of the workers' compensation law is the application of "exclusive" or "sole" remedy: that employees relinquish their rights under common law in exchange for sure recovery under the workers' compensation statutes, regardless of whether an injury is their fault—the concept of "no fault." Employers, in accepting a definite and exclusive liability, assume an added cost of operation (in the form of insurance premiums and claims), which in time can be actuarially measured and predicted.

State requirements and regulations vary widely. Some states require coverage for private employment, whereas others may exempt those with a stipulated number of employees. Many states also exclude certain types of employment because of the nature of the work. For example, in some states, domestic servants, gardeners, and other casual employees are specifically excluded. (However, most jurisdictions permit employers to cover employees voluntarily in an exempted class.)

Minors are covered by workers' compensation in all states. In some states, additional compensation is also provided and penalties are assessed against the employer. Also, many states provide minors with special legal benefit provisions that may override "sole remedy" provisions.

In 1993, the proportion of wages and salaried employees covered by job-injury laws was 87 percent, representing 96.1 million workers, numbers that increase annually, in which virtually all civilian employment in the United States is covered by workers' compensation. The insurance for workers' compensation can be compulsory or elective, depending on a specific state's laws. In states in which it is elective, the employer may choose to accept or reject coverage. However, if rejected, the employer also loses the three common law defenses. For this reason, it is highly unlikely that an employer will reject coverage. Coverage is now elective in only two states: New Jersey and Texas.

Workers' compensation laws have six basic underlying objectives:

1) To provide injured workers with sure, prompt, and reasonable income and medical benefits to work-accident victims, or income benefits to their dependents, regardless of fault

2) To provide a single "exclusive" or "sole" remedy and reduce court delays, costs, and workloads arising out of personal injury litigation

3) To relieve public and private charities of financial drains—incidental to uncompensated industrial accidents

4) To reduce and eliminate payment of fees to lawyers and witnesses as well as time-consuming trials and appeals

5) To encourage maximum employer interest in safety and rehabilitation through appropriate experience-rating mechanisms

6) To promote frank study of the causes of accidents (rather than concealment of fault)—thereby reducing preventable accidents and human suffering.

OVERVIEW OF WORKERS' COMPENSATION INSURANCE

Benefits Provided

Under workers' compensation and the liability for benefits placed on an employer, workers' compensation laws attempt to cover most of an employee's economic loss. Workers' compensation insurance laws specify the types of benefits provided: medical benefits, cash benefits, and rehabilitation.

Medical Benefits. Medical benefits pay all of the costs associated with medical treatment as a result of an injury and are usually provided without a dollar or time limit. Most work-related injuries involve only medical benefits (that is, medical only claims), because substantial physical impairment or wage loss does not occur. These medical benefits represent approximately 30 percent of the total dollars paid out in the compensation system. Medical benefits are unlimited in all states.

Cash Benefits. The purpose of cash benefits under compensation is to replace the loss of income or earning capacity of the injured worker due to the occupational injury or disease. The amount of the benefit and the length of time over which it will be paid are based on the type of disability involved. There are four types of disability:

1) Temporary total disability (TTD)
2) Permanent total disability (PTD)
3) Temporary partial disability (TPD)
4) Permanent partial disability (PPD)

By far, the largest number of cases involve *temporary total disability.* This type of disability implies that although an employee is totally disabled during the period when the benefit is payable, he or she is expected to make a full recovery and return to work, with no lasting impairment. (Permanent total disability generally indicates that the injured worker is totally and permanently unable to be gainfully employed.)

Income benefits for temporary and permanent total disability are expressed as a percentage of the employee's normal wage. Most states use a formula to determine the amount of benefits the person is entitled to, as well as to calculate the maximum and minimum benefits. Some states also limit the total number of weeks and the total dollar amounts of benefit eligibility. The current maximum weekly benefit in most states is 66.6 percent of the injured worker's normal wage. Where there is permanent total disability, most states provide payments for life.

Most awards and most of the money paid out as income are for temporary total or permanent partial disability. Permanent partial disabilities are generally divided into "scheduled" and "nonscheduled" types for the purpose of determining benefits. Scheduled injuries involve the loss of, or loss of use of, specific body members, in which wage loss based on the nature of the impairment is presumed. Loss of a hand, eye, or fingers is compensated according to a predetermined schedule, hence the name. In most states, the actual amount payable is a specific number of weeks of benefits multiplied by the weekly benefit involved. Because disabilities involve current earnings or wage-earning ability, in many states weekly benefit payments for temporary or permanent partial disability of the nonscheduled types are based on the concept of "wage loss" replacement. That is, compensation pays the difference between the wages earned before and after the injury. In some states, nonscheduled permanent partial disabilities are compensated as a percentage of the total disability.

In the event of fatal injuries, the employee's spouse receives a fixed amount or a percentage of the deceased worker's annual income. Usually benefits cease if the spouse remarries. Dependent children also receive death benefits until they reach a specific age (though if the children are handicapped or unable to become independent, benefits may continue throughout their lives). An established amount is also provided to cover funeral costs for fatally injured workers.

Rehabilitation. Workers' compensation programs generally include physical rehabilitation as an integral part of the medical benefits. In most cases, this merely involves therapy to return a worker to a condition in which he or she can return to work.

However, when permanent disabilities are involved and the employee cannot return to his or her original job, rehabilitation may extend beyond medical treatment in the form of vocational rehabilitation and retraining for a different job. An effective vocational rehabilitation counselor evaluates physical capacity, education level, abilities, and aptitudes, with the aim of restoring the injured employee to meaningful and productive employment. Rehabilitation benefits may also include a subsistence allowance and special fund sources to finance the rehabilitation. Vocational rehabilitation must be coordinated with all other aspects of the injury management process. The Federal Vocational Rehabilitation Act is now effective in all states; it includes federal funds to aid states in vocational rehabilitation of the industrially disabled.

Waiting Periods

To reduce the number of small claims, most states impose a waiting period during which income benefits are not payable. This waiting period does not affect medical payments, which begin immediately after an accident. If disability continues for a certain number of days or weeks, most laws provide for income benefits retroactive to the date of the injury. Thus, if a worker suffers merely a minor injury, causing him or her to miss only a few days of work, the worker is eligible only for medical payments. This, in fact, is usually the case: workers' compensation pays for the medical costs, and the employee's lost work time is covered by the normal sick leave benefit provided by the company.

Second Injury Funds

Second-injury (or subsequent injury) funds are another feature of most state systems. These were developed to meet the problems created when a preexisting injury is complicated or worsened by a second injury, producing a disability greater than that caused by the second injury alone. The purpose of these funds is to encourage an employer to hire a physically handicapped worker; in effect, they limit the impact of such compound losses on the employer, as well as equitably allocating the cost of providing benefits to such employees.

Under the second-injury fund program, the employer need only pay compensation related to the disability caused by the second injury alone. The employee, however, will receive a total benefit appropriate to the combined disability. The difference between what the employer pays and what the employee receives is paid out of the second-injury fund. If there were no second-injury funds created by law, an employer might be held liable for compensation due for the total resulting disability, rather than just the injury that occurred while the worker was actually in his or her employment. (For instance, a worker who had lost sight in one eye while working for Company A could lose sight in the remaining eye while working for Company B, thus becoming totally blind, even though, technically, he lost sight in only one eye while employed by the latter company.) Under those circumstances, an employer might be inclined to refuse to hire handicapped persons because of the increased risk. For this reason, second-injury funds are advocated by most jurisdictions.

Compensable Injuries

Workers' compensation was designed specifically to cover the cost of injuries arising out of the work environment, rather than to provide social assistance for accidents that occur outside the workplace. For this reason, certain basic requirements were established to define a "compensable injury," one for which a worker can collect compensation. The primary requirements accepted by all jurisdictions are that the injury must 1) arise out of employment (AOE), and 2) arise during the course of employment (COE).

These requirements suggest, first, that the worker's employment must be the source of the accident or injury (AOE). If no employment relationship exists, the injury is not compensable. And, second, the accident must occur during the time, place, and circumstance of employment (COE). The intent of COE is to restrict compensation to those injuries that are directly tied to job activities. There has been considerable conflict over the precise limits of AOE and COE, and judicial or administrative rulings have tended to broaden these definitions over the years, with each state establishing specific times and places for COE.

Some occupational diseases and cumulative traumas have been disputed, based on these AOE or COE requirements. It is generally agreed that these illnesses, to be compensable, must arise out of employment and be due to causes or conditions characteristic of and peculiar to the particular trade, occupation, process, or employment. For example, a respiratory ailment that can be traced directly to a worker's exposure to harmful dust in the workplace could be treated as an occupational disease, and deafness resulting from repeated exposure to excessive workplace noise could be considered a cumulative trauma.

Ordinary diseases to which the general public is exposed (such as the flu or pneumonia) are specifically excluded. Highly subjective health impairments, such as a stress-induced emotional illness, are often questioned; it is debatable just how much of the stress is due to the work environment and how much to marital, financial, or other personal problems. These subjective illnesses pose a critical concern to workers' compensation administrators. First, the incidence of cumulative trauma cases and awards has been increasing significantly in recent years. Second, medical research has suggested certain links between workplace conditions and medical problems that may not be apparent for many years (as with certain carcinogenic substances, which were not recognized as cancer-causing agents during the worker's period of employment). The disease may not actually develop until years later, when the worker is no longer in the same company or industry.

Employer's Liability

The "sole remedy " or "exclusive remedy" concept in workers' compensation insurance protects an employer against worker liability suits in most states. That is, workers' compensation is an employee's sole remedy, and his or her employer cannot be sued for additional compensation. However, an employer's liability for work-related injuries may still exist in certain circumstances under most state laws. For example, one form of liability in most states can increase an injured worker's benefits to some specified maximum amount if the injuries are caused by "serious and willful misconduct" by his or her employer. These additional benefits, assessed as a penalty, are uninsurable and must be paid by the employer. A typical case might involve an employer with a long history of similar accidents that are controllable through reasonable action by the employer (such as refusal to supply exposed workers with guards or protective equipment). On the other hand, in some states an employee's benefits can be reduced by a specified amount if the primary cause of the injury is found to be the result of serious and willful misconduct by the employee. However, these actions are rare exceptions, and very few cases are ever filed.

Some states also permit an employee to sue a fellow employee if his or her injuries were caused by an intentional assault by that person or if the fellow employee was intoxicated or under the influence of drugs or illegal substances.

In many states, if the compensation law fails to provide a remedy, the injured employee may seek justice in court. These cases may occur, for example, if the employee is ineligible for workers' compensation benefits. This would include cases involving exempt employees, illegal employees, or noncompensable injuries or diseases. It might also include third-party actions, such as are commonly found in contract construction. For example, a general contractor who hires a subcontractor to work on a job site may be sued by the subcontractor's employee if the employee sustains an injury due to the general contractor's negligence. In some states, even though an injured employee receives state compensation benefits, his or her spouse may sue the negligent employer for the loss of the partner's services.

Subrogation

Insurance carriers have the right to subrogate or to sue a third party on behalf of an insured (employer or injured employee) per the insuring contact. This is a common practice in the event of automobile accidents whereby the insurance carrier will sue the auto insurance carrier of the third party who injured the covered employee in order to recover the costs of the claim. Subrogation may also be brought under products liability, in cases in which a product, machine, or process that is found to be defective is a significant factor in an injury. Examples include a poorly designed machine guard, a valve that fails, or a chemical that is determined to cause disease, resulting in injury or illness in the workplace. In these cases, the insurance carrier administers the claims and then pursues recovery, with the understanding that litigation (discovery, depositions, and trials) may take many years. Insurance carriers may perform a cost-benefit analysis of the potential for recovery under such suits, and may elect to settle with a third party or not to pursue subrogation. Each case is evaluated separately. After a period of time, an insurance carrier may reassign the right to sue a third party to the injured employee. Any recovery at a future date would require the insurance carrier to be reimbursed for medical and indemnity claims expenses, with the remaining amount awarded to the injured employee or his or her family, less any legal expenses.

Workers' Compensation Claims Administration

Workers' compensation laws are administered through the state court system, by a special "industrial" or "compensation" commission or board established for this purpose, or by a combination of both. The primary concern in administering a workers' compensation program is to ensure the prompt payment and effective disposition of cases. Without an effective benefit delivery system, many of the problems associated with common law liability actions would remain. An effective workers' compensation administration system should include the following:

- Supervision or monitoring of statutory compliance by employers, employees, insurance carriers, and the legal and medical professionals involved in the program
- Investigation and adjudication of disputed claims
- Supervision of medical and vocational rehabilitation funds
- Management and disbursement of second-injury funds
- Collection and analysis of occupational claims data, and evaluation of the program's effectiveness and efficiency

The more important statutory provisions of administration involve the time limits in which employers must be advised of an employee's injury, the time limits in which the injury claim must be filed, the claim settlement conditions, and regulation of attorneys' fees in litigated cases. All states vary slightly in this regard, and employers must be aware of their state's requirements, forms, posting notices, and so forth. The administering agencies are generally given the power to impose sanctions or penalties on insurance carriers and employers for violations of statutory requirements.

THE COST OF WORKERS' COMPENSATION

Employers are required by law either to obtain workers' compensation insurance or, if financially able, to self-insure. A few states require employers to insure through a monopolistic state fund; other states permit employers to purchase insurance from a competitive state fund or private carrier. Most states permit the option of self-insuring as well.

By far the largest costs associated with workers' compensation are losses. The loss history of a company drives its premium costs. Figure 4-1 depicts the total workers' compensation dollar: losses make up about 80 percent, the premium is about 11 percent, and the remaining 9 percent includes claims administration, taxes, commissions and fees, and related risk management or consulting services.

The premium rates for workers' compensation are developed by a state rating bureau such as those found in Texas or California, or by the National Council on Compensation Insurance (NCCI), which was established specifically for this purpose. The NCCI grew out of a 1915 conference that agreed that rate making for compensation programs could not be handled

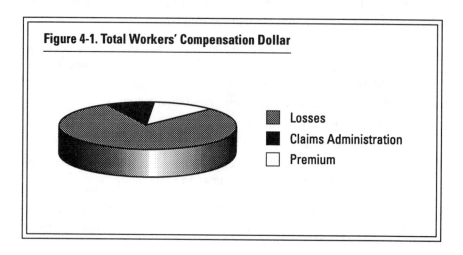

Figure 4-1. Total Workers' Compensation Dollar

Losses
Claims Administration
Premium

by each state separately. The NCCI was established as an independent agency whose primary purpose is to promulgate the premium rates charged by insurance companies.

As established, the NCCI collects data on workers' compensation loss experience incurred by member companies and states. Future rates are developed based on an analysis of this prior injury-cost data. The rates established by the NCCI serve as the standard for a nationwide rate-making procedure approved by the National Association of Insurance Commissioners. The NCCI has established policy language for all but the six monopolistic states, with substantive changes made in 1984 and again in 1992. All policies contain these parts:

1) Workers' compensation insurance
2) Employer's liability insurance
3) Other states' insurance
4) Employer's duties if injury occurs
5) Premiums
6) Conditions

The NCCI's basic manual is standard for all private insurance carriers. It establishes the Council's rules, procedures, and rates applicable to workers' compensation insurance. Because all states now have statutes regulating workers' compensation rates, the NCCI must file annually all proposed rate revisions and all supporting data with these authorities. In most states, public hearings must be held before the rates can be revised.

The workers' compensation premium rate is the amount charged per $100 of payroll. A base rate is established for each major work classification. The work classification system (Four-digit numbers found in the *NCCI Scopes Manual*) was established to group the work operations of all employers who have exposures common to that type of employment. Injury frequency and severity differ considerably between, for example, retail grocery stores and heavy construction firms, and, therefore, a single rate would not be equitable for both. On the other hand, one construction firm's injury experience will be similar to those of other construction firms. Thus, it is necessary to establish rates according to work classification. For example, the work classification code for a retail grocery store is 8006, and this code is assigned to all grocery stores. If the rate for this classification is $3.50, the insured will have to pay $3.50 for each $100 of payroll; the resultant figure will be the premium.

Experience Rating of Premium

Most states have "minimum rate" laws, which make it illegal to charge less than the minimum rate assigned to the work classification. However, a few states have advocated "open rate" laws, most recently California, that permit insurers to charge lower rates for some classifications whose loss experience does not warrant the collection of higher premiums.

Premiums can also be *modified* to reflect an individual employer's loss experience. This modification is calculated by comparing the employer's actual loss experience with the loss experience of all members in the same classification. The loss experience developed for each classification is used to predict future losses. Thus, the insured employer pays a premium based on their expected losses. If the actual losses are considerably different, an adjustment can be made. This process uses an actuarial formula and develops an "experience modification factor," a multiplier used to determine the premium that will be paid.

To develop an employer's experience modification factor, an "e-mod", three full years of prior premium and loss history are analyzed using the actuarial formula developed by the NCCI. The current year is not included in the calculations. For example, if the employer's losses are consistently very low for several years, he or she might develop an experience modification factor of 0.80. If this is applied to the premium, the employer will pay 20 percent less than the average employer in this classification. On the other hand, if his or her losses are consistently much higher than the average, the employer may develop an e-mod of 1.2; thus, the employer might have to pay 20 percent more. Besides being an equitable way of allocating actual losses paid out, this system provides a powerful incentive for the employer to practice accident prevention and create a safer work environment.

Unit Statistical Cards

The unit statistical card (unit stat card) is the instrument that provides the NCCI with the necessary payroll and loss information needed to establish experience modifications. Unit stat cards are prepared by the insurance carrier. The payrolls reported are final audited payrolls by classification for each policy period. Also included are the rates, premium, experience modification used for that policy period, and premium discount applied. The unit stat card is a summary of the exposures and premiums for the policy period indicated on the card.

The same card also includes all the losses incurred (closed claims and open with reserves). Claims are identified by the claim number used by the insurance carrier. The NCCI requires that losses exceeding $2,000 be assigned a claim number. Smaller claims can be grouped together under the corresponding class code. Incurred losses are further separated into indemnity and medical losses. Claims are valued eighteen months after the inception date of the insurance policy and the first unit card is filed twenty months after inception of the policy. A subsequent report is filed every twelve months thereafter to indicate current value of open clams.

Retrospective Rating

Retrospective rating plans for workers' compensation insurance were developed by the NCCI, which carries the major responsibility for overseeing the retrospective rating process, with rules and regulations governing the process. In principle, "retro" rating works to determine an employer's (insured's) premium at the expiration of a rating period based on the insured's actual losses during that period. Such a rating program allows an insured to determine more directly his or her insurance costs through control of his or her own loss experience. A retro plan is not a policy; rather, it is a plan for the ultimate payment of insurance premiums, and its endorsement, which specifies the plan, payment, and audit schedules, is added to the policy. Retro plans can be combined with other lines of insurance coverage and can be written for a period of one to three years.

The concept of an individual employer retrospectively rated premium determined on the basis of *actual incurred losses* is significant for the following reasons:

1) An insured that controls losses is allowed the opportunity to generate a final premium substantially lower than the premium developed under a standard guaranteed cost experience rated plan (that is, prospective rating).

2) An insured that has not historically controlled losses is provided with the *incentive to do so* through an opportunity to reduce significantly the effects of unfavorable experience modifications developed from poor experience in the past.

3) Each insured has the opportunity to earn a reasonable final insurance premium based on its actual losses. Retrospective rating provides more immediate recognition of favorable (or unfavorable)

loss experience than experience rating. There are advantages and disadvantages to retrospective rating plans that all must be weighed carefully before selecting such a plan, but they are one of the best incentives to establish and maintain an effective loss-control program.

Workers' Compensation Deductibles

Workers' compensation deductibles are becoming popular with employers because they allow premium savings and assist in controlling the escalating cost of workers' compensation. Small deductible plans developed by the NCCI are available in all states that have approved these plans. Currently, deductibles range from $1,000 to $10,000, varying by state. By contrast, large deductible plans are available to employers who obtain insurance from carriers whose large deductible plans have been filed and approved by specific states. Large deductibles are usually $250,000 or more. Insureds usually have to meet specific financial requirements to qualify for a large deductible program. Large deductible programs are appealing to large employers because of the advantages of self-insurance without the disadvantages of being self-insured: cost control, significant cash flow, and increased incentive for implementing loss-control programs.

For deductible programs, the reduction in premium varies by state. The workers' compensation deductible is applied in the same manner as a liability deductible. Specifically, the insurance carrier pays the entire claim on a first-dollar basis and then seeks reimbursement from the insured, generally through either a monthly or quarterly billing. Care must be taken to report losses properly to the NCCI per each states requirements.

Unbundling Claims Services

For some employers, the opportunity to "unbundle" claims services from an insurance carrier is desirable, as it provides an opportunity to be more involved in the claims process and reduces insurance expenses. For example, an employer who has purchased a retroactively rated program that includes a $250,000 deductible per claim may want to utilize the services of a third-party administrator (TPA) for the claims within the deductible. The insurance carrier will retain some oversight of the process but, for the most part, will be an insurance funding mechanism, a type of "safety net" for only large losses, and the employer is considered to be in full compliance with laws requiring workers' compensation insurance.

Other Workers' Compensation Insurance Funding Methods

In addition to prospective rating, retrospective rating, and deductible plans, several other methods are available for an employer to fund his or her workers' compensation insurance program. These methods are briefly discussed next.

Self-Insurance. Self-insurance is available to large employers who can qualify for the financial requirements of self-insurance in each state where they do business. Self-insurance suggests that the employer acts as his or her own insurance company, which is true in some respects. Self-insured employers must determine compensability, administer claims, pay losses, and maintain financial security requirements, just a an insurance carrier would. Most self-insureds use the services of a TPA to provide claims services; but some self-insureds maintain an in-house claims staff. Self-insureds usually purchase excess workers' compensation insurance, although an entity can be totally self-insured.

Captive Insurance. Captive insurance programs are established for a specific purpose (a group captive) or as a financial vehicle such as funding large deductibles, usually by an employer (the "parent") of significant premium size ($3 million or more). There are domestic captives and offshore captives. Tax laws have changed in recent years, making the initial reasons for captives less desirable. Regardless of the funding of the captive, monies must be available to pay and administer workers' compensation claims per the statutory requirements. A TPA or in-house claims departments is used, similar to self-insurance.

TPA Services. Contracts are negotiated between a TPA and an employer and usually include a flat charge for handling a medical-only claim and higher fees for indemnity claims. Contracts may be written for the "life of the claim" or a specified number of months, or may require claims to be transferred to a different TPA in the event that the employer changes TPAs. Employers must maintain a funded bank account from which the TPA withdraws funds and pays claims. The fund is replenished on a scheduled basis. There are a number of national TPAs—Alexsis; Crawford and Company; GAB Robisn; Gallagher-Bassett; ESIS; and Frank Gates, to name a few—and many smaller local/regional TPAs. Many are associated with insurance carriers.

STRATEGIES TO REDUCE WORKERS' COMPENSATION COSTS THROUGH LOSS CONTROL

Concern About Cost

After a dramatic rise in workers' compensation insurance rates in the 1980s, which caused great concern to employers and to the insurance industry, both regulations and rates were examined by the states and the NCCI. This examination resulted in a moderation of rates in several states and changes to some of the more liberal applications of the statutes regarding benefits. Workers' compensation fraud, whether committed by an employer, a medical provider, or an employee, is now being aggressively investigated and prosecuted.

However, the continued inflationary trend in medical costs is an area that needs to be addressed because most benefits that are paid in go toward medical expenses. In addition, most insurance executives agree that the following factors have caused workers' compensation costs to increase:

1) High average weekly wage indemnity benefit levels have made the benefits more attractive to injured workers (in other words, they are worth striving to keep).

2) Liberal interpretations of workers' compensation benefits by workers' compensation boards have increased the utilization of these benefits.

3) Attorneys and unions have assisted employees in gaining the maximum benefits possible from the system.

4) The number of recognized occupational diseases (as opposed to specific accidents) is growing and will continue to do so.

5) Many states are experiencing longer duration of claims for disabled workers. This may be an indication of a contentment with benefits and a lack of incentive to return to work.

6) Many employers have not made the necessary commitment to workplace safety and loss control programs because they viewed workers' compensation insurance as a "cost of doing business." This ambivalence, along with a lack of understanding about the workers' compensation system, has lead to "finger-pointing" at the insurance industry, rather than inward at their own lack of participation in the process.

Understanding Indemnity Benefits

Workers' compensation indemnity benefits are not taxed. Therefore, as the average weekly wage benefit levels approach actual wage levels, the incentive to return to work is bound to diminish, because the worker's benefit income may be equal to his or her regular wages after taxes. In addition, more than one member of the employee's family may be working. If one wage earner is drawing compensation benefits and several other members of the household are employed, the family may be as comfortable as or even more comfortable than they were before.

Given current economic pressures and soaring costs, there will undoubtedly be more efforts to reform the administrative agencies that oversee compensation benefits. Ideally, the benefits should be provided only to those who have actually sustained work-related injuries or illnesses. The benefits received by an injured worker should be limited to a level that will encourage a return to work as soon as it is physically possible to do so. This level, in short, should neither create undue hardship nor provide a hidden incentive to remain "disabled."

Cost Control Strategies for Workers Compensation

A number of strategies are available to an employer to reduce or control the cost of claims, sometimes referred to as "cost containment" or "total injury management." All carriers perform several basic cost-containment procedures such as utilization review and bill review on each claim. Following are some of the most effective strategies.

Return-to-Work Initiatives. Return-to-work programs may be known by several names including "modified duty," "light duty," or "transitional duty," but all have the same objective: to return employees to the workplace, among their coworkers, as soon as they are able to perform some or most of their regular job functions. Insurance carriers and nurse case managers have encouraged return-to-work programs for many years; however, some employers wanted an employee back only if he or she was 100 percent recovered, ignoring the indemnity expenses of the claim. A return-to-work program should be custom designed for the workplace and for the injured employee, keeping in mind that individuals have different rates of recovery due to the injury, age, and general health at the time of injury. In a study conducted by William M. Mercer, Inc., for Safeway, the results showed that:

- Modified workers returned to full duty (at maximum recovery) 38 percent sooner
- Modified workers saw a doctor 18 percent fewer times
- Medical payout was 43 percent less

Also, if an employer has a "modified-duty job" available, many states now require that an employee who is released during medical treatment with restrictions must return to work or may loose his or her indemnity benefits.

Medical Case Management. Medical case management is a system for establishing and maintaining communication among injured employees, employers, insurers, and medical providers. Medical case management focuses on active rehabilitation that will expedite the return to work and help an employee achieve maximum medical improvement (MMI). Because case management activities involve review and interpretation of medical treatment, most medical case managers have a strong medical background; in fact, most are nurses.

Although the concept of case management has been around for some time, until recently most insurance carriers and employers utilized it only in "problem" cases, such as when fraud was suspected or a lawsuit was filed. Ironically, the lack of employer involvement was often the reason that the claim became "problematic."

Case management is a tool for managing claims before some complicating circumstance arises. Proactive case management assigns a case manager to every claim that meets specified criteria including lost time claims, amputations, injuries requiring surgery, injuries requiring lengthy physical therapy, and catastrophic claims. Some insurance carriers apply case management, in varying degrees, to every claim. The results of aggressive case management include lower medical costs, lower indemnity costs (less time away from work), and lower legal costs—including fewer lawsuits overall.

Directing Medical Care. Many states now allow an employer to direct some or all of the medical care provided to an employee for a compensable claim. There may be specified time periods for this opportunity, such as thirty to ninety days. Self-insureds can direct all medical care. Obviously, it is important to have a relationship established with a medical provider before an injury occurs.

For the states that allow an employer to direct only one or two visits, it is still important to identify a medical clinic near the workplace, preferably an occupational clinic where the staff is familiar with workers' compensation claims procedures. Even if an employee chooses to see his or her own doctor for treatment, the employer can use the employee's one visit for referral to a specialist or for an intermediate medical exam (IME), which helps to affirm that the employee is receiving timely and proper treatment.

Preferred Provider Networks. A system that has been successful in health care, preferred provider networks (PPOs) are established by an insurance carrier or a TPA to select hospitals and medical providers who will agree to negotiate fees and participate in managed care programs. Some well-known national PPOs for workers' compensation include Corvel, Genex, and CCN.

At the local level, a PPO can be established on the basis of "designated" clinic locations, hours of service, cooperation with local hospitals, and to assist in the implementation of an employer pre- and post-accident substance abuse testing program.

Claim Reviews. Periodic claim reviews are usually held at an employer's office. These reviews provide an excellent opportunity for the employer and the claims adjusters to discuss open claims and claimant progress toward recovery or MMI. The reviews reference a loss run for a given policy period and include "status reports" prepared by the adjuster. The report includes pertinent information about the employee and the injury, a brief history of medical treatment, and a plan of action through claim closure.

Issues discussed at claim reviews include quality of medical treatment, employee cooperation with medical treatment, the availability and use of modified duty, loss-control activities accomplished to prevent recurrence, and a discussion on litigated claims. The status reports also include all medical and indemnity expenses paid and reserved and the "total incurred amount" per claim, along with any anticipated changes in reserves (such as an increase if surgery is indicated in a future doctor visit, following a course of medication, and physical therapy). Participants from the employer should include the risk manager, human resources manager, safety manager, and other interested parties. Participants from the insurance side should include the adjusters on the account, broker/agent claims manager or specialist, and the carrier and broker loss control consultants.

Claim Audits. Claim audits are performed on an annual basis and are used to verify performance standards. Performance standards may be established by an insurance carrier or broker, listing the "best practices" for claims handling and resolution. This includes efficient use of dollars being spent for handling/processing claim payments and related expenses. Claim audits are also used to verify the adequacy of reserves on open claims.

Claim audits take place at an insurance carrier's office or a TPA's office(s) to allow access to the claims files, the adjuster notes (usually on computer), and the adjuster. The audit is usually performed by an employer's risk manager or claims manager and/or by a claims manager/specialist from the broker/agent who represented the employer's interests and advocates for reserve corrections, claim closure, settlements, and so forth.

Following a claim audit, a report is prepared for the employer. The report should comment on the performance standards and make recommendation for improved performance or adherence to best practices. The report also recaps the reserve adequacy of large claims and litigated claims and often includes action items for completion by the carrier/TPA and the employer. If available, savings as a result of any cost-containment initiatives should be captured and reported. The next audit should provide a status report on any prior recommendations or action items until completed.

New Cost-Containment Ideas. Several other cost-containment ideas are being examined by risk management associations and insurance consortiums, including integrated disability benefits and 24-hour coverage.

Loss Control Programs

One of the most important objectives of the workers' compensation system is to prevent or minimize occupational accidents. It is generally agreed that the no-fault system should encourage an honest study of the causes of accidents, rather than the concealment of fault, which tends to prevail under the tort or negligence system. Most workers' compensation insurance carriers offer excellent safety engineering or loss control services to their policyholders, which can greatly aid an employer in improving workplace safety.

It is generally recognized that employers can control their accident costs through an effective loss control program, emphasizing accident prevention and efficient claims administration. Accidents, whether they result in personal injury, property damage, or both, are a major drain on corporate resources. In fact, "direct," or insurable, workers' compensation costs are minuscule compared with "indirect," or uninsured, costs. These uninsured costs include the time wasted by supervisors and other employees, the cost of damaged products or equipment, downtime in processing operations, the cost of first aid, the cost of training a replacement worker, and so on. Uninsured costs represent a far greater cost to the employer and have an enormous impact on the cost of production.

A good loss control effort should involve a partnership between the employer and the insurance carrier. The insurer's loss control representative should be asked to review and analyze existing loss control programs and offer recommendations for improvement. The employer should also take advantage of the supervisory training, industrial hygiene, and other loss prevention services that are available from his or her insurance carrier or broker. If used properly, these services can be extremely valuable in helping to control losses before they occur and the cost of losses after they occur.

To be truly effective, however, the loss control program requires employer involvement—integrating the safety and loss control strategies into all aspects of operations. The following is a sample of business practices that could be positively impacted by implementing loss control strategies.

Management-Supported Safety Program. Traditional safety programs have focused on the regulatory aspect such as OSHA compliance. Traditional safety programs have also been "reactive." Often only after a serious accident has occurred does management enforce safety rules, and even that effort may be short-lived.

Today's workplace is often understaffed and job tasks are both physically and mentally demanding. Employees use technology/computer applications for about 75 percent of all tasks performed, and this is expected to reach 90 percent by the year 2005. There are "tried-and-true" strategies to create and maintain a good safety and loss control program, which safety professionals were aware of, but could not always convince management of their need or value. Safety professionals know that safety should be managed like any other business function. With the emphasis on quality programs in the workplace, safety has been renewed as an integral part of a successful and productive workplace. Elements of the safety program that foster loss control and directly impact workers' compensation are as follows:

- *Job safety analyses* for all jobs at the workplace, including personal protective equipment, ergonomic guidelines, and hazard avoidance steps.
- *Training programs* that are consistent and required on a scheduled basis. Training is critical to new/transferred employees and when processes change or new equipment is added. Whereas regulatory training is often accomplished, hazard recognition training and accident investigation training are often not as frequently or as thoroughly taught as they should be.

- *Supervisory accountability programs* for safety whereby specific safety duties are specified in job descriptions and by company policy. Such programs can include benchmarking, activity and results goals, incentive programs, and bonuses. Supervisors should be held accountable to ensure that new employees receive training, for completion of corrective actions following an accident, and for cooperation with return-to-work initiatives.

- *Safety committees* that meet on a scheduled basis and are actively involved in safety training, accident reviews, workplace inspection programs, and corrective actions following accidents. Safety committees should consist of middle and supervisory management personnel and employee representatives from all shifts and departments. Loss data and pertinent claims information and safety program goals should be shared with the safety committee.

Hiring Practices. How are employees selected and hired? Although the demand for employees fluctuates, it always seems that we need a new employee "now," and sound human resource selection guidelines are not followed. Consider the following strategies:

- Require the applicant to complete the application at the workplace, rather than taking it home. This will enable you to determine whether the applicant is prepared to complete the form and is proficient in written skills.

- Implement substance abuse testing, keeping in mind the adage "If the employer down the street is drug testing and we are not, just whom are we hiring?"

- Provide each applicant with a detailed job description that includes the essential functions and the physical demands of the job.

- After the interview, tour the workplace, showing the applicant where he or she will work.

- After a conditional offer of employment, consider a physical examination designed to the job description. If the applicant is physically unable to perform the tasks, the offer can be withdrawn. Reference the guidelines prepared by the EEOC regarding post offer physicals as defined in the Americans with Disabilities Act (ADA).

- After a conditional offer of employment, if a physical is not given, consider asking the employee whether he or she has had any prior claims or injuries. Again, reference the ADA for the type of questions that are allowed. In the event that the individual has been injured and treated in the past, you can counsel the employee and his or her supervisor to avoid any future injury. This is important, as the employee may have a recognized disability (such as a 10 percent loss of use of the right shoulder from a prior injury), can perform the essential functions, but may not be able to perform other activities on an incidental basis.

- After the first two weeks employment, train and introduce the employee to company safety requirements. Also, provide instructions on reporting injuries and whom to speak to regarding workplace safety and worker compensation issues.

Substance Abuse Programs. Substance abuse is a major factor in workplace accidents. "Substances" refer to both legal and illegal drugs (for example, marijuana, cocaine, heroin, and methamphetamines). Although the use of over-the-counter and prescription drugs has increased, the most serious problem is alcohol abuse. Even though many employers will not tolerate "hard drugs" in the workplace, it is much harder, for a variety of cultural and historical reasons, to get employers and employees to view heavy drinking as an undesirable characteristic or one that should be discouraged. Prescription drugs can also be a casual factor in accidents in the workplace. Despite warnings that persons taking a particular drug should not operate equipment or drive, they often do both.

About 63 percent of the Fortune 1000 companies engage in some type of substance abuse testing (also called drug and alcohol testing). There are three key points for a successful and responsible substance abuse testing program:

1) Establish a substance abuse policy, clearly prohibiting use, possession, or trafficking in alcohol or drugs at work, and specify program parameters.

2) Implement an employee assistance program (EAP). This provides a confidential access to treatment.

3) Implement a drug testing program. The program can include pre-employment, post-accident, random, or site testing procedures. Reference the guidelines as published in the Drug Free Workplace Act or as specified by the U.S. Department of Transportation.

Employee Assistance Programs. EAPs provide employees with access to treatment for a variety of addictive behaviors and mental health ailments. Whether the problems are personal/family or work-related problems, employees who are preoccupied will be less productive and more likely to have accidents. Potential benefits from implementing an EAP include reduced absenteeism, higher productivity, fewer workplace accidents, and lower group health and workers compensation costs. EAPs also offer crisis management and grief counseling in the event of traumatic workplace accidents witnessed by employees and following workplace violence incidents.

Ergonomics Programs. Ergonomics programs are initiated to identify workplace musculoskeletal hazards, establish interventions, and prevent cumulative trauma disorders (CTDs). CTDs are among the fastest growing occupational health concerns, and include lower back strains, shoulder strains, and sprains and repetitive motion injuries/illnesses such as tendonitis and carpal tunnel syndrome. The benefits of ergonomics go beyond the prevention of CTDs and a reduction of workers' compensation costs. As the physical stress related to work is reduced through ergonomic job design, workers' productivity also improves.

Although OSHA legislation on an ergonomic standard has been placed on hold, California has adopted an ergonomic standard, and several other states are considering similar legislation. The basic elements of an ergonomics program include identification of hazards or "stressors" (awkward postures, high repetitions, excessive force, excessive vibration, cold temperatures, machine-paced work); hazard prevention or controls (interventions such as ergonomically designed tools, chairs, workstations); training and education; and a medical management component for CTD claims.

Wellness Programs. Wellness programs supplement an employers health insurance program and include smoking cessation programs, weight management, shift work counseling and sleep management, cholesterol and blood pressure monitoring, lifestyle analysis, and exercise programs. Some employers offer an annual physical to all employees as part of the program. Am injured employee who is in good physical health often recovers faster— an advantage for his or her employer, family, and for workers' compensation cost-control initiatives.

Injury Management Programs. Injury management programs are used to establish a communications process or standard and allows all interested parties to participate in the claim process from the time of injury until the claim is concluded. The process monitors includes the following:

- Injury reporting at the workplace
- Three-point contact by the adjuster

- Clinic for treatment (with a "call back" component to the employer)
- Substance abuse testing (if applicable)
- Accident investigations and reports
- Corrective action completion
- Return to work (or contact with the employee is absent from work)
- Ongoing claim status
- Ongoing medical treatment
- Post claim ADA or other accommodation issues
- Discharge from treatment and claim resolution (settlement or closure)

Insurance Carrier and Broker Services

Insurance carriers and broker/consultants can also provide valuable claims administration assistance to help employers control the cost of the accidents that do occur. This assistance can include developing systems to improve the efficiency of claims reported, providing methods for meaningful accident investigations, and performing timely reviews of an injured worker's case so that management is well informed of his or her condition and situation. Another important service is rehabilitation, which requires a close working relationship between the employer and the insurance carrier's claims department.

Implicit in these services is the partnership between the employer and the insurance carrier. Both have an interest in preventing and controlling losses. The time and money spent by the employer to improve his or her loss prevention and control efforts can result in much lower workers' compensation costs, as well as a more efficient operation and a safer and healthier work environment.

CONCLUSION

This chapter has examined the historical context of workers' compensation law and has provided an overview of how the law is applied for injured workers today. The environment, the laws, the financing mechanisms, and the health care delivery system may all change, but there will always be work and there will continue to be accidents in the workplace. Efforts to better understand workers' compensation and the factors that drive up its costs are part of the skills needed to manage all aspects of work-related injuries. Accidents that are prevented bear no cost, and accident that are not prevented but are managed appropriately cost much less than those that are poorly managed. What is the bottom line? To reduce workers' compensation costs, management must be committed to accident prevention.

REFERENCES

Analysis of Workers' Compensation Laws (1997). Publication No. 0491. Washington, DC: U.S. Chamber of Commerce.

Bregman, R. (Chapter 8: Workers' Compensation Cost Control) and G.M. Nethercut (Chapter 9: Controlling Workers' Compensation Medical Costs) (1995). *The Workers' Compensation Guide, Second Edition*. Dallas, TX: International Risk Management Institute.

Chamber of Commerce of the United States (1997). *Analysis of Workers' Compensation Laws*. Publication No. 0491.Washington, DC: U.S. Chamber of Commerce.

Douglas, J.R (1994). *Managing Workers' Compensation—A Human Resources Guide to Controlling Costs*. New York: John Wiley and Sons.

Malecki, D.S., J.H. Donaldson, and R.C. Horn (1980). *Commercial Liability Risk Management and Insurance, Volumes 1 and 2*. Malvern, PA: American Institute for Property and Liability Underwriters.

Nethercut, G.M. (1995). *The Workers Compensation Guide, Second Edition*. Dallas, TX: International Risk Management Institute.

Scopes of Basic Manual Classifications. (1998). Boca Raton, FL: National Council on Compensation Insurance, Inc.

Trieschman, J.S. and S.G. Gustavson (1995). *Risk Management and Insurance, Ninth Edition*. Cincinnati, OH: Southwestern College Publishing.

U.S. Department of Transportation (1998). Part 382: Controlled Substances and Alcohol Use and Testing. Federal Motor Carrier Safety Regulations. Washington, DC: U.S. Government Printing Office.

Workers Comp: A Complete Guide to Coverage, Laws, and Cost Containment (1997). Dallas, TX: International Risk Management Institute.

Chapter 5

The Changing Workforce

**James P. Kohn, Sharon L. Campbell,
and Leslie C. Ridings**

INTRODUCTION

By the year 2000, many changes in America's workforce will become evident. At present, companies are witnessing an increase of women and nontraditional workers in the workplace. The Bureau of Labor Statistics (http://stats.bls.gov/news.release/famee.nws.htm) reported that in 1996, the number of families with at least one employed person rose by 709,000. Of these 69.2 million families with at least one employed person, 28.1 million had both the husband and wife working. About nine of ten fathers and seven of ten mothers were labor force participants in 1996. At 70.8 percent, the labor force participation rate of mothers was about one percentage point higher than it had been in 1995.

Changes in the composition of the workforce bring awareness to issues such as safety and health, language barriers, and fatigue. Companies are faced with the decision to change their workplace to ergonomically accommodate the nontraditional worker. Language barriers are becoming common in today's workforce, with Spanish becoming an important language. Older workers are making up a larger portion of the workforce each year, especially with baby boomers entering their fifties. "By the year 2005, workers 55 years old and older will number more than 22 million and make up more than 15 percent of the total labor force." (Schatz 1997). What will happen if most of these baby boomers do not want to retire at an early age? Injury-, health-, illness-, and fatigue-related problems will likely occur among these older workers, causing a rise in workers' compensation and health insurance costs within industries. These problems only touch on a few of the issues that the changing workforce will see in the next few years.

Many of these issues are becoming evident in industries today. The following case study is a real-life example of how today's changing workforce is affecting everyone involved.

CASE STUDY:
ADJUSTING TO A DIVERSIFIED WORKFORCE

Bill was nine hours short of completing his masters of science degree in occupational safety management when a great opportunity became available. He was asked if he was interested in a safety internship at a commercial hog farm. Bill was assured that he would have a permanent position waiting for him if he did a good job. Bill jumped at the opportunity for two important reasons. First, he thought that taking the internship would give him real-world safety experience, which he really wanted to obtain before getting his first job. Second, Bill believed it might not be a bad idea to start paying off those college loans that had accumulated over the previous six years.

During the first week at the fifteen hundred-acre farm that employed more than three hundred people, Ted, his boss, and the farm's human resources director gave Bill an orientation. Bill quickly realized during his orientation that Spanish would have been a good subject to take at college. The farm had a sizable number of Hispanic workers, many of whom could not speak English.

The one-week orientation went by very quickly, and Bill knew he had to roll up his sleeves and get to work. The farm had very little in the way of a safety program, and Bill realized that he was going to have to start developing written programs for confined space entry, lockout and tagout, hazard communication, and personal protective equipment (PPE).

Development of the written programs was going very well until he was interrupted by an emergency phone call. Katy M. in the gestation house was injured and Bill was asked to go with her to the emergency room. It turned out that she was struck in the jaw by a feed conveyor lever. Katy ended up missing work for several months because her jaw had to be wired to set the compound fracture.

Bill conducted the accident investigation and discovered that a number of ergonomic and facility design factors contributed to its cause. He discovered that Katy could not reach the feed levers while standing on the ground. Improvising, because all the other employees were busy, Katy climbed approximately two feet above the cement floor with her feet on opposite gates of two hog stalls, in her size 7 boots (she actually required a size 5 work boot). This feed lever was different from all the rest that she had been operating. It required that Katy push it, rather than pull it, to deliver the feed into the sows' trough. As she stated in a letter, written after the accident, "It took all of my 5'0, 97-lb frame to push that lever." The lever never locked in the open position. It kicked back, striking Katy in the jaw and knocking her to the cement floor. Bill knew that once the written programs were completed, he was going to have to look at the way jobs were being performed at the farm. Many female employees were smaller in stature than their male counterparts, and he did not want size differences contributing to any more injuries.

After numerous other interruptions, Bill finally completed the written safety programs. Now it was time to train the employees. A challenging problem faced him for the first time. How was he going to train the Spanish-speaking workers? He talked to Ted about the problem and a plan was developed. In each session in which the Hispanic workers were going to be trained, a bilingual employee would be assigned. Bill would conduct the training in English and the bilingual "translator" would explain the information in Spanish. The translator even shared the jokes that Bill had told. The Spanish post-test indicated that these special employees did understand the information. The training appeared to have worked out well. After his three-month internship was completed, Bill was told that, if he wanted it, he had a job with the company. Bill took the permanent position, but he realized that he still had a lot to learn. His academic classes had prepared him for the basic safety and health knowledge that he needed to be competent as a safety professional. No one, however, had ever mentioned the people-related issues that he faced on the job. Working with female employees and non-English speaking employees had presented Bill with some problems that required creativity to solve. He realized that occupational safety and health would be a career with lifelong challenges requiring lifelong learning.

THE CHANGING WORKFORCE

In 1987, the U.S. Department of Labor published a study conducted by the Hudson Institute titled *Workforce 2000*. The purpose of this study was to determine employer needs and prospective employee characteristics as the United States entered the twenty-first century. The study concluded that two dramatic changes would take place in the American workforce. First, the white male employee population was expected to decrease as women and minorities assumed a larger proportion of the workforce. Second, businesses would face a serious shortage of skilled workers due to a combination of a slowly expanding workforce and the increasing number of jobs requiring advanced training. It appears that some of these predictions have been realized.

In its May 1997 labor statistics report, the Bureau of Labor Statistics of the U.S. Department of Labor noted that nonfarm payroll employment rose and that the unemployment rate was at a twelve-year low of 4.9 percent. The total number of payroll jobs had increased by 217,000; private-sector jobs rose by 151,000. In this same report, the Bureau of Labor Statistics reported that the total number of payroll jobs was at a seasonally adjusted level of 129.4 million. The workforce increased by 1.2 million during the first half of the year, after adjusting for the change in population controls made in January. Both the civilian labor force, 136.2 million, and the labor force participation rate, 67.1 percent, had remained stable. This report also indicated that female and minority employment was up.

So what do these numbers suggest? Just like Bill in our case study at the beginning of this chapter, many safety professionals are witnessing an increasing number of females and minorities employed at their facilities. However, these differences do not totally reflect the entire picture when you study the changing workforce. Safety and health professionals must be sensitive to a changing workforce that includes populations that may not have been employed in as great numbers as before. The reason for these changes is due, in part, to the low unemployment rate and large employment figure reported by the Bureau of Labor Statistics. Part of this changing picture is the result of legislation such as the Americans with Disabilities Act (ADA) of 1990 and the changes in federal unemployment insurance programs. What we are witnessing is an increasing number of individuals entering the workforce that, in years past, would not have been pursuing employment.

Socioeconomic factors have contributed to the changing workforce. As a consequence, today's job applicants include more minorities, women, single parents, immigrants, and handicapped workers (Trunk 1995). In addition to these populations, other groups are present in today's workforce. Illiterate and learning disabled candidates are joining the employment ranks in greater numbers, with companies taking on educational tasks in-house. Part-time or temporary employees are making up a larger percentage of the workforce. And older employees are staying on the job or functioning as consultants to meet the needs of many companies.

This chapter primarily focuses on women and handicapped employees in the workplace. The chapter concludes with a brief discussion of the health and safety issues associated with some of the nontraditional work groups previously mentioned.

WOMEN IN THE WORKPLACE: AN UNIQUE SAFETY SITUATION

According to the Department of Labor Women's Bureau article titled "Hot Jobs for the 21st Century":

> Between 1994 and 2005, employment will rise to 144.7 million from 127.0 million. This represents an increase of 14 percent, or 17.7 million jobs. . . . Women have a huge stake in the current and future job market. Women's labor force growth is expected to increase at a faster rate than men's—16.6 percent between 1994 and 2005 as compared with 8.5 percent for men. This means that women will increase their share of the labor force from 46 to 48 percent.

The number of women working outside the home has increased dramatically within the last few years. In 1954, 35 percent of all women were employed. By June 1984, this figure had risen to 53.9 percent, and by May 1997 this figure had jumped to 60.5 percent. The reasons for this increase are many, including the following:

- Delay of or decision against marriage
- Increased divorce rate
- Increased number of female heads of households
- Federal antidiscrimination legislation
- Inflation
- Increased educational level of women
- Realization of the potential hazard of relying solely on the husband's salary
- Federal welfare-to-work legislation

As a matter of interest, the stereotypical average American family with an employed father, mother at home, and two children applies to less than 7 percent of all families in the United States. The stereotypical American family of the late 1990s consists of two employed parents and one-half a child (Morrison 1990). Both spouses work in 75 percent of American households, up from 51 percent in 1988.

These social changes have made the working mother and the pregnant employee a normal part of everyday life in all occupational environments. Shifts in family structure, changes in the composition of the labor force, and the needs of all workers must be taken into account when a company designs its employment and social policies.

The move into the workplace by women has also involved a substantial number of jobs traditionally held by men. As these traditional barriers fall, more women than ever work in all types of jobs, including construction, mining, and various labor-intensive positions such as fire fighting.

There are specific considerations concerning the safety and health of women in the workplace, not the least of which is the dearth of evidence on potential safety and health hazards associated with places where women generally work. However, job health is not strictly concerned with just the female worker. There has been a great deal of emphasis on the pregnant employee and the health status of the fetus, but these are not the only topics of concern to the employer of working women. Economic and political considerations are also intimately connected to job health.

Employers realize that the health of the workforce—physical, mental, psychological, and economic—influences efficiency and productivity and makes a significant contribution to the bottom line. The following sections explore many of the working woman's considerations and concerns. To improve the work environment, these become the employer's considerations and concerns as well.

HISTORY OF WOMEN IN THE WORKFORCE

The issue of safe and healthy working conditions for women is not new. From the mid-1800s to the present, attention has focused on the special considerations of women in the workplace. In the early 1900s, Dr. Alice Hamilton, one of the most important occupational health physicians in the United States, headed the Women's Bureau. At the turn of the century, many occupational health reformers were women. The organizations they founded, such as the Women's Trade Union League and the Women's League for Equal Opportunity, were active in the fight for improved occupational health conditions for women.

During World War II, when many men were in the armed services, women were called upon to fill the subsequent employment vacancies. After the war, women were removed from mining, construction, transportation, and all types of basic industry. Not until the 1970s and 1980s did they begin returning to these traditionally male areas of employment.

In the mid-1960s, federal laws were passed that granted equal opportunity to and treatment of women workers in the United States. The Equal Pay Act, for example, an amendment to the federal minimum wage law, was passed in 1963. This act stipulated that men and women who work in the same establishment must receive the same pay if their jobs require equal skill, effort, and responsibility.

In 1964, growing out of the civil rights struggles, Title VII of the Civil Rights Act was passed. This act outlawed discriminatory employment practices, including those connected with hiring and firing; wages and fringe benefits; classifying, referring, assigning, and/or promoting; training, retraining, and apprenticeship; or any other terms, conditions, or privileges of employment. The Equal Employment Opportunity Commission (EEOC) is the federal agency that enforces these two laws.

In 1965, President Lyndon Johnson issued Executive Order 11246 to counteract employment discrimination based on race, color, religion, and national origin. In 1967, it was amended to address sex discrimination as well.

In 1978, Title VII was amended by Congress to ban discrimination based on pregnancy. It addressed women affected by pregnancy, childbirth, or related medical conditions. It stipulated the same treatment for all employment-related purposes, including benefits, given to other persons with the same ability or inability to work.

By 1978 at least sixty-five federal laws and orders had been passed on equal employment opportunity. All state protective legislation had been struck down as discriminatory, and women were placed on equal footing with men.

Men's work has traditionally been viewed as the norm. More attention had been paid to their working conditions than those occupations dominated by women and/or those workplaces where most women work. Obvious hazards among healthy workers—such as back injuries from lifting and possible carcinogenic exposures to textile and laundry workers—have not been equally addressed. When women have been considered, the primary topic has been possible dangers to the fetus, should pregnancy occur. This consideration must be broadened in fairness to all workers and employers. Neither sex should have a monopoly on effecting changes designed to produce safer and healthier working conditions.

DEMOGRAPHICS—WHO IS THE WORKING WOMAN?

Women Who Work Outside the Home: Statistics of the Workplace

Women are not only more likely to work outside the home today than in the past, but they also spend more time at work than women did in earlier years. Women have increasingly opted to work both full-time and year-round, partly due to economic necessity, but also due to movement into occupations that require full-time, year-round work. In 1995, of the 57.5 million employed women in the United States, 42 million worked full-time (thirty-five or more hours per week), and 16 million worked part-time (less than thirty-five hours per week). Two-thirds of all part-time workers were women (68 percent).

Many women who work part-time are multiple jobholders. In 1995, 3.6 million women held more than one job. The highest rate of multiple jobholding was among women 20–24 years old and single women—7.3 and 7.2 percent, respectively. Of all women who were multiple jobholders in 1995, those in the 35 to 44 age group were most likely to hold three or more jobs. An important issue is how fatigue impacts the health and safety of employees and their coworkers when they work three or more jobs.

Age and Employment

In 1997, 60 percent of all women in the United States were 16 years of age or older. Of these, 62 million were working or looking for work. More than 70 percent of all women in their twenties to forties are now in the workforce. In addition, approximately 60 percent of women in their fifties are now in the labor force (Morrison, 1990). Even half the nation's teenage women aged 16–19 were labor force participants (52 percent). The women's share of the total labor force continues to rise. In 1995, women accounted for 46 percent of the total labor force in the United States and are projected to comprise 48 percent by the year 2005. Unemployment for all women in 1995 was only 5.6 percent: 4.8 percent for white women; 10.2 percent for black women; and 10 percent for Hispanic women

Women have made substantial advancements in obtaining jobs in the managerial and professional specialties. In 1985 they held one-third (35.6 percent) of managerial and executive jobs and nearly half (49.1 percent) of professional jobs. By 1995 they held 48 percent of all managerial/executive positions and more than half (52.9 percent) of professional occupations.

Education and Earnings

In 1995, of all labor force participants aged 25 years and older, women were more likely than men to have completed high school. Ninety-one percent of female labor force participants held a minimum of a high school diploma, compared with 88 percent for men. A slightly lower percentage of female labor force participants than men were college graduates—27 percent, compared with 29 percent. Wages, however, are not generally on a par with education. Employment and earnings rates rise with educational attainment for both females and males, but earnings are lower for females than for males with the same education.

In 1994, female workers with four or more years of college had an average income above that of men who had some college education with no degree; these average figures were $35,378 and $32,279, respectively. In the same year, the average income of female high school graduates was lower than that of men who had completed fewer than eight years of school; these figures were $20,373 and $22,048, respectively. (Refer to Table 5-1.)

Marital Status

In March 1995, married females with the spouse present had a 61.1 percent labor force participation rate. (See Table 5-2 for the U.S. Department of Labor's break down of female labor force participation by marital status.) The number of families maintained by women alone, however, increased

Table 5-1: Median Income of Persons, by Educational Attainment and Sex, Year-Round, Full-Time Workers, 1994

Level of Education	Women	Men
9th to 12th grade (no diploma)	$15,133	$22,048
High school graduate	20,373	28,037
Some college, no degree	23,514	32,279
Associate degree	25,940	35,794
Bachelor's degree or more	35,378	49,228

Source: U.S. Department of Commerce, Bureau of the Census, *Income, Poverty, and Valuation of Noncash Benefits: 1994.*

Table 5-2: Female Labor Force Participation, by Marital Status, March 1995

Marital Status	Participation Rate
All women	58.9
Never married	65.5
Married, spouse present	61.1
Married, spouse absent	62.0
Divorced	73.7
Widowed	17.5

Source: U.S. Department of Labor, Bureau of Labor Statistics, Unpublished Data, March 1995.

throughout the 1980s. Single-parent families—families with only one parent residing in the household—are continuing to become a larger segment of all families, according to 1995 U.S. Department of Labor statistics. This is especially true for those families maintained by women. They accounted for 14.8 percent of all families in 1980 and 17.6 percent in 1992.

Race and ethnicity are factors when studying marital status and single female head of household employment statistics. For example, in 1992, 3.5 million black families were maintained by women. This represented nearly half (47 percent) of all black families in the United States. Of 57 million white families, 7.8 million were maintained by women; however, this accounted for only 14 percent of all white families. One of every four Hispanic-origin families was maintained by a woman. Sixty-nine percent of female-maintained Hispanic-origin families had children under age 18, compared with 66 percent of black and 58 percent of white families.

Because of the absence of a husband or a second wage earner, women who maintain families are very active in the labor force. Results of the U.S. Bureau of Labor Statistics March 1992 Current Population Survey revealed that single (never married) women with children under age 18 participated in the workforce at a rate of 52.5 percent whereas widowed mothers participated at a 61.4 percent employment rate. Married mothers with absent spouses were employed at a 63.7 percent rate and 80.3 percent of the divorced mothers in the United States were in the labor pool.

Families maintained by women had the lowest median income of all family types in 1991—$16,692—when compared with $40,995 for married-couple families and $28,351 for families maintained by men. [Income is the sum of the amounts received from wages and salaries, self-employment income (including losses), social security, supplemental security income, public assistance, interest, dividends, rent, royalties, estates or trusts, veteran's payments, unemployment and workers' compensation, private and government retirement and disability pensions, alimony, child support, and any other source of money income that is regularly received.] Wages and salaries usually make up the largest portion of a person's income. In 1991, white families maintained by women had a median income of $19,547; for similar black families, $11,414; and for comparable Hispanic-origin families, $12,132.

Divorced female family heads of household are the exception to low salary statistics. In 1991, at least one of every four employed, divorced female householders was working in managerial and professional specialty occupations. Higher educational attainment of divorced women may be one reason for their ability to secure managerial and professional jobs. Divorced women have completed more years of schooling (12.7 years) than other female family heads of household (12.4 years for single and separated female family heads of household, and 12.1 years for widowed family heads of household).

Employment—Where Do Women Work?

During the past few years, with increasing safety and health and antidiscriminatory legislation, employment opportunities for women have increased in breadth and scope. Women in almost every field of employment have obtained jobs.

The majority of women, however, continue to work in traditional professional, clerical, and service jobs. Most of the occupations in which 90 percent or more of the workers are female are in the clerical category, which includes five of the top ten occupations that employ women. Secretaries, cashiers, typists, bookkeepers, and sales workers represent these clerical occupations. Other examples in these female-intensive occupations include nurses' aides, waitresses, and the relatively low-paying, professionally educated, elementary school teachers and registered nurses.

Less traditional female jobs generally offer higher pay. They also tend to differ from female-intensive occupations in providing greater responsibility and more opportunities for advancement. However, studies have demonstrated that although more women now hold supervisory and managerial positions, their occupational status and earnings are generally less than those of their equally educated male coworkers.

According to the Department of Labor, the fifteen leading occupations of women workers in 1996 based on total female participation rates were as seen in Table 5-3.

Many of the women in the managerial/professional category tend to be registered nurses, school administrators, or teachers.

Where Are the "Hot Nontraditional Jobs" for Women in the Future?

The U.S. Department of Labor Women's Bureau defines the term "nontraditional occupation," or NTO, as any occupation in which women comprise 25

Table 5-3. Leading Occupations on Women Participating in the Workforce

Secretaries	5.3%
Cashiers	3.8%
Managerial/professional	3.5%
Registered nurses	3.5%
Sales supervisors and proprietors	3.2%
Nursing aides, orderlies and attendants	2.9%
Bookkeepers, accounting and auditing clerks	2.8%
Elementary school teachers	2.8%
Waitresses	2.6%
Handlers, equipment cleaners, helpers and laborers	1.8%
Receptionists	1.8%
Machine operators	1.7%
Cooks	1.6%
Accountants and auditors	1.6%
Textile, apparel and furnishings machine operators	1.5%

percent or less of total employment. Six major occupational groups fall under this definition for women:

1) Managerial and professional
2) Technical, sales, and administrative support
3) Service
4) Precision production, craft, and repair
5) Operators, fabricators, and laborers
6) Farming, forestry, and fishery occupations.

Congress has passed two acts to help women secure NTO employment. These acts include The Nontraditional Employment for Women Act, effective July 1, 1992, and the Women in Apprenticeship and Nontraditional Occupations Act, effective October 27, 1992. These two acts provide funding for individuals and incentives for employers to encourage females to pursue nontraditional areas of employment.

Table 5-4 presents selected high-paying, fast-growing nontraditional occupations for women as identified by the Department of Labor Women's Bureau.

Table 5-4. Selected High-Paying, Fast-Growing Nontraditional Occupations for Women

Occupation	Women (%)	Weekly Earnings	Total job Openings[1]	Change (%) [2]
Architects	16.7	702	35	17
Police and detectives	15.3	582	416	24
Engineers	8.5	897	581	19
Construction inspectors	8.5	648	28	22
Insulation workers	6.9	485	34	20
Mechanics and repairers	4.3	519	1,950	11
Firefighters	2.6	629	169	16

1 - In thousands between 1994–2005.
2 - Between 1994–2005

What the Demographics Tell Health and Safety Professionals

The demographic information on women in the workforce tells the health and safety professional that women are entering the labor force in increasing numbers. This information also points to some serious safety issues. Women are entering labor-intensive work environments. They are also attempting to raise children and manage family responsibilities at home. Many women are working several part-time jobs to make ends meet. In addition, women are not making the same salaries as their male counterparts. All these factors add up to individuals who may be under a considerable amount of stress while functioning in a general fatigue state. In other words, women employees may be working under personal physical states that are primed for injury and illness. Compound these physical factors with occupational stressors, such as job demands and workplace violence, and it becomes apparent that women are at great risk.

OCCUPATIONAL SAFETY AND HEALTH CONSIDERATIONS OF THE WORKING WOMAN

Women as High-Risk Employees

Traditionally, women have been viewed by our society as physically weak and frail, as well as more susceptible to emotional difficulties and chemical and physical stressors. However, no scientific evidence is available to support any of these generalizations. Except for pregnancy and childbirth, men and women are more similar than dissimilar.

Reported injuries and illnesses for women are on the rise as increasing numbers of women enter the workforce. Injuries include back injuries and repetitive motion/cumulative trauma problems and illness-related problems include reproductive hazards and stress (Weinstock 1994).

Occupational Hazards

According to a study by Kaplan and Knutson (1980), the difference in male and female injury and illness patterns appears related more to different job placements and types of job tasks than to biological differences. In fact, the injury and illness rates of women in nontraditional fields of employment have been demonstrated to approximate those of men.

Some injuries and illnesses experienced by females in the workplace are frequently the result of compound exposures. For example, dermatitis reactions that result from sensitivity to nickel (an allergy) is much greater in women (Plog 1988). This hypersusceptibility may be the result of ear piercing in combination with the handling of coins by cashiers or exposure to metal alloys.

A hazardous exposure for pregnant females in the workplace is an extremely serious issue. An exposure to heavy metals and other substances that may be mutagenic or teratogenic in nature should be avoided. In addition, ionizing radiation exposure is a physical stressor that should be eliminated for pregnant workers.

Heat stress is another physical hazard that creates special hazards for females in the workplace. "Female workers are more sensitive to heat, as are pregnant women, older workers, overweight people, out of shape workers and those taking certain medications, including headache, cold, and flu remedies, tranquilizers, pain medication, antihistamines, and decongestants" (Boyd 1996).

Domestic Violence

Domestic violence is another work-related issue that is a common problem involving females in the workplace. Nearly one million women are the victims of domestic violence each year in the United States. Health and safety professionals must educate themselves about the problem because its effects are brought into the workplace and can cause workplace violence. The U.S. Department of Labor Women's Bureau has reported that domestic violence interferes with a woman's ability to obtain, perform, and keep a job. Nearly three-quarters of the employed battered women who participated in a pilot study admitted that they had been harassed by their abusive partners in person or by telephone. More than half said that they had missed work an average of three days per month because of abuse. A study of Fortune 1000 companies, commissioned by Liz Claiborne, Inc., found that nearly half the corporate leaders surveyed said that domestic violence had a harmful effect on attendance, productivity, and health care costs (Occupational Hazards 1997).

With the participation of women in nontraditional occupations, many ergonomic safety and health problems have developed. Facilities and factories that were built in the past reflect designs that accommodate male anthropometric measurements. "Therefore, equipment, tools, and general plant layouts were designed for the "average Joe." The "average Susie" is dramatically affected by existing workplace reach distances and work heights" (Parker and Imbus 1992). Average Joe's are approximately 5.5 inches taller than the average Susie, contributing to workstation design stressors that affect female injuries and repetitive-motion problems on the job.

Properly Fitting the Worker to the Job

Job Analysis. Job titles and stereotyped job descriptions are often misleading. Therefore, job analyses should be performed and periodically updated, to ensure proper job-worker fit, whether the potential worker is male or female.

Written job analyses should include the following:

- Monitoring the job tasks and the work environment. In this manner, materials and/or activities that might be harmful can be identified. Changes to eliminate or control the hazard(s) can then be recommended.

- Analyzing the circumstances under which the job is performed. Direct observation, with attention to infrequently performed tasks and potential emergencies, provides a definitive picture of the necessary capabilities of a potential worker.

Job Training. All workers, both men and women, must be able to recognize and cope with the hazards of their work environment. They also need to develop the skills to perform their jobs efficiently and safely with the proper tools and equipment. Sound safety and health education and job-training programs, which usually go hand in hand, apply equally to men and women.

Ergonomics. Ergonomics explores the relationship between the worker and the environment and promotes workplace design that is conducive to the physical and emotional well-being of the worker. A correctly designed workplace can increase worker productivity, help prevent occupational disease, and help alleviate stress.

The employment of women in nontraditional roles means that a new group is performing tasks and using tools and equipment that were originally designed for men. Equipment and tools should be selected or designed and arranged, as far as is practical, to be safe and free from health risks. They should coincide with the anatomical and physiological structures of workers. Workers should be classified in terms of their lifting capabilities and strengths through an effective preplacement procedure, not on the basis of sex. Even though it is known that, on average, men are taller and one-third stronger than women, there are many overlaps between the strength scores of men and women. Men also, on average, are thinner than women of similar stature and have broader shoulders, narrower hips, greater leg strength, longer arms, less adipose tissue, and more skeletal muscle than women. What these statements mean, when comparing men to women, needs to be explored in greater depth to understand relative work capabilities more fully. In the meantime, much more data are needed to clarify the capabilities and limitations of women, because most of the current data concern only men.

Tenosynovitis and carpal tunnel syndrome are relatively common among workers who make rapid finger and hand movements. Among hospital workers, back injuries from lifting and turning patients are fairly common. An additional injury risk factor may be introduced if the female employee, after her workday, performs housework with the muscles already tired from occupational activities.

Stress. Currently, a great deal is being written about stress in the workplace. This stress may emanate from two primary sources: stressors inherent in a job, such as deadlines, and non-work-related stressors, such as financial concerns or family problems. The manner in which the worker copes with these stressors is, in part, a function of that person's personality.

Many, if not most, working women have dual roles: that of a worker and that of a wife and/or mother. As role demands increase, a woman may experience role overload—the occurrence of too many demands related to too many roles. This is not uncommon for working women.

Financial concerns, quality child care, the decision to have children or not, responsibility for housework and child care, lack of leisure time, no time to continue education or train for a better job—these are all sources of stress for many women in the workforce. In 1977, the International Labor Office estimated that the typical employed woman worked an average of 75–80 hours per week in all industrialized countries. By contrast, men were estimated to work approximately 50 hours per week. It was also estimated that the employed married woman has 17 percent less free time than the employed married man. A quality of environment survey in 1977 revealed that nearly half of the women surveyed spent an additional 3.5 hours on housework on the days they worked on the job.

For a job to be satisfying and fulfilling, a number of factors must be considered. The physical and psychological conditions of the job, the nature of the work itself, the amount of responsibility and autonomy provided, and the sense of accomplishment after completing a task are all inherent in job satisfaction.

Many women express dissatisfaction with their jobs or occupations because of several types of work-related stress. Some of these stressors include the following:

- Underutilization of skills
- Boredom
- Repetitive, paced work
- Lack of recognition for accomplishments
- Low pay
- Unfulfilled job expectations
- Powerlessness or lack of control over job situation
- Authoritarian, verbally abusive bosses
- Sexual harassment
- Lack of job mobility
- Jobs ranked low in the social structure
- Lack of promotional opportunities

Reproductive Hazards

The issue of reproductive hazards in the workplace is extremely complex. Most current standards for occupational exposures to chemical and physical agents do not consider the effects on reproduction. The teratogenicity (ability to cause birth defects while the embryo/fetus is developing) of most industrial chemicals has not yet been investigated. It is generally believed that the teratogenic ability of a substance depends on the degree of exposure, not just the potential of the substance.

An increasing number of women are being exposed to toxic substances during the childbearing years. The teratogenic effects of a substance can be experienced during the first few weeks after conception, when many women do not even realize that they are pregnant.

It is estimated that 50 percent of first-time mothers are employed at some time during their pregnancies and that 85 percent of women in the workforce will be pregnant at some point during their working lives. In 1983, 30 percent of all babies in the United States (more than 1 million) were born to mothers who were employed while pregnant.

Even though most studies have focused on birth defects caused by a pregnant woman's exposures, a more comprehensive view is needed to identify exposures of both men and women, which are suspected of leading to reproductive failure. Generally, a substance that endangers a fetus or a woman's reproductive capacity also poses a danger to the male reproductive system. There is evidence that a number of substances can cause impotence, lowered sperm counts, infertility, abnormal sperm, and/or adverse fetal outcome after exposure of the male. Chemicals with adverse affects on the male reproductive system include vinyl chloride, dibromo-chloropropane, hydrocarbons, chloroprene, polychlorinated biphenyls (PCBs), anti-neoplastic drugs, anesthetic gases, toluene diamine, organic mercury, and hexachlorobenzene. Another source of danger is radiation. Unfortunately, the exposure history of the father is rarely assessed in relation to the birth of a child.

In a 1992 study conducted by Johns Hopkins University, women exposed to glycol ethers, lead, ethylene oxide, and anesthetic waste gases have shown increased rates of spontaneous abortions (Weinstock 1994). In addition, the National Institute for Occupational Safety (NIOSH) and the Federal Aviation Administration are studying the potential link between flying high northern routes and reproductive health effects. It is believed that cosmic radiation may also be contributing to spontaneous abortions (Weinstock 1994).

In some industries, women of childbearing age have been excluded or restricted from certain types of employment because of concern about liability associated with damage to a woman's reproductive capacity or to a fetus. Various possible solutions have been offered, such as reassignment of responsibilities or telecommuting. In general, such exclusions of women should be permitted only if the reproductive health of women is shown to be in danger of significant harm, as documented by reputable scientific evidence. In such cases, research on the effects on male reproductive systems should also be conducted.

The practice of isolating women who are, or might become, pregnant is inadequate to protect the human population. In addition, such women are often excluded from equal job opportunities and are thereby economically penalized—and men and nonfertile women remain overexposed, in contravention of the purpose of occupational safety and health regulations.

To help reduce the reproductive hazards possibly attributable to the workplace, working conditions must be improved for all employees by reducing or eliminating exposure to all toxins. Occupational and environmental exposures must be more clearly defined and measured. All industrial hygiene practices must be improved and maintained so that exposure is kept to a minimum.

Women are concerned about their reproductive health and the health and normality of their children. They should be informed of all the known and suspected risks of their work environment to enable them to make informed decisions about accepting certain jobs; they must be able to react knowledgeably according to new research findings.

Personal Protective Equipment

When women first encountered nontraditional jobs, there was virtually no appropriate PPE designed for their needs. Therefore, women wore what was available for men and, if possible, modified the equipment. For example, men's hard hats and safety shoes were often stuffed with tissue or socks to provide a better fit. However, women often have to suffer the consequences of poorly fitted equipment:

- Hearing protectors, too large for many women's ear canals, cause irritation and infection
- Respirators that are too large to provide an airtight seal allow inhalation exposures
- Sleeves that are too long are in danger of being caught in moving machinery
- Gloves that extend beyond the fingertips reduce dexterity and are a safety hazard

A number of manufacturers now offer personal PPE for women that provide adequate design, comfort, and fit. It is important for the safety professional to ensure that adequate supplies of several different types of PPE are available. This permits the employee to select the most effective and comfortable equipment.

NONTRADITIONAL WORKERS

Minorities in the Workplace

The insurance industry has significantly boosted its employment of women and nonwhites since affirmative action programs began in the 1970s. According to the Bureau of Labor Statistics, the percentages of women, African Americans, and Hispanic Americans in the insurance workforce continued to rise between 1990 and 1995. However, according to a 1996 survey of 44,000 insurance agencies, by the Independent Insurance Agents of America, managerial jobs largely remain the province of white men (Smith 1997).

Employers are constantly trying to promote diversity in the workplace. A survey conducted by Kossek and Zonia (1993) found that white women and members of ethnic minorities valued employers' efforts to promote diversity in the workplace more than white men did. Groups that had members favoring diversity perceived that insufficient resources are provided for minority groups. Kossek and Zonia also found that groups having more women members than men showed more positive attitudes toward diversity. This study indicated that group characteristics rather than the nature of organization determined attitudes about diversity in general.

The February 1996 issue of *Occupational Hazards* published a study, titled "Inner-City Workers and Injury Risks," about poor and minority workers related to occupational injuries. The study found that poor and minority workers were at an elevated risk of a wide range of occupational injuries, and that relatively few workers received compensation for those injuries. More than one-half of the respondents of an inner-city survey in Philadelphia missed more than three days of work, and 15 percent missed more than a month of work as the result of an occupational injury. Most of the respondents were employed in the health care and service industries, with the remainder in construction, retail, education, transportation and manufacturing.

Diversity and fairness to minorities in the workplace is a major issue facing industries today. Employees' reactions to these issues could directly affect employee productivity as well as overall morale.

Single Parents in the Workplace

Single parenting has always been an issue in the American workforce. It has commonly been perceived that single parents were females. However, today's workforce is beginning to realize that men are now single fathers trying to survive with just one income and no help from the mother. Elbert (1995) discussed the issues that many African-American fathers have faced while raising their children. Whether widowed, divorced, or adoptive, all of the fathers in Elbert's article proved that the stereotype of the noncaring African American father is inaccurate. For example, one father is raising his seven children alone and another has adopted a son. The Joint Center for Housing Studies at Harvard University predicts that the two fastest growing households to emerge by the year 2000 will be single parent and nontraditional couple. The question for workplaces today is, how will the increase in single fathers affect the productivity and morale of the workforce, and will these single fathers face the same discrimination that single mothers have been facing for years?

Illiterate Persons in the Workplace

In the United States, the estimated number of adults (aged 17 and over) that are fundamentally illiterate ranges from 25 million (McGraw 1987) to 80 million (Ford 1992). Although we may believe that this is a problem directly linked to the American educational system, it is interesting to note that proportionally there are an equal number of illiterate persons in the United Kingdom. The percentage of the adult population that is illiterate is approximately 13 percent, and it is not improving. As a matter of fact, it is believed that minority groups will see an increase in illiteracy. The number of illiterate blacks will increase 20 percent while the number of illiterate Hispanics will increase 74 percent (Kovach and Pearce 1990).

The problem is that many adult workers are not totally illiterate: they simply are not able to read, write, or use mathematics well enough to perform their jobs properly. This problem becomes especially important when addressing health and safety training with items such as identification of hazardous materials by reading labels or material safety data sheets. The challenge to safety and health professionals will be to develop health and safety systems that take into consideration the inability of some employees to comprehend, in written form, safety and health information.

Older Persons in the Workforce

As the baby-boomer generation ages, companies are observing higher workers' compensation expenses. "The U.S. labor force is growing older. In the next 10 years the number of workers aged 55–64 will increase by almost 50 percent. Their younger brothers and sisters in the 45–55 age group will grow by 40 percent over the same period" (Mooney 1997). The U.S. Department of Labor reports that injury rates are the same for workers over the age of 35 as for workers under the age of 35. However, the Department of Labor found that the severity rate increases substantially as a worker gets older.

"Using days away from work as a measure of severity, it was found that severity for workers aged 45–64 was almost twice that of the average worker" (Mooney 1997). This study found that older workers took longer to recover from injuries because their injuries tended to be more severe. For example, fractures were three times as frequent in older workers than in younger workers.

Older employees also mean workers with greater sensory and physical limitations. Hearing losses or poorer vision mean that traditional alarm systems or labels on equipment and machinery may have to be changed to meet the needs of this population. Safety professionals should consider how workstations and work processes would need to be modified to accommodate older employees.

SAFETY ISSUES FOR PERSONS WITH DISABILITIES

We have an aging workforce, and an aging population. Because of this trend, and advances in medical care (particularly in trauma cases), more and more persons who would have died in decades past are surviving, usually with some residual disabilities. An estimated 49 million persons are considered by medical professionals to be disabled in some way, and many of these persons are on your company payroll right now. You may even be one of them. And with the ADA in force, more disabled persons are on the job and out and about who would otherwise not be on any payroll, or simply homebound.

Denying employment to a person with a mobility (or any other) impairment because of safety concerns is legally dangerous and not advised; there are almost certain ways that a disabled person can be accommodated if you look hard enough. Under the ADA, unless it would be an "undue hardship," it is the responsibility of the employer to provide any equipment needed to work safely.

You probably do not think of yourself or other persons you know as "disabled," but most likely many persons you know have some limitations that they at least joke about. Many of these limitations have safety implications both on and off the job. This section is equally concerned with on- and off-the-job safety issues, because every reader knows and cares about someone with a disability.

This section is not intended to be comprehensive; even an entire book on the topic would leave something uncovered. After you read the information about the safety implications of some disabilities, think about other ways to apply it. Do you have aging parents? Can you work with the company nurse to identify employees likely to need a little extra safety help?

How can you encourage employees to be honest about what they can and cannot do? Did your spouse break a leg recently? Do people comment that you are not hearing well lately?

As we all age, we will acquire disabilities; how we handle them is up to us. Acknowledging that we need a little extra help and getting it can make the difference between surviving an emergency and dying.

One safety tip applies to virtually every type of disability: notify emergency providers, (company and/or local firefighters) of the locations and special needs of employees and family members who would have difficulty detecting or escaping from an emergency situation. Medical emergency care providers appreciate knowing the usual locations of persons with medical conditions that could cause a crisis (for example, diabetes, heart disease, uncontrolled epilepsy, asthma), especially if those individuals are medically fragile.

The bottom line for workplace safety for persons with disabilities is that they must be able to admit needing assistance or being unable to perform as usual without fear of punishment. If an organization abides by the spirit of the ADA, and willingly accommodates those who need it, it will be rewarded with a safer workplace. When disabled persons feel free to ask for help when they need it, instead of hiding disabilities, they are less likely to cause or become involved in injury-causing events.

The admission of needing help without fear of punishment is also important in family or community situations. Obviously, you cannot fire a family member, but you can express disapproval when a family member takes risks that could lead to injury or death rather than request assistance. Negotiation to ensure that everyone's needs are met is crucial for everyone's maximum functioning.

Sensory Impairments

Loss of Hearing. Noise-induced hearing loss, which damages the ability to hear upper pitches and is the most common type of hearing loss, affects the ability to understand speech and alarms. An estimated 10 percent of the population is hearing impaired, according to Self Help for Hard of Hearing People, Inc. Unfortunately, the average person waits seven years before getting hearing aids, and then everyone else expects that the problem is solved. It is not!

Hearing aids amplify sound; they do not usually clarify it. To understand in a noisy area, assistive listening devices (ALDs) and other communication tools are needed. ALDs are microphones that amplify what the listener wishes to hear; the hearing-impaired person turns off the general microphone on his or her hearing aids and listens instead to what is coming in through the ALD microphone.

Some persons test "normal" with pure-tone hearing tests, but have difficulty understanding speech because of subtle brain damage. Hearing tests should also include a test to see how well a subject can distinguish between different words.

Other communication tools that can be used if a person's hearing loss is too severe for an ALD to be of use include cued speech, sign language, and oral interpreters. Trained interpreters are needed for these types of communications. At meetings, specially trained court reporters can provide real-time captioning, or a fast typist can provide "computer-assisted notetaking." The computer output from a court reporter or computer typist can be displayed on a screen via a liquid crystal display unit on an overhead projector, or the hearing-impaired person can read it on the computer monitor.

When videotapes are used, captioning should be provided as a matter of course. During those seven years before the estimated 10 percent of your employees get help for their failing hearing, a lot of training opportunities are being missed! When performing one-on-one training, writing down the information being communicated is often effective, assuming that the trainee is literate.

Make sure persons who do not hear well understand all important information: safety instructions, job details, machine-operating instructions, or other communications. Pretending can result in injury to them or to someone else. Have hearing impaired workers repeat what they have been told to ensure that they understand how to operate machinery and understand all processes at work. This is not a bad idea for everyone regardless of his or her hearing status.

Other safety tips for the hearing impaired include the following:

- Make sure all employees and family members can hear alarms at work and at home. Persons with deteriorating hearing should regularly test their ability to hear all alarms.
- Obtain alarms with strobe lights for persons who cannot hear alarms, and arrange for a "buddy" to warn them.
- Implement vibrating devices so that deaf employees can be alerted.
- For cars, use devices that flash red lights or turn on signals when emergency sirens are nearby; even normal-hearing drivers often cannot hear outside sounds, especially when the radio is on.

- Put it in writing when you observe that an employee is having trouble understanding when others speak.
- For text telephone (TT) users, contact the local 911 center and arrange to make test calls from time to time so that 911 operators regularly use their TT equipment. The majority of such calls fail to get the help needed unless operators use the equipment on a regular basis.

Loss of Vision. Like hearing loss, vision loss varies from person to person, as does how it affects a person's ability to function. Some common problems include different types and degrees of color blindness, macular degeneration and cataracts (primarily in older persons), and retinitis pigmentosa (RP). Macular degeneration causes a blind "hole" in the center of a person's vision; RP destroys peripheral vision. Cataracts increase the amount of light a person needs in order to see. Some employees will deny a vision problem, and this can lead to safety problems. Everyone should have his or her vision tested regularly and thoroughly, and immediately obtain and wear any needed corrective lenses or magnifiers for safe living and working.

Vision testing is an important component of any safety program. For jobs in which accurate color perception is essential, test employees with the actual colors they need to be able to distinguish. Other safety tips for the vision impaired include the following:

- Make sure that everyone can see any visual alarms at work or home. Turn the visual alarm into an auditory alarm for those who cannot see it.
- Provide any visual assistive equipment for employees to work safely under the ADA, unless it imposes an "undue hardship."
- Make sure that employees can see what they need to. Have them describe what they are looking at if you have any doubts; pretending could result in injury or death.
- If there is no corrective equipment that will compensate for an employee's slowly failing vision, instruct him or her to stop driving, operating hazardous equipment, or doing any activity that is unsafe without good vision before he or she "has to."
- If you rearrange furnishings, equipment, and other items, immediately show blind and visually impaired persons the changes.

Loss of Smell. Gas and chemical leaks and fires are often detected by smell. Smoke and gas detectors with alarms are essential for those who cannot detect odors. It goes without saying that every workplace and home needs smoke and carbon monoxide (which is odorless) detectors, regardless of how keen the occupants' senses of smell are.

Speech Impairment. Speech impairments can range from hoarseness to total loss of voice, or a problem producing words caused by stuttering, cerebral palsy, brain damage, or a host of other problems. Make sure anyone with such difficulties can communicate effectively, and summon aid quickly if needed. A TT is an effective tool for those who can type; signals can be prearranged with emergency providers and others who need to know quickly that an emergency is pending if needed. A caller-ID system can be an effective safety accommodation to identify immediately a caller with impaired speech.

Mobility Impairment. There are many causes of mobility impairments: diseases such as multiple sclerosis and muscular dystrophy, paralysis of the lower extremities or whole body, leg amputation, and coordination problems such as with cerebral palsy, head injury, stroke, and so on. Persons with mobility impairments need to be individually assessed for a good understanding of their capabilities and limitations, and to provide any needed accommodations. Emergency procedures need to reflect what these persons can and cannot do. Persons with progressive conditions that will gradually reduce their capabilities need periodic reassessment.

Any building more than a few stories tall should have evacuation chairs so that individuals who are incapable of walking can get out of the building in the event of a fire or other emergency. Even if no one with a mobility impairment is on the premises, anyone can have a medical emergency rendering him or her incapable of walking. For example, if there is a fire, persons with asthma or heart conditions may need to be carried or otherwise assisted. Without an evacuation chair, those doing the carrying are put at risk of falling and injuring themselves or experiencing and other injuries.

Mental Ability Impairment. Persons with reduced intelligence, emotional problems, and some other mental impairments can make excellent employees, but may need additional supervision and accommodations to reduce stress or job complexity. It is essential that all employees know how to escape from the workplace in an emergency; some employees may need more training and more frequent refresher training than others. Here are some tips:

- Know what the person can and cannot do safely.
- Provide the amount of supervision needed for safety.

- Review safety and emergency procedures frequently.
- Assign a "buddy" for emergencies.

Chronic Illnesses. Many illnesses cause changes in mental and physical abilities from day to day or even minute to minute, including some of the following:

- Diabetes
- Multiple sclerosis
- Chronic fatigue and immune dysfunction syndrome
- Asthma
- Many heart diseases
- Epilepsy
- Migraine headaches
- Arthritis

One key to safety is for a person to understand his or her illness and its effects both on and off the job. Another key is for the person to be honest with himself or herself and others if he or she cannot function as well as usual. Sometimes the affected person is least likely to spot early problems, in which case others need to know what to look for. Assistance or reassignment to another task must be offered in a nonjudgmental way, without any sense of punishment in order to be successful.

Mental Illness. More persons than you would suspect are being successfully treated for mental illnesses, which are often probably biochemical problems in the brain. Sadly, many rightly fear admitting their problem, especially to coworkers or bosses. Therefore, they do not let others help them identify early signs of mental or psychological problems.

The key for workplace safety here is to let all employees know that it is safe to admit a problem and to ask for help when needed. Off the job, open and honest communication about symptoms, and getting appropriate medical care promptly are crucial. However, on or off the job, there should be no tolerance for violent outbursts involving assaults or threats to others, for everyone's safety.

Temporary Disabilities. Injured persons probably face greater safety risks than those who are living with long-term disabilities. They may be inexperienced in handling such aids as crutches or wheelchairs, may not know how to function with just one hand, and may not know what types of assistive equipment are available or how to use such equipment. They are not used to factoring in their temporary impairment in assessing whether they can do something safely, and are often in pain. Pain is distracting, fatiguing, and reduces strength. Pain medications taken for short-term use can impair mental functioning and judgment, too. Workplace safety would be enhanced if safety professionals were notified when injured workers return to work and were able to assess their new safety needs. Off the job, family members and medical professionals can help. The precautions for a particular temporary disability are the same as for the corresponding permanent disability.

Permanent Disabilities and the Use of Assistive Animals. There are many different types of permanent disabilities, and assistive animals are often used to help the permanently disabled. There are seeing-eye and hearing dogs for blind and deaf persons, and "assistive dogs" for those with a variety of mobility problems. Assistive dogs do such tasks as retrieve dropped objects, push elevator buttons, provide stability for unsteady walkers, and fetch desired objects. Some quadriplegics use trained monkeys, and often household pets serve as an owner's ears without anyone realizing it until the animal is gone.

By law, certified assistive animals are allowed in the workplace, with a very few exceptions. These animals are not pets! Certified assistive animals wear an orange harness, leash, or vest, or have some other indication that they are certified. Trying to find an excuse to bar these animals is the same as saying to other employees that you will not allow their eyes, ears, legs, or arms inside.

Resentment or phobias by other employees are not acceptable excuses to bar these animals, either. Phobias are highly treatable; have the affected employee get counseling! Allergies are often cited as another excuse; these, too, are usually highly treatable.

Employee training is also necessary to teach others not to pet or distract working assistive animals. Assistive animals need to be with their disabled owners. Their needs for food, water, and a place to eliminate wastes *must* be accommodated—and they need to be factored into emergency plans as well.

CONCLUSION

Persons with disabilities often have special safety needs both on and off the job. Since aging eventually causes some disabilities for everyone, learning about and accommodating these special safety needs is very important.

The life you save could even be your own! The best experts you have for helping devise reasonable accommodations and solutions to safety problems are often the ones with disabilities themselves. Be sure to consult them early and often, and with a positive attitude.

ADDITIONAL RESOURCES

Thousands of organizations have information on disabilities. Those selected for this list are of the most general interest. *Note:* area codes are changing rapidly; you may need to check if the area code listed here is current. "TT" stands for "text telephone."

AbleData
Newington Children's Hospital
181 E. Cedar Street
Newington, CT 06111
  800-344-5405
  203-667-5405
  Database with over 15,000 listings of adaptive devices for all disabilities.

American Paralysis Association
2201 Argonne Drive
Baltimore, MD 21218
  800-526-3456
  24-hour hotline with referrals and information on rehabilitation and psychological adjustment.

The Arc (formerly Association for Retarded
          Citizens of the U.S.)
500 East Border Street, #300
Arlington, TX 76010
  817-261-6003
  817-277-0553 (TT)
  Advocacy, support, and information group for children and adults with mental retardation and their families.

AT&T National Special Needs Center
2001 Route 46
Suite 310
Parsippany, NJ 07054-1315
  800-233-1222
  800-833-3232 (TT)
  Has products for communication needs for customers with hearing, speech, motion, or vision impairments.

Clearinghouse on the Handicapped
Switzer Building, Room 2319
330 C Street, S.W.
Washington, DC 20202
  202-732-1250
  National clearinghouse group providing information on all types of disabilities.

Computer-Disability News
National Easter Seal Society
5120 South Hyde Park Blvd.
Chicago, IL 60615
  312-667-7400
  Quarterly computer resource newsletter for persons with disabilities.

Direct Link for the Disabled
P.O. Box 1036
Solvang, CA 93464
  805-688-1603
  Nonprofit referral agency with listings of more than 10,000 organizations and community-based resource centers for all ages and disabilities.

Job Accommodation Network
P.O. Box 6123
809 Allen Hall
Morgantown, WV 26506-6123
  800-DIAL JAN (342-5526)
  Free consultant services providing assistance to employers and persons with disabilities.

National Amputation Foundation
12-45 150th Street
Whitestone, NY 11357
718-767-0596
Provides information for amputees and their families.

National Association of the Deaf
814 Thayer Avenue
Silver Spring, MD 20910
301-587-1788
Oldest and largest consumer organization of deaf persons; has information on communication skills and employment rights, among many other areas.

National Clearinghouse on Technology and Aging
University of Massachusetts Medical Center
55 Lake Avenue N.
Worcester, MA 01655
800-433-2306
508-865-3662
Provides a variety of services, including Sensory Technology Information Services, which provides persons with sensory disabilities information on assistive technology and special services.

National Spinal Cord Injury Association
600 W. Cummings Pk #2000
Woburn, MA 01801
800-962-9629

617-935-2722
Has disability information and publications, and local associations throughout the United States.

Self-Help for Hard of Hearing People, Inc.
7800 Wisconsin Avenue
Bethesda, MD 28014
301-657-2248
301-657-2249 (TT)
Has information on hearing loss, communication, assistive listening devices, and alternative communication skills.

United Cerebral Palsy Association National Headquarters
1522 K Street N.W.
Washington, DC 20005
800-872-5827
202-842-1266
Support and advocacy organization; will provide information about the nearest local affiliate.

World Institute on Disability
1720 Oregon Street
Berkeley, CA 94703
415-486-8314 (V/TT)
Public policy center run by persons with disabilities. Uses research, public education, training and model program development to create a more accessible and supportive society for all.

REFERENCES AND SUGGESTED FURTHER READING

Boyd, V. (July 1996). "Dealing with Heat Stress." *Occupational Health and Safety*.

Elbert, A. (June 1995). "Dads on Their Own: Don't Believe the Hype." *Essence*, p. 76.

Ford, D. (November 1992). "Toward a More Literate Workforce." *Training and Development*, pp. 52-55.

"Inner-City Workers and Injury Risks." (February 1996). *Occupational Hazards* 58(1): 24.

Kossek, E., and S. Zonia. (January 1993). "Assessing Diversity Climate: A Field Study of Reactions to Employer Efforts to Promote Diversity." *Journal of Organizational Behavior*, p. 61.

Kovach, K., and J. Pearce (April 1990). "HR Strategic Mandates for the 1990s." *Personnel*, pp. 50-55.

McCoy, F. (January 1990), "B.E. Economists' Report: Standing on Shaky Ground." *Black Enterprise*, pp. 55-60.

McGraw, H.W. (1987). "Adult Functional Illiteracy: What to Do About It" *Personnel* 64(10):38-43.

Mooney, S. (March 1997). "Financial Insights." *National Underwriter,* p. 19.

Morrison, P.A. (April 1990). American Newspaper Publisher's Association Conference, Las Vegas.

Parker, K. and H. Imbus. (1992). *Cumulative Trauma Disorders.* Boca Raton, FL: Lewis Publishers.

Plog, B. (1988). *Fundamentals of Industrial Hygiene.* Itasca. IL: National Safety Council.

Schatz, R.D. (1997). "The Aging of the Work Force," *Working Woman* 22(5):64-66.

Smith, L. (January 1997). "As America Changes, So Does the Insurance Workforce." *Best's Review*, p. 42.

Trunk, C. (October 1995). "Plan Now for Workforce 2000." *Material Handling Engineering*, pp. 113-136.

Weinstock, M. (March 1994). "How Safe Are Women in the Workplace?" *Occupational Hazards*, pp. 68-70.

"Working on Domestic Violence" (January 1997). *Occupational Hazards* 59(1):24.

Chapter 6

Economic Decision Analysis for Safety Professionals

Mark D. Hansen

INTRODUCTION

Companies must choose one of two basic strategies concerning adherence to occupational safety and health regulations: compliance or intentional avoidance. Within each of these basic strategies, there are specific alternatives ranging from *full compliance,* by fully implementing ergonomics, to *extreme avoidance,* by doing nothing (and risking OSHA fines as well as losses associated with workers' compensation).

Although compliance is intended to protect the safety and health of workers, economically struggling companies may not accept compliance and may instead opt for cost-reduction activities even if such action requires avoiding a risk-taking strategy. Regardless of the strategy selected, some health and safety-related costs will occur, and therefore impact a company's (tangible) costs as well as possible product quality, as measured by production rejects and customer feedback. In addition to these tangible factors of cost and quality, there are "vague" or difficult-to-measure factors (intangible) such as production flexibility, public image, and employee morale. Because most health and safety-related decisions involve both tangible and intangible factors, a method of analyzing risks is vital for appropriate decisions to be made regarding the identification of optimal safety and health program actions.

Poor decisions concerning compliance can be costly for an employer, so it is important to analyze costs and related risks factors to arrive at an optimal decision. One method that can be used is the technique referred to as risk mapping by its authors, Kenneth H. Harrington and Susan E. Rose.

This chapter details the risk-mapping approach and summarizes the results of an application at a major chemical company in the United States.

USING RISK MAPPING FOR INVESTMENT DECISIONS

Risk mapping is a tool used to manage risk, optimize resource allocations, and adjust project schedules based on cost and risk information. It combines an order-of-magnitude integrated risk analysis approach with cost data and importance measures. Thus, risk mapping extends safety risk analysis efforts to have true business value to an organization.

The pace of technological change is accelerating rapidly. Change is the only permanent feature of corporate America today. As time is compressed, space is expanded, and industrial and economic activities unfold on a global scale. In this dynamic environment, business enterprises are challenged to continuously create "value." As a result, constant innovation has become a survival skill.

Thus, deriving value from technology has become universally important as a core competency of modern enterprises. However, deriving value from technology is no longer as simple as meeting customer requirements or improving product performance. Value is always multifaceted, often subjective, and occasionally bewildering. Fundamental activities in the process of delivering value include the following:

1) Determining the dimensions of value
2) Systematically identifying and balancing the technological, financial, environmental, and societal risks
3) Establishing measures for value
4) Establishing an organization suited for value creation
5) Identifying and acquiring new technologies
6) Managing the deployment of technologies

Risk mapping was developed with value creation at its core. It is a tool used to establish dimensions and measures of value and to provide a balance among the various aspects of value. Companies have used it for identification of value creation opportunities in a risk management format. That is, by identifying and evaluating an integrated risk profile of safety issues, companies have determined where to invest to achieve the highest value of return on their safety investments.

Overview of Risk Mapping

Value from safety issues most often comes in the form of cost or risk avoidance, but can result in increased productivity, which translates directly to the corporate bottom line. A true value determination must account for both costs and risks. The difference between costs and risk can be summarized as follows:

- Costs are expected expenditures that can be included in a budget of financial forecast for an economic time frame of interest.

- Risks represent expenditures or liabilities that are potential but not expected within the same economic time frame; hence, they are not generally included in a budget or financial forecast. A probability exists that an expenditure or liability will actually be incurred within each time frame of interest. Thus, the expense will be zero if the loss incident does not occur. The expense or liability can be very high if it does occur, and can have a significant impact on a business.

To combine costs and risks, they must both be in the same units of measure. Since costs are generally in monetary units, and decisions are generally made on an economic basis, it follows that risks must also be converted to monetary values.

Risk is defined as a combination of the likelihood of occurrence and the severity of consequences of unexpected loss incidents. To combine risk with costs, the risks are put into units of dollars per year. The "dollars per year" risk measure is thus an annualized liability or loss rate. Eliminating that liability adds value to an organization.

Risk mapping provides risk management and optimized resource allocations based on cost and risk information. It combines an order-of-magnitude integrated risk analysis approach with cost data and importance measures. Thus, risk mapping extends safety risk management efforts to an organization's true business value. The risk-mapping information is stored in a computerized database that interfaces with project management software. By using the risk-mapping tool, decisions can be made in a cost-effective manner based on cost and risk information.

Defining the Scope of Risk Mapping

Risk-mapping methodology may be used to address a wide range of objectives at varying levels of detail. It is important at the onset, however, to define the goal clearly and, therefore, limit the scope as necessary. Examples of applications range from a site-wide risk prioritization, which may include not only performance risk but also the risk of delaying or eliminating a project to a top-level strategic issue prioritization.

To develop an understanding of risk requires addressing three specific questions: What are the hazards? What are the possible undesired outcomes of hazards? How likely are these outcomes to occur? To understand risk, it is essential to view an accident as a sequence of events (see Figure 6-1). A hazard is generally defined as the presence of a material or condition that has the potential for causing loss or harm. An accident scenario begins with an unplanned initiating event, or deviation involving a process hazard. The effects of the deviation are undesired outcomes or consequences and potential harmful impacts. Preventions reduce the likelihood of the deviation occurring, whereas protections reduce the likelihood of the consequences occurring, given that a deviation occurs.

Order-of-Magnitude Methodology

Estimating the risks of safety issues involves determining the likelihood of an undesired outcome and the impact of that outcome, should it occur. To simplify the risk analysis portion of risk mapping, cost and risk parameters are based on an order-of-magnitude basis. Furthermore, to simplify the display and combination of cost and risk parameters, only the exponents of the magnitudes are used. For example, a risk of 100 times per

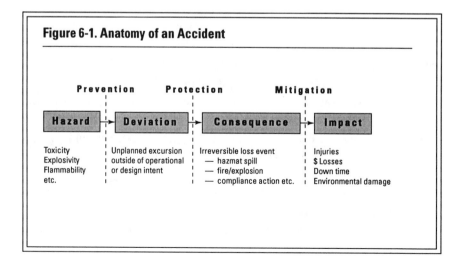

Figure 6-1. Anatomy of an Accident

Prevention   Protection   Mitigation

Hazard → Deviation → Consequence → Impact

Toxicity	Unplanned excursion	Irreversible loss event	Injuries
Explosivity	outside of operational	— hazmat spill	$ Losses
Flammability	or design intent	— fire/explosion	Down time
etc.		— compliance action etc.	Environmental damage

Table 6-1. Order-of-Magnitude Scale Depicting Likelihood of Undesirable Outcome

Magnitude	Times Per Year	Alternate Description
+2	100	Twice per week
+1	10	Once per month
0	1	Once per year
-1	0.1	Once every 10 years, or 10 percent chance per year of operation
-2	0.01	Not expected to occur during facility, but may occur; 1 percent chance per year of operation
-3	0.001	Extremely unlikely to occur during facility life; 1 chance in 1,000 per year of operation
-4	0.0001	Extremely unlikely

year is recorded as a 2, since 100 per year is equal to 102 per year, and only the exponent A2" is used.

The likelihood of occurrence of each undesired outcome, or scenario frequency, is based on the estimated frequency of the initiating event and the effectiveness of the preventive and protective features. An order-of-magnitude scale, as shown in Table 6-1, can be used for capturing the likelihood of occurrence of each undesired outcome.

If an initiating event were expected to occur once every ten years, risk mapping would assign a value of -1 to the initiating event. If there were a 10 percent chance that a particular protection would fail to minimize the outcome of that event, risk mapping would also assign a value of -1 to the protection. The undesired event occurrence frequency is the frequency of the initiating event times the probability that the protection(s) would fail, or

1/10 years x .1 = 1/100 years.

Since risk mapping deals with orders of magnitude, the result can be achieved by the sum of the initiating event frequency and the protection effectiveness, or

$$-1 + -1 = -2, \text{ or } 1/102 \text{ years, or } 1/100 \text{ years}$$

Evaluating the impacts of undesired events includes evaluation of the types of impact to be considered and the severity of each impact type. The risk-mapping approach provides a framework for capturing the wide range of potential impacts that a given scenario might impose, such as worker and public safety, business impact, and social impacts. Table 6-2 gives an example scale for measuring the severity of consequences of undesired outcomes related to facilities handling hazardous materials.

Since impacts are additive rather than multiplicative (as is frequency), combining impact from various impact types in risk mapping is not simple. Impacts must be added and combined in an absolute manner. Thus, if an event had outcomes of medical treatment for workers (severity magnitude 3), exposure above limits for offsite populations (severity magnitude 4), and localized, short-term environmental effects (severity magnitude 4), the event impact calculation would be as follows.

Table 6-2. Measuring the Magnitudes of Severity of Consequences

Magnitude	Cost, Loss or Liability ($)	Effects on Workers	Effects on the Public	Effects on the Environment
7	10 million	Fatality or permanent health effect	Fatality or permanent health effect	Widespread and long term or permanent
6	1 million		Severe or multiple injuries	Widespread and short term or localized and long term
5	100,000	Severe or multiple injuries	Injury or hospitalization	
4	10,000	Lost workday(s)	Exposure above limits	Localized and short term
3	1,000	Medical treatment	Exposure below limits	Reportable spill
2	100	First-aid case	Odor/noise concern	Variation from permit

$$\$103 + \$104 + \$104 = \$1,000 + \$10,000 + \$10,000 =$$
$$\$21,000 = \$2.1 \times 104 = \$104.3$$

In order-of-magnitude terms, this becomes 3 and 4 and 4 = 4.3 or \$21,000.

Risk Determination

Because risk is defined as a combination of the likelihood of occurrence and the severity of impacts of unexpected loss incidents, risk mapping's order-of-magnitude approach allows a simple calculation similar to the frequency determination. In risk mapping, risk is the sum of the frequency and total impact magnitudes. Using the previous equations, the risk calculation would be

$$\text{Frequency} + \text{Impact} = \text{Risk}: -2 + 4.3 = 2.3, \text{ or } \$210/\text{year}$$

An Example Risk-Mapping Application. A U.S. chemical company recently used risk mapping to identify key strategic safety issues. The primary objective was to develop and apply a systematic risk identification process in a cost-effective manner to be used by management on an ongoing basis to assist with risk management decisions. The process involved identifying key strategic issues from a set of high-level potential accidents. The strategy applied included six steps:

1) Identify the issues.
2) Determine impact categories.
3) Develop accident scenarios associated with each issue.
4) Obtain cost and risk information.
5) Determine risk magnitudes.
6) Establish risk tolerability criteria.

Next, each plant site was asked to submit five key safety issues. These key issues were compiled and combined into twenty-one issues that represented the key corporate issues. These twenty-one issues were used as the starting point to identify strategic environmental, health, and safety issues.

The frequency and effectiveness categories established in risk mapping are based on time and therefore can be used in any study. Impact categories and magnitude of impacts are unique to each study. We determined that the following impact categories were important to this company:

- Worker health and safety
- Public health and safety
- Capital assets
- Operational continuity
- Compliance
- Product and service liability
- Ecology
- Society

Based on expected levels of impact, qualitative descriptors were applied that established quantitative levels similar to those in Table 6-2.

The third step in the process included developing a sequence of events for each of the twenty-one issues. That is, for each issue, we postulated initiating events, prevention, protection, and the expected impacts. For each scenario, we estimated the likelihood for the initiating event based on the order-of magnitude method.

The scenario risk was then calculated based on the estimated scenario frequency and relative impact. Next, these risk estimates were combined to reflect the total risk for the given key issue.

The final task was to establish criteria for making risk management decisions. These criteria were used to sort key issues into three categories:

1) Issues whose risk is high enough that action is required regardless of economic return.

2) Issues whose risk is high; however, any risk reduction effort must show an economic return.

3) Issues whose risk is low and warrant no risk-reduction efforts.

These risk tolerability criteria represent the corporate risk aversion. Because senior management makes the decisions that affect the amount of risk to which the company is exposed, a combination of the individual risk aversion of senior management was used to determine corporate risk aversion.

A series of individual interviews were used to measure senior management risk aversion. An estimate of individual risk aversion through a risk tolerability questionnaire based on both economic and human impact was established. Examples of the questions posed are as follows:

What level of annual economic loss from a single type of event at any facility would you consider to be a part of normal operations?

What is the largest human loss you can conceive resulting from a single event over a single plant's lifetime?

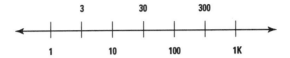

Based on the results of the risk mapping and the risk tolerability questionnaire, the risks associated with strategic safety issues were plotted, forming a risk matrix, as shown in Figure 6-2. The matrix shown plots risk on a log-log scale with indices of frequency and impact. The values of frequency and impact are the order-of-magnitude values used to calculate risk. Levels

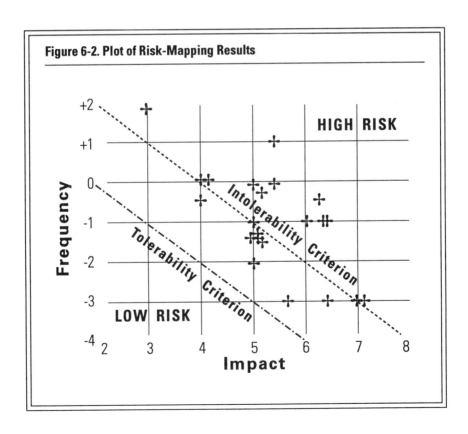

Figure 6-2. Plot of Risk-Mapping Results

of constant risk in the matrix go along the diagonal from the upper left to the lower right with the highest risk in the upper right-hand corner and the lowest risk in the lower left-hand corner.

The lower diagonal line plotted represents the level of tolerable risk. Any event with a risk that falls to the left and below that line warrants no action. The upper diagonal line plotted represents the level of intolerable risk. Events above and to the right of that line are characterized as high risk. Events above the intolerable risk line warrant risk-reduction actions regardless of whether there is a positive economic return associated with the risk-reduction action. The area between the lines represents the area where risk-reduction actions must have positive economic return equal to or greater than other corporate investments.

Of the twenty-one key issues considered in this analysis, fourteen fell into the high-risk area of the plot. These issues must be addressed and the risk associated with them must be reduced even though there may not be a net positive economic return from the investment. Seven of the issues fall into the area between the criteria. These seven issues should be addressed only if risk reduction measures offer a return equal to or greater than other corporate investments.

CONCLUSION

Determining value from safety issues is a process that sets the framework for value creation and that continuously works within that framework to identify the potential of value. Risk-mapping technology is a flexible, cost-effective tool that has been used as an input to a company's investment decision-making process. It provides management with an established risk characterization method to help identify strategic issues, quantify risk, and confirm issues that warrant risk-reduction actions.

In the example application, risk mapping was used to establish a framework for the determination of key issues that have the potential for adding value. Risk mapping was also used to identify criteria for determining when the strategic issue of adding value applies and when other decisions apply to reducing risk. This framework satisfies the initial steps in the value creation process, that is, determining the dimensions of value, identifying and balancing risks, and establishing measures for value.

The next step in the process is the assimilation of the risk-mapping process throughout the organization. If the company plans to establish an organization focused on value creation, it must enable each plant site with the skills necessary to perform a broader search for value creation opportunities. It plans to expand the application to include multiple business units and to identify additional issues. Thus, the company expects to create an environment for identifying and acquiring new value-added technologies.

SUGGESTED FURTHER READING

Alexander, D. (1994). "The Economics of Ergonomics." *Proceedings of the Human Factors and Ergonomics Society Annual Meeting—1994.* Human Factors Society, pp. 696-700.

Canada, J.R. and J. A. White (1980). *Capital Investment Decision Analysis for Management and Engineering.* NJ: Prentice-Hall.

Crandall, R. (1983). *Controlling Industrial Pollution,* Washington, DC: Brookings Institute.

Davis, J.R. (1998). "Effective Decision-Making for Ergonomic Problem Solving." Chapter 11 in: *Ergonomic Process Management.*

Davis, J.R. (1995). "Automation and Other Strategies for Compliance with OSHA Ergonomics." *Proceedings of 1995 International Industrial Engineering Conference,* International Institute of Industrial Engineers, pp. 592-599.

General Accounting Office (1997). *Worker Protection: Private Sector Ergonomics Programs Yield Positive Results,* General Accounting Office, GAO/HSAFETY-97-163.

Grant, E.L., Ireson, W.G., and Leavenworth, R.S. (1990). "Aspects of Economy for Regulated Business." Chapter 17 in: *Principles of Engineering Economy.* New York: John Wiley & Sons.

Hansen, M.D., H. W. Grotewold, and R. M Harley. (1997). Dollars and Sense: Using Financial Principles in the Safety Profession. *Professional Safety* 42(6):36-40.

Harford, J.D. (1978). "Firm Behavior under Imperfectly Enforceable Pollution Standards." *Journal of Environmental Economics and Management* 5:26-43.

Harrington, K. H., and Rose, S. E. (February 1998). *Using Risk Mapping for Investment Decisions,* Mary Kay O'Connor Process Safety Center, First Annual Process Safety Symposium, College Station, TX.

Jones, C.A. (1989). Standard Setting with Incomplete Enforcement Revisited. *Journal of Policy Analysis and Management,* 8(1):72-87.

National Safe Workplace Institute (NSWI) (1992). *Basic Information on Workplace Safety & Health,* p.16.

Owen, J.V. (1991). Ergonomics in Action. *Manufacturing Engineering* 106(6):30-34.

Schelling, T.C. (1983). *Incentives for Environmental Protection.* Cambridge, MA: MIT Press.

Slote, L. (1987). "How to Reduce Work Injuries in a Cost-Effective Way". Chapter 22 in: *Handbook of Occupational Safety and Health.* New York: John Wiley & Sons.

Viscusi, W. K. (1986). "The Impact of Occupational Safety and Health Regulation, 1973-1983." *Rand Journal of Economics* 17(4):567-580.

Zeleny, M. (1992). *Multiple Criteria Decisionmaking.* New York: McGraw-Hill.

Chapter 7

Multinational Organizations

Kathy A. Seabrook

INTRODUCTION

This chapter outlines how successful multinational companies plan and manage their safety and health processes transnationally. The following key areas are addressed:

- What drives international companies to provide a safe and healthful workplace for their global workforce
- Key components of a global safety and health management system
- Planning and implementation
- Managing local safety and health regulatory/legal compliance
- Safety and health performance metrics

THE GLOBALIZATION OF BUSINESS

Multinational companies have been part of the international business landscape for decades. However, the face of the multinational company has changed. In the 1990s companies have focused on international growth, increased global market share, and international revenues. Business recognizes that increasing market share in nondomestic sectors is key to success now and into the future. In the past, the chemical, petrochemical, pharmaceutical, and construction industries dominated the international market. Today, a broader range of companies are now going global. These companies understand that the opportunities for growth and financial results are substantial. As a result of this expansion into global markets, company performance is now significantly dependent on nondomestic revenue growth and market share. Therefore, the global economy has a greater impact on

Copyright ©1998 by Kathy A. Seabrook. Reprinted by permission of author.

the internationally active company. Case in point: the 1998 economic downturn in Asia. This downturn has had a dramatic affect on the business results and stock prices of many U.S. multinationals, as seen in the performance of many U.S. technology stocks.

This trend in globalization is being experienced by businesses no matter what their country of origin. Japanese, Latin American, Canadian, South Korean, European, and other non-U.S. businesses are aggressively seeking new international markets. The trade areas of China, Eastern Europe, Maquiladora-Mexico, Indonesia, Malaysia, and the European Union are wooing companies to relocate manufacturing and distribution facilities. The building of the multimedia super-corridor in Malaysia has produced infrastructure and trade opportunities for every industry from power generation and construction to food, housing, and other consumer product needs. As wealth in China continues to grow, so will consumerism, and the need for a growing infrastructure in regional areas as well as the main cities.

According to Drucker (1989), the key is investment. The World Bank and private investors are creating these trade and development opportunities, and it is internationally oriented companies that have the expertise, knowledge, skill base, capability, and infrastructure to capitalize on these opportunities. The North American Free Trade Agreement (NAFTA), European Union (economic, political, and monetary union) and the World Trade Organization (WTO) have contributed to worldwide economic growth and the globalization of multinational corporations.

Global expansion means a global approach to business. Consistency of quality, name and brand recognition, reputation, and workplace safety and health are critical to business success. This combination of forces is changing the way many companies manage at safety and health within their organizations and with whom they do business.

What Drives an International Company to Provide a Safe and Healthful Workplace for Its Global Workforce?

What is a company's motivation to manage worker safety and health throughout its global organization? After all, it is apparent that in many developing areas of the world, the value of life may not be comparable to your own. This is demonstrated in substandard living conditions in which housing, sanitation, waste treatment, and the availability of food, water, and clothing are not a given. In many of these developing areas, local work environments do not have the same level of health and safety standards as in more developed areas. Specific regional and country politics and government also play a part in understanding this disparity. So, what motivates most multinationals to embrace a global view on worker safety and health? The following are some global safety and health motivators:

- Safety as a corporate value
- Corporate reputation
- Corporate vulnerability
- Local country legal and regulatory compliance
- Competitive advantage
- Contractual requirements for doing business
- Marketability: quality products and services
- Responsibility/duty to shareholders
- Environmental responsibility

Safety as a Corporate Value

Many U.S.-based multinationals have embraced worker safety and health as a *value* throughout their organization. A corporate value is an unchanging philosophy that defines corporate business culture. When worker safety and health is a corporate *priority,* it is negotiable; it changes with the needs of the day, whether management must balance production, quality, or other short-term priorities. When worker safety and health is a corporate value, a formal safety and health management system is developed and implemented throughout all business units. Everyone in the organization is responsible and held accountable for assuring a safe and healthful workplace.

With safety and health as a value, all levels of management understand that worker safety and health is integral throughout the business and manufacturing process, including but not limited to the following:

- Due diligence for potential mergers and acquisitions
- Initial design of a facility
- Facility construction
- Purchase of equipment and raw materials
- Vetting of vendors
- Recruiting and hiring of staff
- Production of finished product for distribution to the customer

Corporate Reputation and Vulnerability

The catastrophic incident at the Union Carbide's Bhopal, India, plant significantly impacted the public image and financial position of that company. The ensuing negative pubic relations, coupled with resulting litigation and damage control costs crippled the company; it never recovered its previous business position. With today's global media, newsworthy events are virtually broadcast live. The public's first exposure to and opinion of a particular company may be how it handles its workplace safety and health and crisis management. Response to and concern for worker safety and health in a crisis may shape the public's perception of a company for a long time. Managing the global news media is a key public relations and loss-prevention measure, and is integrally linked to the perception of an international company's workplace safety and health commitment.

In-Country Legal and Regulatory Compliance

In-country legal and regulatory compliance is one of the main drivers for managing safety and health internationally for many multinationals. Compliance with local regulations is essential to doing business in a local country. Noncompliance with existing workplace safety and health regulations may bring litigation, fines, or potential imprisonment of the company leadership in the local country. In some cases, noncompliance with worker safety and health can even mean the shutdown of production operations. Poor public relations and a questioning of a multinational's ethical principles can significantly impact future business in that country. If a nonindigenous multinational is not trusted or found to have questionable ethics, other local country regulatory agencies may be called in to audit compliance with regulations that go beyond safety and health requirements. This could have a significant impact on business relationships and productivity, as well as regional and local country customer relations. In some countries, there is less tolerance for lack of safety and health regulatory compliance at multinational companies than at their indigenous counterparts. This perception stems from the belief that indigenous companies have limited resources as compared with their large multinational counterparts.

Competition

Customer-driven worker safety and health requirements can be a basis for bidding on contracts as well as a requirement for a company to be on an approved vendor list.

Contractual Requirement

Most companies require a contractor working in their facility to comply with their safety and health management system. In addition, some companies require vendors to provide a safe working environment for their workers and prohibit exploitation of child labor. In the United States, this issue was showcased in the publicity of Kathy Lee Gifford, a known television celebrity and owner of her own clothing line, when, according to the *New York Times,* her company allegedly used vendors that exploited child labor in developing countries. This disclosure had a significant impact on her company's image and also resulted in unexpected costs for public relations and auditing of her company's vendors.

Marketability

The production of quality products and services is a requirement for all successful companies today. A company that receives the International Standards Organization (ISO) 9000 quality management system certification/registration for its global operations is recognized as providing quality products and services throughout the world. In most cases, ISO 9000 is a requirement to do business. Marketability and a competitive advantage go hand in hand with quality. In addition, many companies recognize that a safe and healthful workplace is essential for the manufacture and distribution of quality products and services. This commitment to quality drives the safety and health process in many organizations.

Responsibility to Shareholders

Corporations have a fiduciary responsibility to their shareholders. Compromising a company's reputation due to mismanagement of safety and health could have a negative impact on financial results, affecting shareholder dividends and the company's share price. It could even leave a company vulnerable to a hostile takeover.

Environmental Responsibility

A catastrophic incident as a result of in-plant process safety issues could have a significant impact upon the environment as well as the safety of the workers. A facility's location (that is, its accessibility to rivers, streams, national landmarks) could cause a regional or global outcry affecting the company's reputation. The Sierra Club and Greenpeace are two environmental advocacy organizations with global influence. They have a global network with legal and financial capabilities working to ensure enforcement of companies' responsibilities and financial accountability for environmental incidents. If a company is responsible for an environmental incident, the extent of environmental damage and public outcry will impact the company's ability to retain consumer and shareholder confidence and long-term financial performance, even if the actual environmental damage is contained.

KEY ISSUES: MANAGING SAFETY AND HEALTH IN A MULTINATIONAL ORGANIZATION

The British Standards Institute (1996) defines a management system as a composition and interaction of personnel, resources, policies, and procedures to ensure that a task is performed or that a desired result is achieved. A safety and health management system provides an organized mechanism for integrating the identification, evaluation, control of risk, and the review and continuous improvement of the safety and health process to sustain a safe and healthful workplace for workers in business units throughout the world. The following outlines the components of an effective safety and health management system.

Leadership Commitment

A corporation's business culture drives the effectiveness and results of the safety and health management system. Leadership commitment in the home country is not enough. A multinational company must be prepared to lead domestically and internationally, by demonstrating a commitment not only in policy but in rewards and recognition for positive safety and health results at all levels of the organization throughout the world. Communicating that safety and health is a corporate value integral to the global business process is a key to achieving success. Commitment is demonstrated by establishing and communicating safety and health performance goals and objectives internally as well as publishing those results throughout the company and in the annual report to shareholders.

Organizational Structure

As previously stated, safety and health management must be integral to the business process. Depending on corporate culture and organizational structure, the organizational structure for safety and health may be quite different, but just as effective, company to company.

The key to an effective safety and health organizational structure is that it is invisible. Basically, every individual and job function is responsible and held accountable for safety and health results. This includes all levels of management, all job functions, and all frontline workers. The following list is a sample of job functions that should have specific annual safety and health performance goals with responsibilities for safety and health outlined in the job description for each job function:

- Plant and line managers
- Engineering and design
- Facilities manager
- Risk manager
- Safety and health
- Environmental affairs
- Quality assurance
- Line employees
- Medical personnel
- Human resources
- Legal counsel
- Security

This functional interface provides the frontline manager or supervisor with the authority, accountability, responsibility, and financial resources needed to take correct action wherever and whenever necessary. It also provides a continuous improvement loop with input and feedback from frontline workers and individuals aligned with specific job functions.

In Richard W. Lack's book, *Essentials of Safety and Health Management,* he reviews the management responsibilities of the safety professional, which include the following:

- Assisting line leadership in assessing the effectiveness of the unit safety and health programs.
- Providing guidance to line leadership and workers so that they understand and know how to implement safety and health programs.
- Assisting line leadership in the identification and evaluation of high-risk hazards and developing measures for their control.
- Providing staff engineering services on the safety and health aspects of engineering.

- Maintaining working relations with regulatory agencies.
- Attaining and maintaining a high level of competence in all related aspects of the safety and health profession.
- Representing the company in the community on matters of safety and health.

Exhibits 7-1 and 7-2 outline two examples of company safety and health organizational structure: Corning, Inc. of Corning, New York, and NRG Energy, Inc. of Minneapolis, Minnesota. These two companies have very different organizational structures and both are achieving excellent safety and health results using their corporate-specific safety and health management process. This demonstrates how corporate culture and organizational structure dictate the success of a safety and health management system within an organization.

Resources

Financial and human resources must be allocated at all levels of an organization to achieve desired safety and health goals and objectives. This is especially true internationally, as corporate headquarters and corporate expectations across many time zones and miles may not be the priority of local management.

Exhibit 7-1. Case in Point: Corning, Inc.

At Corning, high-performance work teams operate at all organizational levels and incorporate all job functions within the company. When contemplating a new plant site, process, or change in a raw material, a high-performance work team is formed to address the task at hand. According to Corning's Ron Kitson, CSP, Manager Corporate Safety and Health, "Safety and health is integral to Corning's business process. It is not unusual for many members of the team to identify safety and health implications of a new or change in process." As the team works together to identify the most cost-effective, efficient, state-of-the-art solution, team members begin to recognize how the legal requirements for hazardous material reporting or hazardous waste removal impacts the cost-effectiveness of a given solution.

Exhibit 7-2. Case in Point: NRG Energy, Inc.

At NRG Energy everyone is responsible and held accountable for injury prevention. According to Dan Severson, Executive Director, Occupational Safety and Health, "the goal of the safety and health process is to facilitate an injury-free workplace through proactive actions and activities."

To reinforce proactive safety actions and activities, safety and health activity schedules are developed and performance reviews conducted for all supervisors and mangers. Safety and health activity schedules are aligned with key safety issues identified by corporate safety and health. Safety and health actions and activities include conducting self inspections, job safety analysis, safety and health meetings, developing and implementing safety and health training plans for all workers, and conducting safety observations throughout the supervisor and managers' area of responsibility.

To facilitate these proactive actions and activities, performance reviews are conducted for all levels of NRG Energy leadership worldwide. Performance reviews embrace corporate safety and health goals and objectives and include the following:

1) Demonstrated good supervisor and manager safety and health management practices (for example, communicating the corporate safety and health philosophy and conducting safety and health reviews for all subordinates)

2) A 360-degree safety and health feedback process. Feedback is provided from the immediate superior, subordinates, and safety and health managers.

3) Demonstrated completion of safety program elements (for example, accident investigations and self-inspections)

Severson continues: "This feedback, known in the company as safety and health performance reviews, is intended to define what a good job is, to provide the impetus for, and accountability in doing the right things, to provide feedback on how well the company is doing, and to be the basis for rewarding people for their contributions."

NRG Energy demonstrates how an effectively integrated safety and health management process facilitates an injury-free workplace in operations worldwide.

Planning

Planning for worker safety and health is essential to achieving corporate safety and health goals and objectives worldwide. Planning must be conducted at the corporate, business unit, or division and facility levels, and goals and objectives aligned strategically with corporate safety and health values. Planning should incorporate input from all levels of the company. When all levels and all locations of the company have their own planning process, implementation is much easier. Each business unit or facility owns and agrees to the goals and objectives and the plan or strategy to achieve them. The planning process for global safety and health management includes the following components.

Setting, Publishing, and Communicating Annual Safety and Health Goals and Objectives

Safety and health goals and objectives should be set, revisited, and reworked on an annual basis. The goals, objectives, and performance metrics or criteria should be specific, realistic, measurable, results oriented, and include a timeline. Goals define the desired result (the "what") and objectives are the action plan (the "how to") to achieve the desired result.

A process for publishing and communicating the goals and objectives throughout the global organization should be established. The effectiveness of the safety and health management system will be a result of achieving these goals and objectives. It is important to understand that goals and objectives must be tied to the implementation of the safety and health management system. Accident statistics are only one performance metric. This is especially true when working in the global arena. Underreporting of accidents and incidents is not uncommon.

In countries where English is not the first language, language barriers and terminology such as lost-time or lost-workday cases are misunderstood. In addition, the legal requirements and definitions of lost-time accidents or lost-workday cases in some countries may be different from the expectations of corporate headquarters. Furthermore, the use of accident rates for measuring performance encourages underreporting. Accidents or incidents that are not reported are not investigated, increasing the likelihood for a more serious accident to occur in the future. Effective performance metrics for a global safety and health management system are as follows:

- Demonstration of the commitment to safety and health in the work environment and daily business dealings through company leadership.

- One hundred percent completion of weekly self-inspections.
- Completion of incident/accident investigations for 100 percent of the incidents that occur during a specified time period.
- Conduction of risk assessments for 100 percent of all job tasks at a plant facility.
- Setting of safety and health training schedules for the entire staff. Completion of 100 percent of the training.

Setting Performance Standards

Based on safety and health risk within an organization, best practice hazard control programs should be developed and implemented. Best practice hazard control programs incorporate global good management practices, industry best practices, and local country legal requirements to achieve hazard control results. Examples of hazard control programs include accident investigation, self-inspections, ergonomics, machine guarding, personal protective equipment, and lockout/tagout. (Local country legal requirements may be difficult to obtain, especially in English. The resource/reference section of this chapter may be useful in obtaining legal requirements for some countries.)

The use of best practice standards throughout the organization, no matter the location of the facility, provides consistency and a level of safety and health assurance aligned with corporate goals and objectives. Prior to operating in a country, a company should seek out local safety and health council to determine whether the corporate best practices meet local requirements. Where the local requirements are more strict, they must supersede the corporate best practice. In addition, the new standard should be reviewed and considered for future use as corporate best practice standard. In some countries, safety and health expertise is provided by the legal profession, others through safety professionals, architects, or engineers.

Assigning Responsibility, Authority, and Financial Ability to Meet Goals and Objectives

Goals and objectives must be part of an individual's performance appraisal process. In addition, safety and health responsibilities must be outlined in the individual's job description, particularly if there is not a formal appraisal process in a specific country or business unit. Budgeting must allow for safety and health planning, including cost-sensitive objectives or outstanding action items from previous years.

Designating an Individual to Provide Competent Safety and Health Guidance

Safety and health resources must be provided to achieve corporate-wide goals and objectives; this includes personnel as well as monetary commitment. The nature and extent of safety and health guidance a facility or business unit requires will be based on the risk involved. The safety and health professional must be qualified and properly trained to provide the required guidance.

Setting Policies and Procedures

Policies and procedures should be developed to build in a continuous improvement loop from recommended corrective actions resulting from accident investigations, self-inspections, audit findings, or worker observations. There should be a clear process for monitoring the outstanding items through to completion. Responsibility and accountability for each action item, along with a timetable for completion should be documented for follow-up. Depending on the corporate, business unit or facility organizational structure, the monitoring process could be the responsibility of the facility manager, safety committee, or safety manager.

Processing for Change Management

Written procedures should be developed to manage change in facility processes, technology, equipment, and procedures. As outlined in the CFR 1910.119 (OSHA 1992), change management procedures should address these issues:

- Technical basis for the proposed change
- Impact of the change on worker safety and health
- Modifications to operating procedures
- Time period needed for the change
- Authorization requirements (internal and external to the company) for the proposed change

Workers affected by changes in operations, process, or technology, including process, maintenance, or contract workers, must be informed of, and trained in, the safety and health risks associated with these changes. Training must be completed prior to work being conducted in the area(s) where the change(s) has occurred. In addition, all written procedures or practices must be updated as a result of changes.

IMPLEMENTATION

The key to implementation is in the planning stage. If planning has been well conceived, and if those charged with implementation have been involved in the planning process, then implementation will likely produce positive results. For international companies, two aspects of the implementation process can pose challenges between the corporate and international business unit or facility: communication and cultural differences. To meet the communication challenge, corporate leadership and the safety and health manager must define an effective system to manage safety and health process across

- Various local business cultures
- Time zones
- Geographical distances
- Language barriers
- Cultural norms, customs, and religions

Overcoming communication challenges requires patience and the use of several communication methods, the most effective being in person, face-to-face communication. Developing a rapport or relationship with individuals in global business units will ease many a future miscommunication and provide a foundation for a positive business relationship. Using fax and e-mail is also effective; these technologies make working across different time zones and vast geographical distances very efficient.

The second challenge is cultural differences. Culture is real; it is who we are as human beings, individually and collectively. Culture defines our experiences, perceptions, values, beliefs and behaviors. Therefore, the way in which a safety and health goal or management system is implemented in one country may not be the same as in another. Trompenaars (1994) provides a scenario in which a pay-for-performance philosophy may or may not be an accepted way of doing business.

> The internationalization of business life requires more knowledge of cultural patterns. Pay-for-performance, for example, can work out well in cultures where this author has had most of his training: the U.S., the Netherlands, and the UK. In more collectivist cultures like France, Germany, and large parts of Asia, it may not be so successful, at least not the Anglo-Saxon version of pay-for-performance. Workers may not accept that individual members of the group should excel in a way that reveals the shortcoming of other members. Their definition of an outstanding individual is one who benefits those closest to him or her. (5-6)

Trompenaars's example demonstrates the importance of working with international business units and facilities during the planning stage. Listening to their implementation concerns and focusing on the best way to achieve safety and health results is essential. This provides an understanding between corporate and the business unit to work out the best solution to achieve safety and health goals within a particular country.

PERFORMANCE MEASUREMENT

Safety and health goals and objectives define desired results. Planning and implementation provide the means to achieve those desired results, and corporate audits and individual performance appraisals provide a measurement tool to determine whether the desired results are being achieved.

As discussed in the "planning" section of this chapter, accident statistics are not the best measurement criteria for safety and health performance. Accident statistics do not prevent injuries and illnesses; they are not proactive, and in some instances they are not accurate. If a corporate goal is to reduce accidents by 20 percent, and actual results are only a 6 percent reduction, there is no way to know what went wrong to cause the accidents. Accident statistics indicate *where* and *when* an accident occurred. Instead, measurement criteria based on components of the safety and health management system provide information on what causes accidents and why they continue to occur. The performance metrics given in the "planning" section are proactive, planned for, implemented, and easily measurable. If current safety and health goals are not being met, here are some questions to ask that may highlight why:

- What are the objectives or "how to's" to achieve the safety and health goals and objectives?
- What aspects of the management system are not being implemented or may not be working in a particular operation or country?
- Is the corporate value of safety embraced within the local business unit or facility?
- Is local leadership committed to corporate and local safety and health goals and objectives?
- Has the facility had a recent corporate or local safety and health audit? What were the audit findings? Have they been completed? Do outstanding audit findings impact accidents that have occurred?

One final note on performance metrics. Many international companies believe competition between global business units is good and that publishing worldwide accident results motivates leadership to better manage safety and health. However, this is not always the case; in fact, publishing accident results can produce a work environment in which accidents are not reported. Nonreporting will often occur in order to maintain the *appearance* of a quality safety and health program.

CONCLUSION

Managing safety and health in an international company takes leadership commitment to make safety and health a value, which is communicated and demonstrated in the actions and behaviors of everyone in the company. To be effective, a safety and health management system must be aligned with the corporate culture and the organizational structure of the international company. Thoughtful preplanning, implementation, and performance metrics must be identified and reflect components of the safety and health management system. The components of the safety and health management system are drivers for continual improvement of safety and health results throughout the worldwide organization. As we enter the next millennium, the globalization of business will continue apace with the key role of the safety and health professional in managing an effective global safety and health management system.

ADDITIONAL RESOURCES

Directorate V-F
Commission of European Committees
Batiment Jean Monnet L29 20
Luxemborg
  Fax: + 35 243 13 4511
  Directorates General V (DG-V)—
  Department of the European Union (EU),
  which oversees employee industrial
  relations and social affairs.

European Commission office
2100 M Street N.W. (7th floor)
Washington, DC 20037
  Tel: +1 202-862-9500

Fax: +1 202-429-1766
Provides all information on the EU

Eurosafety
Industrial Relations Services
18-20 Highbury Place
London N5 1QP, England
Tel: + 44 (0)171 354 5858
Fax: + 44 (0)171 359 4000
A quarterly periodical providing the
"state of play" on European safety
directives, as well as articles related to
the judicial interpretation of those
directives.

European Community's Health and Safety Legislation, Vol. I
ISBN 0 442 31651 8 (USA)
Van Nostrand Reinhold Inc.,
115 5th Avenue
New York, NY 10003

Croner's Health and Safety Directory
Croner Publications Ltd.
Croner House, London Road
Kingston upon Thames
Surrey KT2 6SR, England
Tel: + 44 (0) 181-547-3333
Fax: + 44 (0) 181-547-2637
Provides names, addresses, and telephone and fax numbers of safety and health organizations by country, throughout the world.

Conducting International Meetings: A Meeting Planning and Resource Guide
Greater Washington Society of Association Executives
Fax: +1 202-833-1129
Provides all you need to know for conducting meetings outside the United States: handling planning logistics, planning for security, transportation and shipping, and bringing international meetings to the United States.

U.S. Chamber of Commerce, International Division
1615 H Street, N.W.
Washington, DC 20062-2000

REFERENCES AND SUGGESTED FURTHER READING

Axtell, R.E., compiled by the Parker Pen Company (1993). *Do's and Taboos around the World, 3rd edition.* White Plains, NY: The Benjamin Company.

Brake, T., D. M. Walker, and T. Walker. (1995). *Doing Business Internationally.* Burr Ridge, IL: Richard D. Irwin Inc.

British Standards Institute (1996). *British Standard 8800: Guide to Occupational Health and Safety Management Systems.* London, England: British Standards Institute.

Drucker, P. F. (1989). *The New Realities.* New York: Harper and Row, Publishers, p. 229.

Lack, R. W. (1996). *Essentials of Safety and Health Management.* Boca Raton, FL: CRC Lewis Publishers, p. 27.

National Safety Council (1994). *Accident Prevention Manual for Business & Industry, 10th ed.* Vol. 1: Administration & Programs. Chicago: National Safety Council.

Occupational Safety and Health Administration (August 1992). Process Safety of Highly Hazardous Materials Standard, 40 CFR 1910.119.

Trompenaars, F. (1994). *Riding the Waves of Culture: Understanding Cultural Diversity in Business.* Burr Ridge, IL: Irwin Professional Publishing.

Chapter 8

Safety Cultures and the Behavior-Based Model

**Thomas R. Krause, John H. Hidley,
and Stanley J. Hodson**

INTRODUCTION

Since the mid-1980s, safety professionals have shown a growing inter-est in behavior-based approaches for improving the safety cultures at their sites. This chapter presents the fundamentals of the behavior-based model, a report on critical success factors for implementing the model, and case history material from sites using the model. The following section presents an overview of the behavior-based model as it is applied at many sites by maintenance crews and others.

OVERVIEW OF THE BEHAVIOR-BASED MODEL

At the core of the behavior-based model presented here are four linked steps:

1) Identify safety-related behaviors that are critical to performance excellence.
2) Gather data on work group conformance to safety excellence.
3) Provide ongoing, two-way performance feedback.
4) Use accumulating behavior-based data to remove system barriers to continuous improvement.

As its name indicates, behavior-based safety focuses on human fac-tors. In this context, behavior simply means an observable act; there is no connotation of good or bad. Behaviors—observable actions—are the proper upstream focus for safety for two reasons: 1) At-risk, task-related behaviors are the final common pathway for almost all incidents; and 2) most at-risk behaviors commonly at a site are supported by the culture of the site.

Copyright ©1998 by Behavioral Science Technology, Inc. Reprinted by permission of author.

Taken together, these two points convey an important message for conventional wisdom at all levels of an organization. In effect, the first reason says, *Don't blame conditions alone;* and the second says, *Don't blame employees.* Stated positively, behavior-based safety engages personnel at all levels of an organization to reduce rates of at-risk behavior and raise the rates of identified safe behaviors.

Identifying Critical Behaviors

At most sites the task of identifying the core cluster of critical behaviors is carried out by a steering committee guided by a consultant. Steering committees usually have management input and are composed primarily of wage-roll personnel. This group reviews the site's incident reports (including near misses) for the past two to three years. It identifies the cluster of at-risk behaviors that served as the final common pathway in the most serious and/ or most numerous incidents. In the course of this behavior-based review, it is common for the steering committee to discover a set of twenty to thirty behaviors that accounts for 90–95 percent of recent incidents. Furthermore, the wage-roll steering committee members—those most familiar with the daily risks of the job—sometimes identify additional behaviors that may not be implicated in incident reports but that they know are critical to worker safety. The committee members then define each of the identified behaviors in operational terms and categorize them for inclusion in a data sheet. The operational definitions might focus on areas such as pinch points, line of fire, eyes on path, and three-point contact on ladders or stairs or scaffolding. The completed data sheet is used to train site personnel as observers, who then gather data on work group performance of the identified behaviors.

Gathering Data

Typically, most of the critical at-risk behaviors deal with shortcuts, temporary conveniences, or with systems issues that prevent safe behavior. Trained observers use a data sheet developed by their peers to measure the rate at which work groups perform the identified critical behaviors either safely or in an at-risk manner. The operational definitions guide the observers as they sample or measure performance. The categories of the data sheet also have examples to help calibrate the observers. This calibration produces several important benefits. First, it ensures that the data are objective and accurate. Second, fluency in the data sheet also means that workers from different trades can observe each other because they have a new common vocabulary for safety. Finally, the categories of the data sheet give everyone (observers and observees) a shorthand method of referring to critical safety behaviors. While many sites train supervisors in behavior-based observation procedures, the observer corps at most sites is composed largely of wage-roll personnel. They regularly observe their peers, and then provide performance feedback.

Providing Ongoing, Two-Way Feedback

Typically, immediately after an observation the observers speak with the observed personnel. The observers inform coworkers about the critical behaviors they are performing safely and the ones they are performing in a way that puts them at risk for injury. In the case of at-risk behaviors, the observers question their coworkers to determine why they are engaging in at-risk behavior. In the course of these discussions, the observers may uncover system barriers to safe performance. For instance, for the line-of-fire category, the observers may be determining whether workers stand in the way of relief valves or bleed-off points when they are working on pressurized gear. This category extends to helpers or associates. Sometimes workers may be careful not to position themselves in the line of fire but fail to direct a helper or other associate to stand out of the line of fire while they are working together. When observers see coworkers using at-risk behavior under this category, they inform and then ask why they are exposing themselves or their associates to at-risk line-of-fire behavior.

During such discussions, the observees may say that they didn't recognize or register the exposure in the practice under observation. In this case, the observer shows them the at-risk aspects of their performance so that they understand it. Often the observees say that they knew about the at-risk behavior but forgot or became distracted, and that they will avoid such risk in the future. However, they may also say that because of the way a particular piece of equipment is engineered and/or installed, they do not see an alternative to performing the at-risk line-of-fire behavior that was observed. In the comments section of the data sheet, the observer records coworker suggestions and remarks about system barriers that favor at-risk behavior. In addition to this verbal feedback, the data gathered by the observers is analyzed by computer software, and reports and charts of work group performance are printed and posted as documented feedback.

Removing Barriers to Continuous Improvement

Using the comments and observation data, site personnel can target areas for improvement. For instance, the observation data may show that performance of three point-contact is very high (97 percent safe) and that good progress is being made on pinch points (up from 70 percent safe to 82 percent safe), but that line of fire is running at 65 percent safe. Therefore, site personnel would flag this as an area where an "accident is just waiting to happen." The written comments of the observers can go a long way toward showing the number and kinds of remedy needed. Action to address at-risk line-of-fire behavior might proceed along several lines:

1) A sizable group of new-hires is still having trouble recognizing this at-risk behavior. More training is called for in their case. The steering committee agrees to deliver that training over the next month in coordination with supervisors in charge of crew-safety meetings.

2) In most cases crewmembers recognize the at-risk behavior after the fact, but they are still having difficulty "internalizing" the safe behavior. The observers agree to focus their observation and feedback on line of fire for two months to reinforce crew performance of the identified safe behavior.

3) Three of the site's seventeen pumps have been installed in such a way that surrounding equipment makes it very difficult for workers to avoid being in the line of fire while doing routine maintenance on those pumps. An announcement about the three pumps is added to safety meeting agendas. The steering committee meets with the engineering staff to brainstorm corrective measures for the three pumps. Engineering staff uses behavior-based walkthroughs and data to fine-tune the equipment improvements.

Outcomes. Using the behavior-based approach, many companies have engaged personnel at all levels to address human factors and put safety performance on a more solid footing. If companies have been experiencing the safety/accident cycle, this approach brings better continuity to their efforts. If they have been stalled on a performance plateau, this approach helps them to make a baseline shift toward continuous improvement.

IT IS MORE DIFFICULT THAN IT SOUNDS

Because the concepts and procedures of the behavior-based model are clear and straightforward, applying the model can seem deceptively straightforward. When experience managers are first introduced to this model, they often respond with statements such as, "This is not anything new in particular, but the configuration is different—it brings many important pieces together in one approach." On the other hand, when behavioral scientists read this application to safety, they often say, "Why of course there is a natural fit," without realizing how many organizational realities need to be dealt with in the course of making this "natural" fit.

In fact, it can be difficult to implement any significant behavioral and cultural change initiative. In safety, for instance, all of the systems that produce injuries are stable, with their roots firmly embedded in the culture of a site. Change is not easy, and it is resisted with vigor and even ingenuity. Relations between labor and management often have an implicit rule prohibiting change. One of the primary purposes of culture is to provide stability by avoiding change—even otherwise admittedly "good" change. Most sites have a history of failed change efforts, failures that create skepticism and even cynicism.

On the other hand, while these challenges are serious, implementation efforts characterized by the seven critical success factors method make impressive gains. As implementation of the change effort begins, someone outside of the site organization needs to give guidance that encourages leadership at the site level. The guidance can come from a consultant within the corporation or from an outside consultant who has experience with multiple implementations. Understanding the day-to-day aspects of how business organizations work, however, is a critical factor because the goal is continuous improvement, and that goal necessarily involves adaptation of the change method. The challenge is to adapt the method to be used to fit the culture in which it is to function. Fostering such an adaptation allows the installation of a mechanism for change.

The Driving Mechanism for Continuous Improvement in Safety

The driving mechanism for continuous improvement in safety is the proper use of modern scientific methods coupled with employee involvement. In establishing this mechanism, management is faced with the daunting question, "How can we maintain in our organization the focus of effort that is required for continuous improvement?" All kinds of change are stressful to people—even changes that bring improvement. Consequently, resistance to change is natural. This hard fact of life is true in every management field, but it is more emphatic in those fields that do not have well-structured support mechanisms such as measurement, feedback, continuous training, and cultural values and systems. Until recent developments, safety was a prime example of a field handicapped by a lack of adequate structure.

In a market economy, it is not difficult to maintain a focus on production. The culture in any manufacturing plant supports this focus. Managers and supervisors at all levels support it. Measurement systems are in place to provide frequent, often daily, feedback on small variations in production performance. Everyone understands the priority of production. However, safety is a performance area for which it is much more difficult to maintain focus. Most companies lack cultural and systems support, and as a result, managers have traditionally attributed variations in safety performance to variations in "awareness." Of course, over the years, engineering, facilities, and maintenance have received attention and have provided some safety improvement. But, any supervisor or team leader knows that incidents come primarily from the "human element." The workers least at risk for injury are those who are familiar with critical task-related behaviors but who have not yet become set in their ways. This optimal training phase of adaptive readiness to safety is very much a human element, but merely identifying it is not enough. The management question is, "How can we maximize this level of readiness for the workforce as a whole?" Or, "How can we minimize variability of performance?"

In business and manufacturing, the concept of variability first became a subject for management in relation to the quality of goods and services. With the advent of the quality movement, a new field has been sketched. This new field aims to understand and manage variation by involving employees in teams that problem solve for continuous improvement. The science and statistics at the core of total quality management (TQM) are solid and dependable, but implementation itself has met with variable success because the cultural and feedback mechanisms for quality are not securely in place. In the final analysis, the success or failure of quality initiatives does not depend on the brilliance or truth of the insights of Deming, Juran, and others. Whether in safety or in quality, the ultimate success of these methods rests with leadership.

The driving mechanism for continuous improvement in safety is the proper use of behavioral and statistical science coupled with employee involvement. The proper use of these methods to manage safety hinges on two factors: 1) scientific measurement and management of all employee levels of workplace behavior; and 2) the involvement of all employees in this ongoing feedback and problem-solving process. The reason to focus on behavior is that when an incident occurs, behavior is the crucial, final common pathway that brings other factors together in an adverse outcome. Therefore, ongoing, upstream measurement of the sheer mass of these critical at-risk behaviors provides the most significant indicator of workplace safety.

SEVEN CRITICAL SUCCESS FACTORS FOR IMPLEMENTING THE BEHAVIOR-BASED MODEL

As interest in behavior-based safety grows, it is important for practitioners of this method to communicate the factors that make a behavior-based safety initiative effective through all its important phases, from start-up to implementation to maintenance for long-term continuous improvement.

Defining the Seven Success Factors

In studying behavior-based safety practitioners and their implementations, seven factors were identified as critical to the success of their behavior based safety processes. Inadequate attention to any one of these factors can mean significantly impaired success or even failure:

1) Use a process blueprint.
2) Emphasize communication and buy-in.
3) Demonstrate leadership (management and labor).
4) Assure implementation team competence.
5) Use action-oriented training.
6) Use data for continuous improvement.
7) Provide appropriate technical resources.

Interestingly, in a parallel review of the TQM literature, a search for lessons about success and failure factors in TQM repeatedly turned up the same factors that are critical to the success of behavior-based safety at a site. Figure 8-1 shows the relative frequency with which each of the seven factors appeared in the TQM literature.

A comparison with the TQM literature is relevant because, like behavior-based safety, TQM is a profound undertaking of organizational change with far-ranging potential benefits and similarly extensive associated direct and indirect costs. Much of what is being attempted in TQM parallels that of behavior-based safety. In a rigorous study of award-winning

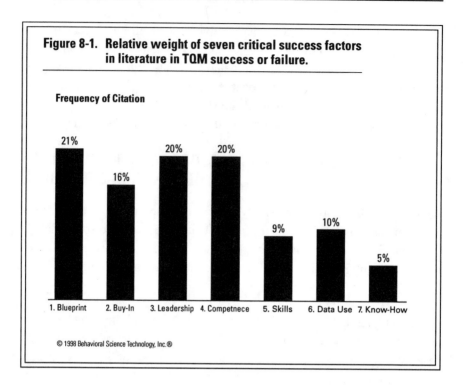

Figure 8-1. Relative weight of seven critical success factors in literature in TQM success or failure.

© 1998 Behavioral Science Technology, Inc.®

TQM initiatives, Hendricks and Singhal (1995) found that all 230 companies in their sample significantly outperformed matching companies that did not have award-winning quality initiatives. This was true whether the TQM companies were large or small, or whether they earned TQM awards before 1986 (early in history of this application) or later as TQM methods matured. The bad news is that the successes, while striking, are relatively rare. Some observers estimate that only 25 percent of U.S. initiatives achieve success (Spector and Beer 1994), and after reviewing a variety of surveys of TQM success, Hawley (1995) concluded that "the data strongly suggest that TQM's overall success rate is so low that [for most organizations] better odds could be achieved by increased efforts directed at business as usual. Typical TQM is a practice that will probably fail."

The important point is that TQM is only worth doing if you are prepared to do it right; then it is truly worthwhile. The same is true of behavior-based safety. In both cases, doing it right means paying close attention to each of the seven success factors.

Factor 1: Use a Process Blueprint—A Structured and Rigorous Implementation

It is important to have a well-thought-out blueprint for the entire implementation sequence. This involves defining a well-structured approach and implementing it rigorously. Structure means knowing the necessary steps and rigor means executing them meticulously. *Structure and rigor do not mean inflexibility or rigidity.* There is plenty of room for creatively adapting the proper principles to the particular needs of the organization. There is room for creativity. However, there is no room for sloppiness, fooling around, and blindly trying "this and that." As in making a major structural modification in a plant, implementing an organizational change-effort requires thorough planning and self-discipline. As Figure 8-1 shows, this is also the most frequently appearing success factor in the TQM literature.

Characteristics of a Structured Approach. In successful change-efforts the implementation has a very clear plan, a path forward with specific steps and a predefined sequence of events and timeline.

Characteristics of Implementation Rigor. Sites using a structured approach do not improvise with the structure of the plan. Rather than pursuing every possibility, they steadfastly pursue precisely those steps that lead to their objective.

Factor 2: Emphasize Communication and Buy-In

Buy-in and communication deal with how well the change-effort is "marketed" internally to an organization. Buy-in starts when top leadership, management, and labor analyze safety performance and jointly commit to the need for improvement. The purpose of communication is not only to explain but to engage supportive activity at all levels of the organization. Throughout an organization with a successful change-effort, personnel are highly committed to the initiative.

This buy-in is critical to the success of the initiative; ultimately, people will not do what does not make sense to them. A corollary is that in the long run, it does not help to get buy-in for things that do not make sense. So buy-in has two important aspects: 1) people become actively engaged in the behaviors that are needed for the initiative's success; and 2) the initiative itself makes a kind of hard-headed sense (as opposed to having a merely ideological or idealistic appeal).

An effective implementation blueprint includes a "marketing" plan that attends not only to what to market but also to how and when. For instance, one-size-fits-all training that is given too early and that is primarily abstract or conceptual can backfire by generating enthusiasm about improvements that cannot materialize overnight (see Factor 5).

Factor 3: Demonstrate Leadership (Management and Labor)

Factor 3 addresses the central importance of leadership's orientation and demonstrated and appropriate support for the change process. For example, do plant managers expect to be regularly involved in the behavior-based progress, or do they assign all oversight to staff members? At sites represented by organized labor, do the union presidents and other officers understand what their support really means and are they willing to provide it? There is nothing wrong with a union taking a cautious approach, especially at a site that has failed to support its own initiatives in the past, but a cautious approach does not mean taking a wait-and-see position. Rather, it means making sure that the site makes the commitments needed to ensure the success of the safety change-effort.

In over six hundred implementations throughout the United States, Canada, and the United Kingdom, and in France, Jamaica, Mexico, South America, South Africa, Australia, and the Philippines, we have found in surveys and assessments *that the single best predictor of success at a site is the willingness to address issues.* Note that the predictor is not the absence of issues to address, but the willingness to address them. Sites that seem to have a good chance of success because they start with few obvious issues can nonetheless do very poorly if their leadership is unwilling to address those issues. Companies that seem to face insurmountable difficulties because they have many issues (adversarial relations between labor and management, old plant and equipment, loss of market share and resulting pressures to downsize) can nonetheless achieve remarkable success when leadership is willing to raise and solve these and other issues.

Management Activities. Successful behavior-based safety initiatives have managers who monitor the change-effort, holding the implementation team accountable. These managers may also make observations, serve on the implementation team, attend training, and function as knowledgeable spokespersons for the change-effort. The latter duty often involves promoting the change-effort beyond the bounds of the site to the site manager's boss, to the corporate headquarters, the community in which the site resides, and to other companies that are interested in behavior-based safety. Finally, managers of successful change-efforts encourage benchmarking and networking.

Feedback. Effective managers and supervisors ask for and use feedback on how effectively they are supporting the process. Some have established routine and formal mechanisms for subordinates and hourly employees to give them input about the process and their support of it.

Labor. The role of labor leadership, whether union or non union, is to actively pursue safety excellence for its own sake. Unions do this when they are careful not to make safety activities a bargaining chip during contract negotiations. Labor leaders also help ensure this factor when they do not insist on seniority as the sole criterion for safety committee membership, allowing the best-suited personnel to serve on the implementation teams. (See the next factor for more on this important subject.)

Disciplinary Action. Both labor and management make important leadership contributions to disciplinary action. They both protect the integrity of the fact finding versus fault-finding aspect of behavior-based safety by making sure that the data produced by the system are not used for disciplinary purposes. Behavior-based safety is not a process for dealing with gross violations of rules or enforcing compliance with government regulations; it is a system to help employees manage risk and exposure more effectively.

Integration. The lesson here is that although change-effort starts outside the normal structures of the organization, it cannot succeed in the long term and bring continuous improvement if it is left to stand alone. Both management and labor are responsible for integrating the new initiative. Successful change-efforts are strongly integrated into the daily functions, procedures, and structures of the organization. This requires an active effort to find ways to interface and integrate change efforts with existing organizational structures. Furthermore, successful change-efforts become part of the common language of the organization. At their most successful, these initiatives help the organization create a new common language. Finally, once they are integrated, successful change efforts also evolve with the organization.

Factor 4: Ensure Implementation Team Competence

Ensuring the competence of the implementation team primarily deals with the identification, recruitment, and training of the change-effort steering committee or team and its leader or facilitator. This group is crucial to success. The answers to the following questions will determine a group's success:

- Are the members and the facilitator individuals whom supervision and the general population trust and respect?
- Do they have good people skills and task competence?
- Do they engage others, creating a sense of pride in community around the change-effort so that people will want to participate?
- Are they competent communicators?
- Do they use many communication channels?
- Are they organized?
- Do they follow up on action items?
- Do they take the technical expert's advice or put their reasons not to on the table and deal with them?
- Do they have a clear path forward and are they working that path forward systematically, thoroughly, and rigorously?

Successful implementation teams are "believers" in the change-effort. Since these teams know that resistance to change is natural and inevitable, they do not take it personally. They also recognize and appreciate the *limits* of their role—which is to identify *for* management the existing barriers to continuous improvement. These teams must have, at every turn, a close working relationship with managers, from supervisors to plant managers and their staffs. In effect, the steering committee and facilitator are themselves "in the middle," between their coworkers and the plant management personnel. Successful change-efforts identify and train wage-roll personnel who function well in this new role. In most cases, these new roles require intensive real-world training that is carefully matched and paced to the emerging responsibilities and tasks of the change-effort itself.

Factor 5: Use Action-Oriented Training

One of the marks of successful change-efforts is that personnel are trained for their roles and responsibilities immediately before they are expected to carry them out. This approach is a training version of just-in-time delivery of skills coaching. Furthermore, this is the most assured approach to optimize transfer of training from the classroom or seminar to the shop floor. The emphasis throughout is on *skills* acquisition. Skills training provides trainees a structured, safe environment in which to practice the new skills that the change-effort requires for its success. Skills training is not the same as conceptual training; conceptual training is easier to accomplish and requires fewer resources on the part of the organization. Skills training takes time, and itself needs to be timed. It needs to be implemented shortly before the skills are to be used. A common error in skills training is to overemphasize technical tools at the expense of interpersonal skills, and to assume that by teaching the concept of the skill, people will develop the skill itself. Successful skills training always involves follow-up feedback on performance.

Factor 6: Use Data for Continuous Improvement

Successful change-efforts continually measure their own processes and results by putting in place systems for gathering information about the change-effort itself and about its effects. In the case of behavior-based safety, such a system involves a computerized data-tracking system. Once a behavior-based safety initiative is on-line, it generates a data stream that is so rich in both scope and detail that it cannot be analyzed and reported effectively without computer software.

Furthermore, successful companies measure their change-effort in an evolving way because the change-effort itself is evolutionary. During start-up companies measure start-up activities, and as the initiative achieves on-line status, they measure action planning and other safety improvement activities. As companies ready their systems for removing barriers to safe behavior, they use prioritizing and other decision techniques to identify their most important or pressing problem-solving opportunities. These practices are confirmed in the history of TQM initiatives. Rigorous process measurements and performance feedback are a common characteristic of TQM initiatives that have survived four to seven years. Those process measurements cover all important aspects of the respective TQM change-efforts, including support by management and leadership.

An effective change-effort surfaces many barriers to improvement, some with high priority and some with lower priority. Successful companies pay careful attention to this matter and quickly address items that

- Have strategic importance
- Affect the customers of the change-effort
- Are historically recognized significant issues
- Affect the success of the change-effort itself

Directing the change-effort to these ends requires personnel who are skilled in analyzing, problem solving, planning, and communicating. Identifying, recruiting, and training these personnel is handled under factors 4 and 5.

Factor 7: Provide Appropriate Technical Resources

Providing the proper technical assistance influences many aspects of the change-effort, but this is an area where management may drop the ball at the very beginning and thereby create problems for the entire implementation. Some managers wrongly conclude that because behavior-based safety is an employee-driven effort, wage-roll personnel should be left alone to select the outside resource. Wage-roll representation, and even predominance, on the selection committee can be a very good first step. However, successful sites make sure that the committee clarifies and shares its vision of what it is looking for and is adequately educated on the techniques needed to make valid apples-to-apples comparisons and ratings of what the various technical resources are offering.

The central aspect providing proper technical assistance is *know-how that is site-specific*. Appropriate technical resources provide solutions for *site-specific* needs. Negatively stated, this means not just telling a site what to do and leaving it up to the site to carry out the committee's recommendations, it means working closely with the site to figure out the very best way to accomplish the site's objectives.

Appropriate technical resources provide *know-how*. In the context of change-efforts, know-how means good judgment about exactly which changes are possible and when to press for them. It also means that when the resource makes recommendations, the recommendations should be sensible and practical. Judgment is something we acquire through well-considered experience. Persons who are very intelligent may nonetheless have poor judgment in areas in which they are not experienced. Furthermore, even persons who are knowledgeable about the theory of behavior-based safety may still not have good judgment about how to advise a particular company to make the needed changes in its ongoing procedures and practices.

Lessons about Success Factors

The growing interest in behavior-based safety is justified by the important gains it has helped companies make. Central to those gains are the seven factors presented here. Although there is no single silver bullet for companies that are setting a course for an injury-free culture, these seven factors have proven their staying power.

THE CRITICAL-MASS APPROACH TO AT-RISK BEHAVIOR

On first acquaintance with the behavior-based approach, the single most important thing to understand is that it focuses on the sheer mass of at-risk behaviors at a facility. The at-risk behaviors in question are the work practices of the facility that are necessarily interwoven with management systems, including safety systems. This statement emphatically does not mean that an injury is the employee's fault. Nor does this statement contradict the diagnosis of quality improvement personnel that 85 percent of the problems with quality are due to poor management practices. In the statement that "most accidents are caused by at-risk behavior," the type of cause referred to is known as the final common pathway. Those critical behaviors are "produced" by management systems of the site and, therefore, decreasing them is the key to accident prevention.

For example, a worker may be feeling pressured by the production schedule, and at the same time, perhaps he or she is preoccupied with a daughter's illness. Under the circumstances, if the worker gets hurt the cause is almost always related to the worker doing something at risk in response to the situation, some action such as trying to clear jammed equipment without first turning it off. The work configuration may encourage or even require that this behavior occur. In other words, production pressure, the work design, and family worries are important variables in the situation. By definition the variables are continually changing. The common thread among almost all incidents is the observable at-risk behavior of reaching into moving equipment. This behavior is a critical behavior—so called because it is a behavior that makes a critical difference in whether or not a worker gets injured while using the equipment in question.

The statistics of the work environment make it clear that a very large number of at-risk behaviors and/or conditions precede every accident. To use an image, this swarm of preexisting at-risk behaviors and conditions is in the air, like water vapor just waiting to precipitate out as a thunder shower or like an avalanche primed to happen. Focusing on behavior is crucial because an instance of at-risk worker behavior is like the small sound that touches off an avalanche. These small causes that precipitate large effects are causes that provide a final common pathway for many preceding causes to come together. This is the type of cause that at-risk employee behavior represents. In virtually all accidents, employee behavior provides the last link and common pathway for an accident to happen. However, the at-risk behavior at issue is a part of the management system, implicitly either encouraged or condoned by management. Therefore, to blame employees is counterproductive.

Instead of blame and fault finding, the behavior-based approach identifies critical safety-related behaviors, measures the sheer mass of them, and manages their frequency. This includes analyzing all the cases of at-risk behavior and developing action plans for performance improvement. This concept is new in safety circles, but it has gained a solid foothold in quality improvement efforts, and has been known and used in science for a century. To picture the linkage between at-risk behavior and injuries, it is helpful to think of the relationship between the critical mass of a radioactive substance and its explosiveness. For instance, in the case of a piece of uranium of critical mass, no one has any idea precisely which unstable atom will touch off a chain reaction resulting in an explosion. Just as there is a randomness or unpredictability about individual atoms in a mass of radioactive uranium, there is a randomness and unpredictability about individual employee behaviors at a particular facility. On the other hand, the activity of the whole mass of uranium is statistically very predictable. And in the same way, the overall safety performance of an entire facility is statistically very predictable; at a given level of at-risk behavior, there will be explosive events—accidents are going to follow.

Given what physical science knows about radioactive uranium, the people who manage it are very careful to store it in quantities well below its critical mass threshold. Given what behavioral science knows about behavior, leaders responsible for safety do something similar: they identify the accident threshold (the critical mass) of their facilities by using applied behavior analysis to identify critical behaviors. Once they have identified the critical behaviors, they measure frequency through observation and they provide feedback for improvement. The data gathered from observation is used to develop improvement targets. This core linkage—percentage of safe behavior up, injury frequency down—has been demonstrated in a variety of business and industrial applications.

In each facility, the rise and fall in the level of safe behavior is a function of various factors: management system, workforce, physical plant, machinery and processes of production, the product itself, and so forth. As the frequency of at-risk behavior increases, the likelihood that injuries will occur increases. Many first aid injuries generally occur prior to a more serious injury, and so forth up to fatalities, as shown in Figure 8-2. The challenge is to identify the critical behaviors of a particular facility and then to track and manage safe behavior levels. This proactive management of safety performance is steered by indicators in advance of even first aid accidents. To steer by injury levels even as relatively benign as first aid accidents is to give up management control and to invite the fluctuations of the accident cycle Figure 8-3 shows the relationship between a facility's thresholds of acceptable injury rates and management's changing response to the rates: When the recordable rate rises above a facility's upper limit, management acts to drive the rates down; when the rates fall below the acceptable limit, management loses focus on safety, and the recordable rate rises again. In this cycle, the management action for improvement follows fluctuations in the injury rate. The accident cycle is a familiar fact of organizational life in which an organization's safety effort is based on response to

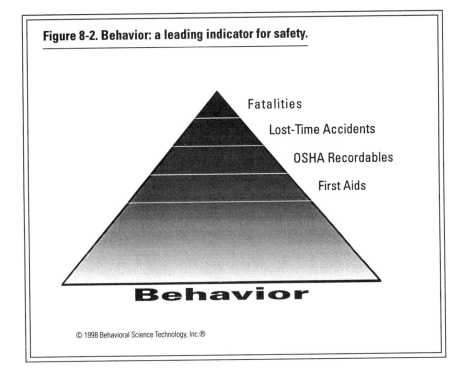

Figure 8-2. Behavior: a leading indicator for safety.

Fatalities

Lost-Time Accidents

OSHA Recordables

First Aids

Behavior

© 1998 Behavioral Science Technology, Inc.®

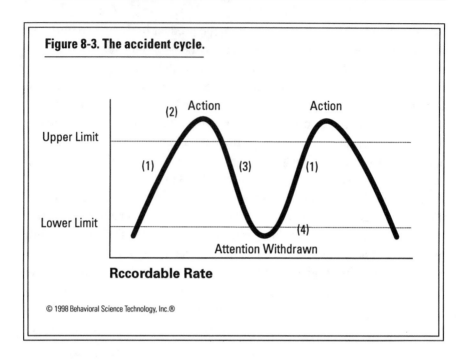

Figure 8-3. The accident cycle.

© 1998 Behavioral Science Technology, Inc.®

injuries. This approach is reactive rather than proactive, and it produces an intermittent, marginal solution to the problem. By waiting for recordable accidents or injuries to trigger its response, the reactive approach proceeds as though continuous improvement were impossible.

Whether hidden or explicit, the premise of reactive safety management is that staying even is all that is possible. Behavior-based safety management focuses on the strategic, long-term needs of a company by perceiving safety as a long-term product of the organizational system rather than as an accidental by-product of unknown origin. It is also important to distinguish the behavior-based approach from other approaches that emphasize attitude or culture change over behavior change. As its name implies, the behavior-based approach improves safety attitude and culture by identifying and then managing a change in the behaviors that are critical to safety in a given facility.

Understanding the Relationship of Behavior to Attitude and Culture

Most people naturally register the sequence of attitude change followed by behavior change, but behavioral science has shown that behavior change also causes a change in attitude. Because attitude has the power to affect behavior, people are tempted to focus on it in the first place. Another factor that inclines people to focus on attitude rather than behavior is that at first glance attitude seems simpler to handle. A list of necessary behavior changes can have so many items that without the benefit of the behavior-based approach there is no obvious way to work with all the items systematically. The important point for safety management is that this attitude-based approach introduces an indirect connection. Although everyone agrees that an improvement in safety-related behavior is what is needed, conventional wisdom falsely concludes that since the behaviors are so numerous, desired behavior change has to be achieved through attitude change. Therefore, the call goes out for a new batch of safety posters and a round of meetings urging people to change their attitudes about safety. Because behavioral science provides a method for measuring and managing the mass of behaviors, it cuts through the indirectness, pointing out that if behavior change is what is needed, there is no need for a detour that addresses attitudes first to get at behaviors later. Behavior can be managed directly. There are two powerful reasons to focus on behavior first: 1) Behavior can be measured and therefore managed, whereas attitude presents measurement problems; 2) a change in behavior leads to a change in attitude.

Although programs emphasizing attitude change make a strong appeal to our common sense, such programs are flawed because they overlook these two points. Everyone agrees that a good safety attitude is important. Given the importance of measurement in the management process, it follows that attitude change is very difficult to manage. To actually manage a process of continuous safety performance improvement solely by means of attitude change would require some form of attitude monitoring and control, which is not feasible even if it were desirable. In actual practice, attempts at changing safety culture by changing attitude invariably suffer from a lack of control.

The key here is that culture cannot be systematically changed by focusing on attitude change. For instance, when asked, "What are we doing to improve safety attitude, and how do we know whether it works?" a prudent manager would be hard pressed to say anything more than, "We try many things—it is hard to tell what works and what does not;" or, "We are giving motivational talks to all employees, and we think they are having a good effect."

But when asked, "Is the incidence of at-risk body placement on the increase or the decrease at our plant?" a behavior-based safety management process can give a definite answer, indicating by how many percentage points the incidence of the specified behavior is up or down over the last measurement period. The objectively observable character of behavior makes it amenable to both measurement and management, addressing the first weakness of attitude-focused programs. The second strength of behavior-based management is that changed behavior causes a change in the attitude of the workforce.

This principle has been demonstrated in the workplace on numerous occasions. For instance, during the assessment effort at one facility, applied behavior analysis revealed that one of the behaviors associated with a significant number of accidents and/or injuries was at-risk body placement in relation to task. Here was a prime candidate for the facility's behavior-based inventory, one that can be targeted as a leading indicator of how safe or at risk the workplace is, in advance of any accidents at all. The rationale is straightforward—the more frequently workers use at-risk body placement in relation to task, such as standing in the line of fire, the more likely it is that an accident will happen. Conversely, other factors being equal, the less frequently somebody stands in the line of fire, the less likely an accident will happen. Using a data sheet that incorporates the operational definitions of such critical behaviors, trained observers make random samples of workplace behavior, producing a measure of the facility's level of safety performance.

At the facility in question, a common sequence of events took place. Work crews typically took their at-risk behavior for granted. Within the safety culture of this facility, there was the unspoken assumption that at-risk body placement was simply part of being an efficient worker—a team player with hustle. Within six to eight months of experience with the behavior-based safety process, however, their behavior and their attitude changed dramatically. The workers came to understand the basic concepts of the process. They realized that the observers were careful to keep their observations objective and accurate. They also realized that management was careful not to distort the observations with disciplinary action or punitive measures. The result was that the same workers who used to believe standing in the line of fire was acceptable came to regard it, and related at-risk body placement, as irresponsible behavior by a mature team player.

Ensuring Employee Involvement

The behavior-based approach to accident prevention builds in a highly significant workforce involvement. From the outset, site-wide input is used to draw up and define the behavior-based inventory for the facility. Properly administered, this process means that crews become interested in their safety performance curves relative to their own past ratings and relative to the ratings of other crews. Whereas in the past supervisors may have felt they were always having to nag their workers about safety, they now find that the workers initiate discussions about their safety performance ratings. When workers achieve several periods of unbroken improvements, they are proud of themselves. Then when the graph of the workers' performance shows a decrease in the percentage of safe behaviors for their crew, they want to know exactly which critical behaviors are responsible for the dip in their good record.

The "secret" of this kind of participation and involvement is that behavior-based safety management avoids personalities by focusing on something that is objectively measurable. Besides, motivational speeches are not likely to have long-term effects, especially if system factors remain constant. Although it seems a bit paradoxical at first, it is because the behavior-based approach appreciates attitude change that it does not waste time and resources ordering, exhorting, or "motivating" people to change their attitudes. This is a fruitless exercise because people cannot obey such orders, even if they want to. It is a fact of human nature that people cannot be commanded in attitudinal matters. People cannot be commanded to change their attitudes any more than they can be commanded to hope or to believe or to like something. They can, however, manage not to stand in the line of fire while they are working. Therefore, the first step in developing this safety process is to make an accurate analysis of the safety-related behaviors that are actually occurring in an organization.

Antecedent-Behavior-Consequence (ABC) Analysis

An essential tool of safety management is discovering and addressing the roots of accidents. All safety efforts that work—whether or not they are consciously behavior-based in their orientation—are effective because they influence employee behavior. Equally crucial is that most organizations have what amount to very strong behavioral incentives that favor at-risk behaviors. Applied behavior analysis helps an organization assess the factors that are really driving its safety efforts. This basic tool of applied behavior analysis is known as ABC Analysis, and it provides the powerful foundation of behavior-change technology. In terms of this analysis, an antecedent is an event that triggers an observable behavior. A consequence is any event that follows from that behavior. An example is a ringing doorbell (antecedent), which we answer (behavior) to see (consequence) who is at the door. Common sense tends to identify the antecedent—in this case the doorbell—as the most powerful stimulus to behavior—in this case answering the door. And of course the antecedent is important. However, applied behavior analysis demonstrates that consequences are more powerful determinants of behavior than are antecedents.

To see the truth of this discovery, suppose a situation where the doorbell rings repeatedly and repeatedly there is no one at the door. Perhaps the bell is malfunctioning, or pranksters are ringing the bell and then running away from the door. In such a case, the behavior of answering the door to see who is there is frustrated for lack of the expected consequence. In fairly short order, one would stop "automatically" answering the door. As soon as the ringing doorbell no longer reliably signaled the presence of a caller at the door, it would no longer elicit in us the behavior of going to the door to see who was there. Taken by itself, the antecedent (the bell) does not directly determine the behavior (answering the door). Instead, antecedents elicit certain behaviors because they signal or predict consequences. The goal of ABC Analysis is to discover which antecedents and consequences are influencing a particular behavior. Once these factors are known, they can be changed; and when the antecedents and consequences change, behavior changes. Where safety performance is concerned, this means finding out which antecedents and consequences are actually driving behavior in the workplace.

In a nutshell, ABC Analysis involves the following principles:

- Both antecedents and consequences influence behavior, but they do so very differently.
- Consequences influence behavior powerfully and directly.
- Antecedents influence behavior indirectly, primarily serving to predict consequences.

Many well-intended safety programs fail because they rely too much on antecedents—things that come before behavior—safety rules, procedures, meetings, and so on. All too often these same antecedents have no powerful consequences backing them up.

In the ABC Analysis of real-world situations, one finds that most behaviors have a cluster of consequences that follow from them. For instance, the following is a brief but representative list of consequences commonly cited by workers regarding failure to wear hearing protection:

- Greater comfort when they are not wearing the protective equipment
- Greater convenience in not having to locate the protective equipment and put it on
- The possibility of hearing impairment

Each of the natural consequences in this list is like a plus or a minus, competing among the others to determine what behavior a worker will exhibit the next time he or she does a job that requires hearing protection. The antecedents in the environment—the sign that reminds the workers to wear hearing protection, the supervisor who mentions it now and then, the high-pitched or high-intensity sound of the machinery itself—have a less direct influence over the workers' behavior than the consequences that follow from their behavior. With this in mind, it is important to note that the first two consequences listed push the worker toward not wearing protective equipment for hearing, whereas the last consequence pushes the worker toward wearing it. In addition to discovering that consequences are stronger than antecedents, behavioral science research has also found that in the competition of consequences to control behavior, some consequences are stronger than others.

Soon-Certain-Positive: The Strongest Consequence. Three features determine which consequences are stronger than others:

1) *Timing:* A consequence that follows soon after a behavior influences behavior more effectively than a consequence that occurs later.

2) *Consistency:* A consequence that is certain to follow a behavior influences behavior more powerfully than an unpredictable or uncertain consequence.

3) *Significance:* A positive consequence influences behavior more powerfully than a negative consequence.

These three rules mean that the consequences having the most power to influence behavior are those that are simultaneously soon, certain, and positive. By contrast, the weakest consequences are the ones that are late, uncertain, and negative. Because they are so common, and often so ordinary, examples of the first kind of consequence can seem trivial and not worth noticing. But that is precisely the point; such consequences are so common because they are so powerful in spite of their seeming triviality. The previous example about protective equipment for hearing is a case in point. A safety program that tries to motivate the use of personal protective equipment solely by stressing the possibility of hearing loss is relying on the weakest kind of consequence—one that occurs slowly or eventually (later), if it happens at all (uncertain), and that is negative. Common sense may dictate that it is illogical to risk something as serious as hearing loss for some minor but immediate convenience or comfort, such as not having to locate and wear protective equipment for hearing. The fact is that in the give-and-take of everyday situations, most people continually take such risks for precisely these small but immediate, certain, positive outcomes. Whether it is common sense or not, it is human nature to behave this way. From a long-range point of view, clearly a safety effort that relies primarily on the possibility of hearing loss to influence workers to use protective equipment for hearing is a safety effort that is going to sustain hearing loss.

Consider two managers, one who wears hearing protection when it is required, and one who does not. What difference between these two is most likely to explain the difference in their safety-related behavior? Not their nature, and not the antecedents. They have the same human nature, and they are both aware of the antecedents of their behavior—that is, they both know that they should wear protective equipment when it is required. One manager wears protective equipment for hearing and the other does not and may not even think about it. The difference in their behavior is most likely the result of different sets of consequences for their behavior. The manager who is careful to wear hearing protection most likely receives strong, positive consequences for behaving in an exemplary fashion. This manager is part of a safety culture that defines compliance with safety measures as good performance and consistently gives positive feedback for it. The safety culture of the other manager is very likely one that offers no support or feedback for paying attention to hearing protection in particular, and for guiding safety-related behavior in general.

The two managers have different attitudes about safety because their attitudes internalize and reflect their respective safety cultures, and their safety cultures differ in what they reward and what they punish. The first manager feels the importance of safety and experiences a sense of leadership when setting a good example. This manager's attitudes about safety have become an internal source of soon-certain-positive consequences for safety-related behaviors. The difference between these two managers highlights the importance of the safety culture as the primary source of consequences that build the attitudes that, in turn, guide and reinforce safety behavior.

Between the most powerful consequences (soon-certain-positive) and the weakest ones (late-uncertain-negative), there is a range of consequences. For instance, among the consequences of not wearing protective equipment, comfort and convenience were the most powerful, whereas the possibility of hearing loss was the weakest. This is a clear-cut case. Oftentimes, however, each of the consequences of a particular behavior is a mixture of strong and weak effects.

Many safety programs are oriented toward penalties and punishments for violating safety measures, much like getting a traffic citation for speeding. Such a program can have some effect in those very rare situations in which being caught and cited soon is very certain to follow as a consequence of infringement of the rules. However, even in this case, the management effort would be wasteful because it would be spending its resources on delivering negative consequences. Thus, even if a negative approach did not have unhelpful side effects, it would be less effective than a positive approach because, dollar for dollar, negative consequences are less powerful in their impact on worker behavior than positive consequences. However, there are always side effects, and behaviorally trained managers know that the punitive approach presents several serious problems. The usual effect of such a program is to teach people not to get caught. As a practical matter, in most facilities it is as prohibitively expensive to ticket every safety infraction as it is for the police to ticket every driver who exceeds the speed limit.

Finally, a safety effort that relies solely on the threat of some accumulating physical debility to motivate workers is like a stop-smoking campaign that expects smokers to quit smoking simply because they have been given information about associated lung problems. ABC Analysis brings clarity to these matters by insisting on a simple fact: when a facility's safety effort is not working it is because the consequences in favor of safe behavior are weaker than the consequences in favor of at-risk behavior. The function of ABC Analysis, therefore, is to understand a facility's most stubborn safety problems. (See Case History 2 for an application of this approach.)

ABC Analysis of Failure to Wear Respiratory Protective Equipment. Behavior analysis of antecedents and consequences (ABC Analysis) is a tool that can be used in several areas of behavior-based safety management. First, ABC Analysis illustrates the problem that safety management faces in most facilities; there are many soon-certain-positive consequences in place that favor at-risk behavior. This problem makes ABC Analysis a good tool for introductory presentations of the behavior-based approach. Second, ABC Analysis is valuable during accident investigations because it brings a clear understanding of the tangle of consequences that elicited the behavior precipitating the accident. And third, ABC Analysis serves the function during the problem-solving sessions of using observation data to select improvement targets. ABC Analysis has three steps:

- Step 1: Analyze the at-risk behavior.
- Step 2: Analyze the safe behavior.
- Step 3: Draft the action plan.

See Figure 8-4 for an overview. In Step 1. the at-risk behavior under analysis is stated in objective, observable terms (see Figure 8-5). This requirement is very important. From the outset, ABC Analysis is focused on behavior—failure to wear respiratory protective equipment—not on attitude or blame. This nonjudgmental approach characterizes discussions and interviews with workers, and from these a list of the triggers or antecedents of the at-risk behavior is developed. The antecedents represent the things that workers are aware of as triggers of the at-risk behavior. In cases in which workers fail to wear respiratory protection, they are typically aware that the protection is unavailable, or perhaps it is available but inconveniently so. Workers are also aware that there is peer pressure not to wear the protection and that when they feel rushed, they are much less likely to wear the equipment. Workers may say that whether they wear respiratory protective equipment depends on the time of day or shift. They may also note that even supervisors sometimes do not wear protection when it is required, and hence are not good models of safe behavior in this matter. In addition, workers may say that they have never really been trained to take the protective equipment seriously. And, finally, workers often admit anticipating that no one will reprimand or penalize them when they do not wear respiratory protective equipment. Each of these points is then entered in the following list of antecedents of at-risk behavior:

- Availability
- Peer pressure
- Rushing
- Time of day
- Modeling

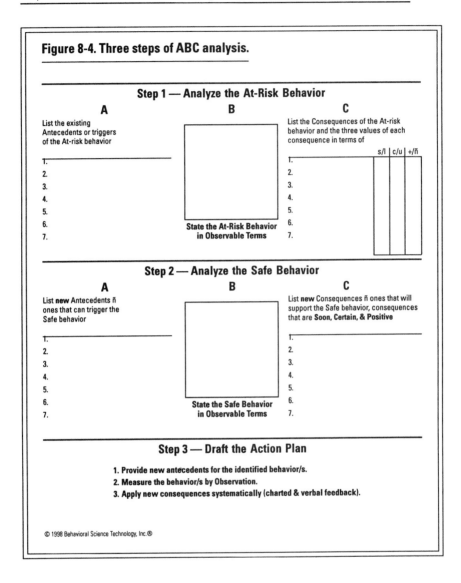

Figure 8-4. Three steps of ABC analysis.

Step 1 — Analyze the At-Risk Behavior

A

List the existing Antecedents or triggers of the At-risk behavior

1.
2.
3.
4.
5.
6.
7.

B

State the At-Risk Behavior in Observable Terms

C

List the Consequences of the At-risk behavior and the three values of each consequence in terms of

| | s/l | c/u | +/ñ |

1.
2.
3.
4.
5.
6.
7.

Step 2 — Analyze the Safe Behavior

A

List **new** Antecedents ñ ones that can trigger the Safe behavior

1.
2.
3.
4.
5.
6.
7.

B

State the Safe Behavior in Observable Terms

C

List **new** Consequences ñ ones that will support the Safe behavior, consequences that are **Soon, Certain, & Positive**

1.
2.
3.
4.
5.
6.
7.

Step 3 — Draft the Action Plan

1. Provide new antecedents for the identified behavior/s.
2. Measure the behavior/s by Observation.
3. Apply new consequences systematically (charted & verbal feedback).

© 1998 Behavioral Science Technology, Inc.®

- Lack of training
- Anticipation of mild consequences

The consequences are then listed. These are the consequences that the workers are aware of as following from their behavior when they fail to wear respiratory protective equipment. Workers typically note such things as convenience, comfort, or time pressure, saying that it saves time to work

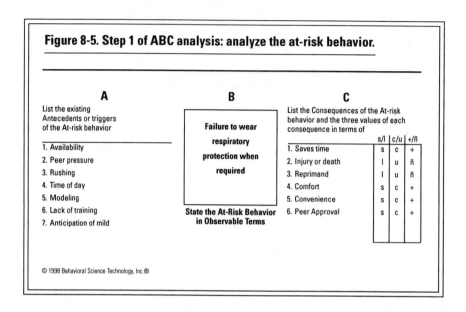

Figure 8-5. Step 1 of ABC analysis: analyze the at-risk behavior.

A	B	C
List the existing Antecedents or triggers of the At-risk behavior	Failure to wear respiratory protection when required	List the Consequences of the At-risk behavior and the three values of each consequence in terms of

(Column C values below, with headers s/l, c/u, +/ñ)

	s/l	c/u	+/ñ
1. Availability			
2. Peer pressure			
3. Rushing			
4. Time of day			
5. Modeling			
6. Lack of training			
7. Anticipation of mild			

State the At-Risk Behavior in Observable Terms

	s/l	c/u	+/ñ
1. Saves time	s	c	+
2. Injury or death	l	u	ñ
3. Reprimand	l	u	ñ
4. Comfort	s	c	+
5. Convenience	s	c	+
6. Peer Approval	s	c	+

© 1998 Behavioral Science Technology, Inc.®

without respiratory protection. Workers are aware that they may receive a reprimand and that injury, or even death, may be a consequence of their behavior. However, a worker aware of peers favoring a kind of bravado in this matter, would be "rewarded" for showing a disregard of respiratory protection. These consequences represent pressures on the worker, either for safe behavior or for at-risk behavior. The following are consequences of failure to wear respiratory protective equipment (see also Figure 8-4):

- Saves time (soon-certain-positive), strongest
- Injury or death (late-uncertain-negative), weakest
- Reprimand (late-uncertain-negative), weakest
- Comfort (soon-certain-positive), strongest
- Convenience (soon-certain-positive), strongest
- Peer approval (soon-certain-positive) strongest

Each consequence that is listed is also analyzed in terms of the soon-certain-positive scale to determine which ones are in fact controlling the behavior. In this list, the consequences in favor of the at-risk behavior—saving time, comfort, convenience, and peer approval—are the strongest, whereas the consequences in favor of safe behavior—possible injury or death, and reprimand—are the weakest. This balance of consequences in favor of at-risk behavior is extremely representative. It is quite common for the natural consequences in the workplace to favor at-risk over safe behavior. Step 2 of ABC Analysis addresses this fact. See Chapter 5 for the application of this problem-solving technique during work group action planning for continuous improvement.

The important general principle at the heart of this chapter is that the work group focuses on the mass of behaviors that make up its performance, paying particular attention to the identified at-risk behaviors discovered to have the lowest percent-safe values during the immediate period of peer-to-peer observation. On the basis of this observation of its own performance, the work group produces a critically important perspective on the state of safety at the site. As an illustration, think of the flashing patterns of Christmas tree lights, a common sight during the holidays in many countries. Picture a 30 or 40-foot tree in a shopping mall or some other public place, decorated with multiple systems, each with hundreds of lights and each of the systems blinking in different patterns. Anyone standing near the foot of the tree would see individual lights blinking on and off, and might get some sense of the pattern in the immediate foreground. But for someone standing back far enough to see the whole mass of blinking lights at a glance, the patterns would not only be visible but inescapable.

In a further correlation, imagine that the lights represent safe behaviors blinking on and off, on and off, at a rate of thousands of behaviors per second. Suppose that each system of lights has a different color of bulb, and that each system represents a different kind of safe behavior. This imagined tree representing the safety performance of an entire site at all moments of the day is like the vision that guides the work group engaged in behavior-based safety. Each individual observer sees only a small part of the tree. However, since the behavior-based inventory has been developed carefully, the individual observers know that the lights they are seeing are part of a large system of critical factors for continued safety and well-being at their site. Once the observer corps is calibrated so that it uniformly registers, tallies, and comments well on what it observes, simple math yields a birds-eye view of the whole picture. This is the same as for the holiday tree. A well-calibrated observer corps registers and records a sample of workforce behavior that is representative of performance as a whole.

The outcome of this process is a responsive and statistically valid measurement of the safety performance of a facility, in advance of any injuries or accidents. Since it was developed with input from hourly employees, the behavior-based observation process has credibility with them. Therefore, feedback from the measurement process becomes a powerful consequence. The workforce receives this feedback in the form of discussion with the observer, posted charts and graphs, and in reports at safety meetings. Supervisors also learn that they will be judged rationally and fairly on their safety performance. The measurement process puts in place a mechanism that positively delivers the strongest possible consequences for continuous improvement in safety.

PUTTING THE PRINCIPLES OF BEHAVIOR-BASED MANAGEMENT TO WORK

It is not the point of behavior-based safety management to change human nature, but rather to change the safety culture, to use the nature of behavior in favor of safety instead of against it. This amounts to devising consequences for safety that are soon, certain, and positive, which is the implementation effort, and it is a bit like priming a pump. The initial soon-certain-positive consequences in favor of safe behavior build new attitudes toward safety. These new attitudes, in turn, become the source of both broader and more finely tuned attention to safety, bringing new soon-certain-positive consequences to bear on a facility's safety performance. In this way the positive and continuous improvement process is established in a facility, becoming a safety mechanism.

One of the most powerful consequences at work in any organization is peer feedback, which offers immediate, certain, positive feedback to the worker from a peer. (Note that the employee-driven, behavior-based safety process engages this peer dynamic and realigns it for safety.) Peer feedback is accepted differently than supervisory feedback, which has different connotations. A supervisor, too, may inadvertently encourage or maintain at-risk behavior by praising production performance even though the supervisor saw workers' at-risk behaviors while they were hustling. The effect of this praise is further compounded when the workers know that the supervisor saw them taking risks. Finally, the most common consequence in favor of at-risk behavior is simply that most at-risk behavior is not even observed, let alone noted and addressed. The following case studies illustrate some of the outcomes of peer-to-peer ongoing observation and feedback.

Case History 1:
Behavior-Based Safety Outcomes at a Paper Mill

The following case history shows the predictive relationship between observed percent-safe and injury rates. The subject of this study is a wood products lumber mill that began planning for implementation of its behavior-based safety process in 1993. With a union-represented workforce of approximately 425 employees, the mill prides itself on the level of employee involvement it maintained from the beginning. Union officials played an active role during the search phases of consultant review and selection—accompanying mill managers on an extended visit to another site using the behavior-based safety process. Subsequent to that trip, collaborative decision making on the part of management, union, and wage-roll employees at the mill was a key element for securing involvement at all levels. This involvement led to high-quality group engagement. Employees at the paper mill adapted the behavior-based approach in their own safety initiative. Their effort focuses on improving safety through coworker observations of everyday task performance. These observations provide data for work groups to use in the development and implementation of action plans for improvement.

Reductions in Injury Rates

Before implementing a behavior-based safety initiative, the mill had a recordable rate of 16.8, above the industry average of 15.9. Other case histories given in Chapter 13 show results with companies starting below their respective industry averages. Figure 8-6 illustrates a significant step change when mill employees began peer-to-peer observations with feedback and data analysis for improvement targets. The mill realized a 47 percent reduction in injuries in the first two years of their process [$t(54) = 2.84$; $p < .01$]. Figure 8-7 shows this improvement over a two-year period of time. Not only was the average for the post-implementation period significantly lower than the preceding two years, but the recordable rate showed continuous improvement. After observations started, the recordable rate decreased from more than 10 to less than 5. This trend was highly significant ($R = -.47$; $p < .01$).

Relationship between Behavior-Based Process Indicators and Injury Rates

One predictor of the recordable rate is the percent-safe score derived from behavior-based observations. Figure 8- 8 shows that as the percent-safe

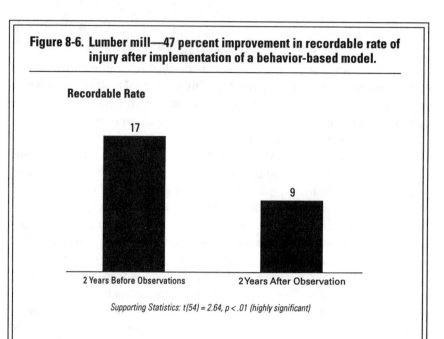

Figure 8-6. Lumber mill—47 percent improvement in recordable rate of injury after implementation of a behavior-based model.

Recordable Rate

17

9

2 Years Before Observations   2 Years After Observation

Supporting Statistics: t(54) = 2.64, p < .01 (highly significant)

© 1998 Behavioral Science Technology, Inc.®

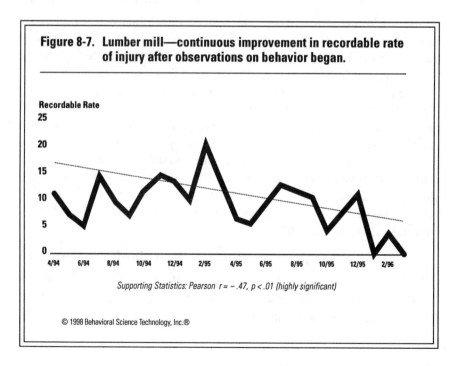

Figure 8-7. Lumber mill—continuous improvement in recordable rate of injury after observations on behavior began.

Recordable Rate

25

20

15

10

5

0

4/94   6/94   8/94   10/94   12/94   2/95   4/95   6/95   8/95   10/95   12/95   2/96

Supporting Statistics: Pearson r = − .47, p < .01 (highly significant)

© 1998 Behavioral Science Technology, Inc.®

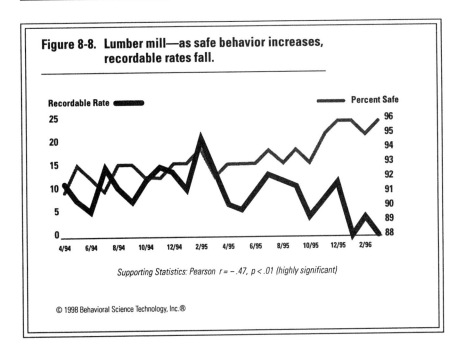

Figure 8-8. Lumber mill—as safe behavior increases, recordable rates fall.

Supporting Statistics: Pearson r = − .47, p < .01 (highly significant)

© 1998 Behavioral Science Technology, Inc.®

rate increased in the mill, recordable rates decreased. This correlation is statistically significant ($R = -.47$; $p < .02$). Another predictor of the recordable rate at the paper mill is the contact rate. This contact rate is specific to the behavior-based safety process as it is presented in this book. Figure 8-9 shows that the higher the contact rate, the lower the recordable rate. This correlation approaches significance ($R = -.31$; $p < .07$).

By focusing on behaviors critical to safety performance, this paper mill achieved a steady decrease in recordable injuries. Ongoing peer-to-peer observation and feedback about safety performance has become part of the mill's culture: it's what employees talk about. Data collected during observations is successfully used for the development and implementation of action plans. The relationships between these activities and injury rates were significant: high levels of observation activity and percent-safe behavior correlated with low recordable rates. Finally, in a reassessment of the safety system at the paper mill, wage-roll employees, management, and union officials demonstrated a high level of interest and involvement in safety and prided themselves on their achievements. Scores on the safety survey (Chapter 6) administered before implementation and then two years later reflected statistically significant increases in safety perception:

1) The scale measuring involvement rose from 88 percent to 93 percent.

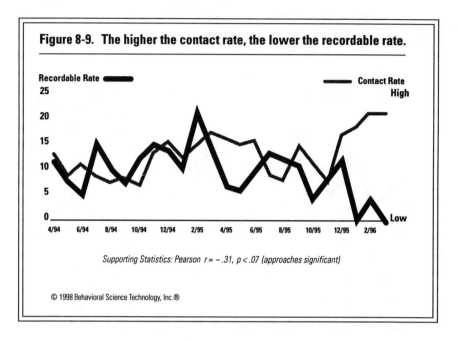

Figure 8-9. The higher the contact rate, the lower the recordable rate.

Recordable Rate ━━━   ━━━ Contact Rate

Supporting Statistics: Pearson r = − .31, p < .07 (approaches significant)

© 1998 Behavioral Science Technology, Inc.®

2) The scale for facilities maintenance rose from 76 percent to 82 percent.

3) The score for the overall percent-safe response rose from 73 percent to 78 percent.

Once it has been established, the behavior-based continuous improvement safety process represents a closed loop Figure 8-10. During implementation, however, a facility's efforts advance through the stages of the process—identifying critical behaviors, training observers to measure those behaviors and to give and receive feedback about them, and problem solving to remove barriers to continuous improvement. After these stages, the loop closes with further adjustments of the process, and the facility is on its way to establishing a mechanism for continuous improvement. The following case history illustrates different kinds of problem solving within the behavior-based model.

Case History 2:
3M Company Behavior-Based Initiatives Remove Barriers to Continuous Improvement

This case history reports on behavior-based safety efforts at three 3M facilities:

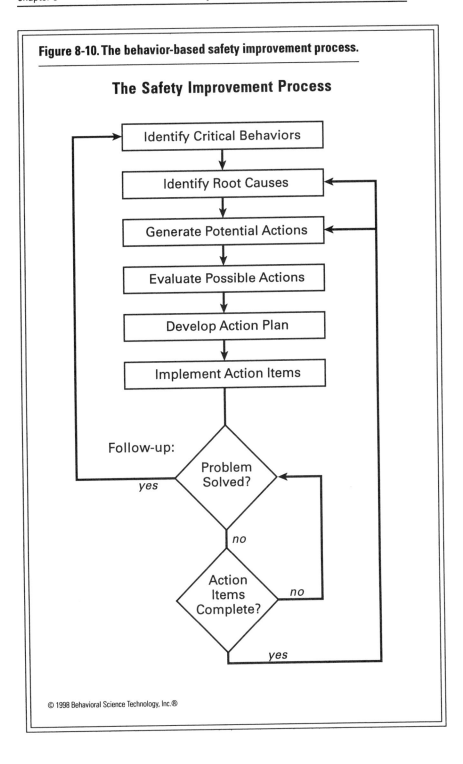

Figure 8-10. The behavior-based safety improvement process.

The Safety Improvement Process

Identify Critical Behaviors

Identify Root Causes

Generate Potential Actions

Evaluate Possible Actions

Develop Action Plan

Implement Action Items

Follow-up:

Problem Solved?
yes

no

Action Items Complete?
no

yes

© 1998 Behavioral Science Technology, Inc.®

1) The STAR (Stop Taking a Risk) safety initiative at 3M's film plant in Decatur, Alabama.

2) The SAFE (Safety Awareness for Everyone) initiative at 3M's specialty chemicals plant also in Decatur, Alabama.

3) The CHAMPS (Changing Habits Affects Motivation, People, and Safety) safety initiative in Greenville, South Carolina.

At each site, the behavior-based safety team uses its computer software to identify barriers for removal as work group follow-up items. This approach allows the safety teams or steering committees to pursue the many smaller but important safety remedies that do not require supervisory decisions or direction. With this method, the teams can reserve full-fledged action planning for barrier removal that needs group coordination and management input. As a result of their streamlined follow-up procedures, the 3M steering teams have earned the trust and support of wage-roll personnel, who see their action items being followed up, and of supervisors, who see their crews participating actively to improve work group safety performance.

Barrier Removal Is Crucial

While the adaptation of the behavior-based model varies from site to site, each implementation involves the four signature activities of behavior-based safety:

- Step 1: Identify critical behaviors.
- Step 2: Gather data on those behaviors.
- Step 3: Provide ongoing performance feedback.
- Step 4: Remove barriers to improvement.

Barrier removal is a crucial activity because at all sites a portion of identified at-risk behaviors are beyond the control of individual workers. As soon as a site's behavior-based observers begin gathering performance data, they also collect detailed information about barriers. Those barriers need to be addressed or the safety change-effort may lose credibility with site personnel. That loss of credibility can finally cripple the implementation effort.

Levels of Barrier Removal

After the observer data sheets and comments are entered in a site's computer database, the 3M team facilitators make one of the three responses shown below. The level of response is determined by the degree of control that observees have over whether or not they use the site's identified critical safe behaviors:

1) Is the critical safe behavior within the control of the individual observee? If yes, reinforce with *observer feedback.*

2) If no, is the remedy in the control of the behavior-based safety team and/or workgroups? If yes, do *follow-up* item(s).

3) If the steering committee cannot pursue a remedy without management input, then initiate *action planning.*

The facilitators produce a computer tracking report of all *follow-up* items by department and then they forward a copy to each work group to handle. When an *action plan* is called for, the facilitators discuss it with the appropriate supervisor. And the 3M observers are trained to handle their own *feedback* assignments.

Figure 8-11. 3M plant follow-up or action items for first three quarters of 1997.

Star Follow-up or Action Plan Items

	IDENTIFIED	COMPLETED	IN PROCESS
Building #15	46	18	28
Building #19	27	21	6
Boiler house	49	25	24
Converting	71	38	33
D1 Makerline	14	8	6
D3 Makerline	9	4	5
D4 Makerline	19	6	13
D5 Makerline	29	15	14
D7 Makerline	22	15	7
D8 Makerline	38	31	7
D9 Makerline	17	7	10
DSR / Reclaim	5	1	4
Maintenance	26	7	19
Packing	54	41	13
Q.C. Lab	24	24	0
Warehouse	5	4	1
Totals	**455**	**265**	**190**

© 1998 Behavioral Science Technology, Inc.®

Observer Feedback for Enabled Behaviors. A critical safe behavior is within the control of individual workers when the behavior is *enabled.* A safe behavior is enabled when it is triggered by a site's system antecedents and reinforced by the system consequences. When observers see personnel who are not using an enabled safe behavior, feedback is called for. For example, STAR observers in the warehouse noticed that contrary to standard operating procedure, some personnel were lifting heavy rolls of film into high bins instead of putting them into easier-to-reach lower storage bins. The identified safe lifting behavior was enabled by the site, but the workers themselves were choosing to perform the at-risk lifting behavior. Therefore, the action called for was observer feedback to remind warehouse personnel to use body mechanics that did not put them at risk for injury.

Follow-Up for Difficult Behaviors. Between enabled and nonenabled behaviors are *difficult behaviors.* A difficult safe behavior has some system antecedents and consequences working for it, and some working against it and for its related at-risk behavior. Beyond the control of the individual employee, removing this kind of barrier is within the control of the steering team and the work group. This category is primarily composed of fix-it items. Although each of these barriers is relatively small, removing them can quickly add up to important gains.

Action Planning for Nonenabled Behaviors. A safe behavior is *nonenabled* when the site's system antecedents and consequences favor the related at-risk behavior. To intervene on behalf of a nonenabled safe behavior often calls for an action plan. For example, production equipment may be designed, installed, or maintained in a way that requires the workforce to perform at-risk behaviors. Addressing a barrier of this kind at 3M Decatur requires an equipment change request (ECR), and an ECR automatically involves supervisory input and work group coordination across shifts and/or areas. Careful and rigorous *action planning* is well suited to formulating and tracking this level of barrier removal.

A Follow-Up for Cut-Resistant Gloves

3M procedure calls for film line operators to wear Kevlar gloves when they are cutting film to start a new roll. STAR observers noticed that workers on the film line either were using no gloves or were carrying two pairs of gloves and alternating between them. The reason? The Kevlar cut-resistant gloves do not have good gripping qualities. So although the available gloves protect the workers' hands from cuts, they make film handling more difficult. Out of professionalism, some workers preferred the "hassle" of carrying leather gloves for their gripping quality, and switching to the Kevlar gloves when they were about to cut the film. However, others reacted by using no gloves at all.

The STAR facilitator took on this follow-up item himself, researching the matter with other 3M film plants. He called the CHAMPS facilitator at 3M in Greenville, South Carolina, who had just the gloves that STAR needed: Kevlar cut-resistant gloves with leather palms sewn into them. Figure 8-11 shows the totals of STAR follow-up and action planning items for sixteen departments at the 3M film plant for the first three quarters of 1997. As the department work group members continue to make gains on their items, they call the STAR office to ask for more to do, which pleases their supervisors.

As follow-up successes like the gloves become more common, work groups and their supervisors know they can reserve their action-planning efforts for larger or longer-term projects such as the SAFE action plan reported here. For example, during a recent monthly review of its behavior-based data, members of the SAFE team focused on at-risk behaviors associated with changing condenser heads on some rooftop equipment. The manual-control hoist in place on the rooftop made the safe behaviors *nonenabled* by forcing personnel to climb over pipes and condensers to activate the hoist mechanism. At a meeting with supervisors and production managers, the SAFE facilitator and a steering team member presented a possible remedy: installing a remote-control hoist. The resulting action plan called for the SAFE team member to price remote-control hoists with vendors, and to put the best bidder in contact with the supervisors. The facilitator met with her site's product manager to budget the hoist upgrade. From start to finish, the action plan took ten weeks to complete. As a result, the identified safe behaviors have been upgraded from nonenabled to fully enabled. At $7,000 the new hoist came in under budget; it eliminates a cluster of seriously at-risk behaviors and allows personnel to perform the condenser change-out task faster—all signs of a good action plan.

During the same period at the 3M film plant in Greenville, South Carolina, the site's CHAMPS team was engaging supervisors to play an active role in observer motivation. Given the importance of the observer function in the behavior-based model, it is important to motivate observers to achieve adequate numbers of high-quality observations. With the support of the plant manager, in November 1996, the CHAMPS facilitator handed over observer motivation to 3M Greenville's supervisors. The supervisors took on that responsibility because the CHAMPS team made it worth their while by pursuing the same aggressive approach to follow-up items reported above of sister sites in Decatur, Alabama. Beginning in April 1996, as work groups used the data from CHAMPS observers to clean up many barriers to safe behavior, their supervisors began to appreciate those gains. Figure 8-12 shows the history of this development at the 3M Greenville site. Supervisor involvement is seen as critical for long-term success of the behavior-based model.

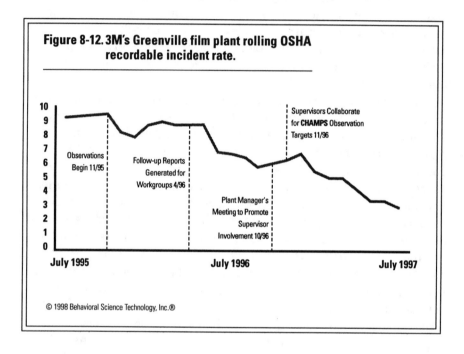

Figure 8-12. 3M's Greenville film plant rolling OSHA recordable incident rate.

© 1998 Behavioral Science Technology, Inc.®

On the basis of those achievements, in November 1996, the 3M Greenville plant manager informed plant supervisors that the CHAMPS process was one of the site's primary safety assets, and that he wanted them to actively use the behavior-based reports and other data and to support the CHAMPS observers on their crews. With that direction, the supervisors began coming to the plant manager to find out more about how to "get with the program." Realizing that the supervisors were important internal customers or clients for the CHAMPS safety "product," the facilitator developed for the supervisors a plan that spelled out the primary supervisory roles and responsibilities in connection with CHAMPS. From January through September 1997, the effect was a new period of continuous improvement.

Contrary to the expectations of many people, the behavior-based model works well at both union and nonunion sites. The 3M sites we have presented have nonunion workforces. The following case history shows what union and management can accomplish with behavior-based safety.

Case History 3:
ARCO Chemical Personnel and Contractors Make Gains with Behavior-Based Safety

In 1991 per sonnel at the ARCO petrochemical plant in Channelview, Texas, launched a behavior-based safety initiati ve. ARCO personnel named their employee-dri ven safety process FOCUS, whic h means *Focus on Changing Unsafe Situations.* That safety ef fort has helped the site ac hieve important results—an OSHA recordable rate that has impro ved steadily from 8.0 in 1 990 to 0.39 in 1   996 (Figure 8-1 3). L ocated 20 miles east of Houston, the Channelview plant produces propylene o xide, styrene mono-mer, eth ylbenzene, phenyl eth yl alcohol, meth yl ter tiary butyl ether , alk yl alcohol, and butanediol. The site is a nonunion facility with a total popula-tion of appro ximately nine hundred. Almost seven hundred are ARCO personnel and the remaining two hundred are per sonnel working on-site

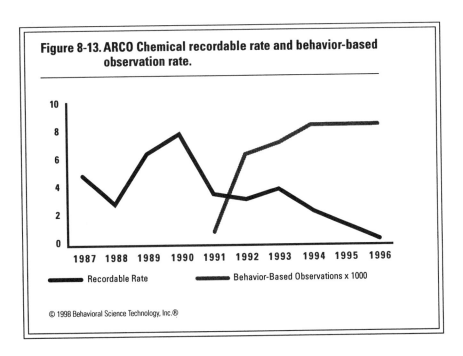

Figure 8-13. ARCO Chemical recordable rate and behavior-based observation rate.

Recordable Rate   Behavior-Based Observations x 1000

© 1998 Behavioral Science Technology, Inc.®

for independent contractors. As the observation rate improved, the ARCO Channelview recordable rate fell.

A Contractor Rate of Zero

The behavior-based safety effort at the site includes the independent contractors. The largest of the contracting companies is Austin Industrial, whose employees take care of all pipe fitting and general services onsite. In 1996 Austin Industrial had a record-setting OSHA recordable rate of zero. Each unit or department at the site is represented on the steering committee that directs the behavior-based safety effort. Austin Industrial personnel also serve on the steering committee.

Peer-to-Peer Observation and Feedback

To date, the steering committee has trained 558 ARCO personnel and 54 contractors as behavior-based observers. The observers measure work group safety performance using an inventory of operationally defined safe and at-risk behaviors. Company personnel develop their own inventory by analyzing the site's incident reports. By using this system, personnel are able to identify and define the cluster of twenty to thirty safety-related behaviors implicated in most incidents at the site within the past several years. Those are the behaviors that put persons at risk for injury. The behaviors in question often fall into categories such as line of fire, eyes on path, pinch points, and pre-job inspection. By observing for those behaviors, the observer corps at the site collect performance data and give their coworkers verbal feedback on how they are doing. The observer coaching effort has been a major undertaking of the safety committee, requiring dedication and commitment. The effectiveness of the coaching and of the observers is shown in actual figures. As the number of observations gained momentum, the plant's OSHA recordable rate dropped steadily.

Currently, there is one coach per shift in each unit or department of the site. The duties of the coaches include leading and coaching coworkers in behavior analysis, assuring closure of action plans, updating the work group on observation data, and acting as a liaison between the behavior-based steering committee and shift workers.

Adding Ergonomics Behaviors

In 1996 ergonomics-related critical behaviors were identified and added to the site's data sheet for ongoing observation. The steering committee is in charge of developing the observer training materials for those new items. The ergonomics items are also being added to the site's refresher training materials.

Examples of Action Planning

In a plantwide activity in 1996, the steering committee conducted behavior analysis on two areas of exposure to risk: chemical contact and failure to wear coolvests. A coolvest is a garment with pockets for ice packs. Many tasks at the site require workers to wear full-body protective "slicker suits," which put the wearers at risk for heat exhaustion given the heat and humidity of the work area.

For each of these identified exposures, the steering committee produced a list of five most-recommended action items. All items were followed up and closed out during 1996 with significant improvements in the related behaviors. Progress on these plans was then communicated to the entire plant.

Next Targets

These are the steering committee's next important goals:

- Extend observer training to all personnel in the operations and maintenance areas by the close of 1998.
- Include the administrative and clerical areas in the behavior-based safety effort.

In addition, the site is planning to offer refresher training for observers. The steering committee is putting together the training matrix to make sure that the training covers all relevant new material in development, such as the ergonomics items.

The steering committee is also looking for ways to make more effective use of its computer software to produce graphs and charts that clearly indicate the correlations between behavior-based data and incidents. As part of this new push toward more effective communication of performance targets and achievements, in the first quarter of 1997, the steering committee began publishing a newsletter, and is now placing its reports in two other ARCO publications.

A Piece of the Safety System

The ARCO facilitator of the steering committee and the safety effort say:

> Here at ARCO Chemical, behavior-based safety is a piece of the safety system. Until a company takes this piece as seriously as everything else they're doing, they are missing one of the spokes in the wheel. Industrial safety is a big job and we can't afford to ignore anything that makes a real contribution. The steering committee, coaches, and observers are making that kind of contribution—and they're proud of it.

Case History #4:
Ultramar Diamond Shamrock Petroleum Reduces Injury Rate and Completes Turnaround with Record-High Performance

In mid-1994, site personnel at Ultramar Diamond Shamrock Petroleum in Ardmore, Oklahoma, launched their behavior-based safety process. From 1990 through 1993, the refinery had an average recordable rate of 10.96 (Figure 8-14), well above the petroleum and coal products industry average of 6. Figure 8-15 shows the step-change improvement in safety performance and consistency achieved by the site in 1994, when refinery employees began behavior-based observations and feedback. At the end of 1995, the refinery had realized a highly significant 55 percent reduction in injuries. And for 1996 its OSHA recordable rate continued its trend of improvement.

As seen in Figure 8-14, from 1990 through 1993, the refinery's average recordable rate was 10.96. By the end of that period, the site had reduced its rate by 55 percent, and the end of 1996 was still showing continuous improvement. Figure 8-15 shows that in 1994, as the behavior-based safety initiative took effect, the refinery achieved a baseline shift or step change in its performance along with a marked increase in consistency.

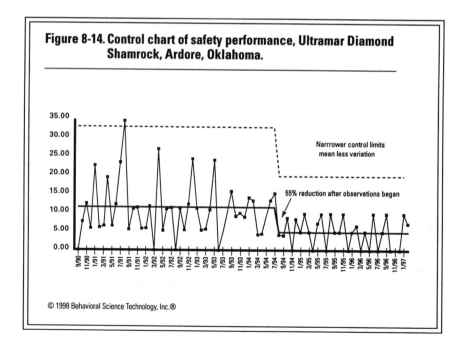

Figure 8-14. Control chart of safety performance, Ultramar Diamond Shamrock, Ardore, Oklahoma.

© 1998 Behavioral Science Technology, Inc.®

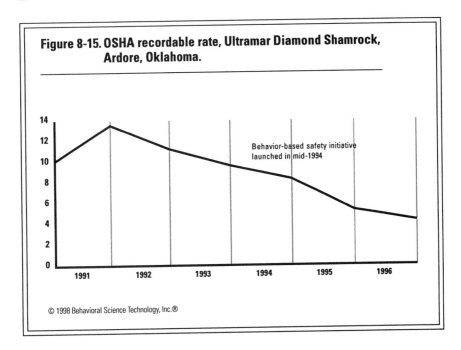

Figure 8-15. OSHA recordable rate, Ultramar Diamond Shamrock, Ardore, Oklahoma.

Behavior-based safety initiative launched in mid-1994

© 1998 Behavioral Science Technology, Inc.®

Plant Turnaround—A Special Application

In addition to this high level of general safety performance, the site's behavior-based safety process was used in an intensive special deployment during its 1996 plant shutdown and turnaround. With numerous contractors working throughout a plant, and site personnel doing tasks that are new to them, turnarounds are often associated with a "rash of incidents" that cause a spike in the safety numbers and drag down annual performance. However, in 1996, during Ultramar Diamond Shamrock's 30-day turnaround, the personnel at Total Petroleum, an Ultramar refinery in Ardmore, Oklahoma, used behavior-based safety to set a new standard in safety.

Total Petroleum processes 77,000 barrels of crude oil daily. The 180 wage-roll personnel at the site are represented by the International Union of Operating Engineers Local 670. The site began planning for implementation of its behavior-based safety process in 1993. Refinery personnel named their initiative WASP, short for Workers Awareness Safety Process. By reviewing incident reports, site personnel were able to identify and operationally define the cluster of at-risk behaviors implicated in the majority of the site's incidents.

Site personnel trained in behavior-based observation and feedback use track those at-risk behaviors as leading indicators of safety performance. Such at-risk behaviors include the following:

- Pre-job inspection
- Line-of-fire
- Body mechanics
- Eyes-on-work
- Walking/surface
- Lifting-and-bending

Along with peer-to-peer feedback on these behaviors, the observers also provide comment reports that a steering committee uses to do action planning for continuous improvement.

The following are recent action plans developed from behavior-based data include:

- Giving focused feedback on lifting and bending and demonstrating proper lifting techniques in the plant
- Installing chain operators on valves that are 20 feet or more above the plant's floor, changing worker behavior of climbing in pipe racks
- Paving roads and installing gutters to divert rainwater runoff to operating units (for both safety and environmental reasons)
- Improving lighting as needed

Behavior-based action plans are a critical element of the continuous improvement mechanism because of their role in removing barriers to safe working conditions. Action plans are evaluated qualitatively according to their number, quality, and the extent to which they address behavioral issues.

Successful Turnaround

The 30-day turnaround at Total Petroleum in January and February 1996 saw many more contractors onsite than usual. It was also a time of exceptionally high activity for the site's behavior-based safety process. Turnaround is always a safety challenge because of a lower level of knowledge and alertness. People new to a site are working in many areas of it, and many veteran workers are doing tasks that are new to them. As in the past, the site brought in extra safety professionals for the turnaround. However, they could not watch everyone all the time, so the behavior-based team recruited volunteers to do observations on both site personnel and contractors in the plant. Consulting with other users of behavior-based safety, the WASP team put in place the following strategy: when a contract company first came onsite, the WASP team would meet with the company's safety departments to explain the behavior-based approach and how it would help them. Using this approach was successful because as one WASP facilitator said, "Once they understood we were trying to help them, they had no objections, and they let us give their employees a 30-minute orientation on behavior-based safety."

Then after WASP observers first began observing the employees of the contractor company, they generated a report and explained to the employees what the information indicated about their performance of the critical behaviors. The safety professionals and other contractor personnel then used that information in their own safety meetings. The result was that by the end of the 30-day turnaround, the site experienced only one lost-time accident and a few first aid incidents. Other users of behavior-based safety have recorded similar outcomes.

CONCLUSION

In the mid-1980s, both union and corporate observers were being prudent in taking a wait-and-see attitude about behavior-based safety, which was new then and had not yet produced long-term results. But this technology has passed the trial stage and is here to stay. This is a matter of record at hundreds of sites—as many union as nonunion. By the mid-1990s, behavior-based safety initiatives were well established throughout the United States and Canada. In addition, the approach is making headway in the United Kingdom, and there are some ongoing efforts in Latin America, Jamaica, Australia, and the Philippines.

Since June 1997, more than 200,000 workers at more than 500 sites benefit from behavior-based safety initiatives. Since 1995 an ongoing study of injury reduction at sites implementing behavior-based safety has been conducted (Behavioral Science Technology, Inc. 1997). Of the more than 160 companies that have submitted injury data for the study, 74 provided data sets that met all three criteria of the study:

1) The site had been using behavior-based safety for at least one year.

2) The data covered only departments or areas using behavior-based safety.

3) Both baseline and follow-up data were available for the site.

The sites in the study population are a 50/50 mix of union and nonunion. This same ratio holds for the overall population of sites using behavior-based safety. Furthermore, the average reduction in injury rates at union and nonunion sites is the same. For all 74 sites studied, the average, cumulative annual reduction in injury rates from year-one baseline is 27 percent for the first year, 42 percent for the second, 50 percent for the third, 60 percent for the fourth, and 69 percent for the fifth (Figure 8-16).

The bar chart in Figure 8-16 shows reductions for all seventy-four organizations; the line chart shows reductions for the top 10 percent of those organizations.

Although the behavior-based model does not offer a quick fix, companies that implement it with the critical success factors presented here have made long-term improvements in their safety performance.

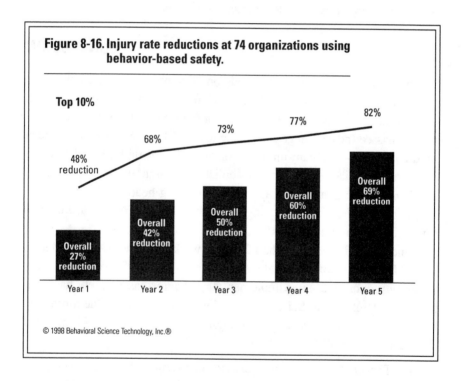

Figure 8-16. Injury rate reductions at 74 organizations using behavior-based safety.

© 1998 Behavioral Science Technology, Inc.®

REFERENCES AND SUGGESTED FURTHER READING

Behavioral Science Technology, Inc. (1998). *Ongoing Studies of the Behavioral Accident Prevention Process Technology, Third Edition.* Ojai, CA: Behavioral Science Technology, Inc.

Deming, W. E. (1996). *Out of the Crisis.* Cambridge, MA: Massachusetts Institute of Technology, Center for Advanced Engineering Study.

Hawley, J. K. (1995). "Where's the Q in TQM?" *Quality Progress* 28(10):63–64 .

Hendricks, K. B. and V. R. Singhal (1995). "Firm Characteristics, Total Quality Management, and Financial Performance: An Empirical Investigation." [Unpublished paper].

Krause, T. R. (1995). *Employee-Driven Systems for Safe Behavior—Integrating Behavioral and Statistical Methodologies.* New York: John Wiley & Sons.

Krause, T. R. (1997). *The Behavior-Based Safety Process—Managing Involvement for an Injury-Free Culture.* New York: John Wiley & Sons.

Spector, B., and M. Beer (1994). "Beyond TQM Programmes." *Journal of Organizational Change Management* 7(2):63-70.

Chapter 9

Determining Safety and Health Program Requirements

O. Dan Nwaelele

INTRODUCTION

Few programs run successfully by spontaneous desire, but most do by calm and prudent planning. Consequently, planning becomes the foundation of an effective program. Safety and health programs are no exception. The specific elements incorporated into a safety and health program must be chosen judiciously through careful assessment of workplace hazards and the risk of exposure. Currently, there is no formal standard governing minimum requirements on the national level. This chapter focuses on the need for a formal program and minimum program requirements, including the following:

- Safety policy
- Clarification of management and employee responsibility
- Employee involvement
- Hazard assessment
- Hazard prevention/control
- Information dissemination and training
- Program (auditing) evaluation
- Contractor safety requirements

THE NEED FOR A FORMAL PROGRAM

Within the safety and health profession, there is no argument about the need for formal safety and health programs. However, safety and health professionals rarely have the final say when business decisions are made.

Traditionally, management reserves the right, as it should, to make the final decisions that affect its business. Unfortunately, many managers are not aware that a formal accident prevention program is a means to manage their safety and health obligations and related liabilities and costs. The following stories exemplify the need for and show where careful planning and implementation could have made a difference.

---

2 Workers fall to deaths in Dome

"Two men working on the round-the-clock repair effort on the Kingdome ceiling were killed last night when a crane gave way as they prepared to sandblast over right field. A third man was seriously injured when the crane's basket plummeted more than 250 feet and crushed the cab of the crane. A worker for one of the companies who has been at the Kingdome for the past month said the crane that collapsed had been performing like others, 24 hours a day, with workers changing in eight-hour shifts. Although cranes are supposed to be inspected by operators every eight hours . . . , a co-worker on site late last night said the crane broke at the base of the jib where the boom and the jib connect. The crane has a 120-ton capacity, which means it is able to lift up to 240,000 pounds. 'This crane is used to holding all that weight, but cranes aren't made for that kind of shaking that was going with this sandblasting,' the worker said." (*Seattle Post Intelligencer,* August 18, 1994)

"County officials, faced with paying more than $11 million in repairs and compensation to the Mariners, have been under pressure to reopen the Dome for the Seahawks' first regular-season home game Sept. 18 against the San Diego Chargers. Dick Sandaas, the county's new interim director of the Kingdome, said all plans will be re-examined. 'No more work, no more repairs until the investigation is complete', he said. 'We've had a tragedy here, folks. Let's think about the families." (*Seattle Times,* August 18, 1994)

Work resumed at the site after the county hired a new contractor with heavy construction experience. Funds were authorized for safety improvement. A safety engineer was assigned to coordinate site safety. More important, for the first time a site-specific safety work plan was developed and implemented. Elements of the safety work plan included the following:

- Engineering controls
- Communication
- Routine evaluations of the cranes for structural integrity by a licensed professional engineer with extensive industry experience
- Personal protective equipment
- Other administrative controls

A few weeks later, the life of another worker was saved when his fall-restraint system (full-body harness and lanyard) prevented him from falling from the top of the Kingdome roof.

The Seattle community was reminded of how deadly the lack of planning can be only four months later, when tragedy struck again. Four firefighters lost their lives a few blocks from the Kingdome in January 1995. The following was published regarding the fire:

Report: Fire chiefs ignored safety

"The Seattle Fire Department's former training and safety chief has told state investigators the agency was so fearful of criticism that safety sometimes took a back seat to image. Stewart Rose said top fire department administrators censored safety reports and then banned them altogether so they would not be available to the media or state investigators. He also said the agency's top safety officer . . . was considered a 'snitch.' 'The organization has no desire to be safe, and most of the things they take credit for aren't happening. They're on paper but not realistic,' Rose said in an interview last August with investigators from the state Department of Labor and Industries.

"Rose's blistering comments were released yesterday along with a 2-inch-thick stack of documents gathered during Labor and Industries' investigation into claims by former Safety Chief Jones that he was demoted for criticizing department practices. . . .

"Labor and Industries announced yesterday that it has concluded that the city violated state law by retaliating against Jones." (*Seattle Times*, April 20, 1996)

Unfortunately, these scenes are replayed too often throughout the United States, nearly thirty years after the Occupational Safety and Health Act was passed. No doubt, progress has been made in reducing workplace deaths, injuries, and exposure to toxic substances through the efforts of federal agencies, states, employers, employees, and employee representatives. However, work-related deaths, injuries, and illnesses continue to occur at alarming rates. This imposes a substantial economic burden on employers, employees, and the nation in terms of lost production, wage loss, medical expenses, compensation payments, disability, and reduced quality of life. Additionally, in the ill-fated OSHA Reform Act of 1993, the 103rd Congress found that

- Employers and employees are not sufficiently involved in joint efforts to identify and correct occupational safety and health hazards.
- Employers and employees require better training to identify safety and health problems.
- Mandatory regulation is necessary to protect employees from health and safety hazards but federal agency standard setting has not kept pace with knowledge about such hazards.
- Enforcement of occupational safety and health standards has not been adequate to bring about timely abatement of hazardous conditions or to deter violations of occupational safety and health standards.

According to Lincoln Nebraska Safety Council, in a survey of 143 national firms, it was found that firms that did not have a written safety program experienced 106 percent more accidents than firms that did. Companies that ignore the business necessity for a written safety program face liabilities including the following:

- Higher insurance premiums
- Regulatory fines
- Product liabilities (third-party) lawsuits
- Cost of recruitment of replacement workers
- High turnover rate
- Criminal citations for company management

No one can be satisfied when a life is lost regardless of the external pressures, organizational politics, profit motives, or personal aspirations.

Consequently, management's obligation for employee safety and health must be paramount in all that it does. The need for a written safety program cannot be overemphasized. The foregoing problems, and many more, are directly attributable to the lack of effective written safety and health programs that address accident prevention, hazard communication, and, most of all, education in many companies and agencies, especially the smaller ones. The question is no longer whether it is important to have a written program but to determine the requirements of an effective program. Let us look at the elements of an effective program.

SAFETY POLICY

Safety policy is an executive action that requires commitment for success, not just support in principle from top management. It is a statement of goals, objectives, and operational principles that govern the organization. Safety policy must be created or approved at the highest possible level—president, chief executive officer, or chairman of the board. It is the first evidence of top management's commitment to employee safety and health in writing.

A good safety and health policy establishes the value system for the company. A good policy must

- Create mutual understanding
- Create or enhance a corporate culture that is conducive to high creativity, productivity, and profitability
- Strengthen the corporate identity
- Allow middle management and employees the freedom to carry out their safety and health responsibilities as an integral part of their work
- Emphasize the need for all employees to support safety and health activities throughout the company
- Promote employee quality of life
- Encourage participation by all

The following is an example of a safety policy currently in use by one of the nation's high-technology companies:

Safety and Health Policy

The safety and health of all employees shall be paramount to all operations of <Company name> at <Location>.

It is the policy of < Company Name> to furnish a workplace free of recognizable hazards, and to comply with all applicable safety and health laws, regulations, and ordinances.

The **Management and Employees** of <Company name> will maximize our efforts to:

- Eliminate or minimize exposure of employees to safety and health hazards.
- Work collaboratively with public and private health and safety organizations to solve health and safety issues.
- Increase operational reliability through a system of internal auditing.
- Foster management and employee support through the commitment of resources to eliminate safety or health hazards.
- Encourage and support communication regarding health and safety issues between employees and management at all levels.

Chairman

Clarification of Roles and Responsibilities

To be effective, a safety program must assign responsibilities and authorities to pertinent members of the organization. In general, management is responsible for establishing and /or supervising the establishment of

- A safety policy
- A safe and healthful working environment
- An accident prevention program including a hazard communication program
- A training program to improve the skill and competency of all employees in the field of occupational safety and health
- A safety and health committee
- A policy of using personal protective equipment (PPE)

Employees are responsible for

- Coordinating and cooperating with all other employees in an attempt to eliminate accidents
- Studying and observing all safe practices governing their work
- Offering safety suggestions that may contribute to a safer work environment
- Applying the principles of accident prevention in their daily work and using proper safety devices and protective equipment as required by management
- Promptly reporting to their supervisor each industrial injury or occupational illness, regardless of degree of severity
- Properly caring for all PPE

The establishment of more specific roles and responsibilities enhances the operational efficiency of the safety program. Therefore, the following (Nwaelele 1994) are offered for consideration:

Specific Responsibilities

Chief Executive Officer or President. The chief executive officer or president is ultimately responsible for the health and safety program. Consequently, as the management's representative, he or she must

- Establish the responsibilities of managers, supervisors and others for managing safety and health at the workplace
- Provide managers and supervisors with the authority, access to relevant information, and training in safety and health commensurate with their safety and health responsibilities
- Appoint a manager to control the recording-keeping responsibilities
- Constantly provide continuing support for the program.

Managers. Managers are responsible for

- Providing adequate funding for health and safety training for employees in their supervision
- Requiring all employees within their supervision to comply with all health and safety policies, procedures, and practices
- Ensuring that projects under their control have adequate built-in health and safety precautionary measures

- Ensuring that employee health and safety measures have top priority in budgetary considerations
- Providing a supportive culture for employees, especially when health and safety is an issue
- Seeking and providing feedback to the safety coordinator

Safety Coordinator. The safety coordinator is responsible for developing and implementing appropriate health and safety policies and practices including, but not limited to

- Coordinating the overall health and safety program
- Auditing the safety program
- Knowing the current legal requirements concerning regulated substances
- Providing managers, supervisors, and other appropriate personnel, with updated safety information
- Incorporating regulatory changes, issued by OSHA and other agencies, by notifying the management team, project managers, and other appropriate personnel
- Ensuring that all contract documents include adequate language for health and safety provisions
- Determining the required level of protective apparel and equipment
- Ensuring that appropriate health and safety related training are adequate
- Seeking ways to improve the safety program
- Ensuring that employee training programs are regularly updated
- Developing and maintaining a database of employee training and medical surveillance programs
- Managing the health and safety records

Supervisor. Each supervisor is responsible for implementing the health and safety program for his or her area of responsibility including, but not limited to

- Ensuring that each employee under his or her supervision knows and understands the safety program and follows all safety rules
- Ensuring that protective equipment is available, in working order, and used properly

- Ensuring that appropriate training for working with hazardous materials has been provided according to standards
- Providing regular maintenance and housekeeping inspections
- Ensuring that leads understand their responsibilities for the health and safety program
- Reviewing all accident reports with group leads and seeing that corrective action is taken immediately
- Filing complete and concise accident reports with the safety coordinator in a timely manner
- Ensuring that a copy of the health and safety program is on-site and that copies of applicable federal, state, and local regulations are at the job-site office
- Keeping abreast of current legal requirements concerning safety on construction sites

Group Lead. Each group lead has responsibility in his or her respective area, for implementing the program including

- Ensuring that each employee under his or her supervision knows and understands the safety program and follows all safety rules
- Ensuring that protective equipment is available, in working order, and used properly
- Ensuring that appropriate training for working with hazardous materials has been provided according to related standards
- Reviewing all accident reports with employees
- Assisting with the investigation of accidents
- Discussing safety and maintaining personal contact with each employee on every operation
- Ensuring that employees do not commit unsafe acts and that no unsafe conditions exist on his or her job site

Contractor. Each contractor shall be responsible for

- Submitting for evaluation its comprehensive health and safety plan during preconstruction meetings to the safety coordinator
- Requiring each subcontractor/consultant to provide, for submission to the safety coordinator, a complete health and safety plan during the pre-award meetings
- Performing work in a safe manner and observing established health and safety practices at all times while on company property

- Informing their employees of their obligations to identify potentially hazardous conditions, or changes in procedures that may create hazardous conditions, and to report those conditions immediately to management

Employee. Each employee is responsible for

- Performing work in a safe manner at all times
- Observing established health and safety practices at all times
- Identifying potential hazards or changes in procedures that may create hazardous conditions
- Reporting observed hazardous conditions immediately to the safety organization via his or her immediate supervisor
- Reporting any job-related injury, illness, or property damage to the supervisor
- Promptly seeking treatment for any job-related injuries or illness
- Observing all hazard warnings and signs
- Keeping aisles, walkways, and work areas clear
- Knowing the location of fire/safety exits and evacuation procedures
- Keeping all emergency equipment such as fire extinguishers, fire alarms, fire hoses, exit doors, and stairways clear of obstacles
- Reporting to work sober and staying sober on the job
- Refraining from fighting, horseplaying or distracting others
- Remaining in his or her work area unless otherwise instructed
- Operating only the equipment for which he or she is authorized and properly trained
- Using safe procedures with all equipment that he or she is authorized and trained to operate

Employee Involvement

Current federal safety and health regulations do not specifically address employee involvement in the development and/or management of a safety program. However, many states including Alaska, California, Connecticut, Florida, Minnesota, Montana, Nebraska, Nevada, New Hampshire, North Carolina, Oregon, Rhode Island, Tennessee, and Washington have specific regulations requiring employee involvement in terms of safety committees. For instance, Washington State safety code [WAC 296-24-045(1)] states that "all employers of eleven or more employees shall have a designated safety committee composed of employer-selected and employee-elected members."

It is prudent for management to involve employees in the development and administration of the safety program. Here are some of the reasons:

- Demonstrates management's commitment to safety
- Actively seeks employee contribution and participation in solving safety problems
- Creates a culture of trust in the workplace
- Encourages open communication across the board
- Boosts employee morale
- Controls costs by reducing workers' compensation claims
- Enhances perception of management care and understanding

The following are some of the functions of a safety committee:

- Overseeing the implementation of the safety program at its site
- Reviewing health and safety inspection reports and helping to correct unsafe conditions or practices observed
- Reviewing accident investigations and reports to help identify causes and preventive measures
- Evaluating the effectiveness of the health and safety program
- Establishing and maintaining a safety culture in the workplace
- Conducting peer safety training
- Communicating safety information to other employees
- Preparing minutes of each meeting to be filed and kept for at least one year in the health and safety office

HAZARD ASSESSMENT

Hazard assessment, also called job hazard analysis (JHA) or job safety analysis (JSA), describes the process of systematically identifying and evaluating each job task, and pinpointing the hazards associated with it. This is the foundation from which effective safety and health programs are built. Some of the OSHA's new (performance) standards such as Personal Protective Equipment (29 CFR 1910.132) and Process Safety Management of Highly Hazardous Chemicals (29 CFR 1910.119) mandate employers to conduct hazard assessments to determine whether hazards are present or likely to be present. It is anticipated that most new standards would incorporate the hazard assessment as a requirement. In fact, OSHA considers hazard assessment so vital an activity that it developed pamphlet 3071, "Job Hazard Analysis," to guide employers.

Hazard assessment is best done concurrently with other traditional industrial engineering analyses, such as design review; flow process analysis; and motion and time studies, which take into account the time and movement of each part of the body.

The Concept of Hazard Assessment

Hazard assessment is a technique used to help identify what combination of equipment/tool placement, engineering and administrative controls, behaviors and PPE in an operation are safe and correct. It begins with data collection to define the system, and proceeds through the identification of hazards and the analysis of causes and effects. Hazard assessment is a form of task analysis that breaks down an operation into activities of workers. The assessor identifies the hazards associated with each activity in the operation and for each activity describes how to do the task correctly and safely.

Who Should Conduct the Assessment?

The assessors should be thoroughly familiar with the operations being assessed and knowledgeable about safety hazards associated with the operations. Although a safety professional is most appropriate, alternatives include supervisors of the work area and members of the safety committee. A team that includes the safety professional, the supervisor, and employees' representatives works best. No matter who does the assessment, preassessment training in hazard recognition is essential to ensure that the assessor conducts an effective walkthrough.

Conducting the Assessment: Data Collection

Perhaps the most important task of the assessor is the collection of accurate data about the substances present, the processes and equipment, plant (facility) layout, topography, and demographics of the facility and surrounding community. There are a number of resources from which to collect data:

- Design review of proposed facilities
- Review of as-built-drawings for older facilities
- Equipment specifications
- Process and instrumentation drawings for process plants

- Maintenance records
- Physical and chemical characteristics of all substances used or planned to be used in the facility
- Flow process charts
- Standard operating procedures
- Emergency management procedures
- Logic diagrams
- Motion and time studies
- Feasibility studies
- Safety minutes
- Training records

Each step of every job done or planned including any movement of any body part (reaching, bending, standing, and so on) must be documented. One possible means of breaking a job down into the tangible steps is the use of a flow process chart.

Once the data are collected, an analysis of the hazards associated with the facility begins. Many formal techniques have been developed for the systematic analysis of complex and not- so- complex systems. Some of the most commonly used methods are as follows:

- What-if/checklist
- Hazard and Operability Study (HAZOP)
- Failure mode and effects analysis (FMEA)
- Fault-tree analysis

What-if and Checklist Analyses. The what-if and checklist analyses are actually two separate methods, commonly combined because they compliment each other. The assessor goes over the elements of the job or process in line-by-line details, asking what-if questions. For instance, an assessor may ask what would happen if, at a certain step in an operation, the pressure builds to full scale? This is like a brainstorming exercise—highly unstructured, but effective in developing creative potential failures and consequences. Some answers to some of the what-if questions are found on failure and consequence checklists developed based on experience of the operation.

Hazard and Operability Study. The HAZOP is one of the most common and widely accepted methods of systematic qualitative hazard analysis. The technique is useful for both new and existing facilities. It can be used to analyze an entire facility, a unit of a plant or operation, or a piece of equipment. It helps assessor(s) and plant personnel identify possible hazards in the plant including those caused by equipment malfunction or human error. The HAZOP relies on the information usually gathered as part of data collection and the expertise and judgment of the assessor(s).

The main objectives of the HAZOP (Lipton and Lynch 1994) are to

- Identify areas of an operational or project design that present unacceptable risk to humans and property;
- Identify design features that may make an operation hazardous;
- Ensure that all recognizable hazards associated with an operation or project are evaluated;
- Identify and evaluate any pertinent design information that may not have been available or known to the design team.

Most successful HAZOPs follow established problem-solving model protocol:

- Definition of scope, purpose, and objective
- Selection of the team
- Collection of data
- Analysis of the data
- Write-up of the report with recommendations
- Follow-up

Failure Mode and Effect Analysis. As the name implies, FMEA is a technique or method of analyzing failure modes and effects or consequences of those failures on other systems or components of the system being analyzed. FMEA is an inductive procedure that analyzes the potential failure of specific equipment or components and traces the effects of such failure on the whole system. Typically, FMEA uses a spreadsheet form to log data during the analysis. Any good system safety book has sample forms, or the assessor can create one.

Fault-Tree Analysis. Fault-tree analysis is another method of analyzing or assessing hazards. It is based on a boolean logic concept developed in 1962 by H.A. Watson at Bell Telephone Laboratory. This method is used to analyze how and why a disaster or an accident could occur. It is a deductive process that pictorially starts with the end result (the accident or disaster) and works backward to find the initiating event or combination of events (cause or causes) that would lead to the undesirable outcome. If the probability of the potential initiating events is known or could be estimated, the probability of the undesirable event can be calculated. Therefore, fault-tree analysis is a quantifiable method. Many good texts have been written on fault-tree analysis. The reader is encouraged to peruse some of the system safety texts and get some training on application before using this method as a primary method of assessment.

Hazard Assessment Evaluation

The effectiveness of the hazard assessment in determining a safety program's requirements can be evaluated by answering the following questions (Moran 1996):

1) Has hazard analysis been performed to identify, evaluate, and control associated hazards?

2) Is the analysis appropriate to the complexity of the process or operation?

3) Does the analysis provide a clear understanding of the following:

   a. The hazards and risks associated with the operation?

   b. The engineering and administrative controls applicable to the hazards and their interrelationships?

   c. Identification of any previous incident that had a likely potential for catastrophic consequences in the workplace?

   d. Facility siting?

   e. Human factors?

   f. A qualitative evaluation of a range of the possible safety and health effects of failure of controls upon employees in the workplace?

4) Is the analysis an orderly, systematic approach and follow one or more of the following recognized methodologies:

   a. What-if?

   b. Checklist?

   c. What-if/checklist?

   d. HAZOP—hazard and operability study?

   e. Failure mode and effects analysis (FMEA)?

   f. Fault-tree analysis?

   g. Other?

5) Was the hazard analysis conducted by a team of persons with expertise in safety, unit operations, or other disciplines knowledgeable in the area assessed?

6) Did the team include at least one employee with experience and knowledge specific to the operation that was (or is being) evaluated?

7) Did the team include one or more individuals knowledgeable in the specific hazard analysis methodology used?

8) Is there a written and approved hazard analysis report describing the team's findings and recommendations?

9) Is there an established system and schedule to address in a timely manner, the findings and recommendations in the hazard analysis report that includes:

   a. Documentation of what actions are to be taken and a schedule of when those actions are to be completed?

   b. Documentation of actions taken?

   c. Communication to operating, maintenance, and other personnel whose work assignments are in the facility and who are affected by the recommendations or actions?

   d. Assurance of satisfactory implementation of the recommendations as soon as possible?

HAZARD PREVENTION/CONTROL

Once the hazard assessment is completed and the hazards are identified, it becomes necessary to prioritize the hazards based on their seriousness. The prioritization should be done judiciously by management using employees with experience and expertise in hazard assessment and risk ranking. The result of the ranking becomes the guiding document to develop safety and health programs to mitigate the hazards. The program must be in a written format; manuals seem to work best for most companies. Specific program elements must reflect the priority established during the hazard and risk ranking process.

Again, it must be emphasized that hazard assessment is the bedrock of any effective program to prevent or control workplace hazards. No program should be started without completing a comprehensive hazard assessment of the facility or operation first. To do otherwise would be irresponsible.

Information Dissemination and Training

No matter how good the hazard assessment and formal accident prevention programs, if employees are not aware of the programs, all efforts are in vain. Programs are not worth the paper they are written on without strong communication and training elements. Pertinent safety and health information about operations must be passed on to all employees at all levels at all times. Safely speaking, safety communication channels must be kept open across the board.

The first line of defense against any hazard is engineering control. However, this may not always be feasible; therefore, training must be an integral part of the safety program. The OSHA recognizes the need for training to ensure employee protection. Consequently, it mandates training in many of its standards, including Confined Space [29 CFR 1910.146(g)(1)], Process Safety Management [29 CFR 1910.119(g)], Hazardous Waste Operations and Emergency Response [29 CFR 1910.120(e)], and Hazard Communication [29 CFR 1910.1200 (h)]. Simply put, excellence in safety performance cannot be achieved without ongoing safety education.

Each employee must be trained so that he or she has the understanding, knowledge, and skills necessary for the safe performance of his or her duties. Training should be provided

- Before the employee is first assigned to perform a task that might expose him or her to a hazard
- Before there is a change in the employee's assigned duties
- Whenever there is a change in the operation that presents a hazard about which an employee has not previously been trained
- Whenever the employer has reason to believe either that there are deviations from the procedures or that there are inadequacies in the employee's knowledge or use of these procedures

Elements of a training program vary. However, training program elements should cover the following as a minimum:

- Operation of the facility or unit
- Specific duties of each person involved in the operations
- Hazards associated with the operation, including information on the mode, signs or symptoms, and consequences of operational failure

- Proper use of equipment required to operate safely including, testing and monitoring equipment, communication equipment, PPE, and emergency equipment
- Importance of communication between employees and management
- Conditions under which employees must evacuate the facility
- Emergency management procedures

Program (Auditing) Evaluation

An audit, in this context, is defined as a systematic, documented, objective examination involving analysis, tests, and confirmation of procedures and practices leading to verification of a facility's compliance with legal requirements, corporate policies, and/or accepted practices. The established program must undergo periodic audits to ensure that it is achieving stated goals. The fact is that programs become obsolete as regulations change or as technology improves. Consequently, the safety and health program must be continuously evaluated for improvement opportunities. The frequency of auditing depends on a number of factors including the following:

- Complexity of operation
- Stability of operations
- Employee turnover
- Rate of accidents/incidents including near misses
- Size of facility or operation
- Technical support

However, in no case should an audit be conducted less than annually. An established checklist can be helpful in conducting an audit (see Nwaelele 1997).

Contractor Safety Requirements

Many employers have fallen victim to third-party lawsuits as a result of inadequate contractor safety requirements and management. Many well-meaning employers develop and implement some of the best safety programs for their own employees, yet turn a blind eye to contractors working on their sites. This, of course, is a mistake!

The majority of contractors are involved in designing and building activities. They design and install structures that the employees of the site's owner have to live with. Shouldn't owners pay more attention to how the contractors deal with safety issues relating to construction? The reality is that many companies undergo several expansions, or modifications involving some level of construction throughout their existence. The construction industry employs only about 5 percent of the workforce, yet posts the highest fatality rate of any major industry. According to Washington state's Department of Labor and Industries, in 1996 the construction industry "accounted for 25 percent of the state's work-related fatalities" (*Leading Edge* 1997).

Workplace accidents and injuries are no longer the sole responsibility of the primary employer. The courts in Washington State have made some striking rulings. Businesses and owners are well advised to pay attention. On March 29, 1990, the Washington State Supreme Court issued its ruling in *Stute v. PBMC* (1990), which held that a general contractor could be held liable for an injury to a subcontractor's employee that occurred as a result of a Washington Industrial Safety and Health Act violation. Since that decision, the Washington Courts of Appeals have also extended the rule to include an upper-tier subcontractor (*Husfloen v. MTA Construction* [1990]) and owner/developers (*Weinert v. Bronco National Co.* [1990]). In Weinert, the Court of Appeals held that the owner/developer held a position so comparable to that of general contractor that the owner/developer was responsible to all employees on the worksite. On January 7, 1991, in *Doss v. ITT Rayonier* (1991), Court of Appeals extended the rule in *Stute v. PBMC* to impose potential liability to a landowner whose independent contractor failed to comply with safety and health regulations.

There is no doubt that general contractors, owners and developers, and landowners may be held liable for safety violations caused by employees of contractors doing work for them. It is therefore incumbent on the facility owner to ensure that its contractors comply with safety regulations. There is no guidance to employers or owners on how to enforce the safety of contractors, except through contract requirements. Therefore, employers or owners must include contractor safety requirements as part of their contract documents. Additionally, they must strictly enforce the provisions of the contract that relate to safety. Appendix A is an example of contractor safety requirements. It is always prudent to have a lawyer review all contract documents to ensure their lawfulness.

CONCLUSION

There is no substitute for proper planning when it comes to determining the requirements of an effective safety and health program. Although materials in this chapter, especially hazard assessment methods, may seem cumbersome and perhaps very costly, they are fundamental in determining the hazards associated with the work-site. No successful program can be developed or implemented without a careful hazard assessment process. Management has the ultimate responsibility for the safety and health of "all workers" on its site, irrespective of their primary employer. This awesome responsibility should not be ignored. The consequences are too great. With genuine management commitment and employee involvement, any operation can develop and implement a site-specific safety and health program that would minimize third-party liabilities, while increasing employee confidence and morale, improving the quality of products or services, increasing productivity and profits; and lowering insurance premiums.

Contractor Safety Requirements General Conditions

General

The contractor shall be responsible for initiating, maintaining and supervising all safety precautions and programs in connection with the work. The contractor shall comply with all applicable laws, ordinances, rules, regulations and lawful orders of any public authority bearing on the safety of persons or property or their protection from damage, injury or loss.

Protection of Persons

The contractor shall take all reasonable precautions for the safety of all employees on the Work and all other persons who may be affected thereby. The contractor shall designate a responsible member of his organization at the project site whose duty shall be the prevention of accidents. Except as otherwise stated in the contract documents, if the contractor encounters on the project site material reasonably believed to be hazardous, the contractor shall immediately stop work in the area affected and give notice of the condition. Work in the affected area shall not be resumed without written direction by the owner.

Employee or Operator Safety

The contractor shall be solely and completely responsible for conditions of the worksite, including safety of all persons and property, during performance of the work. The contractor shall maintain the worksite and perform the work in a manner that meets statutory and common law requirements for the provision of a safe place to work and that does not pose any safety risks to operators of the plant or other workers and visitors in the facility. This obligation shall apply continuously and not be limited to normal working hours. That the owner's safety engineer conducts review of the contractor's performance does not and shall not be intended to include review of the adequacy of the contractor's safety measures in, on or near the site of the work. The contractor shall comply with the safety standards and provisions of applicable laws, building and construction codes, and the Federal Occupational Safety and Health Act of 1970 (OSHA), including all revisions, amendments and regulations issued thereunder, and the provisions of other state or local agencies with jurisdiction, including all revisions, amendments and regulations issued thereunder. In case of conflict between any such requirements, the more stringent regulation or requirement shall apply.

The contractor shall maintain at the worksite office or other well known place at the worksite all materials (e.g., a first aid kit) necessary for giving first aid to the injured, and shall establish, publish and make known to all employees procedures for ensuring immediate removal to a hospital or a doctor's care, persons, including employees, who may have been injured on the site. Employees shall not be permitted to work on the site before the contractor has established and made known procedures for removal of injured persons to a hospital or a doctor's care. The contractor shall ensure that its employees have valid, effective first aid cards.

The contractor shall prepare a *written safety program* demonstrating the methods by which all applicable safety requirements of this contract will be met. The contractor shall ensure its subcontractors have a written safety program or formally adopt the contractor's safety program. The contractor shall designate a qualified safety officer who shall be responsible for proper implementation of the safety program. The contractor shall submit a copy of its safety program to the safety engineer as required in the specifications. The safety engineer's review of such program shall not be deemed to constitute approval or acceptance thereof.

The contractor shall conduct a monthly safety meeting with all subcontractors and others on the site performing work hereunder to discuss general and specific safety matters. The contractor shall provide written notice of each meeting to the safety engineer. The contractor shall provide the safety engineer with a record of each meeting, including a sheet on which each attendee signed in and a list of the matters discussed.

The contractor shall conduct weekly safety meetings ("tool box talk") with employees of the contractor and subcontractors. The contractor shall provide written notice of each meeting to the engineer. The contractor shall provide the engineer with a copy of the sheet on which each attendee signed in and a description of the safety topics discussed at the meeting.

There is no acceptable deviation from these safety requirements, regardless of practice in the construction industry. Any violation of OSHA, or other safety requirements applicable to the work shall be considered a breach of this contract.

Protection of Work

Unless otherwise agreed, the contractor shall bear the risk of loss to the work and all materials and equipment to be incorporated therein until substantial completion.

Protection of Other Property

The contractor shall perform the work so as to protect from damage private and public property not scheduled for repair, replacement, or removal and shall ensure that interference with the use of such property is minimized. The contractor shall promptly repair or replace property damaged by or resulting from the performance of the work, including without limitation structures, pavement, vegetation, and utilities.

The contractor shall promptly repair and/or replace all damage to any property referred to in this Article caused in whole or in part by the contractor, any subcontractor, or anyone directly or indirectly employed by any of them, or anyone for whose acts any of them may be liable and for which the contractor is responsible, except damage or loss attributable to the fault or negligence of the contractor.

These obligations are in addition to the contractor's obligations under any part of the contract.

The contractor shall protect and be responsible for any damage to his work or material throughout the contract time and shall hold the owner harmless from any damage or loss to that Work or material.

Health and Safety Specification[1]

Part 1   General

1.01 Description

A. This section specifies procedures for complying with applicable laws and regulations related to safety and health of the worker and the public. It is not the intent of <insert company name> to develop and/or manage the safety and health programs of Contractors or in any way assume the responsibility for the safety and health of their employees. It is required that all Contractors adhere to applicable federal, state, and local health and safety standards.

B. This section describes the Accident Prevention Program, which is a subset of the Safety Program defined in the General Conditions.

1.02 References

A. Comply with and enforce on-the-job site current applicable local, state, and federal health and safety standards including, but not limited to, the following:

Reference	Title
29 USC 651 et seq.	*Federal Occupational Safety and Health Act*
29 CFR 1910	*OSHA General Health and Safety Standards*
29 CFR 1926	*OSHA Construction Safety and Health Standards*

1.01 Definitions

A. A hazardous substance is defined as follows:

1. A substance classified as "dangerous waste," in accordance with any standard, or that in sufficient quantities would be classified as "dangerous wastes"

2. A solid waste, or combination of solid wastes, which because of its quantity, concentration, or physical, chemical, or infectious

---

[1] Source: Nwaelele, O. Dan, *Your Safety and Health Manual,* Government Institute, 1997.

characteristics may (1) cause or significantly contribute to an increase in mortality or increase in serious, irreversible, or incapacitating reversible illness; or (2) pose substantial present or potential hazard to human health or the environment when improperly treated, stored, transported, or disposed or otherwise managed

3. Polychlorinated biphenyls (PCBs), polynuclear aromatic hydrocarbons (PAHs), explosives, radioactive materials, and other materials designated as hazardous by regulatory agencies having jurisdiction over such matters

B. A contaminated substance is defined as follows:

1. A substance containing materials in sufficient quantities such as hydrocarbons, PCBs, diesel fuels, gasoline, heavy metals, solvents, and other types of fuel oils present in the soil, water, or air

2. An element, compound, mixture, solution, or substance designated under Section 102 of CERCLA

3. A hazardous waste having the characteristics identified under or listed pursuant to Section 3001 of Solid Waste Disposal Act (i.e., RCRA) except those suspended by an act of Congress

4. A toxic pollutant listed under Section 307(a) of the Federal Water Pollution Control Act (FWPCA)

5. A hazardous air pollutant listed under Section 112 of the Clean Air Act

6. An imminently hazardous chemical substance or mixture with respect to which the EPA administrator has taken action pursuant to Section 7 of the Toxic Substance Control Act

C. Confined space is defined as follows:

1. It is large enough and so configured that a person can bodily enter and perform assigned work.

2. It has limited or restricted means of entry or exit.

3. It is not designed for continuous employee occupancy.

D. Permit-required confined space is a confined space that has one or more of the following characteristics:

1. It contains or has potential to contain a hazardous atmosphere.

2. It contains material that has potential for engulfing an entrant.

3. It is shaped inside such that someone entering could be trapped or asphyxiated.

4. It contains other recognized serious safety or health hazards.

1.01 Submittals

A. Submit the following to the Safety Engineer:

1. A formal Accident Prevention Program for the company. This program shall outline the anticipated hazards and safety controls necessary to safeguard the Contractor's employees, the public, and employees of the owner. It shall be specific to the job and site and shall meet federal, state, and local jurisdictional requirements. The program will be reviewed for compliance with this section prior to the start of work.

2. Revisions. Revise the accident prevention program prior to the start of work to accommodate changes requested by the company and/or regulatory agencies or jurisdiction. Post a copy of the accepted program at the Contractor's job site office and at each of the subcontractors' offices. Three additional copies shall be posted at the Engineer's Office.

3. Health and safety equipment and/or training material as specified in this section.

1.01 Quality Assurance

A. Ensure that subcontractors receive a copy of this specification section. The Contractor is responsible for ensuring compliance with the Accident Prevention Program.

B. Coordinate with the Safety Engineer to obtain approval to disconnect or reconnect utilities.

C. Coordinate with the Safety Engineer regarding the shutdown and safety tagout/lockout of pressurized systems, electrical, mechanical, pneumatic, hydraulic, etc., systems, and other equipment and utilities.

D. Maintain good housekeeping in work areas in accordance.

E. Ensure that all health and safety submittals are reviewed and approved by a Certified Safety Professional (CSP) and/or Certified Industrial Hygienist (CIH).

F. Provide a qualified (CSP or CIH) health and safety supervisor, with responsibility and full authority to coordinate, implement, and enforce the Contractor's accident prevention program for the duration of this contract. The name and telephone number of the safety supervisor shall appear in the Accident Prevention Program.

1.01 Special Considerations

A. This paragraph describes certain minimum precautions for consideration in developing an Accident Prevention Program. It supplements the regulatory requirements of the General Conditions. Failure to comply with health and safety regulations will result in work suspension until adequate health and safety measures are implemented.

1. Hazard Communication (29 CFR 1910.1200).

    a. Provide a written Hazard Communication Program and emergency management plan addressing the potential hazardous substances on site.

    b. Prior to commencing work, provide a list and corresponding Material Safety Data Sheets for hazardous chemicals to be used on site. If no hazardous chemicals are to be used, provide a statement to that effect.

2. Confined Space (29 CFR 1910.146).

    a. The nature of work under this contract may expose workers to permit-required confined spaces having possible explosive, toxic and oxygen fluctuation conditions.

    b. Prior to execution of work in confined spaces, submit a written confined space safety program that meets Requirements 29 CFR 1910.146.

3. Underground Construction. Provide a written program detailing how employees and visitors on the site will be protected from the dangers of underground construction. Such program shall include the following as a minimum:

    - Air monitoring
    - Ventilation
    - Illumination
    - Communications
    - Flood control
    - Mechanical equipment
    - Personal protective equipment
    - Access and egress
    - Rescue
    - Hazardous classification
    - Gassy operation
    - Haulage
    - Electrical safety
    - Hoisting

- Use of explosives, if applicable
- Fire prevention and protection
- Emergency procedures, including evacuation procedures and check-in/check-out systems
- Designated person
- Emergency lighting
- Ground support
- Pneumatic and hydraulic safety

4. Other Site Safety Considerations. Supply to the company for review, prior to commencing work on this Contract, a comprehensive written Accident Prevention Program covering the Contractor's activities on site. As a minimum, the program shall include the following:

- Respiratory Protection
- Accident/Injury Reporting
- Emergency Plan (SARA Title III—Community Right-to-Know)
- Crane Operations, Rigging, and other Overhead Lifts (WAC 296-155-525)
- Excavation and Trenching
- Personal Protective Equipment
- Fall Restraint and Fall Arrest
- Fire Safety and Prevention
- Signs, Signals, and Barricades
- Material Handling, Storage, Use and Disposal (WAC 296-155-325)
- Floor Openings, Wall Openings and Stairways
- Hand and Power Tools
- Welding and Cutting
- Electrical
- Ladders and Stairways
- Scaffolding
- Tagout/Lockout
- Temporary Buildings
- Dangerous Waste Management Program
- Asbestos and other Carcinogens
- Demolition
- Hearing Conservation
- Vehicles and other Motorized Equipment

5. Special Hazards.

   a. *Lead:* Exposure to lead is a safety concern at this site. Submit an accident prevention program demonstrating the methods by which applicable health and safety requirements will be met. This program should include the following as a minimum:

   - Exposure monitoring
   - Respiratory protection
   - Engineering controls
   - Administrative controls
   - Personal protective equipment
   - Housekeeping
   - Personal hygiene practices
   - Medical surveillance
   - Medical removal protection
   - Employee training
   - Signs
   - Plans for transpiration and disposal of lead-based paint debris
   - Decontamination equipment and procedures including washing facilities
   - Recordkeeping

b. Hazardous Waste (29 CFR 1910.120): Promptly suspend work and notify the Engineer of unusual conditions including oily soil found on project site. Work shall remain suspended until authorized by the Engineer to resume.

c. *Process Safety Management (WAC 296-67)*: This company shall comply with the provisions of the Process Safety Management regulations, 29 CFR 1910.119. This project may affect the regulated processes. Submit a written Process Safety Management Plan. At a minimum, the plan should cover the following:

- Process safety information
- Operating procedures
- Management of change
- Contractors or subcontractors
- Review of hot work permits
- Training

- Process hazard analysis
- Safe work practices
- Pre-startup safety review
- Mechanical integrity
- Emergency planning and response
- Compliance audits

c. *Work Zone Traffic Control:* Exposure of employees to traffic is a concern on this project. Submit work zone traffic control plan demonstrating the methods by which applicable safety requirements will be met.

d. *Fall Protection:* Work activities on this project may expose employees to fall hazards. Contractor must provide a written Fall Protection Plan for each fall hazard encountered throughout the project.

1.01 Utilities

A. Take appropriate precautions in working near or with utilities and dangerous substances during the performance of work in order to protect the health and safety of the worker, the public, property, and the environment.

B. Such utilities and dangerous substances include, but are not necessarily limited to, the following:

1. Conductors of

   a. Petroleum products

   b. Toxic or flammable gas

2. Natural gas pipelines

3. Electric conductors

Part 2 Not Used

Part 3 Execution

3.01 Safety and Health Compliance

A. Occasionally, <insert company name> will audit the Contractor's Accident Prevention Program. The company reserves the right to stop that portion of the Contractor's work that is determined to be a serious health and safety violation. On-going work that is considered a safety or health risk by the Engineer shall be corrected immediately.

B. Ensure that necessary air monitoring, ventilation equipment, protective clothing, and other supplies and equipment as specified are available to implement the Accident Prevention Program.

C. Notify the Safety Engineer immediately of accidents resulting in an immediate or probable fatality to one or more employees or the public, or which result in hospitalization of two or more employees.

D. Complete the Monthly Contractor Injury Summary Report (see Attachment).

3.02 Accident Prevention Program Revision

A. In the event that the Company, regulatory agencies, or jurisdictions determine the Accident Prevention Program or associated documents, organizational structure, or Comprehensive Work Plan to be inadequate to protect employees and the public, do the following:

1. Modify the Program to meet the requirements of said regulatory agencies, jurisdictions, and the company.

2. Provide the Engineer with the revisions to the program within 7 days of the notice of deficiency.

MONTHLY CONTRACTOR INJURY SUMMARY REPORT

CONTRACTOR: _____

MONTH: _____   CONTRACT NO.: _____

OSHA RECORDABLE CASES

WORK GROUP	NUMBER OF CASES	
	Month	Year to Date
Hourly employees		
Supervisory personnel		

LOST-TIME ACCIDENTS

WORK GROUP	NUMBER OF CASES	
	Month	Year to Date
Hourly employees		
Supervisory personnel		

TOTAL HOURS AT <INSERT COMPANY NAME> PROJECT SITE

Month	
Year to Date	

INCIDENT AND SEVERITY RATE

Date since last lost-time accident: _____

No. of hours worked since last lost-time accident: _____

$$\text{Incident Rate} = \frac{\text{No. of OSHA Recordables} \times 200{,}000}{\text{Total Hours Worked}}$$

$$\text{Severity Rate} = \frac{\text{No. of Lost Workdays} \times 200{,}000}{\text{Total Hours Worked}}$$

Rates	Month	Year to Date
Incident		
Severity		

Submit completed form by the 10th day of each month to the Safety Engineer.

REFERENCES

Business Publishers, Inc. (1994). Comprehensive Safety and Health Reform Act of 1993.

Haines, T. W. and H. Gupta (August 18, 1994). "Two Workers Killed in Fall; Kingdome Closed Indefinitely," *Seattle Times*.

Nwaelele, O. D. (1994). *Health and Safety Risk Management*. Rockville, MD: Government Institutes, Inc.

Nalder, E. (April 20, 1996). "Report: Fire Chiefs Ignored Safety," *Seattle Times*.

Penhale, E. and R. Jamieson, Jr. (August 18, 1994). "2 Workers Fall to Deaths in Dome," *Seattle Post Intelligencer*.

Chapter 10

Total Quality Management and Planning

F. David Pierce

INTRODUCTION

Total quality management (TQM) runs counter to many traditional American ways of thinking. This fact has certainly made the true application of total quality challenging in American organizations and businesses. It has also added to the numerous misconceptions concerning what total quality applies to. Nevertheless, due to many successful applications, the power and positive influence of total quality is without question. For applications in nonline, specialty type functions such as safety and health, total quality offers the most significant management philosophy change and most promising avenue for success.

Being composed of many interwoven concepts and focuses, total quality management brings high participation and holistic management to safety and health programs. Additionally, through applied strategic planning, goal and objective setting, and dynamic measurement, traditional reactive management is replaced with proactive practice. Program evaluations and benchmarking offer internal and external learning that raises program effectiveness. When interlocked with continuous improvement, total quality can transform struggling or average safety and health programs into world-class examples.

CASE STUDY: TRANSFORMATION OF SAFETY AND HEALTH MANAGEMENT

In the early 1990s, a western American company was following the Deming fad, attempting to move from traditional command and control management

(Theory X) to a participative approach. This company had selected total quality management as the model it would follow. Hiring an eastern management-consulting firm to guide its efforts, the company began mapping the transformation and the first phase of management training. The management team was immediately divided by the complete change in management philosophies. To many team members, the firm's paradigms of what management practices were successful were too strong to invite their query into total quality. Others, mostly those who were younger and who had little investment in the traditional management ways, saw total quality as the opportunity for which they had always hoped. But the impasse between management camps was deep. After many bitter squabbles, the battle lines became so entrenched that the two sides became opposed and unmoving. The management structure was near stalemate and the business started to be impacted. The total quality initiative was put on the back burner in the hope that the battle lines would soften with a cease-fire.

The safety and health director of this company saw total quality as the answer to his frustrations and the inefficiency of the safety program. Moving out on his own, he continued to implement total quality management in his program and department. Over a period of three months, as his department continued its studies and conceptual imagery, a strange transformation occurred. The department began to work more as a team. Creative thought became the norm. "Can do" attitudes replaced the pessimistic "can not do" roadblocks. Team members began to find new and successful ways of getting line management support and commitment for safety and health and getting production workers to become involved and to participate.

"Extinguishing fires" became rare whereas, before the move to total quality, they had been daily, and even hourly, occurrences. Support and funding issues that had seemed impossible dreams became realities through upper-management commitment. Others in the company began to notice. What had happened to the safety program and department? What had transformed them? Others started asking because they wanted the same results in their programs and departments. Learning that it was total quality began a resurrection of the total quality effort in the company and had a major impact in lowering the anxiety of those who had so fiercely resisted it before. Now they had proof. Now they could see that it worked.

CONVERSION TO TOTAL QUALITY

The previous case study has many important points. First, total quality is a powerful concept that works. More important, for the use of safety professionals, it works extremely well in safety and health programs. Second, total quality is so counter to the traditional and historically successful ways we have managed in America that successful implementation is racked with misconceptions and roadblocks. Adopting total quality is certainly not for the fainthearted or those who merely want to stick their toes in the total quality waters to see how it feels. And third, conversion to total quality certainly does not require an organization-wide effort to be successful. All it requires is a leader in that organization who sees the value total quality brings and who has the courage to do some pioneering.

With all the misinterpretations and misuses of total quality in today's world, it is valuable to discuss what total quality is *not*. First, total quality is *not* management by objectives (MBO) revised. MBO was a management concept that tried to bridge the gap between management by luck and management by plan. Although successful total quality programs do utilize extensive planning and performance measurement, total quality takes them a big step farther.

Total quality also is not new generation Theory Y management. In this theory of management, management serves the employees and allows them to accomplish improvements in the workplace. It is true that successful total quality programs have high employee participation, but management is far from subservient. In a total quality program, management must lead the efforts.

A newer misinterpretation is this: TQM = ISO 9000. ISO 9000 is a long list of quality specifications. Total quality, on the other hand, is a process, a way of thinking and doing business. Total quality is *not* ISO 9000.

Total quality is also considered only a manufacturing concept, with no applicability to either service industries or staff functions. Total quality is a process; it can be applied to any business or industry or to any function, including staff functions, production, or even running a Boy Scout troop. Total quality is, therefore, *not* just a manufacturing concept.

So what *is* total quality? Total quality is a systematic, highly participative process that identifies inefficiencies and improves them. In reality, total quality is no more complex than that. Adding more specifics to total quality only limits it to a specific application or focus. Total quality is bigger than any one application or focus. Total quality is a process that is fueled by a different way of thinking. In a holistic sense, it can be defined even more concisely. Total quality is a highly participative process for improvement, which is why there is so much confusion over total quality. It is too simple! Because of its power, it is natural to make total quality more complex, but in reality, it is not.

So, if total quality is so simple, what makes it so important and powerful? Total quality is important because it is the way of the future. Total quality is powerful because it changes cultures, personal and business, and the process by which improvement is achieved continually refuels itself by the challenges, participation, and successes. Look at all the examples of companies that have implemented total quality programs. Look at their successes—higher production, lower costs, reduced inventories, higher quality, high employee morale, higher profits. It is difficult to argue with examples such as Hewlett Packard and Motorola. Would they be as successful today if they had not risked their futures on total quality? Maybe they would have, but most likely not. Sure, market, good people, and dumb luck might have produced some positive results, but not at the magnitude nor the sustained performance that total quality has produced. Powerful? Yes it is. To most of America's companies and endeavors, however, especially in the staff and service sectors, total quality is still an untapped resource.

Is total quality useful in a safety and health program? Ideally, you cannot do better than having a safety and health program in a total quality organization. Set in this environment, the traditional frustrations and roadblocks that have haunted safety and health practitioners for years simply disappear. Even when implemented in a traditional organization, a total quality safety and health program can provide significant rewards, simply because it works better. As it has been said many times, no form of management or management process is more advantageous for safety and health than total quality

ROOTS OF TOTAL QUALITY THINKING

Where did "total quality" come from? Like most concepts, it has many roots. The most important one, however, was an American: W. Edwards Deming. After World War II, Deming openly offered his new manufacturing concepts to American industry and was rejected. At that time, American companies represented the strength of the world's manufacturing. They had no reason to change what had been so successful: "If it ain't broke, don't fix it!"

But Deming did not give up. Frustrated, he went to Japan to teach his ideas. After the war, unlike American industry, Japan had nothing to lose and everything to gain. The Japanese learned so well that their contribution to total quality needs to be separated from Deming's. Anyone who has studied the Japanese culture knows that they have a "learning culture." The Japanese can learn so well that they improve things to near perfection. That is just what they did with Deming's ideas. They expanded them, refined them, and perfected them.

European industry also helped mold what we call *total quality* today. Europeans, especially their industries, have long had a "quality is everything" attitude that pervades their society. Name brands such as Mercedes, Porsche, Hasselblad, Rolex, and BMW stand as standards for quality. In fact, all manufacturers, including the Japanese, compare their product quality to these name brands. Unfortunately, Europeans mistakenly held that a knowledgeable consumer would always pay an inflated price for high quality. The Japanese destroyed that belief with quality at reasonable costs.

Total quality was not totally born abroad. Some of its development and processes are homegrown. For example, during the 1970s, self-managed teams swept America. At the time, this concept did not pack the wallop that was anticipated. It fell notably short of the mark and, in most industries and facilities, was quickly discarded. However, one of the side effects of self-managed teams took root and is an integral part of the total quality process—employee participation, input, and ownership. What started out in American industry as a good idea with limited application changed face and provided a building block for total quality today.

TOTAL QUALITY AND TECHNOLOGY

Total quality has also been significantly impacted by the world's technological revolution. In the past, product research, development, design, and, finally, manufacture took years—many years. In our fast-paced, highly competitive world, this same process today takes only months. In some cases, it takes only weeks to go from concept to consumer. The technological revolution has had an enormous impact on the development of total quality. This speed demands that communication and accuracy become woven into the total quality fabric. Just like today's concept-to-manufacturing cycle, total quality has had to become very fast without loosing quality or value.

In fact, total quality today is a blend of many cultures, thoughts, industries, people, and revolutions. It will continue to evolve as these impactors change. Total quality is a dynamic concept, not a static one that is frozen in time. So, total quality tomorrow will certainly be much different from what it is today.

TOTAL QUALITY MANAGEMENT

Studying successful total quality programs, there are five common, specific program concepts and components. Specific applications or specialties may not require all of them. These concepts include common focuses, dynamic people and management concepts, strategic planning that includes goal setting and measurement, benchmarking and evaluations, and continuous improvement.

Common Focuses

Total quality programs have common focuses. They focus on customers, waste, time, and excellence. Focus, after all, is a measure of intensity. Focus is not casual and it is not an awareness or knowledge of something. It is an intense concentration, an obsession. It requires extreme, all mind-and-body dedication and commitment. Look at any NFL football team at the beginning of the season. What is everyone on the team, on the sidelines, in the front office, and in the stands concentrating on? The Super Bowl! Each team player is convinced that with a dedicated, full-team effort, they can and will make it to the Super Bowl. And given that chance, they will win it all. It is not a maybe. Like Mike Ditka recalled from the Chicago Bear's Super Bowl year in 1985, "From the first moment in training camp, everyone on the team *knew* they were going to win the Super Bowl." That is focus!

Customers. In the focus area of *customers,* it is more than merely knowing who your customers are; it is a total focus on them, what they need and what they expect. Customer focus is one of the things that separates business superstars from also-rans. Look at McDonald's and Disneyland as excellent examples of customer focus. Walt Disney put it this way, "We want to do our job so well that they will want to come again, and bring a friend." That is customer focus. Customer focus goes much deeper than just external customers. Each organization, program, or business also has internal customers. Internal customers include management, employees, and departments that take semi-finished goods from your department. In a total quality program, internal customers are no less important than external ones.

Wastes. Focusing on *wastes* requires the same level of concentration and dedication as all of the others. Do not just focus on the reductionistic definition of the term. Wastes are not only what is thrown away or disposed of. Wastes include anything that is inefficient, wastes resources, or generates waste, such as product rework, inventory, disposal costs, too much labor, injuries, illnesses, and having to walk too far to do a job. Wastes are inefficient and drive costs up. Good total quality programs must focus on wastes and their elimination.

Time. Successful total quality programs focus on *time.* Time applies to waste, but because it is so important, it is given a focus of its own. Here, the important time element is called *cycle time.* Cycle time is the amount of time measured from the start of an activity to the finish or completion of the activity. It can be all-inclusive, such as the time it takes to make a product from raw material until it is finished and shipped to the customer. Cycle time can also be a portion of the process, such as the time it takes to complete a task.

Why is cycle time so important? Obviously, because the longer it takes to do something, the more it costs. There are two important areas of wasted time: Q-time and set-up time. *Q-time* is nonproductive time. It is the time spent waiting for parts, equipment availability, inspection, paper or directions, approvals, or decisions. Q-time is waiting time in which no productive work is done. It is caused by something in the system, administration, or workplace. Just as important is *set-up time.* Unlike Q-time, people remain busy, either setting up or changing over some operation or piece of equipment. Changing machine dies and setting up a machine for precision work are good examples of set-up time. Q-time and set-up time are two main extenders of cycle time, and they affect both product and operation costs.

Excellence. Successful total quality programs have to focus on *excellence.* Excellence is the road to tomorrow. Anything short of excellent is second rate. Excellence, of course, can be determined only by your customers and only by comparison. As Lee Iacocca said, "We want to be the best. What else is there?" Excellence is a continual pursuit. It is also the outcome of the total quality process. Like the other focus areas of successful total quality programs, it requires total concentration and commitment to the pursuit of excellence. There can be no compromises. Nothing else will do.

What Focus Brings to the Organization

How are these aspects of total quality applicable to service and staff functions such as safety and health? Let's look first at the focus on customers. Who are your customers and how important are they to your program? Historically, if you had asked a safety or health practitioner, "Who is your customer?" you would have heard, "The employee." In a total quality program, that definition is expanded. The employee is still one of the highest prioritized customers of any safety and health program. Today's definition, however, includes management at all levels, other staff positions, matrix-reporting positions, and those who have chosen you for benchmarking. What is the ranking of priorities? Each will have his or her time at the top of your priority grid. Overall, however, each has pretty much the same priority—high.

Do you know what your customers want or expect from your service? Do you regularly ask them what they want or expect or do you just assume that you know?

The focus of American industries on waste is new. One does not have to go back very far to remember a near-sterile Lake Erie, smokestacks belching black smoke, and hazardous waste disposal via barges and the open sea. However, in total quality businesses today, the definition of waste has been expanded. It includes the following:

- Wasted inventory
- Production
- Rework
- Manpower
- Inflated prices for materials or services
- Storage
- Too much material transportation
- Injuries and illnesses
- Inefficient manufacturing processes
- Stops in production
- Imperfections in material quality
- Returned products
- Product liability issues
- Loss of employees

Focusing on waste allows wasted effort and costs to be eliminated or minimized. So, it allows a business to be more productive and make more profits.

How is this focus on waste different in a safety and health program? It is not! Focusing on wasted effort, expenses, and related factors allows the program to be more productive and avoid costs that will ultimately improve the business' profit margin. It is important to note that aside from consultation, safety and health programs generate no revenue to the business. Therefore, cost efficiency is a key focus. What kinds of wastes can a safety and health program focus on? A short list would include the wasted effort in doing a routine task such as dispensing safety glasses or approving chemical purchases. It would include wasted costs of safety equipment such as gloves or respirators and wasted time reacting to recurring problems such as accidents and communication problems. It would also include unproductive efforts to locate something. Get the idea? Obviously, an analysis begins with knowing where your time goes. This, naturally, deserves to be the next waste in our discussion.

Time is probably the greatest resource that is wasted. After all, what is your number one cost of doing business? People! Inefficient use of people costs money—lots of money. That is why total quality organizations focus on time. How much time does it take to make a product, do a task, or perform a function? This can be determined by breaking the time into both value-adding time—time when something is actually being done to further a product's manufacture, and non-value-adding time—time spent waiting, setting up equipment, transporting material, putting parts into and taking them out of stock, getting needed tools or equipment, and starting up or shutting down equipment. Eliminating or minimizing nonvalue-adding time increases the efficiency of this most costly resource—people.

Is focusing on time different for a safety and health application? Not at all. People are still our most valuable resource. And, like it or not, anyone who practices safety and health has a limited amount of time to work with. Total quality safety programs focus on wasted time in both routine and nonroutine functions. How long does it take to do a safety inspection, complete and review an accident report, approve a new chemical product, measure an employee's exposure, do a physical examination, do a monthly report, have a problem-solving meeting, file records and reports, ready a shipment of hazardous waste, provide safety training? Starting from the tasks that happen most frequently or take the most time to do, and analyzing them for steps and then for value-adding and non-value-adding time allows a function or task to be streamlined and made more efficient. Process mapping is an excellent tool here. Efficiency wastes less time and allows everyone to spend their time on more worthwhile activities, such as proactive safety and health functions.

For too long, the American industry accepted the paradigm that products were supposed to break and, thereby, require repair. It was not until more forward-thinking players such as the Germans and the Japanese came along that our thinking changed. And these established players are continuing to improve as are new players entering the game. Excellence is being defined and redefined each day. If excellence is the only road that assures tomorrow, the focus on excellence in a total quality business is a matter of life or death. Products must be made better, with higher quality at competitive prices; last longer; provide more no-cost features; and require fewer repairs. The customer expects it more today than ever before, and will expect it even more tomorrow.

How about focus on excellence in safety and health programs? For too long, we have been focused in the wrong direction when it comes to excellence. We have focused on our injury rates, lost workdays, and workers' compensation costs. These are the wrong indicators of excellence because too many things outside the safety and health program influence them. For example, using injury rates as a measure of excellence will place good and bad programs equal, even if different injury severities are involved, or the numbers are being altered. If workers' compensation costs are used as an indicator of excellence, a program can significantly worsen simply because a new doctor who is ignorant of occupational needs and practice or workers' compensation rules comes in, or because a new surgical procedure is in vogue (for example, Carpal Tunnel Syndrome surgery). These traditional indicators are very poor measures of safety and health program quality. They cast too many "shadows" that draw attention away from what makes a program excellent. What should a safety and health program focus on to move toward excellence? It should use measures such as customer surveys, multifaceted program evaluations, and benchmarking. These are true indicators.

Dynamic People and Management Concepts

Total quality programs have dynamic people and management concepts. People, after all, are the most important resource to a successful organization or business. Looking over the history of management and the resultant management theories, the importance of people has cycled up and down. Theory X management, for example, considered people as necessary resources, much like raw materials. People were there to do work, and work was to make a product or perform a service—nothing more, nothing less. A good worker was a productive worker, a happy employee. And a good worker was a secure employee—at least, as long as he or she performed at the expected productivity and quality level, and provided that the managers ran the business well. It was a "your world-my world" separation concept. Management ran the business and made the decisions. Workers operated machines, inspected product, shipped product, unloaded raw materials. They simply were not included in the decision-making process. They were not even asked!

Theory Y management came along, and in the companies that used it, the importance of people made a quantum leap to the opposite end of the continuum. Under Theory Y, people were everything! Managers became known as "huggers and kissers," catering to employee needs, keeping them happy. The concept was different—very different from Theory X. Here, employees had all the answers. Management was there to coordinate and serve. As a matter of historical perspective, there have been successful businesses in both theaters—Theory X and Y. In a total quality environment, neither works.

Theory Z management, also commonly known as Japanese Management (although it is not totally), is better described as team management or participative management. Originally described in Japanese industry, this type of management uses team dynamics to solve problems and make decisions. It has been very successful. One of the disappointing aspects of Theory Z management, however, is the devaluation of the individual. Teams are the only recognized entity. To "belong," you have to be part of a team. Recognition and rewards are focused on team performance and successes. Americans, who have long admired individual effort and heroics, have had trouble adapting strictly to this management style, primarily due to this individual devaluing. America has an entrepreneurial society that strongly holds that individuals who risk and succeed get ahead. Successful total quality programs need to take the synergy and dynamics that are available within the Theory Z school of thought, blend it with a modification of the Theory X management that is now called leadership, and mix them with the entrepreneurial concept of individuality. Total quality uses people concepts, group and individual, as a dynamic tool.

Concepts of Participation, Empowerment, and Ownership. Within these people concepts utilized by total quality programs, three notable terms stand out: participation, empowerment, and ownership. *Participation* is a deep concept. It not only speaks of involvement at all levels of an organization, top to bottom, but also includes involvement in most of the areas that impact the organization. For participation to occur, a number of interrelated needs must be met. Management creates some of those needs, some by the organization. These needs include opportunity, knowledge and training, time, commitment, resources, support, encouragement, and responsiveness. Because participation is a deep concept, it must be fostered, nurtured, and supported in order to be sustained.

Empowerment, to a great extent, is a measure of the autonomy awarded to an employee or group of employees. Autonomy is a function of trust and challenge. In his book, *The Game of Work,* Coonradt (1985) set a special, large area in the middle of his "field of play," which was his way of describing an employee's work. This area defines the concept of empowerment very well. In this middle area, employees do their jobs, independent of being overly managed. Coonradt called this area the G.O.M.B. or Get Off My Back Zone. That is what empowerment is: letting an employee do his or her job without over- or undermanagement and control. In reference to purpose, Peter Drucker, in his seminal book *Management,* stated, "The purpose of an organization is to enable common men to do uncommon things." That is empowerment!

Ownership is accountability, recognition, and reward. This is the people concept that keeps employees focused as appropriate, and measures their progress and success. Effort, together with success, should that occur, is recognized personally and publicly. Reward is very personal. Whatever rewards are used, they must be selected and made appropriate to the employee and the organization's culture.

Team Building. In American organizations, a lot of progress has been made in people concepts over the past ten years or so. There are still *wide* ranges in the application of these people concepts, however. They range from not-at-all to fully ingrained in the organization's culture. Using these people concepts, classical definitions of worker and management roles are being challenged and discarded. Long-standing walls between management and labor are being torn down. Communication pathways are being redefined. Critical aspects of these people concepts include trust, active communication, autonomy, involvement, sharing, accepting responsibilities, supporting other team members, and working together.

One valuable and widely used people concept is team building. Far from the traditional definition, which is synonymous with working team or work crew, today's team transcends the horizontal and vertical aspects of an organization. Team members come from different departments, trades, staff units, different levels of management, and from outside the organization. The teams themselves are dynamic. They come together to solve a problem or provide a task. Once completed, the team dissolves only to have the members become involved in new teams and new missions.

There is a common but destructive belief in many safety programs. "If I, as the safety and health professional, do all the work and have all the knowledge, it is my program and I am irreplaceable." It is called "turf." In a total quality safety and health program, the use of dynamic people concepts challenges the basic premises of turf. By educating and including other people in all safety and health activities, there are both positive and negative aspects. If people are insecure about his or her abilities and engages in turf, letting go may be too threatening. This would be viewed as a significant negative. If, however, they are secure with themselves and want to be able to accomplish more, it is overwhelmingly positive. Through using people concepts such as team participation, line responsibilities can be more firmly entrenched and reinforced. Additionally, the number of projects becomes disassociated from the amount of available time. The safety and health knowledge level is significantly increased in the organization. And, the number of safety and health disciples in the workplace can be enhanced.

Strategic Planning

Total quality programs use strategic planning concepts as a tool to measure their progress and plot their pathway to success. Strategic planning incorporates three interrelated levels: strategic, long term, and short term. Strategic planning is what has to be accomplished now and in the future to achieve set goals, such as being the best. Strategic plans change as the organization makes progress and as the standards change. Because of that progress, strategic plans change frequently. Long-range plans are more concrete, are usually progressive, and are aimed at achieving your strategic plan. Long-range plans usually cover a period of three to five years. Short-term plans are bite-sized pieces of the long-range plan. They are more detailed and usually cover one year of operation. The strategic planning process is critical to a successful total quality program. Without it, any success is forced to happen by chance, or accident.

Strategic planning depends on setting goals and objectives, and measuring progress toward them. Objectives not only complement the planning process but also identify the "have to's," the "need to's," and the "want to's" that should be accomplished. Objectives by themselves mean little to the success of an organization. Objectives must be accompanied by measurement. Measurement allows everyone to know whether progress is being made, whether the path needs correction, and when the objective has been achieved. Objective setting is important, but measurement is critical.

Traditionally, how many safety and health programs operate by strategic planning? From my observations, I would guess that the percentage would be significantly less than 5 percent. Historically, safety programs, especially, could best be described as "management by fire drill." Managers are so busy "putting out the fires" that they never get into planning what they *should* be doing. A strategic planning process must manage the total quality safety and health program. Preferably, the strategic plan is derived by a team and complements the organization's plan. To a total quality safety and health program, strategic planning is not a choice, but a necessity.

Do traditional safety and health programs use goal and objective setting and measurement in their strategic planning process? Taking away the traditional measurement approaches we have discussed that do not work, injury rates, lost workdays, and compensation costs, my guess, again is, less than 5 percent. Objective setting in most safety and health programs is a day planner. Measurement of accomplishments is also almost totally absent in the safety and health fields. This is a strong statement. As with most realities, the truth often hurts. Setting objectives and measuring progress is critical to a total quality safety and health program.

Benchmarking and Evaluations

Total quality programs use benchmarking and evaluations. Benchmarking is an improvement tool that front-running organizations, companies, or programs use as models. Improvement objectives are based on reaching those models. Benchmarking also includes a process of measuring that improvement. Benchmarking is a very valuable tool that avoids, to a great extent, the need to "reinvent the wheel." Model programs and organizations usually invite benchmarking because it "shares the wealth." It is also a real ego boost.

Program evaluations are systematic tools for assessing program growth, areas where improvement can be made, or where programs are strong. Benchmarking without program evaluations is like icing without the cake.

Through program evaluations, good programs can become great programs. Through benchmarking, great programs can become industry leaders.

Factually, most businesses and organizations do not benchmark nor do self-evaluations. However, successful total quality companies always conduct self-evaluations. Why is benchmarking so important? First, it is hard to know how to improve without a model. Benchmarking can provide that model. Second, it is very costly to invent everything. Benchmarking is a learning tool that avoids the need to "reinvent the wheel" at each company. Just as important, it also can help you learn from the mistakes others have made.

Is benchmarking important to a total quality safety and health program? If the program is focused on excellence, then it is very important. Again, how many safety and health programs today use benchmarking? I would guess that almost none do! Why? There are two reasons. First, successful safety programs have historically been kept "close to the chest." There is a mystique about them, a mystique that seems to be driven by superstition. Second, it is a personal reason. For the program team to ask someone else why the program was successful might be an admission that the team did not know what it was doing. In a word, pride! In many total quality companies today, staff functions such as human resources are beginning to benchmark across to human resources programs at other companies. It has not trickled down to safety and health programs to any extent yet, but it will.

Program evaluations, however, have existed in safety and health programs for many years. They have existed in many forms, have had different applications, and have used countless different formats and levels of detail. In short, program evaluations have existed to meet the perceived needs of the safety and health program itself. Those who foolishly feel that minimal self-evaluation is necessary use minimal or no evaluations. Performing programs that find greater value in evaluation use detailed, team-oriented self-evaluations at regular intervals. Some use program evaluations that are free-form, asking basic questions to guide group discussion; others use program evaluations that are literally small books containing extreme detail. The purpose of program evaluations, of course, is to provide information for improving programs. This is an extremely fundamental concept of total quality.

Benchmarking is an optional exercise once a highly evolved or improved status is reached. Some programs use benchmarking because they wish to reach a higher level. To any competent safety and health program, especially those aspiring to total quality, program evaluations are *not* optional. They are too fundamental to total quality. If benchmarking is not pursued, fine. If, however, program evaluations are unthinkable, quit thinking about total quality!

Continuous Improvement

Total quality programs use the concept of continuous improvement. This idea runs counter to traditional American industrial thought of "just throw money at it." Continuous improvement is not just an employee empowerment issue. It also focuses on small, low, or no-cost improvements. Classical American industrial thought says that to improve a process, a new process, assembly line or automation is necessary. These improvements are expensive and are not always successful. Continuous improvement takes ideas from the "grassroots" level of an organization and implements those small improvements. Such improvements could be as simple as modifying a hand tool, elevating a work surface, increasing the lighting, moving a storage cabinet closer to a worker, or rearranging equipment to smooth out product flow. Implementing small changes has several benefits:

1) Major costs are controlled.
2) Employees are involved in the improvement process.
3) There is little or no interruption in production.
4) Employee morale and enthusiasm is built.
5) Major capital expenditures may be completely eliminated or changed so that they can be more effective.

Total quality programs believe in continuous improvement. In total quality companies, it is a way of life. On the other hand, to most safety and health programs today, continuous improvement is a foreign term; it simply has no meaning. The concepts and tools of continuous improvement— *Kaizen,* a Japanese word—are just as applicable to safety and health programs as they are to manufacturing. Successful total quality programs should have the following:

- Common focuses (customers, wastes, time, and excellence)
- Dynamic people concepts (participation, empowerment, and ownership)
- Strategic planning using objectives and measurement
- Benchmarking and program evaluations
- Belief in continuous improvement

CONCLUSION

The following simple, step-by-step checklist can help you place total quality concepts into your safety and health program:

1. Have you established the correct program focuses on customers, waste, time, and excellence?

2. Have you determined, written down, and shared a mission statement about why your program is important to the organization?

3. Do you know which direction to go for making your program successful?

4. Have you developed a strategic plan for your program?

5. Have you developed a short-term or annual strategic plan?

6. Have you identified ways to measure your program's performance and the successful implementation of your plans, and are your measurements there for everyone to see?

7. Have you performed a formal evaluation of your program and imbedded that process in your annual schedule?

8. Have you used any benchmarking or partnershipping opportunities to improve your program?

9. Do you know your processes in detail and have you identified ways to continuously improve them?

10. Have you implemented any program(s) that increase employee participation and ownership in your program?

11. Have you implemented any program(s) that seat true line management responsibility for the safety and health of their workers?

Total quality offers significant advantages for managing any company, organization, or program. For nonline, specialty-type functions such as safety and health, total quality can offer considerably more. Not only can it provide the mechanism for a successful program on a continual basis, it can remove most of the traditional frustrations associated with the safety and health profession. The perceived differences in total quality application—manufacturing versus service—exist *only* in one's paradigms! That is not to say that this perception is not important. Quite the opposite. Because of this supposed conflict of interest, it is important, first, to recognize the fact, second, to open your mind to a new way of thinking, and, third, to actively challenge your paradigms and those of sponsoring or upper management.

Additionally, the total quality focuses align our thinking with emerging philosophies in American business. These focuses are critical to the success of America in the global economy. But by aligning our focuses with these emerging philosophies, we place ourselves inside the value-providing segment of our organization. This alone is a significant deviation from our traditional role and one that will serve us well into the future.

REFERENCES AND SUGGESTED FURTHER READING

Barker, J. A. (1989). *Discovering the Future: The Business of Paradigms*. Burnsville, MN: Charthouse Learning Corporation.

Christopher, W. F. (1993). *Vision, Mission, Total Quality: Leadership Tools for Turbulent Times*. Cambridge, MA: Productivity Press.

Coonradt, C. (1985), *The Game of Work*. Salt Lake City, UT: Shadow Mountain Press.

Drucker, P. F. (1993). *Management Tasks and Practices*. New York: Harper Business.

Pierce, F. D. (1995), *Total Quality for Safety and Health Professionals*. Rockville, MD: Government Institutes, Inc.

Pierce, F. D. (1997), *Managing Change for Safety and Health Professionals: A 6-Step Process*. Rockville, MD: Government Institutes, Inc.

Chapter 11

Benchmarking

Arthur Billington and
Andrew Sorine

INTRODUCTION

Benchmarking is a very powerful tool for companies that wish to improve the effectiveness of their existing safety efforts, or for companies that desire to reduce costs associated with operational errors. Benchmarking is so powerful because you can apply processes outside of your industry to your specific company. Benchmarking can be defined as "a process for rigorously measuring your performance (safety) vs. 'best-in-class' companies, using the analysis to meet and exceed the best-in-class." Benchmarking safety management practices and safety competencies allows a firm to measure internal alignment. Internal alignment is the degree to which members of an organization coordinate their efforts to achieve company safety goals. Many companies have invested money and time in compliance-driven programs only to find limited success. Why is it that safety problems or unsafe acts and conditions persist even though companies put "safety first"? Traditionally, safety programs have been relied on to address the many nonsafety issues that underlie performance deficiencies. However, programs, by their nature, are short-lived and therefore can cause misalignment. It is our opinion that safety issues continue to reappear because management practices and principles have not been applied to the implementation of safety programs. Many firms have never evaluated what safety activities are in place, or why. They have never developed a concise vision or mission statement for safety, or formulated an action plan to achieve this vision. The "big picture" is missing, so individual safety activities appear to be disjointed, segmented, and unconnected to the company's business goals and mission. No wonder managers do not want or know how to manage safety, and safety "services." Safety becomes misaligned with the goals and purpose of the organization.

Benchmarking a company's safety program to established "best practices" gives executives and managers a way to evaluate their program. It provides a means of comparison to a theoretical model that comprises safety activities and the best practices associated with those activities that excellent safety companies use. Alignment can be achieved by reviewing individual safety activities as part of a comprehensive safety program, permitting company executives the opportunity to see the big picture. Benchmarking, according to David T. Kearns, former CEO of Xerox, "is the continuous process of measuring products, services, and practices [safety] against the toughest competitors or those recognized as best in class or industry leaders."

Beyond knowing how a firm stacks up against excellent safety companies regarding the implementation of safety activities, a firm must know how effective it is in implementing those activities. Safety management practices, like any other management practice, must be in place to guarantee the continuous smooth implementation of safety activity best practices. Safety management practices, deployed with skill and consistency, will have an impact on business processes and will influence the company's safety culture. Strategic misalignment for safety management can occur within an organization when strategic or cultural differences develop over time as the result of safety management competencies not being effectively used by those responsible for implementing safety.

CASE STUDY: BENCHMARKING A COMPANY'S "BEST PRACTICES"

A metal container manufacturing company was experiencing an increase in workers' compensation costs. The frequency of accidents was up, and the severity of employee injury was impacting the bottom line of the division. The company did have a safety program, with the accompaniment of compliance program manuals and video training programs. However, the company could not understand why they where experiencing accidents since they did have a safety program and safety committee in place.

Management considered a benchmark study and approval was given to analyze the existing safety program and to evaluate the program's effectiveness to promote error-free performance. A baseline was established from which all future progress would be measured. A team visited the facility with the intent to interview those responsible for safety, tour the facility to view first-hand how the safety program was implemented, and speak with employees regarding how they felt about safety. First, the project team interviewed those individuals with accountability and responsibility for safety within their job. Second, the team toured the facility not to identify unsafe conditions, but to develop an opinion of how safely employees performed and how supervisors interacted with their department. Also, during the tour the team stopped at random and asked employees how they felt about safety. Employees were asked to determine how important safety was to management and their supervisors. They were also asked if other employees followed the established safety rules for the plant. Last, the team reviewed job descriptions, performance reviews, minutes of safety committee meetings, accident investigations, current written safety compliance manuals, and safety training records. The results of the information gathering gave the project team enough background and understanding of the current safety climate to ask the appropriate questions during the roundtable discussion that followed.

At the roundtable, persons directly or indirectly involved with the safety program sat down at one table to discuss "How safety functions at _____." The facilitator of the roundtable served as a moderator and drew information out of several sources for the purpose of identifying accountability, performance measurement, and consistency of implementation. Each of the twenty safety activities identified as critical to safety program effectiveness were discussed and compared to established National Safety Management Society "best practices" (refer to the list of safety activities chart found in Figure 11-1). The purpose of the roundtable was to gather several persons, all acquainted with the safety program, to discuss their perception of the program. The group format was used, which permitted individuals from different job functions important to the program implementation to have a chance to express their opinion and be part of the discovery process. No one person knows how safety truly functions or does not function within a company. The group process is used to broaden the perspective and provide different experiences to develop a realistic snapshot of program effectiveness. As the group participants work their way through the questioning, and interaction among the participants begins, they discover the strengths and weaknesses of the current program.

Figure 11-1. Gap Analysis

Describe Program Procedure/Strategy	Is It Documented	Who Is Responsible	How Is It Communicated/Implemented	How Do You Verify/Measure/Conformance/Performance

Function: 1.0 Management

CURRENT PRACTICE	BEST PRACTICE "BENCHMARK"	GAP
1.1 Established Goals/Objectives	Corporate and individual objectives are well defined and measured. Accountability is defined. Senior management evaluates attainment of objectives. Objective is to always to exceed regulatory requirements. Loss Control objectives line up with business objectives and are included in business plan. A mission statement has been established and distributed.	Score
1.2 Assigned Responsibility/Accountability	Loss Control activities are established within all job categories both management and employees. Authority and responsibility for these activities are well defined within the organizational structure, as well as for individual performance evaluations. Loss Control functions smoothly because of clear understanding in the assignment and carrying out of tasks.	Score
1.3 Performance Measurement	Well defined loss control objectives are established within individual performance goals. Results are measured and used to evaluate performance within all levels. Both result measures and activity measures are used to measure performance. Performance measures for groups are charted and used to guide continuous improvement at the organizational level. Individual goals line up with company objectives.	Score

Figure 11-2. Safety Performance Gap Analysis

LOSS CONTROL MANAGEMENT PRACTICES

FUNCTIONS	Baseline 1	2	Compliance 3	4	Leadership 5
1.0 MANAGEMENT					
1.1 Established Goals/Objectives					
1.2 Assigned Responsibility/Accountability	▓	▓			
1.3 Performance/Measurement	▓	▓			
1.4 Organization/Committee	▓	▓			
1.5 Recognition/Visibility	▓	▓	▓		
1.6 Enforcement	▓	▓			
1.7 Hiring Procedures					
2.0 CONTENT					
2.1 OSHA Targeted: Written Programs					
2.2 Standard Operating Procedures	▓	▓	▓		
2.3 Planned Inspections/Observations	▓	▓	▓	▓	
2.4 Accident Investigations					
2.5 Employee Training					
2.6 Supervisor Training					
2.7 Cost Containment					
3.0 ASSESSMENT					
3.1 Regulatory Requirements	▓	▓	▓	▓	
3.2 Employee Perceptions					
3.3 Internal Accident Analysis					
3.4 Suggestion System					
3.5 Publicity/Awareness	▓				
3.6 Audits					

Legend

Rank	
1	No Attempt
2	Some Attempt to Implement
3	Documented and Implemented
4	Integrated and Monitored
5	Performance measures – Best Practice – Continuous Improvement

The result of the information gathering and roundtable discussion is a *gap analysis* (Figure 11-2). The gap analysis summarizes, on a chart, the comparison of current safety program practices to "best practices." The difference between the actual and the ideal "best practice" is the gap in safety performance that needs to be addressed to improve results and effectiveness. This gap in performance can then be addressed in an *action plan,* which comprises specific goals along with activities required to accomplish those goals.

Incorporated into the roundtable, along with the evaluation of current practices, is a discussion of what safety management practices exist to ensure continuous and smooth implementation of each safety activity. A score of 0 to 5 is assigned to each activity based on positive responses for the following safety management practices:

- Formal documentation
- Consistency (say what you are going to do and do what you say)
- Assignment of accountability and responsibility
- Communication
- Measurement of performance and feedback of results

This score is used to prioritize needs and to assist in the allocation of resources in the completion of the action plan.

The case study firm discovered during the individual interviews and roundtable discussion that management commitment was present, but that responsibility and accountability for safety performance had been limited. As a result, there was little awareness of safety within the plant, and employees did not feel involved or accountable for safe performance. The gap analysis documented this conclusion with low scores for 1) established safety goals and objectives, 2) supervisor training, and 3) employee awareness and safety perceptions. Activities with low scores uncovered a common weakness in assignment of accountability and responsibility, communication, and performance measurement. The project team assembled an action plan to address closing the gap between the best practice and the current practice in place for each of the twenty safety program activities evaluated. Prioritization of resources and the establishment goals critical to safety effectiveness were selected from those with low scores.

BENCHMARKING PROCESS

There must be an understanding of what benchmarking is and its relationship to target settings. Since benchmarking involves setting new directions, its relationship to targets should be understood. This relationship should give a better understanding of where benchmarking fits into the overall planning scheme.

Benchmarking is a new way of doing business: it forces an external view to ensure correctness of objective setting. Benchmarking is also a new management approach: it forces constant testing of internal actions against external standards of industry practices. In addition, benchmarking promotes teamwork by directing attention to business practices necessary to remain competitive, and removes the subjectivity from decision making.

Benchmarking is an objective-setting process. Benchmarks, when best practices are translated into operational units of measure, are projections of a future state or end point. In that regard, their achievement may take a number of years to achieve. More important, benchmarks may indicate the direction that must be pursued rather than the specific operationally quantifiable metrics that are immediately achievable. A benchmarking study may indicate that costs must be reduced, that customer satisfaction levels must be increased, or that return on assets must be increased. In addition, the concentration on best practiced supports the general direction that must be pursued, with specific insights into how the benchmarks can or should be attained. The conversion of benchmarks to operational targets translates long-term actions into specific activities.

Targets are more precise, but their quantification should be based on achievement of a benchmark. Furthermore, a target incorporates what realistically can be accomplished within a given time frame (usually one yearly budget cycle or business plan horizon). The consideration of available resources, business priorities, and other operational factors convert benchmark findings to a target. Targets can show progress toward benchmark practices and metrics. The significant difference between a complete benchmark definition and a target is that a carefully conducted benchmark investigation will not only show what the benchmark metric is, but also how it will be achieved.

Benchmarking Phases

It is important to have a general understanding of the generic phases of benchmarking and their rationale. The process starts with a planning phase and proceeds through analysis, integration, action, and finally maturity.

Planning

- *What to benchmark?* Every function has a product or an output. These are priority candidates to benchmark with a view to improving performance.
- *Whom to benchmark?* World-class leadership companies or functions with superior work practices wherever they exist are the appropriate comparisons.
- *Data sources and data collection.* A wide array of sources is available, and a good starting point is a business library. An electronic search of recently published information dealing with the area of interest can be requested.

Analysis

- *Measuring the gap.* It is important to have a full understanding of internal business processes before attempting comparison with external organizations to provide the baseline for analyzing best practices.
- Projecting the gap. Whether negative, positive, or parity, these categories provide an objective basis on which to act and to determine how to achieve a performance edge.

Integration

- Progress should be reported to all employees. On the basis of the benchmarking findings, a vision or end-point picture of the operation can be developed.
- Performance goals should be revised.

Action

- Specific implementation actions, periodic measurements, and assessment of achievement should be put in place.
- People who actually perform the work should be responsible for implementing the benchmarking findings.
- Companies should stay current with ongoing industry changes by continuously benchmarking and updating work practices.

Maturity

- Maturity is achieved when best practices are incorporated in all business processes and the benchmarking approach is institutionalized.

FIVE TYPES OF BENCHMARKING

The benchmarking process can take many forms. The term means different things to different people, and it is important to understand what kind of benchmarking will best suit your application. The important thing about the benchmarking process is to make sure that you set up your process in such a way as to support the overall strategic plan for product development—for quality, safety, marketing, and customer service. The safety management benchmarking process should complement and support other plans, not cut across them.

The mainstream benchmarking literature discusses five kinds of benchmarking: internal, competitive, industry or functional, process or generic, and collaborative.

Internal Benchmarking

Internal benchmarking refers to benchmarking against internal operations. In most large companies, there are similar functions in different business units. One of the simplest (and least expensive) benchmarking exercises is to compare these internal operations. The objective of internal benchmarking is to identify the internal performance standards of an organization. The advantages of internal benchmarking are that 1) it encourages the sharing of information and opens up communication processes; 2) you can obtain immediate gains by identifying your own internal best practices and transfer those to other parts of the organization; and 3) it gives you practice before you start external benchmarking. The disadvantage of benchmarking is that it fosters an internal view. It is all too easy to ignore that other firms have the edge on you if you are concentrating on outperforming internal rivals.

Competitive Benchmarking

Competitive benchmarking means benchmarking against your direct competitors in the market you serve. For example, Coca-Cola might benchmark against Pepsi, Apple against IBM, or General Motors against Ford. The objective is to compare companies in the same markets that have competitive products or services or work processes, and identify how your organization can beat that competition. The advantage of competitive benchmarking is that it's fun to see how your competition does things and how you measure up. In addition, you have the comfort of benchmarking something that is (at least on the surface) directly comparable with your own company's processes and products. The disadvantage is that it's very hard to obtain the kind of detailed information you are seeking without either engaging in espionage or breaking trade-practice laws.

Industry or Functional Benchmarking

Industry or functional benchmarking means benchmarking against leaders in an industry, such as manufacturing, or in a function, such as human resources management. You can benchmark against others in the same industry as your company that may have the same type of products or services, but these companies may not be competitors in the same market. For example, a motor cycle manufacturer might benchmark against a car manufacturer, or a rental car company against a taxi company or limousine service. The big advantage of industry benchmarking is that it is easier to obtain willing partners, because the information is not going to a direct competitor. The disadvantages are cost and the fact that the most popular companies for benchmarking are beginning to feel a bit exploited, which may limit access to them.

Process or Generic Benchmarking

Process or generic benchmarking means breaking your company or functions into processes and benchmarking these various internal functions. For example, the human resources function can be broken into a number of processes including recruitment, training, and dispute resolution. The claims management function in an insurance company can be broken into processes such as claims creation, investigation, and assessment. You do not have to benchmark processes in similar organizations or even in industries similar to your own. The point is to focus on excellent work processes wherever they can be found rather than on the business practices of a particular organization or industry. The advantage of doing process or generic benchmarking is that this is often when the breakthrough ideas for change are generated. Process benchmarking has the potential of revealing the best of best practices. The disadvantage is that it is difficult to do. This type of benchmarking requires careful preparation, open minds, creative application, and commitment from senior management. It is also expensive in time, effort, and dollars. Many companies, however, believe that the payoff outweighs the investment.

Collaborative Benchmarking

Collaborative benchmarking is when a group of organizations agree to develop a benchmarking project that will allow direct comparisons within the group, often with a goal of developing a best practice model that all the participants can use to improve their processes. The difference between this and other forms of benchmarking is that the partnership is formed before measures are determined and data collected. The group shares the results openly among themselves, but does not usually share with outsiders.

WHAT TO BENCHMARK

Identifying what is to be benchmarked, or the benchmark outputs, is often one of the most difficult steps in the process. However, there is a way to arrive at benchmarking outputs in a logical, well thought-out manner. The first step in determining what should be benchmarked is to identify the output of the safety function or activity to be benchmarked. The development of a clear mission statement detailing the reasons safety exists within the organization, including typical outputs expected by users is fundamental. Next, the function's broad purpose should be broken down into specific activities to be benchmarked. Activities should be documented to the level of detail necessary for analysis of key tasks and performance measures— "best practices."

When deciding exactly what to benchmark, there are two areas that are "benchmarkable": performance indicators and management procedures that drive the performance indicators. Performance indicators include training costs, recruitment costs, absenteeism, turnover, or accident rate. The management practices (competencies) that drive safety activities are documentation, consistency, accountability, communication, measurement, and feedback.

Most of the benchmarking surveys that are carried out concentrate on benchmarking performance indicators. Your company and many other companies collectively form a database, and responses to questionnaires are aggregated. You then get an individual report for your company showing your performance against the database as a whole, with helpful information about how you measure up. The second type of benchmarking concentrates on process. Naturally this is harder to codify into a questionnaire.

Benchmarking process usually involves a detailed examination of how the processes or activities are presently performed, and then looks at benchmarking partners that you have specially selected to find out how they perform the process or activity. Then you decide what scope there is for improving the way you perform the process or activity so that it moves closer to the "best practice."

An awful lot of benchmarking, especially of the second type (benchmarking process/activity) is about intangibles. Employee satisfaction, morale, and motivation are very difficult to measure, especially if a good result means the absence of adverse events. For instance, fire protection processes are effective if no fires occur, but how do you tell how many fires have been prevented? One of the best things about benchmarking is that it enables you to get a better fix on the intangibles. Since people carry out most processes, and most services and activities are totally dependent on people, benchmarking plays an important part in working out measures for productivity and effectiveness.

It is typical for a manager of a service, especially an internal service such as safety management, to say something like, "You can't measure what we do—but trust me, we do good things." These days, that is not enough. Safety managers and other providers of internal services are expected to show what they contribute and how they add value to the organization, by providing detailed cost-effective analyses for any new initiatives. Benchmarking safety is basically the same as benchmarking anything else, but three types of benchmarks that are particularly appropriate. They are:

1) **Broad Measures of Performance.** These focus on the overall level of an organization view of safety management, examining broad productivity measures such as incident rate, accident costs per employee, frequency and severity ratios, workers compensation costs, and developed experience modification factors.

2) **Safety Management Activities.** These focus on highlighting which safety activity is being deployed. Comparisons can be made with "best practices" across companies. Safety activities that have been identified as critical to the effectiveness of a safety plan are:

   • *Safety management activities*—to include established goals, assigned responsibility and accountability, performance measurement, safety organization, recognition, enforcement, and hiring procedures.

- *Program content activities*—to include written programs, standard operating procedures, planned inspections, accident investigation, employee training, and cost containment.

- *Assessment activities*—to include employee perceptions, accident and behavior analysis, feedback, and audits.

3) **Safety Competencies.** These focus on safety management skills, knowledge, and/or abilities of staff and those responsible for safety and measure how effectively the safety activities are being deployed.

Determining whether something should be benchmarked depends upon how important the process is in the internal customer chain or in satisfying end users, customer needs or government requirements. The following are some critical questions that need to be asked:

- How significant is the problem to be benchmarked in relation to other areas where benchmarking resources could be directed?

- Will your customers (employees) notice the difference if you implement best practices for this business process? If so, will they change their behavior significantly enough to make a visible impact on the results of the organization?

Deciding on what to benchmark is the critical beginning of the benchmarking process and it is important that it be done carefully. If you decide on the wrong items to benchmark at the beginning of the process, you will waste valuable time and resources on items that may be best left alone while the real problem areas of the company continue to get worse and cost the organization money.

HOW TO QUANTIFY SAFETY PERFORMANCE

The next step in evaluating safety program effectiveness is to measure performance. The traditional method of measuring performance has been to compile and compare accident statistics and accident trends—all of which are reactive. Accident statistics may give useful information regarding the frequency, severity, and location of past accidents, but provide very little insight into how to prevent future operational errors. To be truly proactive we need to identify those "upstream" practices that will predict success. The activities or programs themselves do not improve performance; it is the safety management practices that are implemented and embedded within a company culture enhance the likelihood of effectiveness. The safety management practices that have proven to be critical to excellent safety performance are as follows:

- Formal documentation—know what you are going to do
- Consistency—have the will to do it
- Accountability/responsibility—know who will do it
- Communication—say what you are going to do
- Measure results—show how well we do it
- Feedback—build on what we did correct and help us to improve upon our weakness

No one particular safety management practice will make all safety programs effective simply by its presence; it is the accumulation of the safety management practices together that make programs effective for the long term. At present, we do not know whether different combinations of management practices are more important than others. This could be the subject of another study. The influence of the management style that embraces employee empowerment and input into the process also has proven to be significant. It is difficult to measure "empowerment" or any cultural type assessment, but the presence of such a culture does contribute positively to safety performance.

Safety performance is scored on a scale from 1 to 5. Each of the twenty critical safety activities associated with effective safety programs is evaluated. The higher the score the more safety management practices are present with that particular activity. A higher score, also means there is a greater likelihood that safety efforts will be effective, not simply put down on paper to fulfill a regulation. One point is given for each safety management practice: documentation, consistent implementation, communication, accountability and responsibility, and performance measurement and feedback. A point is awarded if it has been determined that the safety management practice meets the criteria established for the activity evaluated. The scoring is an all or nothing methodology. The role of the auditor who conducts the benchmarking study is to gather information from the persons interviewed. Questioning would follow this sequence of logic:

- Describe to me the activity and/or program within _____ Company.

- Show me any documentation.
- Who is responsible to implement? Is it part of anyone's job description?
- Would employees know about this activity, could they describe it, and would they know who was responsible if asked?
- Is the implementation of the activity/program consistent among different departments and areas of the company, or does it rely solely on the initiative of specific individuals?
- Do you measure performance? How do you measure performance? Is the measurement quantifiable?
- Do you give feedback regarding performance? How do you give such feedback? Is it part of a person's performance review? Is it used to continuously improve the safety process? If yes, how? Is both positive and negative reinforcement given when required?

The actual quantification of safety performance should be a group activity and completed with all stakeholders in the safety program present. A group is preferred because to develop a true "snapshot" of how a safety program is functioning, it takes several individuals with different perspectives and experiences to develop a realistic interpretation. Our experience with such groups is that when open discussion is encouraged and several persons with knowledge and interaction with safety activities are promoted, a more accurate picture of safety is developed, because safety is multifaceted and impacts many areas and individuals. No single person can interact and be intimate with so many activities. With the input of several persons, the entire picture can be developed and a true picture of performance developed. Remember, all programs have strengths and weakness. We must build on the strengths and improve upon the weaknesses to continuously improve.

GAP ANALYSIS

The last phase of benchmarking safety programs is to develop a *gap analysis*. The gap analysis is the accumulation of all the work completed so far. This analysis quantifies the difference (gap) between what is and what is the "best in class." The result is an *action plan* with specific activities and actions that are required to meet and exceed the best in class.

The first phase of a benchmarking project should determine which functions (activities) are to be benchmarked. The second phase should identify key performance variables (practices) to measure. The National Safety Management Society's "best practices" will simplify and speed your benchmarking project. Twenty critical safety activities required for effective program implementation have been identified. The best practices have also been identified and aligned with each activity. The third phase of a benchmarking study should compare and evaluate the company's current activities to established best practices. This third phase results in a gap analysis.

A gap analysis is the report that summarizes the project findings and establishes a baseline evaluation from which to measure future progress. The gap analysis slices the current safety program and associated activities into several elements. Each element of the safety program is then analyzed and effectiveness evaluated by performance of safety management practices. The gap analysis dissects the safety program undergoing benchmarking into the following elements:

- The twenty critical activities required for an effective safety program, in place and functioning
- Best practices for each of the activities in place
- Safety performance by scoring the implementation of safety management practices

The gap analysis puts into one table all the evaluation and analysis completed during the benchmarking study. The gap analysis puts the current program's practices next to the established "best practice benchmark." This side-by-side comparison allows the auditor to identify current program strengths and those areas in need of improvement or current program weakness. The gap analysis presents the findings of the auditors if the required safety activity is present and functional within the benchmarked company. Current safety practices associated with each critical activity are quantified and put down on paper. The table used places "current practice" in the left-hand column and the "best practice benchmark" in the middle column. The right-hand column is used for comments regarding the activities that need to be implemented or included with the current program to meet or exceed "best-in-class."

The "gap" column also has a section for entering a score. This score is the number developed by evaluating the safety management practices used to implement the critical safety program activity. The numerical rating given to each critical activity is to be used to help the auditor prioritize action and allocation of scarce resources to improve the current program. The "gap" column also is to be used to develop specific actions that are required to be taken to implement "best practices" within the benchmarked company. This series of action is called the "action plan." The action plan not only addresses what practices need to be improved upon to meet or exceed best practices but also prioritizes the actions. Those activities that need improved safety management practices are the action that will improve effectiveness faster and will impact the bottom line faster.

Chapter 12

Safety in Design: A Simple Perspective for the Designer of Machinery

Robert N. Andres

INTRODUCTION

Effective safety and health management begins long before the worker appears in the shop or a machine or process is put into use. There is a critical fact the astute manager must remember: It is less costly and more efficient to correct safety and health hazards at the design stage, before they become part of the workplace. The aim of effective management should be to ensure that safety and health hazards are, for all intents and purposes, eliminated before they exist.

The management role in design starts with a corporate policy that clearly indicates that safety is to be built into the job and the product. Each firm has its own management philosophy, its own way of viewing safety and health, consistent with the firm's underlying objectives. No single organizational plan would be useful or appropriate for all firms, but one sound approach, whether specifying equipment for purchase or offering equipment for sale, is to establish a corporate committee to develop the design requirements of the firm.

DESIGN RESPONSIBILITY OVERVIEW

Safety begins during the product design phase. This rightly implies that the designer who ignores or minimizes the application of safety principles may be, in whole or part, responsible for many equipment-related injuries. Even if the designer does not actually commit errors in design, he or she may still inadvertently omit factors needed to protect users of the equipment.

Copyright ©1998 by Robert N. Andres. Reprinted by permission of author.

Every machine or process creates some hazards during its life cycle, which in turn pose risks of harm, either directly to people, or indirectly through damage to the environment. As these machines and processes become more sophisticated, it is necessary to take due diligence to *identify and control all the hazards.* Safe design may well end up being a complex multi-disciplinary effort, involving many people and functions on whose input and cooperation the designer must rely. Design defects involve a host of considerations, including the following:

- Material stress factors
- Human behavior and other factors
- Construction materials
- Safeguarding measures
- User needs
- User evaluation
- Foreseeable emergencies and unplanned maintenance
- Accident or crash worthiness
- Proper operating instructions
- Codes, standards, and accepted practices
- Manufacturing specifications
- Product service experiences and facilities
- Properly trained users
- Design envelopes
- Financial constraints

This list presents only a few of the inputs the designer must considered. It is plain that the designer cannot know everything; therefore, a coordinated, supportive effort involving many types of expertise is needed.

SUPPORTING ROLES IN THE DESIGN PROCESS

To describe extensively the many supporting roles, functions, and disciplines in good design is beyond the scope of this chapter. A few of them, however, are so essential that some discussion is required. In particular, the senior manager should be aware that all of these individuals play a valuable role in safe design:

1) **Designers**—who will determine the parameters and prepare the specifications and drawings

2) **Managers**—who must oversee that the process is properly carried out

3) **Supervisors**—who, for both the user and manufacturer, have more intimate knowledge of installation, setup, and operational requirements

4) **Safety personnel**—who, by virtue of their training, have insight to safety issues and risks

5) **Maintenance personnel**—who have insight to maintenance requirements at all levels

6) **Safety committees**—the members of which can provide valuable insight to a variety of issues and are often most attuned to the behavioral aspects of safety

7) **Loss control administrators**—who are most familiar with the history of accidents and injuries and can equate economic costs

8) **Field installation personnel**—who understand the limitations of the machine or process.

9) **Sales personnel**—who may understand how the equipment may fit with the user's requirements better than anyone

10) **Users of the equipment or process**—who can give insight to their requirements and field experience with similar machines and processes.

11) **Safety consultants**—who can provide input from a broad array of experience with similar equipment or processes in various settings, often including accident and injury specifics, and an objective viewpoint that transcends internal politics.

The design of equipment and products is essentially evolutionary—always changing as new materials, processes, and applications are developed. Designers must constantly acquire wider experience and knowledge to keep abreast of requirements. Their position is critical, since it is a key to eliminating problems and forestalling unsafe situations. In addition to doing design work, designers must assess safety and health hazards, be knowledgeable about safety and health regulations, know the environmental demands of the workplace, and be aware of the various engineering tools, controls, and equipment that can solve their problems.

To be approached properly, equipment and process design must include the concept of *risk assessment and analysis* from the onset. Risk assessment, which dictates the design and inclusion of appropriate safeguards, can be significantly improved by knowing loss histories (from loss control personnel or legal counsel), maintenance requirements (from internal and user maintenance personnel), anticipated use and foreseeable misuse of a machine (from users, field installation and maintenance personnel, and sales personnel). These factors are often overlooked.

RISK ASSESSMENT AND EVALUATION METHODOLOGY

The heart of safety in the design process is the determination of risk of injury. Risk is the product of the severity of injury that could occur as a result of exposure to a hazardous event and the probability of that occurrence.

For years, many in our profession have taken the position that, because a hazard exists, it must be "safeguarded"—regardless of the exposure. Some have even gone further to require hazards to be safeguarded to the "highest level"—regardless of its potential to cause injury (an argument often used by attorneys for the plaintiff). Fortunately, the paradigm has shifted! Safety professionals are realizing that we have been spending valuable resources guarding hazards posing little or no risk while ignoring those outside the immediate scope of responsibility. The purpose of the "risk assessment and evaluation" steps outlined in this chapter is to assure that *appropriate and cost-effective* steps are taken to reduce risk to an acceptable level.

Risk assessment can be simple, if a logical step-by-step iterative process is followed. For purposes of this discussion, let us assume that the end product is to be an industrial machine. Figure 12-1 shows the steps in the iterative process of risk assessment and reduction.

The limits of the machine or process will determine what tasks can logically be performed. For example, a power press is defined by its physical size and weight, tonnage capability, size of the bed, its stopping time, and so on. Tasks exceeding these limits create a much greater risk of a hazardous event. The limits establish the framework of operation for determining what tasks can be reasonably performed, and what types of use and misuse are foreseeable.

Determine the Tasks to Be Performed

The tasks involved with a machine or process are not limited to setup and operation. Rather, the designer must envision all the tasks that will be done with respect to the machine or process, including the following:

Figure 12-1. The Iterative Process to Achieve Tolerable Risk

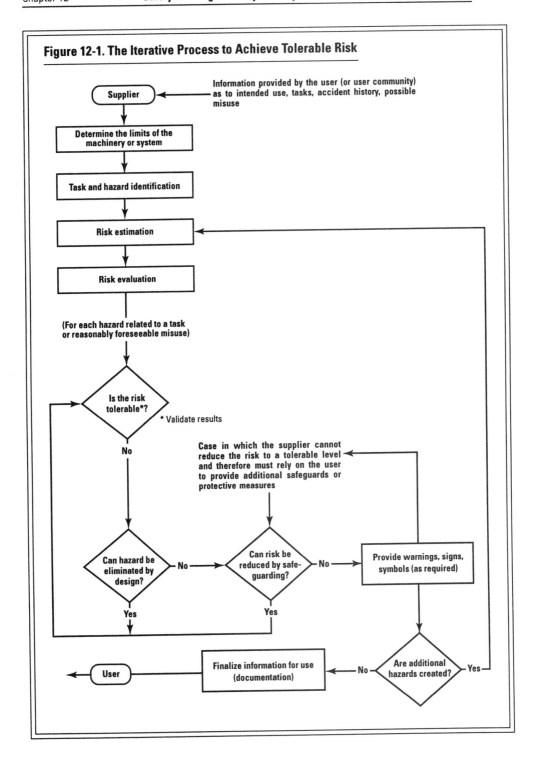

- Packing and transportation
- Unloading and unpacking
- Preparation of the area
- Provision of services, such as air and electrical
- Commissioning, including handling and filling of fluids and connection of services
- Setup, including installation of tooling, fixtures, and accessory items
- Operation, including foreseeable use and misuse
- Maintenance, including planned and unplanned maintenance
- Tooling changes
- Housekeeping
- Decommissioning, including safe removal and disposal of fluids
- Removal, including packaging
- Disposal and transportation

Almost every mishap can ultimately be traced to a human error, although it may not be an operator error; an error may have been caused by the designer, the production worker, the maintenance worker, the supervisor, or anyone even remotely connected with the process, including the worker two thousand miles away who carelessly put the wrong label on a container.

Too often the human worker's limitations receive little attention during the design phase. Consequently, managers are left in the position of having to fit the worker to the machine. They are urged to instruct, train, or motivate the worker to avoid the hazard when the hazard should have been eliminated from the equipment in the first place. Fitting the worker to the task is a poor approach, particularly in light of the fact that minor changes in equipment can greatly improve performance and reduce accidents. In short, human considerations can radically affect the safety, efficiency, and reliability of the best-designed process or system, and must be given top priority in the risk assessment process.

Identify the Hazards Associated with the Tasks to Be Performed

The most essential step in safe design involves identifying exactly the hazards posed by the equipment. All hazards are a product of some form of energy or toxicity. The designer must first recognize a hazard before he or she can successfully deal with it. Here are some of the things to look for:

- **Mechanical hazards**—due to machine parts or work piece shape, relative location, mass and stability; mass and velocity, limits in mechanical strength; accumulation of energy; hazards from springs, liquids or gases; the effects of vacuum; and hazards that may result in crushing, shearing, cutting or severing, entanglement, entrapment, puncture, abrasion, fluid injection, striking, or impact.
- **Electrical hazards**—due to contact of persons with live wires or components, either directly or as a result of a fault condition; persons approaching parts with high voltage; electrostatic phenomena.
- **Thermal hazards**—which may result in burns, scalds, or damage to health from exposure to a hot or cold environment.
- **Noise and vibration hazards**—which may result in interference with communication, hearing loss or other physiological disorders, neurological and vascular disorders, and physical trauma.
- **Radiation hazards**—from low frequency, radio frequency, microwaves, infrared, visible and ultraviolet light; alpha, beta, gamma, neutron, and X-ray production; electron or ion beams; and lasers.
- **Ergonomic hazards**—from unhealthy and unnatural postures and movement or excessive effort, inadequate placement of controls and displays, or poor lighting.
- **Material and substance hazards**—caused by harmful fluids, gases, mists, fumes, and dusts; fire and explosion hazards; chemical, biological, or microbiological (viral or bacterial) hazards.

Keep in mind that such hazards may be presented by or associated with the following:

- Unexpected startup caused by failure of the control system
- Restoration of energy supply after interruption
- External influences
- Software error
- Environment and surroundings
- Human error

There are several commonly used design analysis techniques to ensure systematic identification of all probable hazards. The two major approaches are generally classified as (1) fundamental and (2) technical. The fundamental approach consists of a study of all possible hazards that may exist. They are first examined qualitatively, to be sure that they are actually hazards, and then quantitatively, to calculate the overall level of severity of injury that may result from a particular hazardous event. This approach is sound, whatever the new product or process.

The technical approach involves a careful study of as many accidents as possible in order to identify the hazards that had a role in the accidents. The findings are then applied to the design of new equipment and processes or the redesign of existing systems. The technical approach nearly always involves some version of failure mode and effects analysis (FMEA) and fault tree analysis (FTA). Since the average person has little occasion to deal with either of these techniques, a few words of explanation may be appropriate here:

- FMEA looks at the possible failures of each component of a product or system and determines the effect of each failure. It is equally useful when a system involves many parts, several failure modes, and moderately complex designs.

- FTA organizes the cause-and-effect relationships of an accident or event, deducing the various combinations of failures and errors that could result. By using statistical information, it can also determine the probability of an accident. FTA works well when there are many parts, many modes of failure, a complex design, and a sophisticated operation. It can also give information for making decisions, choosing alternatives and allocating resources. The objectives of this analysis are to:

    1) Detect and address potential accident situations during the design phases.
    2) Perform such analysis systematically until every activity has been covered by a hazard evaluation.
    3) Perform such analysis whenever a process or equipment is changed.

Establish Hazard Levels

All hazards are not equal; they vary by degree. Although it is essential that all hazards be *identified,* it is not always necessary that they be *addressed.* The designer must keep in mind that hazards posing a threat of exposure to people or the environment—in any reasonable manner—must be evaluated.

Hazards are generally "rated," either qualitatively or quantitatively, according to the *severity of consequence*—measured in terms of the severity of injury and the economic damages caused by exposure to a hazardous

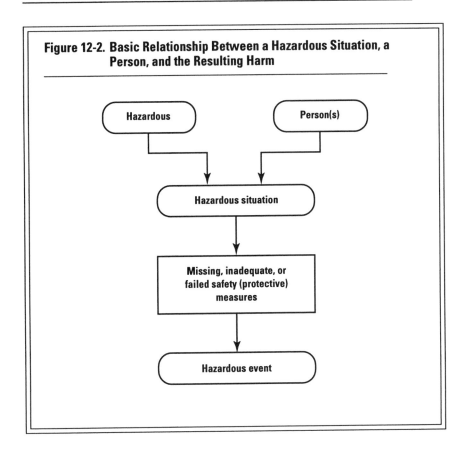

Figure 12-2. Basic Relationship Between a Hazardous Situation, a Person, and the Resulting Harm

event. All machine accidents have associated costs, both direct and indirect, that must be considered in determining the severity of consequence.

One example of this rating is derived from Mil-Std-882C. Assuming, for our purposes, that we concerned only with the effects on people, this example establishes four qualitative hazard categories relating to the effects that could result from a hazardous event (See Table 12-1):

Determine the Elements of Probability

Once the hazard level has been determined, the other factor of risk is the probability of occurrence of a hazardous event. There are several elements to consider in this analysis:

- **The frequency of exposure.** This is probably the most important single element of probability. As an example, a rabbit crossing a

Table 12-1. Qualitative Hazard Categories and Their Effects

Hazard Rating/Severity of Harm	Effects of Exposure
Catastrophic	Death or seriously debilitating long-term injury, such as multiple amputation, coma, or permanent confinement
Critical/Serious	Permanent and nonreversible injury that significantly impacts the enjoyment of life and which may require continued medical treatment
Marginal/ Moderate	Permanent and nonreversible minor injury that does not significantly impact the enjoyment of life, or a reversible injury—either of which requires medical treatment
Slight	Reversible injury requiring only simple medical treatment with no confinement

busy highway (the hazard) has a much better chance of survival if he does so rather infrequently, rather than several times per hour.

- **The motivation to be exposed to the hazard.** This often-underestimated element is strongly determined by the tasks to be performed, and is the basis for determining foreseeable use and misuse of the machine during its life cycle. For every hazard, the question must be, does anyone have a reason to be exposed to it—and to what extent?

- **The possibility of avoiding exposure.** This element, present in the European models, is dependent on the task being performed, the speed at which the hazard presents itself, and where bodily parts are placed. It also may depend upon the skill and perception of persons associated with the machine—behavioral factors that are difficult to define especially when the history is sketchy.

- **The duration of exposure.** The probability of a hazardous event is determined, in part, by the *amount of time* a person is exposed to a hazardous situation. As an example, the rabbit running quickly across the busy highway has a much lower risk of being struck than the much slower possum, or the even slower turtle. (The turtle, however, uses PPE to compensate somewhat for his lack of good sense with respect to busy highways).

- **The extent of exposure.** This element, explored in the UK BS5304 standard, presumes that less bodily exposure lowers the potential of the hazardous event. This is true, to some extent, but the minimized exposure also may cause a modification of the hazard by reducing the severity of injury that may result from the hazardous event. As an example, the person with full bodily exposure in a press is far less likely to avoid a malfunction of the press and will also sustain a more serious injury than a person hand-feeding parts to the same press.

Of all the elements, *frequency* is the easiest to identify, and the most significant. But in models in which this element is used exclusively in lieu of *probability,* such as that shown, the designer must recognize that the other elements, particularly motivation and avoidance, can modify the result.

In the following matrix, used for determining the level of risk, frequency is denoted as:

- High or frequent—greater than once per shift.
- Occasional—less than once per shift, but greater than once per month.
- Remote—less than once per month, but likely within the year.
- Infrequent—so unlikely that it can be assumed that occurrence may not be experienced during the life of the machine. The military generally looks at a probability of one in a million as being in this category.

Establish a Level of Risk

Using the severity of consequence and the probability of occurrence, the level of risk can be determined with some degree of accuracy using a model. The designer, however, must keep in mind that this is only a model, which takes into account subjective inputs and gives a subjective output, which, in turn, must be interpreted and modified by the subjective application of "non-safety-related" items such as:

- Public perception
- Productivity
- Economic costs—direct and indirect
- Politics

Over the years, many "models" have been developed to assess risk levels based on an analysis of the elements of risk. British Standard BS

5304 uses an interesting "Lego block" model that determines level of risk by assigning values to the various elements and using the resultant sum of the values. The most popularized model is derived from EN 954, a European Community standard for electrical component performance requirements. This model determines risk using a decision-tree method. Most domestic users are anxiously awaiting the release of ANSI B11.TR3, a technical reference that will provide guidance and harmony with EN 1050, the European Normative on "risk assessment."

It appears that the *most model* for domestic use has evolved from Mil-Std-882C, as shown in Figure 12-3. This matrix allows the risk level to be determined using the two basic inputs, and, for all intents and purposes, establishes a "tolerable risk" as a condition in which either the *severity of consequence* is negligible, or the *probability of occurrence* is so remote as to be considered nearly zero. (The military considers this, in general, to be a likelihood of less than one in a million.)

Three similar models are shown in Figure 12-3: one assumes that harm is not avoidable; one assumes that it is possible to avoid harm, and the last assumes that there is a high probability of avoiding harm. Note the differences in the outcome. These choices are shown only because, under the circumstances, the designer and users are the same person. In most instances, however, the assumption that harm is not avoidable under normal circumstances, is the only position that the cautious designer can take. In most cases, the designer has no control over the user's training, perception, and awareness, the environment, or maintenance of the system.

Apply the Hierarchy of Safety Measures

If the application of the risk model indicates that the risk is acceptable, and the results can be validated, the job is complete. If the risk is not acceptable, it is necessary to apply safeguarding measures in accordance with the safety hierarchy—in this order of preference:

- Eliminate the hazard.
- Modify the hazard to reduce the energy or toxicity.
- Safeguard the hazard using a guard (barrier) or safety device.
- Warn of the hazard.

The purpose of safeguarding measures is, simply, to reduce the risk to an acceptable level. The application of guards, safety devices, or warnings will do nothing to modify the severity of consequence—only the probability of occurrence.

Figure 12-3. Matrices Showing Difference in Risk Outcome Using Avoidability

Not possible to avoid or limit harm

Frequency and Duration of Exposure	Severity of Harm			
	Catastrophic	Critical	Marginal	Negligible
High	Level A	Level A	Level C	Level D
Occasional	Level A	Level B	Level D	Level E
Remote	Level A	Level C	Level D	Level E
Improbable	Level B	Level D	Level E	Level E

Some possibility to avoid or limit harm

Frequency and Duration of Exposure	Severity of Harm			
	Catastrophic	Critical	Marginal	Negligible
High	Level A	Level A	Level C	Level D
Occasional	Level A	Level B	Level D	Level E
Remote	Level B	Level D	Level E	Level E
Improbable	Level B	Level D	Level E	Level E

High possibility to avoid or limit harm

Frequency and Duration of Exposure	Severity of Harm			
	Catastrophic	Critical	Marginal	Negligible
High	Level A	Level B	Level C	Level E
Occasional	Level A	Level C	Level D	Level E
Remote	Level B	Level D	Level E	Level E
Improbable	Level C	Level D	Level E	Level E

Application of Safeguards

The determination of a safeguarding method depends on the nature of the operation. The *safety hierarchy* tells us that physical guards are the logical first choice for guarding many hazards, but both electronic and mechanical devices may also serve in this regard.

Keep in mind that the purpose of the safeguard is to reduce the risk posed by a hazard to an acceptable level, and that it does so by reducing the probability of a hazardous event. This means that

> The greater the initial risk (or the more reduction of risk required) the greater the required *level of performance* that must be demanded from the safeguard, and, where applicable, the machine safety control system. As an example, where a barrier is selected as the safeguard, a low-level risk may require only a simple post and chain, whereas a high-level risk may dictate a locked steel guard secured with special fasteners and protected with cross-checked, dual-channel, specially keyed interlock switches. In other words, the selection of the safeguard performance level should be commensurate with the risk— to avoid wasting resources.

Table 12-2 is an example of how risk can be addressed with appropriate safeguards.

Are Additional Hazards Created by the Safeguarding Measures? There is a simple axiom in the safety community: "The safeguard, in itself, shall not pose a hazard." The risk assessment procedure must be repeated after the application of safeguards to ensure that any introduced hazards are addressed.

Are All the Risks "Tolerable" Yet? After application of all protective measures at the design stage, the residual risk must be assessed. If the risk is still not acceptable after all practicable measures have been taken at the design stage, the designer must effectively communicate the nature of the residual risk to the user. The designer can assist the user to provide additional safeguarding or take alternative protective measures to ensure that the residual risk is acceptable. Such measures include the following:

- Additional safeguarding specific to the use
- Additional warnings
- Training and safe work practice
- Personal protective equipment

If you were to take everything that has been said in this chapter and distill it into one sentence, that statement would be: "Design the machine so that your loved one, given the current status of training and communication, could do the tasks required during the life cycle of the machine, and still be assured of coming home every night without injury."

Table 12-2. Appropriate Safeguards to Address Risk

Risk Level From Chart	Suggested Performance Criteria
E Negligible	No safeguarding required
D Low	• A barrier that provides awareness and minimal protection against inadvertent exposure to the hazard. Examples are a post and rope, swing-away shield, or movable screen. • Electrical, electronic, hydraulic, or pneumatic safeguards that utilize industry-grade components in a single-channel configuration. • Accompanying signage to use "NOTICE" or "CAUTION" legend.
C Moderate	• A barrier that provides substantial protection against inadvertent exposure. Examples are a fixed screen or railing, lathe chuck guard, movable barrier with interlocking pieces or manifest evidence of misuse. • Electrical, electronic, hydraulic, or pneumatic safeguards that utilize single-channel configuration with a high safety factor, or simple redundancy—either of which can be checked for proper operation. • Accompanying signage to use "CAUTION" legend.
B High	• A barrier that provides substantial protection against intentional exposure; if movable, interlocked per system criteria. Examples are a full barrier guard not readily adjustable or removable. • Electrical, electronic, hydraulic or pneumatic safeguards and control systems that utilize redundant (dual-channel) configuration using safety-rated components—with self-checking on startup or cycle. • Accompanying signage to use "WARNING" or "DANGER" Legend.
A Very High	• A barrier that provides substantial guarding, affixed with special fasteners or lock; if movable, interlocked per system • Electrical, electronic, hydraulic, or pneumatic safeguards and control systems that utilize continuous self-checking to ensure continuance of redundancy. • Accompanying signage to use "DANGER" or "EXTREME DANGER" legend.

SUGGESTED FURTHER READING

Andres, R. N. (June 1997). *Machine Safety–A Global Approach.* Proceedings of the 36th Annual Professional Development Conference, American Society of Safety Engineers, Des Plaines, IL.

Department of Defense. Military Standard System Safety Program Requirements (MIL-STD-882C). Washington, DC: Department of Defense.

Manuele, F.A. (1993). *On the Practice of Safety.* New York: Van-Nostrand Reinhold.

Pilz, GmbH and Co. (March 1996). *A Guide to the Machinery Safety Standards, Vol. 1, 5th Edition.* Northants, England: Pilz GmbH and Co.

Scientific Technologies, Inc. (1997). *Safebook 2.* Publication 44512-0350. Fremont, CA: Scientific Technologies, Inc.

U.S. Department of Labor–OSHA (1996). Concepts and Techniques of Machine Safeguarding. Publication 3067. Washington, DC: U..S. Government Printing Office.

Staffing and Developing the Safety Function

Brett Carruthers

The true leader inspires in others self-trust,
guiding their eyes to the spirit, the goal.
— Bronson Alcott

INTRODUCTION

Congratulations, you are now responsible for corporate safety of a multidivisional, multilocation chemical manufacturer. Or maybe you lead the safety efforts of a regional retailer. Or perhaps you are responsible for safety at a single manufacturing location of precision-machined goods. The staffing and developing of the safety function in each of these cases will vary significantly; however, the approach and strategy to this end is similar.

Each situation for staffing and developing safety professionals is like a glove: no "one size fits all." In the ideal world, each position would be filled by a degreed and experienced safety professional who has earned the Certified Safety Professional (CSP) designation. Now back to reality. The individuals filling the various safety roles within an organization will have diverse backgrounds, education levels, and experiences. This is where the challenges begin.

Fulfilling your responsibility to manage safety is no easy task. Along with leading the organization's safety efforts, the challenge to nurture, develop, promote, and staff a team of professionals presents additional challenges. Developing and implementing a successful plan must be carefully devised. This plan cannot be implemented until a thorough study of the organization's landscape is undertaken and understood. The following items must be examined and comprehended:

- Organization philosophy
- Organization goals and objectives
- Strategic initiative

- Organization culture and value system
- Current position holders' strengths and weaknesses

Each of these factors will influence how positions are staffed and how the individuals are developed. Another factor that will impact staffing is "high-profile" or "high-risk" facilities. Plant managers at these locations will typically demand individuals who are more experienced and educated. They may also be less flexible when it comes to moving personnel to other positions.

In determining development needs, search first for common ground. What competencies must all individuals possess? Do the incumbents have these competencies? After this common ground is determined, the critical success factors must be determined. What skill sets and competencies must individuals develop to further themselves to reach the next step in their career progression? This is where most development plans fail. Adequate time is generally not allocated to define fully these critical success factors. Consequently, individuals often lack the necessary skills to be promoted and progress within the organization. This leads to frustration, poor performance, and high turnover rates.

This chapter presents information on 1) safety professional self-assessment and career planning, and 2) corporate safety succession planning and development. First and foremost, self-assessment is a means for safety professionals to assess their strengths against competencies demanded by the safety profession. Once these strengths are assessed and compared to current job demands, a plan to exceed present demands and achieve advancement in one's career can be charted. Corporate safety managers can then utilize these competency assessments as a means for internal staff-succession planning, safety professional development, and career pathing.

THE CHALLENGE

Today's safety professionals are stretched in many directions. Priorities are ever changing. Organizations demand more, accomplished with less. All of this in a business culture in which downsizing, re-engineering, and rightsizing are the buzz words in corporate boardrooms.

Industrial Safety and Hygiene News' "12th Annual White Paper Report on U.S. Industry and Health Practices" (Kohn, Timmons, and Bisesi 1995) presents a picture supporting the stress experienced by many safety professionals. Here are several highlights of this report:

- Only 9 percent of respondents reported Environmental Health and Safety (EHS) staff expansions in 1996—down from 19 percent in 1995 and 14 percent in 1994.

- Seventy-nine percent of respondents reported that EHS budgets were maintained or increased—down from 94 percent in 1991 and 90 percent in 1992.

- Eighteen percent of respondents reported that they would look for another job in 1996—up from 5 percent in 1993 and 14 percent in 1994.

Bill Rhodes, a recruiter with Bench International, Ltd., stated, "There are just fewer chairs for the players to sit on" (Kohn, Timmons, and Bisesi 1995, 8) He sees no end to industry's downsizing fervor. Another recruiter stated, "I hate to say it, but a lot of safety and industrial hygiene people will have to find something else to do" (Kohn, Timmons, and Bisesi 1995, 8).

Finally, *Industrial Safety and Hygiene News'* survey revealed an increase in worries regarding job security among safety and industrial hygiene professionals. Twenty-four percent of survey respondents stated that maintaining their job was a major concern—up from 20 percent in 1995 (Kohn, Timmons, and Bisesi 1995). Is the glass half empty or half full? The optimists look to the future with enthusiasm and challenge. The pessimists? . . .

MANAGING YOUR DESTINY

Taking control of your career begins by taking a "hard look in the mirror"—taking personal responsibility for your future. This is best summed up by Jeff Vincoli, safety and health manager at McDonnell-Douglas, in a recent *Industrial Safety and Hygiene News* article: "If my destiny is in question, I'm going to be in charge of it" (Burke and Johnson 1995, 24). This is further reinforced in the same article by personnel recruiter Dan Brockman, who stated, "Your job security is in your skills, not with any one employer."

Times have changed. Safety professionals must be aggressive in the struggle for survival in today's competitive job market. A safety manager who is departing industry to become a consultant says, "I hate to say it, but it's almost every man for himself in safety these days" (Burke and Johnson 1995, 24). Larry Hansen, a safety consultant, has a few valuable pieces of advice:

It's time for aggressive self-study and self-development. Go back and take a college course in finance; read books by the leaders of the quality movement. You've got to take the initiative. The careers people have today will change more quickly than you think. Forget planning for the next five to ten years; think in one to three year cycles. Changes are happening that fast (Burke and Johnson 1995, 24).

Another key to improving your destiny is through certification. In the safety and health professions, the two most commonly awarded certifications are the Certified Industrial Hygienist (CIH) and the Certified Safety Professional (CSP). Certifications are beginning to become the standard for midrange career positions (five to fifteen years of experience) and are almost a requirement for senior positions (more than fifteen).

Certification is an important building block that carries a long-term personal commitment. Both the CIH and CSP certifications carry ongoing maintenance requirements (continued personal development) to maintain the certification. Again, Vincoli sums up the importance of professional certifications, "certification is a career building block, not the end all and be all" (Burke and Johnson 1995, 24).

Understanding business and how one's organization does business is another key to controlling one's destiny. Does this mean earning a Masters in Business Administration? No. But strong business awareness is certainly a strength that begins to separate the "cream of the crop." Again, Hansen provides valuable insight to safety professionals, "You must have a financial orientation, and understand how safety and health relate to the business mission" (Burke and Johnson 1995, 24).

Understanding human nature and human behavior are two skills that can be intangibles in controlling one's destiny. Dale Carnegie consultant Marita Mugler recommends continued development of human relations skills. Mugler states, "While you're at it, work on your attitude. Knowledge is your ticket into the game, but without a strong attitude—enthusiasm, determination, commitment, optimism—you won't survive" (Burke and Johnson 1995, 25).

Finally—read, read, read, and read some more. Look beyond books on the profession. Staying abreast technically is important; however, expand your reading list to such publications as the *Wall Street Journal, Forbes,* and other similar business publications. (See Appendix A for a list of books that savvy safety professionals are reading).

DEFINING SAFETY PROFESSIONAL COMPETENCIES

To control one's destiny, the competencies needed for one's "toolbox" must be defined. Defining the "tools" needed for success as a safety professional are constantly changing. These tools will vary based on the safety professional's education, experience, and responsibilities. In addition, the safety professional's position and organizational culture and values will influence the skills that must be developed.

A survey of safety professionals conducted by Indiana State University (ISU) (Kohn, Timmons, and Bisesi 1991) helps to define safety professional competencies. Two key questions were asked by this survey that help define these competencies:

- How do you rate the proficiencies most important to health and safety program success?
- Which areas need to be developed in order to achieve career advancement?

The five most important skills necessary for safety program success were as follows:

1) Hazard recognition
2) Verbal communications
3) Written communications
4) Safety training
5) Management ability

The six areas where development was needed in order for career advancement were as follows:

1) Management ability
2) Computer science
3) Industrial hygiene
4) Ergonomics
5) Hazardous materials
6) Fire Science

A recently completed and published survey by Lon R. Ferguson (1995), of Indiana University of Pennsylvania, focused on the "appropriateness of major content topics in baccalaureate safety curricula." While these may, on the surface, appear to be unrelated, the following survey question results help define safety professional competencies:

1) Content topics based on perceived importance
2) Content topics based on use in current job
3) New knowledge or competencies required of safety professionals in the future

For, "content topics based on perceived importance," the following were in the top 50 percent of responses:

- Verbal communications
- Accident causation and investigation
- Written communications
- Measuring safety performance
- Safety and health regulations
- Safety management
- Safety training
- Environmental safety and health
- Design for engineering hazard control
- Ergonomics
- Computer applications
- Industrial hygiene
- Ethics
- Fire safety
- Risk management
- Behavioral safety
- Hazardous materials

For, "content topics based on use in current job," the following were in the top 50 percent of responses:

- Verbal communications
- Written communications
- Safety and health regulations
- Safety management
- Environmental safety and health
- Computer applications
- Behavioral aspects of safety
- Measurement of safety performance
- Accident causation and investigation
- Workers compensation
- Recordkeeping
- Ergonomics
- Safety training
- Hazardous materials
- Risk management
- Industrial hygiene
- Fire safety

For "new knowledge or competencies required of safety professionals in the future," these were the responses:

- Human behavior and performance
- Computer applications
- Total quality management
- International safety practices
- Learning theories/training
- Financial aspects of safety
- Task/process analysis
- CSP preparation
- Risk management
- Salesmanship
- Diversity in the workplace
- Conflict management
- Ergonomics

Although this latest survey offers more subjects than the ISU survey, there are continued parallels. In addition, several of the key points discussed in the previous section on managing destiny are supported by the results of Ferguson's (1995) survey.

SKILL/KNOWLEDGE ASSESSMENT

When skill and knowledge levels are assessed, different levels of ability will result. By prioritizing competencies that need development and establishing a plan of action focusing on achievable goals, safety professionals can improve their personal and organizational value.

In rating personal competencies, two measures must be taken: 1) priority (as it relates to current job position), and 2) development level (of an individual).

Priority

The first competency rating must be based on the importance to the current job position. This can be based on an "A, B, C" or similar rating. This A, B, C rating system translates as follows:

A = essential to job/position success
B = moderately essential
C = nonessential

Development Level

The development level rating assesses the safety professional's current level of development. Each individual will have varying competency levels.

These levels of development can have various titles, but cover the spectrum from novice to experienced master. They should be evaluated less by the time an individual has been working at a particular competency than by the depth and breadth of an individual's demonstrated knowledge and ability. These are the four development levels used for this portion of the assessment:

1) Novice
2) Bright upstart
3) Prominent prospect
4) Experienced master

Novice. A person is probably at the novice level in a competency if he or she

- Has not had the task responsibility long
- Has little prior task specific experience
- Seldom anticipates problems
- Does not improve faster by trying harder

Bright Upstart. A person is probably at the bright upstart level in a competency if he or she

- Has had the task responsibility for less than a year
- Has some task-specific experience
- Occasionally anticipates problems
- Improves gradually by trying harder

Prominent Prospect. A person is probably at the prominent prospect level in a competency if he or she

- Has had the task responsibility for at least one year
- Has had at least nine months of task-specific experience
- Frequently anticipates problems
- Improves immediately by trying harder

Experienced Master. A person is probably at the experienced master level in a competency if he or she

- Has had the task responsibility for two or more years
- Has more than one year of task-specific experience
- Consistently anticipates problems
- Has little room left for additional improvement

ASSESSMENT OF COMPETENCIES

The following is a solid list of competencies that safety professionals can use to assess themselves, or that can be used as a measurement tool for various corporate planning purposes. The list can be further expanded or reduced as appropriate.

Interaction competencies:

- Oral/verbal communication
- Written communication
- Influence and persuasion skills
- Meeting leadership skills

- Human relations
- Resource planning
- Training skills
- Leadership

Administrative competencies:

- Economics/budget management
- Workers' compensation management
- Planning and objective setting
- Problem-solving skills
- Time management

- Computer skills
- Claims management
- Organization skills
- Recordkeeping

Technical competencies:

- Practical safety law
- Regulatory requirements and interpretation
- Safety procedures
- Security management
- Boiler and machinery
- Chemical information and interpretation
- Occupational health
- Preventive maintenance
- Construction safety
- Hazardous material management

- Environmental
- Ergonomics
- Process safety
- Industrial hygiene
- Behavioral safety
- Fire safety
- Product safety
- Property protection
- Emergency response

Now comes the challenge. From the assessment, prioritize the competencies that have the greatest impact on your current job situation. Focus on two to four development areas where you will get the "biggest bang for the buck." Develop an action plan that is objective and uses a variety of conventional and nonconventional methods to meet the desired outcomes. Here are some examples of conventional development methods:

- College courses; company-sponsored training; correspondence courses
- Outside seminars
- On-the-job training
- Reading

Here are some examples of nonconventional development methods:

- Computer learning—distance learning
- Observing and modeling
- Researching
- Practicing
- Consulting

Safety professionals have much to gain and even more "to lose" by not planning for the future. Chart a course for tomorrow and follow the winds of success to better opportunities and personal growth.

CONCLUSION

Safety professionals must assume control of their testing and development. There are many avenues and means to pursue personal growth, select the ones that are available and fit your goals and needs. Your investment in self-development will pay significant dividends.

Industrial Safety and Hygiene News' Savvy Safety Professional's Reading List

- Stephen R. Covey's books.
- W. Edwards Deming's books.
- Alvin Toffler's books.
- B. F. Skinner's *Walden Two,* MacMillan Publishing Co., 1948; and *Beyond Freedom and Dignity,* Alfred Knopf, 1971.
- *The Art of the Long View,* by Peter Schwartz, Doubleday, 1991.
- *Being Digital,* by Nicholas Negroponte, Alfred Knopf, 1995.
- *The Care of Souls,* by Thomas More, Harper Collins, 1992.
- *Empires of the Mind,* by Denis Waitley, William Morrow, 1995.
- *Images of Organization,* by Gareth Morgan, Sage Publications, 1986.
- *In the Age of the Smart Machine: The Future of Work and Power,* by Shoshana Zuboff, Basic Books, 1988.
- *Job Shift,* by William Bridges, Addison Wesley, 1994.
- *Leadership and the New Science,* by Margaret Wheatley, Berrett-Koehler, 1992.
- *Liberation Management: Necessary Disorganization for the Nano-second 90's,* by Tom Peters, Ballatine Books, 1994.
- *The New Paradigm in Business,* edited by Michael Ray and Alan Rinzler. Jeremy Tarcher, publisher, 1993.
- *The Portable MBA Series,* Wylie Publishers, 1990–1994.
- *Workplace 2000,* by Joesph H. Boyett and Henry P. Conn, Dutton, 1991.
- *The World in 2020,* by Hamish McRae, Harper Collins, 1995.

Source: Burke and Johnson (1995, 26).

REFERENCES

Burke, A. and D. Johnson (October 1995). "What Does the Future Hold for Careers in Safety and Health?" *Industrial Safety and Hygiene News* 29:23-28.

Ferguson, L. R. (November 1995). "Baccalaureate Safety Curricula: A Survey of Practitioners." *Professional Safety* 40:44-48.

Kohn, J. P., D. L. Timmons, and M. Bisesi (January 1991). "Occupational Health and Safety Professionals: Who Are We? What Do We Do?" *Professional Safety* 35:24-28.

————. (November 1995). "12th Annual White Paper Report on U.S. Industry Safety and Health Practices." *Industrial Safety and Hygiene News* 29:1-34.

Structuring Resource Protection Technology Strategy

Anthony Veltri

INTRODUCTION

A company's ability to perform competitively tends to be based on protecting and using organizational resources productively. One way companies can improve business performance is to rethink the manner in which they have organizationally structured a strategy for resource protection technology. These functions are commonly identified as environmental affairs, safety and industrial hygiene, fire prevention, security protection, emergency/disaster preparedness, and risk/loss control management. Currently, senior-level executives, responsible for positioning functions within the organizational chart of the company, are not taking full advantage of the strategic and technical tools developed by these functions.

Many resource protection functions are organizationally misarranged, inadequately grouped, and out of position, making them unable to contribute effectively to business performance. As a result, it becomes extremely difficult for managers of these functions to leverage their management strategy and technologies in ways that go beyond traditional concerns for regulatory compliance to more modern concerns for enhancing eco-safe business performance. However, among some senior-level executives, there is a growing interest in the role these functions play in helping companies to perform competitively. This interest is occurring primarily at two business levels. At the executive level, conceptual interest in a company-wide structure that 1) prepares, protects, and preserves organizational resources, 2) enhances compliance with regulatory requirements, and 3) uses resources in ways that create added value is economically appealing. At the operating level, the practical notion of a structure that assists design and process engineers in 1) assessing and counteracting risk to resource problems at an early stage in the process life cycle to prevent later risk burdens, and 2) profiling cost burdens and profitability potential associated with their products/technologies/ processes sends a powerful message that they are a main source for improving a company's competitive performance.

The purpose of this chapter is to present conceptual guidelines and key benefits for structuring a resource protection technology strategy into a single source organizational department. The intent is not to take away the autonomy created by directors and managers of these functions. Rather, the intent is how to group directors and managers in ways that senior-level executives can leverage the collective competencies and capabilities of these functions to mount an aggressive and sustained attack on the risks to resource problems that confront the company, while adding competitive value and maintaining compliance with regulatory standards.

STRUCTURING RESOURCE PROTECTION TECHNOLOGY STRATEGY

Recent efforts in structuring resource protection technology strategy have been focused on enhancing structural fit (that is, assuring congruency between the functions strategic intent, mission, initiatives, and the long-term business and regulatory performance standards of the company). Any modern attempt at structuring a resource protection technology function should begin with a well-formulated strategy.

The following approach to devising strategy and structure is offered (see Figure 14-1) because it enables senior-level executives and resource protection technology specialists to think in a systematic way during the strategy formulation and structuring process while avoiding duplication and misallocation of resources. This method is organized around a two-stage procedure. Stage 1 concentrates on developing a strategic intent, mission statement, and initiatives that set the direction necessary for guiding the performance of the function. Stage 2 focuses on constructing an organizational structure that welds the resource protection technology strategy in a manner that gains broad consensus and support throughout the organization.

Figure 14-1. Stages and Steps for Devising Resource Protection Technology Strategy and Structure

---

Stage 1. Strategy Formulation

Steps   1. Review corporate competitive performance strategy

2. Envision a strategic intent

3. Construct a statement of mission

4. Identify initiatives (developmental and reform)

5. Develop projects that support the initiatives

Stage 2. Organization Structure

Steps   1. Concentrate on assuring a structural fit

2. Determine a logical process for taking direction

3. Assure appropriate positioning arrangement within company chart

4. Think of ways of to integrate vertically and laterally

Stage 1: Strategy Formulation

The first step in devising strategy for the resource protection technologies function is to study the company's business plan or competitive performance strategy and determine ways to assure congruency between the resource protection technology strategy and the company's business strategy. Assuring congruency sends the correct signal to senior-level executives: that the resource protection technology strategy is in compliance with the company's business performance needs and expectations. Attention should be focused on identifying those statements appearing in the company's business road map that indicate the need for a strategy to sustain business performance, to protect organizational resources, and to use resources productively. This type of study provides the greatest opportunity for the resource protection technology strategy to justify its organizational existence.

The second step focuses on developing a strategic intent (vision statement). Strategy formulation needs to begin with a strategic intent (vision) of what the desired future for the company's resource protection technology strategy should be. Strategic intent basically means the envisioning of building and sustaining a unique set of competencies and capabilities that enhance an organization's ability to sustain long-term business performance. The vision statement can be developed by requesting the senior-level executive team and a sample of organizational stakeholders to envision what they would like the resource protection technologies function to become. It seems reasonable to believe that a member of the senior-level executive team and a few organizational stakeholders, based on their firsthand experiences, their secondhand information, or their own creative insights, would possess information or well-founded beliefs about a vision for the company's resource protection technology strategy. The challenge is to effectively tap this store of creativity, knowledge, and judgment from these individuals. Another way of developing the strategic intent is to base the vision statement on the technical knowledge, experience, and intuition of the resource protection technology specialists. Specialists keep abreast of research developments in the field and could recommend cutting-edge strategy and tactics to pursue. Both methods are usually products of information gathering that can be performed by interview or by survey-type techniques. However, these are the criteria that must be used to judge the value of the information:

1) The perceived degree for sustaining the business performance of the company
2) The protection and use of resources productively
3) The impact on cost, manufacturing yield, and logistics
4) The enhancement of compliance with regulatory requirements

The third step in the strategy formulation process is to construct the mission statement. A mission statement describes the ambitious long-term strategic purpose of the function and is intended to guide decision making and operating-action capabilities of all stakeholders. The statement of mission is the cornerstone on which initiatives (developmental/reform) are developed and the organizational structure is constructed. The basic question that should be answered in order to develop a resource protection technologies mission statement is, "What is our strategic purpose?" or, "What should our purpose be?" The following competitive business performance concept is provided to help guide the process of developing a mission statement: "Long-term economic and regulatory performance within the enterprise derived from cost-effective use of resources" is recognized as being applicable to any business. Such a statement has an overarching relationship to the various core strategic business units and operating functions, and serves as a guideline statement for the resource protection technologies function to develop a statement of mission that is congruent with the intent of the firm's competitive performance strategy. For instance, " Optimal preparation, protection, and preservation of enterprise resources" is an example of a mission statement that describes the congruency between the function's mission and the firm's competitive performance strategy. The following rationale illustrates this congruency.

A firm's ability to compete and stage for long-term business sustainment depends on three factors: first, the attractiveness of the industry in which it is located; second, establishing a competitive advantage over competing industries; and third, using assets and resources effectively. The resource protection technology function should concentrate on the third factor. Because assets and resources are used to compete and stage for long-term business sustainment, they need to be prepared for risk, protected against danger, and preserved from loss. The resource protection function can play the lead role in preparing the organization's resources and must possess a statement of mission for guiding decision-making and operating-action capabilities in this important area of business concern. History has proven repeatedly that designing protection strategies at an early stage in the product/process life cycle tends to prevent future problems, roadblocks, and expenses.

The fourth step is to identify initiatives (development and reform) that support the accomplishment of the mission. You identify initiations by 1) reviewing results from assessment phase (that is, analyzing the gaps between strategies); 2) selecting developmental and reform initiatives for closing the gaps; and 3) determining specific projects that related to each initiative. This step is principally based on the results extrapolated from the assessment phase and is intended to close the gaps between an organization's current resource protection technology strategy and what it should be.

Closing the gap between an organization's current strategy and its future strategy requires perceiving and structuring a set of initiatives—developmental and reform—and mobilizing resources and commitment to those initiatives. This requires making a series of coordinated decisions that transform the way the organization thinks and performs relative to resource protection. These initiatives are examples of ways of shaping a firm's sense of direction and creating the architecture needed for preparing, protecting, and preserving company resources. The following initiatives are examples of such decisions:

- Initiative I: Rethink how resource protection technology strategy is formulated and directed.
- Initiative II: Reconstruct the organizational structure of the function with emphasis on 1) integrating vertically and laterally and 2) changing the positioning location arrangement within the company's organization chart.
- Initiative II: Modify the financial tools used in 1) profiling activities that drive environmental, safety and health, fire and disaster costs; and 2) determining the fiscal allocations needed for preparing the organization to deal effectively with risks to resource problems.
- Initiative IV: Improve the process for influencing and preparing senior level executives and design and process engineers to take part in resource protection technology activities.
- Initiative V: Rethink and revise the technical tools, competencies, and capabilities related to emergency preparedness; mitigation; response; business recovery and resumption; risk-ranking methods; environmental, safety, and health costing; up-front process design; mass/energy balancing; and life-cycle analysis; and find ways to offset existing compliance costs through innovation.
- Initiative VI: Refine methods for measuring, evaluating, and benchmarking environmental, safety, and health management performance.
- Initiative VII: Consider developing and installing management information systems to improve decision-making and operating-action capabilities and for setting priorities concerning environmental, safety and health, security and fire protection, emergency/disaster performance, and risk and loss management issues.
- Initiative VIII: Consider establishing the means for designing and conducting research on resource protection technology operations.

Stage 2: Organizational Structure

There are a variety of ways to structure resource protection-type functions. However, each approach should be influenced by its strategy and how other functions/departments need to integrate with the resource protection function. Any company's resource protection function should reflect its concern for maintaining compliance with regulatory requirements and in creating added value to the company's business performance. However, resource protection technology functions tend to evolve and be structured in unpredictable ways, and sometimes are thwarted in being an asset to the company due to their business strategy and/or in the organizational structure of the company. The purpose of organizing is to weld the resource protection technology strategy and structure together in a manner that gains broad consensus and support throughout the organization. The approach used for structuring strategy and selecting the positioning arrangement for the function within the organizational chart of the company tends to fall into one of four developmental levels, characterized as follows:

1) **Level 1: Reactive.** No formal organizational structure exists. The company relies primarily on the services of a small group of environmental, safety, industrial hygiene, fire and emergency management specialists, and collateral duty personnel positioned in various departments to provide direction.

2) **Level 2: Adaptive.** The structure is organized around a small group of environmental, safety, industrial hygiene, security and fire protection, emergency management, and risk and loss management specialists. This group is positioned and dispersed in the organization's core business units (that is, manufacturing, quality control, transportation, engineering, and maintenance) to provide technical advice on resource protection efforts. The function takes its direction from core business unit executives and from recommendations of external consultants.

3) **Level 3: Active.** A formal organizational structure exists. The function reports and takes direction from a senior-level executive and is composed of a team of specialists who create and fulfill needs and expectations, develop policies, and provide strategic and technical solutions regarding ways to protect and use resources productively.

4) **Level 4: Dynamic.** The approach is to integrate efforts internally (vertically and horizontally) within the overall organizational chart and to integrate externally with outside strategic alliances and agencies. Functions at this level are structured in the business strategy process of the firm and are organized to remain constantly tuned into the needs of both internal and external constituents while providing

strategic management and technology-transfer services. The function takes its direction from research findings, audits, special assessments, studies, and task forces organized by senior-level executives.

Recent efforts in structuring management strategy have been focused on structural fit (that is, assuring congruency between the strategic intent, mission statement, initiatives, and the organizational structure) (see Figure 14-2). As previously mentioned in the strategy formulation phase, any modern attempt at structuring resource protection technology strategy should begin with a strategic intent (vision) statement that describes the desired future for the function's performance. Next, conditions must be arranged to close the gap, between the current performance and desired performance, by identifying initiatives that focus efforts on a mission statement. Then, an organizational chart can be drawn for providing a pictorial representation.

Constructing the organizational structure after formulating the strategy tends to enhance structural fit. Some of the worst mistakes in designing for structural fit have been made by imposing a concept of an ideal resource protection technology function without attending initially to the strategy formulation process. When these structural fit concepts are mismatched, and when the wrong strategy is formulated, the organizational structure will tend to be ineffective and inefficient. Although there is no universally agreed-upon framework for structuring resource protection technology strategy, the following organizing principle should be used: *structure follows strategy*. That is, strategy determines structure. Structural fitness, therefore, is the glue for bonding structure and strategy.

Positioning Arrangement

A dilemma commonly encountered by senior-level executives is determining the optimal organizational location for positioning the environmental, safety and industrial hygiene, security and fire protection, emergency/disaster preparedness, and risk and loss management functions within the overall organizational chart of the firm. This task is particularly difficult when one discovers that 1) there is no organizational location that is universally accepted and sufficient for positioning these functions, and 2) that there are no research studies or conceptual models to review that could facilitate decision making in this area. In general, there is a lack of literature in this area; however, the use of the seven positioning principles presented

Figure 14-2. Concept of Structural Fit: An Application to Resource Protection Technologies Management Structure

(Organizing Loop)

STRATEGIC INTENT ....?

Describes the future state (vision) for the organization's resource protection technology function.

as ultimately desired accomplishment.

MISSION STATEMENT ....?

Describes the ambitious long-term strategic purpose of the function.

for guiding decision-making and operating actions.

INITIATIVES ....?

Describes ways and means for closing the gap between the organization's current level of performance and its desired level.

for transforming the function and the organization.

ORGANIZATIONAL STRUCTURE ....?

Provides a pictorial representation of what the function looks like, how the flow of executive action and communication takes place and how individuals are welded together in a common mission, while allowing various individuals and groups to focus their attention and expertise on a specific structure of strategies and tactics.

for coordination of work.

REPLICATE MODEL WITH BUSINESS UNITS/FUNCTIONS ....?

for integrating decision-making and operating actions.

(Reorganizing Loop)

subsequently may help guide resource protection specialists to promote to senior-level executives an optimal organizational location. Based on numerous discussions with senior-level executives; management consulting firms; environmental, safety, and health managers; and university professors, I have found that improved organizational positioning arrangements tend to be influenced by

1) The actual degree of contribution a function makes to the competitive and regulatory performance standards of the firm

2) The perceived capability of the function's formulated strategy to contribute to the long-term competitive and regulatory performance standards of the firm

3) The strategic fit of the function's formulated strategy with the competitive performance strategy of the firm

4) The capability of the function's organizational structure to integrate vertically and laterally across all business units within the company; provide technology transfer services; and arrange conditions that facilitate organizational learning

5) The capability of the function to create and sustain a management information system that improves decisionmaking and operating-action capabilities within the company

6) The capability of the function to apply research and development methodology for influencing competitive and regulatory performance

7) The perceived ability of the function to be strategically opportunistic (that is, the ability of the function to be focused on the long-term, while remaining flexible enough to solve day-to-day problems and recognize new opportunities)

These positioning principles are essential to deciding where to locate the resource protection technology function. However, they do not uniquely identify a particular solution. Based on the concept that structure follows strategy and location positioning charting of the function follows structure, Figure 14-3 gives an example of an organizational location and positioning arrangement that should be favorable to the accomplishment of the resource protection technology mission.

SAMPLES OF SYNERGIES AND ALLIANCES

The following are samples of key synergies and alliances that might develop as a result of merging environmental affairs, safety and industrial hygiene, security and fire protection, emergency and disaster preparedness, and risk and loss management into a single source organizational unit:

Figure 14-3. Enterprise Statement of Mission

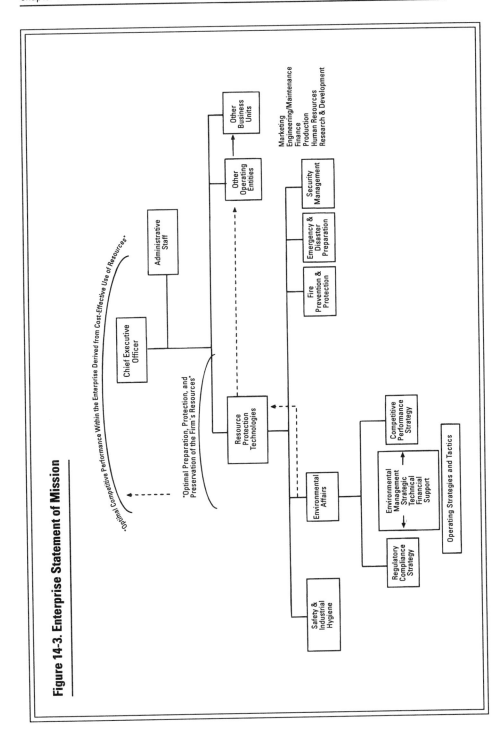

Key Synergies

1) Support of the company's competitive performance strategy by having a company-wide strategy specifically concerned with resource protection, use, and preservation and a company-wide structure in which client departments will no longer have to find their way through the maze of fragmented units to resolve their risk to resource problems.

2) A single source (point of contact) department consisting of an integrated group of multidisciplinary functions that are concerned with the preparation, protection, and preservation of the company's resources, while enhancing compliance with regulatory requirements.

3) Consolidation and coordination of resources for conducting risk to resource research, training, inspections, investigations, assessments, measurements, and evaluations. A variety of cross-functional training opportunities exist for personnel already employed in these functions. Personnel can work together and be more productive in new groupings when they share a common mission, strategy, and structure, and analytical tools for problem solving.

Key Alliances

1) An internal organizational alliance with senior-level executives and design and process engineers to protect and use resources more productively and for leveraging resources in ways that create distinctive competencies and capabilities.

2) An external organizational alliance with regulators, by sending the correct signal of the company's commitment to excel in regulatory requirement matters by having a department that is principally concerned with resource protection and regulatory compliance.

3) A "shield of creditability" against adversaries who might challenge the company's strategic intent of protecting and using resources productively, and an increased likelihood that insurers will minimize the denial of claims due to an inadequate resource preparation, protection, and preservation strategy, structure, and practices.

CONCLUSION

The guidelines in this chapter are essential to deciding where to position resource protection functions. However, they do not uniquely identify a particular solution. I believe that failure to follow these guidelines will only absorb the energies of the function's participants without producing any significant output to the organization.

These guidelines are not intended to "clear the air" for the structural approach I describe, but are intended to make it clear that there is no organizational structure that is universally sufficient for structuring resource protection strategy. The organizational structure ought to accommodate the messy dilemmas resource protection managers face. Therefore, each company should customize its own approach, using the guidelines identified in this chapter as the building blocks. Whatever the chosen organizational approach, constant change in resource protection structure will be necessary to achieve the particular strategy set out to achieve. The positioning structure described in this chapter is useful only for companies in which the strategy formulated drives the organizational structure and positioning arrangements.

Chapter 15 ———————————————————————————

Budgeting for Safety

Mark D. Hansen

INTRODUCTION

Safety, like any other important corporate function, must have and be accountable for a budget. Budgeting is an important part of the planning function and must be completed as early as possible in the planning process to maximize its effectiveness. Before implementation, the budget must be developed and agreed to by all concerned parties.

Budgeting requires the safety professional to identify the costs of doing day-to-day safety. Before you can determine what safety items should be purchased, the current safety equipment and budget should be analyzed. Included in this chapter are the current costs for safety equipment in use, such as safety glasses, prescription safety glasses, goggles, rubber and other boots, rubber and other gloves, acid suits, flame-resistant clothing, hard hats, hearing protection, or emergency medical equipment. Current budget items also include the cost for capital equipment, such as sprinkler systems, fire-fighting equipment, training courses, upgrades identified from insurance audits, and others. Once you have identified where all of your safety expenditures are going, try to determine a more efficient way to spend your budget dollar. For example, some employers use contractors to maintain first aid cabinets at a monthly cost. By ordering the medical supplies by mail and stocking the first aid cabinets yourself (which will increase your visibility), this cost can be minimized.

Next, illustrate control of costs. When you treat your department like a business, your desire to control costs will be apparent. It is always a good idea to plan to finish the project on schedule and under the estimated cost. It is also important to let upper management know that you were able to return some of the allocated funds. If you do not market yourself, nobody else will.

Another way to visibly illustrate cost control is to include non-capital expenditure items in your safety budgeting. Items that include little or no financial support could include working with engineers to help design a safety interlock or a machine guard, or performing field evaluations yourself rather than budgeting extra manpower expense. Coupling an extensive safety budget with such clearly beneficial low-cost actions may help your department avoid being labeled as wasteful or indulgent.

Use the standards to your advantage in budgeting. Once you have identified your costs and you have shown that you are controlling these costs, the next step in safety planning is to tie the budgeted items to an applicable standard. There are standards established for every industry sector, by agencies such as OSHA, MSHA, EPA, ANSI, NFPA, NEC, UL, and the National Electronics Manufacturing Association. There are also industry-specific standards required for the petrochemical industry, set by agencies such as the American Petroleum Institute, the Chemical Manufacturers Association, the Synthetic Organic Chemical Manufacturers Association, and the American Institute of Chemical Engineers. The aerospace industry uses military standards such as MIL-STD-882, MIL-STD-1574, MIL-STD-454, Air Force Regulations, and Air Force Occupational Safety and Health standards.

The standards spell out almost everything required to make facilities safe. Compliance with these standards is the reason that many safety and health managers are hired. Use them!

In summary, to budget responsibly, you must do several things. First, identify costs. Find out where money is being spent. Is it being wasted? Is it accomplishing your goals? Are you getting the best value for you dollar? Once you have answered these questions, take action to illustrate cost control. Illustrate your fiduciary responsibility. This may include changing vendors or rescheduling calibrations for instruments. The bottom line is to do as much as possible with your budget dollar without sacrificing the value you are receiving. Finally, tie budget items to compliance standards. Show that everything you are doing is directly linked to compliance. If upper management elects not to approve the budgeted items, remind them of the specific compliance reference for that item. Make those managers with fiduciary power *accountable* for their decisions: provide a sign-off list that physically requires top management to sign or initial for every budgeted safety item (on the proverbial bottom line) whether to spend the money or accept the risk of not spending the money. Figure 15-1 gives an example of budgeting for Level A suits for HAZMAT.

Figure 15-1. An example of budgeting for Level A suits for HAZMAT

Example:

Budget Item #6. HAZMAT Level A Suits for Emergency Response Team (OSHA 29 CFR 1910.120(H)&(c)(5). $ 2,360.00

Tychem 10000 Level A Suits (4 @ $500.00)	$ 2,000.00
Training Suits (4 @ $90.00)	$ 360.00

In the case of a hazardous chemical spill, company policy is to have a minimum of two Emergency Response Team (ERT) members respond and two to stand by as backup. The Level A suits provide the minimum requirements for responding to such an emergency. The use of training suits provide our ERT members with the opportunity to become familiar with the chemical suits prior to an incident and to minimize panic in an emergency situation. Each ERT member should be trained in these suits and self-contained breathing apparatuses (SCBAs).

Implement as Stated Above: Accept Risk of Criteria 2, Freq 2 = 2:
Concur: HR_Date_MGR_Date_PRES_Date_Concur: HR_Date_MGR_Date_PRES_Date
1910.120 Title Hazardous Waste Operations and Emergency Response

(a) Scope, application, and definitions . . .
(iv) Operations involving hazardous waste that are conducted at treatment, storage, disposal (TSD) facilities regulated by 40 CFR Parts 264 and 265 pursuant to RCRA; or by agencies under agreement with U.S.E.P.A. to implement RCRA regulations; and
(v) Emergency response operations for releases of, or substantial threats of releases of, hazardous substances without regard to the location of the hazard. (2) Application. (i) All requirements of Part 1910 and Part 1926 of Title 29 of the Code of Federal Regulations apply pursuant to their terms to hazardous waste and emergency response operations whether covered by this section or not. If there is a conflict or overlap, the provision more protective of employee safety and health shall apply without regard to 29 CFR 1910.5(c)(1).
(ii) Hazardous substance clean-up operations within the scope of paragraphs (a)(1)(i) through (a)(1)(iii) of this section must comply with all paragraphs of this section except paragraphs (p) and (q) . . .

(b) Safety and health program.
Note to (b): Safety and health programs developed and implemented to meet other federal, state, or local regulations are considered acceptable in meeting this requirement if they cover or are modified to cover the topics required in this paragraph. An additional or separate safety and health program is not required by this paragraph.
(1) General.
(i) Employers shall develop and implement a written safety and health program for their employees involved in hazardous waste operations. The program shall be designed to identify, evaluate, and con-trol safety and health hazards, and provide for emergency response for hazardous waste operations.
(ii) The written safety and health program shall incorporate the following:
(A) An organizational structure;
(B) A comprehensive workplan;
(C) A site-specific safety and health plan which need not repeat the employer's standard operating procedures required in paragraph (b)(1)(ii)(F) of this section;
(D) The safety and health training program . . .

(5) Personal protective equipment (PPE) shall be provided and used during initial site entry in accor-dance with the following requirements:
(i) Based upon the results of the preliminary site evaluation, an ensemble of PPE shall be selected and used during initial site entry which will provide protection to a level of exposure below permissible exposure limits and published exposure levels for known or suspected hazardous substances and health hazards and which will provide protection against other known and suspected hazards identified dur-ing the preliminary site evaluation. If there is no permissible exposure limit or published exposure level, the employer may use other published studies and information as a guide to appropriate personal pro-tective equipment.

Although most safety professionals may have little or no formal training in budgeting, and although their experience is likely to be limited, it is important that they participate in this effort. They are the safety experts that know what is required for an effective safety effort and what is desirable but optional. Budgeting without the safety professional's input will result in a budget that does not adequately reflect the needs or best interests of the safety function or the company.

The level of experience is paramount when developing a budget. Experienced persons may be able to work more quickly and correctly than inexperienced ones. However, the higher salary of the experienced person may offset the difference in the budget. In addition, there is always some degree of inaccuracy in budgeting when one considers only the hours required to perform a given task. If the task is performed by a more highly compensated professional than budgeted, the hours can be correct but the dollar cost higher than expected.

TIME FACTORS IN BUDGETING

Budgeting for various tasks should be assigned to the most qualified person in the safety function. This person might not be the safety manager, but rather the individual responsible for a specific project or task. A vital component of planning for this individual is to plan effectively for a task's completion, factoring in the financial budgetary factors required at each stage. The budget must be linked to certain milestones in order to make it trackable. If there are no milestones, the budget becomes difficult to spread out correctly over the entire duration of the project, and even harder to control once it is in place. Setting the minimum completion time should be tempered by the knowledge that there are some things beyond the control of the safety program that can impact the schedule and cause a task to take longer than first planned. For this reason, it is usually wiser to budget for average time to completion rather than minimum time.

PRACTICAL REASONS FOR BUDGETING

Budgeting is one of the most important tasks performed by safety professionals. Without enough funding, any job is almost impossible to accomplish. With overbudgeting, unnecessary tasks are performed and the costs of the safety program are higher than necessary. When the costs of a program are higher than budgeted for, the cost overrun can be traumatic for everyone. Tasks must be taken out or cut back, personnel may be laid off, managers must spend time and resources to defend their efforts, and the program suffers. Cost overruns can also adversely affect the company's image and that of the people involved.

DEVELOPING THE OPTIMUM BUDGET

The best budget is one that allows all necessary tasks to be accomplished with the appropriate persons at the correct time. The budget matches the skills of the individual who is performing the task with the time and materials needed to complete it. The optimum budget must allow for revision and necessary research to develop documentation.

It must also be foremost in everyone's mind that a budget is a planning document for the proper disbursement of resources. It deals not only with personnel and dollars but also with the outlook of the company. The approved budget reflects the perspective of the company's management.

METHODS OF DETERMINING A BUDGET

There are two major types of budgets: fixed and variable. The more common one is the fixed budget. It plans for a specific period of time and is updated and compared with actual results at the end of that period. The time period may be lengthened or shortened, depending upon the needs of the company.

The variable budget provides for the possibility that the actual costs may not be the same as the planned costs. This provision allows for restructuring of the budget to align it with actual costs. The variable budget obviates the need for contingency budgets but requires a very good accounting system.

For the purposes of this discussion, the fixed budget is considered, but the concepts hold true for the variable budget as well. There are several methods to determine the budget that should be allocated for safety. Three methods are briefly discussed.

One method that can be used to determine a budget is to define program elements or tasks. If a task can be defined, it can be properly budgeted for; if it cannot be defined, it cannot be adequately covered in the budget. This method requires significant up-front detail in describing the total program in manageable steps.

A second method for establishing a safety program budget is based on a specific percentage of the total company budget. A variation of that method is to allocate a percentage of the engineering, operations, or administration budget to the safety program budget. This is an easy way to budget, but it is not as good as other methods, because it does not determine how to perform any one task or define the cost of accomplishing it. This method requires little effort to develop the budget but is useless in identifying where resources will be used.

Level of effort on an as-needed basis is a third way that safety may be budgeted. This is the "I'll call you when I need you" method, but it rarely allows the company to perform the tasks correctly and on time. This means that no matter what the tasks are, the allocated budget does not change to accommodate them. There is no measurable effort in terms of significant accomplishment.

Of these three methods, the task definition one is the best one to identify and allocate where to spend and control company resources. It also provides an easy way to identify the problem area when an overrun or underbudget situation occurs. For this reason, the task definition method of budgeting is discussed at length.

Task Definition Budgeting

The first step required when using the task definition method of budgeting is to decide which tasks are to be budgeted. In any safety effort, there are certain tasks that must be completed to ensure that the product and the workplace have the best level of safety. To ensure that the programs that the safety function must implement are fully covered, it is probably best to divide the tasks.

An effective comprehensive safety program can be broken into four major categories:

1) Workplace safety
2) Environmental safety
3) Product safety
4) Administration

These major categories can be further broken down into subcategories, which can be called "safety programs." The various safety programs necessary for any company could include the following:

1) Safety management

2) Industrial hygiene

3) Hazardous materials/waste

4) Radiation safety

5) Hearing conservation

6) Safety training

7) System safety

8) Electrical safety

9) Construction safety

10) Personal protective equipment

This is not a comprehensive list. It does not include the subcategories of each of these programs, nor does it indicate the order of priority or importance. Prioritization is dependent on the product or services of the company and should also be considered when developing a task definition budget. All these factors may not be pertinent for any one company, but many of them will be appropriate for most organizations. There may also be other tasks, not identified, that must be budgeted for and controlled. The basic requirements for any safety program depend on what the company needs for a safe and productive workplace and to produce a safe and effective product. As a first step in the budget process, the safety professional must determine what resources will be required for each task in the overall safety program.

Each of the previously listed programs requires that certain tasks be performed in order to reach its goals. The tasks serve as the basis for the budget. Identifying the scope for these tasks is the hardest part of budgeting.

DEVELOPING THE BUDGET FOR SAFETY

Several considerations must be kept in mind when budgeting for safety. Among the most important are the personnel assigned and the equipment, materials, and time required for completing a task. These are closely interrelated and have a direct effect on the budget.

Personnel budgeting commonly takes two forms: one portion of the budget is the time, in worker-hours, needed to complete a task; the other is the dollars required to pay for the level of skill required to perform the task. Therefore, identification of the skill level required to do each task is very important to the budgeting process.

When the budget is developed, considerations include the type of personnel needed and the difficulty of the task. Two items to consider when discussing personnel are availability and level of experience. Commonly, safety functions consider only the worker-hours needed to complete a task, but this leads to a false sense of security when assigning personnel. If a lower-level person is designated in the budget but a higher-level person performs the tasks, the budget may go into overrun. Conversely, if a lower-level person performs tasks that require a higher level of skill, the time needed to complete the task may exceed the allocated budget and impact the program schedule.

The availability of highly skilled personnel to perform the safety tasks can be a major concern. If the tasks are complex, highly skilled personnel will be required to accomplish them in a timely fashion. However, the skilled personnel may not be available due to priorities and requirements of other programs. The personnel necessary to service a program must be identified early and an effort made to secure their services for the task.

Outside services such as periodic physical examinations, laboratory services, and hazardous waste disposal costs must be included in the budget. The calibration of instruments must also be specified. In addition, some type of contingency budget should be included to ensure that unexpected or unplanned tasks are completed. Contingency funds should be used only in an emergency.

It is very important to budget the total task. A common mistake is to budget only the personnel necessary to perform the task and to budget the equipment and materials for the total program or department separately. Figure 15-2 is a form that outlines the items for each task.

Budget reduction in a company is always a possibility. Contracts expected may not materialize, the product may be late, or a lawsuit may require more resources than expected. The task definition method allows the budget to be reduced by eliminating or postponing specific tasks.

Sometimes, however, budgets are often cut by a certain percentage, with no reduction in the number or scope of the tasks. With a well-defined task definition budget, the effort can be reduced as appropriate. This allows the safety professional to obtain concurrence from company management to delete tasks and reduce budget requirements.

During development, the budget should be spread over the length of each specific safety program. Most companies budget annually; however, various safety programs are often for shorter durations. This budget spread

Figure 15-2. Budget Planning Form

TASK BUDGET

ASK _____

EXPECTED DATE OF COMPLETION _____

DATE OF FIRST BUDGET REVIEW _____

DATE OF SECOND BUDGET REVIEW _____

ITEM	MINIMUM	MAXIMUM	AVERAGE	QUOTE	1ST REVIEW	2ND REVIEW
Manpower (number)						
Skill Level (years experience)						
Equipment List ($)						
Materials List ($)						
Other List ($)						
Time to Complete (man-days)						
Contingency ($)						

should cover the entire period that safety will be involved and should be linked to specific milestones to ensure that the coverage is complete and correct. The budget should be spread as needed to accomplish specific tasks. The budget should put the extra personnel where and when it will be needed. Figure 15-3 shows a typical budget spread linked to product design reviews, manufacturing, and testing, for a system safety program. Note that most of the effort is expended in the early stages of the program. This is known as "front loading" of the system's safety effort and is considered to be the most effective method.

Other types of ongoing safety programs have a different budget spread. The cost peaks relate to activities relevant to performing various tasks within the program. Figure 15-4 shows an industrial hygiene program whose milestones refer to specific air and noise monitoring activities. Note that it is

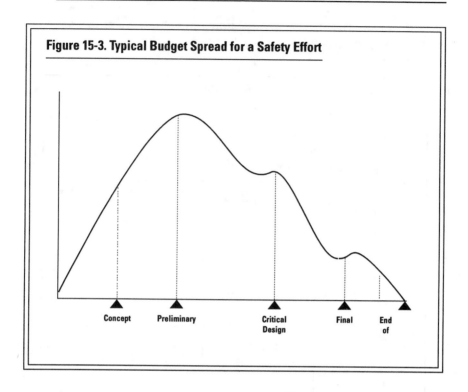

Figure 15-3. Typical Budget Spread for a Safety Effort

Concept   Preliminary   Critical   Final   End
                                  Design   of

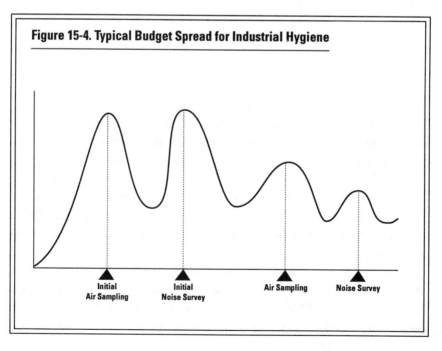

Figure 15-4. Typical Budget Spread for Industrial Hygiene

Initial   Initial   Air Sampling   Noise Survey
Air Sampling   Noise Survey

also front-loaded. This is common with increased costs to acquire equipment and the training necessary to use the equipment properly.

Although the budget is normally developed on the basis of worker-hours needed to complete a task, worker-days should be used, where possible. An excellent measure should be based on a 40-hour week, or at least on an 8-hour day. It is difficult to use only part of a worker's time and still keep that worker active and interested in the program. If a person is working on several programs, his or her effort is diluted, making that person less effective.

Each task to be accomplished should have a basis for the budget estimate. Developing a basis for the estimated budget can be a difficult and frustrating task. There are several methods for developing a basis for the budget estimate, but this chapter deals only with the methods used with task definition budgets.

The ideal basis for a budget estimate is history. The best approach if possible, is to use the actual data from another program on the hours expended to complete tasks similar to those of your program. However, the estimate will usually be based on the judgment of the individual who is preparing the budget, because there are little or no available documented historical data. Even when historical data are available, they are generally not well documented or are not presented in an easy-to-use format.

The engineering judgment method of justifying the budget is always subject to interpretation. However, if a task is broken down into the smallest possible parts, budgeting becomes much more straightforward. The judgment of the individual may be questioned, but if the tasks are well defined and defendable, there is little room for argument.

There must be a task description and a basis of estimate for each task. The task should be described in detail so that a nonsafety person can understand it. The description should be as specific as possible, and should list the requirements and state where they may be filled. It should show the task to be performed, why it is needed, and the scope of the task.

It has been suggested that a fault-tree approach could aid in this effort, that is, top-down approach, which breaks down the task into its necessary parts and then subdivides those parts. Figure 15-5 is an example of this approach. When discussing the safety tasks to be budgeted, keep in mind that each program is different and that there are various ways to reach a budget total.

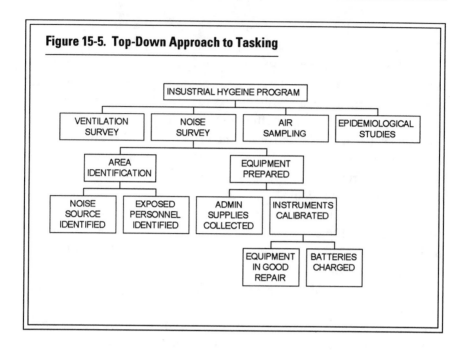

Figure 15-5. Top-Down Approach to Tasking

BUDGETING THE TOTAL EFFORT

Although task determination can be done in a top-down manner, all budgeting must be done on a bottom-up basis. To determine where a company will spend its resources, it must be clear what the company will receive for that investment. Many tasks are required in order to have an effective and complete safety department. Each of these tasks needs a budget to ensure that it can be accomplished effectively. In the previous section on task definition budgeting, the safety tasks were broken down into four categories. These subcategories are also very broad and should be broken down into further subcategories. These subcategories are also very broad and should be broken down further into subcategories to be effective in the budgeting process.

WORKPLACE SAFETY

An effective workplace safety program needs several subprograms to be effective. These subprograms should include those required by law and those that make good sense in doing business on a day-to-day basis. These subprograms are briefly discussed next.

Eye Protection

For an eye protection program to be effective, a survey of eye hazards must be made. Once the eye hazards are determined, controls must be established. When the controls are in place, personnel must be trained to use them. The procedure is typical of most safety-related programs. The cost of replaceable materials, such as safety glasses, goggles, and face shields, must be considered. The time required to monitor, train personnel, and document the program must be established. Outside services needed to provide prescription safety glasses, fit nonprescription glasses, and so on, should be specified in the budget. One important effort to include is the time needed to identify and control those eye hazards that can be eliminated by a good engineering effort.

Industrial Hygiene

Industrial hygiene programs provide the quantitative evidence needed in court to prove that hazardous conditions in the plant are properly controlled. These programs provide the best method to identify a potential problem early enough to correct it before someone is injured or an OSHA citation is issued. Accurate, independent evaluation of the work site can be performed effectively and efficiently by an in-house industrial hygienist or by an outside consultant. Either is appropriate, depending on the desire of company management; a highly paid, well-qualified, full-time industrial hygienist may not be justifiable. However, the convenience of being able to pick up the phone and get an industrial hygienist on the scene immediately may be well worth the investment. The outside consultant cannot provide the instant service that a company may require or desire, but it can provide the service in a planned mode very effectively.

The budget for an industrial hygiene program must include the equipment necessary to perform the required tasks. Such equipment typically includes air sampling pumps, noise survey equipment (such as sound level meters and noise dosimeter), gas detectors, radiation survey meters, ventilation survey equipment, and heat stress equipment.

The cost of sample analysis also must be included in the budget. Very few companies can afford the cost of setting up a comprehensive, accredited industrial hygiene laboratory; therefore, most send their samples to a certified laboratory. Costs vary according to the complexity and the time necessary to run a sample.

Radiation Safety

The basis for all radiation safety programs is the monitoring of personnel exposure. Personnel dosimetry is usually handled by an outside firm that provides radiation monitoring badges and processes them for a set fee. The cost of this service varies with the number of badges used or how often they are processed. Some companies desire a radiation safety effort large enough to justify a full-time health physician on staff, but most rely on the safety department to provide this support.

To ensure that there is no contamination from radioactive materials, area surveys are performed on a regular basis. These surveys usually require portable monitoring equipment and wipes. The equipment requires regular calibration and the wipes must be analyzed, usually by an outside service on a fee basis. The monitoring equipment may be leased from the firm doing the calibration or purchased and sent out for calibration on a scheduled basis. The time and materials necessary for these surveys can be substantial.

Voluminous documentation is needed to manage a radiation safety program. Records of exposure must be requested from other companies on new employees if they have had occupational exposure to radiation at your facility. Responses to these requests must be handled quickly.

Regulatory audits require an extensive amount of time to provide the agency with the necessary data. Since these audits occur on a regular basis, they should be planned and their cost incorporated into the budget.

Hearing Conservation

With the knowledge that noise-induced hearing loss can result in a disability claim, it becomes imperative to reduce noise to acceptable levels and provide protection to employees when noise reduction is not possible. A hearing conservation program identifies sources, gives a pre-placement audiometric examination for all employees, provides periodic audiometric testing for employees working in high-noise areas, fits and issues hearing protection devices, and provides ongoing noise surveys and evaluation of all work sites. Audiometric examinations are usually conducted by an outside firm, but can be conducted inside with the proper equipment. Wherever these examinations are conducted, they require properly calibrated equipment used by a trained and certified audiometric technician.

Hearing protection comes in various sizes, shapes, and materials, and the cost of this equipment can vary dramatically. A program that includes custom-fit ear protection devices is more expensive than one using off-the-shelf materials. However, the former may be preferable because of greater acceptance by employees.

Sound-level monitoring equipment can be expensive, and the complexity of the equipment depends on many workplace factors. The commonly used sound-level meter combined with personal dosimetry provides excellent results in most environments. But if there is a need for octave band analysis or for measuring impact noise, the cost rises quickly. Calibration by in-house laboratories is possible, but usually not feasible.

Safety Training

An important element in any comprehensive safety program is training. Safety training should be designed to prevent accidents and teach the employee the hazards associated with his or her job.

When budgeting for training, several factors must be considered. The instructor's time in teaching the class must be taken into account. If an outside instructor is desired, the procedure is fairly simple: a fee is paid and a service is rendered. For inside service, one must account not only for the time needed to present the material but also for the instructor's preparation time.

The hazard communication requirement brings into sharper focus the need for a safety training program. The law now requires the training of employees in the hazardous nature of their jobs. This training is ongoing and must be done on a periodic basis.

Refresher training is necessary for employees to remain up-to-date on new developments and for recertification. If the company has a require-ment that all electricians be cardiopulmonary resuscitation certified, then there is an annual requirement.

The materials necessary for a training class must also be considered. These include handouts, slides, quizzes, and samples. The widespread use of audiovisual aids, such as films and videocassettes, has opened a new per-spective on training. These films can be purchased and shown over and over, with little instruction required. The technique does require specialized equip-ment, which may be a necessary added expense to include in the budget.

Some organizations have developed safety training correspondence courses to train employees in specific hazards and operations. These can be very effective and can be tailored to a specific location. However, such courses can also be expensive and require much updating and research time.

Construction Safety

New construction projects for facility renovation or for new buildings and facilities are usually well planned. The safety effort required for this activity must also be planned. The problems encountered on construction sites are sometimes unique and different from those in a manufacturing effort. A consultant or a safety engineer specifically trained in construction safety may be necessary for a large, complex activity. In either case, the cost must be budgeted.

If the company is the prime contractor and provides all of the personnel and materials, the safety department must be alert to potential problems involving personnel who are not familiar with construction sites.

Regulatory safety inspections of the site to ensure the maintenance of worksite safety are needed. These inspections should also ensure that personnel use proper protective equipment, such as head, eye, and hearing protection devices. This equipment may have to be acquired separately, and its cost should be included in the budget.

Construction safety orders have different requirements from general industry safety orders and may require some research by the safety department to understand them fully. The time necessary to do this research should be included in the budget.

Subcontractors may cause specific safety problems due to the increased use of poorly trained personnel or inadequate equipment. The safety department must budget extra time for each subcontractor to ensure that the safety requirements on a particular job are understood.

A review of drawings is very important to the safety of a construction project. The safety engineer can identify particularly hazardous operations and can ensure that he or she is present to make certain that such operations are conducted in the safest manner possible.

Forklifts and other vehicles used on construction projects have specific requirements that must be met, including rollover protection, seat belts, and backup alarms. The safety professional should ensure that these vehicles have this equipment before use.

A common safety problem on construction sites is inadequate or inappropriate scaffolding. Safety department personnel are required to ensure that the scaffolding meets the regulations. Cranes and other lifting devices also have special inspection and maintenance requirements of which the safety professional must be aware. The time needed to conduct the inspections and to solve any problems must be considered in the budget.

Crane Safety

Cranes and hoists need special handling. They require routine inspections by operators and also by another individual who is qualified to inspect them. Nondestructive testing may be needed periodically to ensure that this equipment is safe to operate, which may require the use of a consultant or an outside service.

Respiratory Protection

Most manufacturing companies have some personnel who must wear respirators on a periodic basis. The costs of a respiratory protection program must be listed in the budget. All personnel who wear a respirator must pass a physical examination. This examination can be conducted by a company physician or by a contract physician.

Training in the limitations of the respirator and proper respirator selection must be provided. An employee who uses a respirator must be provided with one that fits his or her face. Testing for qualitative or quantitative fit must be provided whereby the individual wears the respirator in a test atmosphere. Depending on the type of testing, the equipment can be expensive. Different sizes and types of respirators must be available for fitting and testing of personnel. Consumable materials for the testing and the cleaning of equipment after testing must be included in the budget.

ENVIRONMENTAL SAFETY

The impact of industry on the environment comes in many forms: noise, hazardous materials, oil spills, ionizing and nonionizing radiation, or thermal energy. The sources of environmental impact must be identified and controlled, and the various programs needed to carry out this task must be established and used by the safety department.

Hazardous Materials

Materials used by the company must be identified, their hazards noted, and controls placed to ensure that such materials do not harm personnel or the environment while they are in the plant. The safety department must be a part of the purchasing activity to ensure that the hazardous materials coming into the plant are identified before arrival. This ensures that proper storage facilities exist and that there are procedures for safe use of the materials. For many manufacturing companies (and, in some states, all companies), a

Material Safety Data Sheet must be on file for all hazardous materials received in a plant, or one must be developed and provided to the users. The maintenance of a hazardous materials inventory is one very effective way of knowing what materials are in the plant and what hazards they pose. This inventory must be updated periodically. The labeling of the chemicals is an essential part of any hazardous materials program, and it must tell users what hazards exist and how they can avoid them.

A hazardous materials spill plan must be developed and used. This plan can require a significant amount of equipment, materials, and personnel. Personnel must be trained to respond correctly to different types of spills. Another approach is to hire an outside contractor to respond to spills. This approach can be very costly, but is sometimes well worth the expense because it does not require the company to train its own personnel and maintain the equipment.

Hazardous Waste

Every company, regardless of its size, must be concerned with the proper disposal of hazardous wastes. Large companies typically have full-time staff to do nothing but manage hazardous waste programs. Smaller firms, usually classified as "small generators" because of the small quantity of waste they create, use the safety department to manage the program. The mandatory use of the "manifest system" requires that all waste be identi-fied and a chain of custody established for it. This manifest system and the cradle-to-grave responsibility for waste require a significant amount of time and effort, which must be budgeted.

Annual reporting of waste usage to federal or state regulatory agencies can take a significant amount of time and documentation. If hazardous waste is stored at a company site for any time period, it may require a significant amount of contingency planning. Planning is needed for the proper response to a spill of hazardous waste in order to prevent it from becoming a problem for personnel or the environment. These plans may require a consultant who is an expert in their preparation, or much work by in-house personnel. Such plans should be updated annually or as significant changes occur. Their development, implementation, and documentation must be budgeted.

Oil Spill Control

If a company stores a significant amount of oil at its facility, it may be required to have a spill prevention, containment, and countermeasure (SPCC) plan. This plan, mandated by federal regulations, requires extensive research and documentation. It involves inspections of tanks and other storage vessels, and documentation of these inspections. If a spill occurs, the SPCC plan directs how the company will respond to the cleanup and how it will provide funds for cleanup costs. Documentation and implementation of SPCC plans can be extensive and expensive.

PRODUCT SAFETY

An effective product safety program must identify all the potential hazards in the product or service, warn the user that they exist, and follow the product from cradle to grave.

System Safety

An effective system safety program covers the product from concept through disposal. The safety engineer should review the drawings, prepare a preliminary design and operations hazard analysis, ensure that the labeling and operations manuals describe the hazards, and provide documentation that is appropriate and detailed. This review can require a minimal effort for many of the products that have a good history, but can be extensive and time- consuming for new product lines. Analytical testing may be required to verify information in the hazard analysis. Stress analyses may have to be run, and various types and levels of inspection performed, to ensure that the product is made as designed. The effective system safety program requires that the safety engineer review the data and ensure that they meet the safety requirements. Time for consultation with legal personnel, design engineers, operations personnel, and repair and maintenance personnel must be budgeted to ensure that all aspects of the product are evaluated.

Marketing

Marketing activities can cause a significant problem for effective safety programs due to misleading or inaccurate advertising. The product safety engineer should review the marketing literature to ensure that it does not mislead the user or show the product used in a manner in which it was not intended. He or she should also review the disclaimers for faulty or inaccurate statements. This review requires time and documentation, as well as coordination with marketing, advertising, and legal personnel. The time and documentation needed for such review must be budgeted.

Recall

If a product is recalled, the safety engineer must have some method to identify all the users of the product. This requires the maintenance of a database of users or distributors. Before a recall occurs, the safety professional should be consulted, or he or she should consult the legal department or corporate counsel, on the proper wording of the recall.

Litigation

All current and pending litigation should be included in the budget. Time for personnel to give depositions, testify in court, or perform research activities should be budgeted. If there is a settlement out of court, a budget for the costs of this settlement should be included.

ADMINISTRATION

One common mistake made in the budgeting process is failure to budget for the administrative part of the program. The following subsections discuss several items that should be considered when budgeting for the administrative tasks in a safety department.

Safety Management Program

The safety management program must include the tasks necessary to supervise, evaluate, and counsel personnel assigned to the safety department. These individuals can include nontechnical personnel, such as secretaries and clerks, and technical personnel, such as industrial hygienists, health physicians, and safety engineers. The program should include the costs of acquiring a new person, either for replacement or for growth of the department. The tasks associated with acquiring a new person should include advertising, recruiting, interviewing, hiring, orientation, and training. Each of these tasks should be broken down into its subtasks. For advertising, that means the work of writing the job description, reducing it to an advertisement, and getting it approved. These tasks could also be broken down into smaller tasks, depending on the difficulty of getting an advertisement approved by the company.

The costs associated with planning, organizing, and using new products must also be included in the budget. A hazardous process has a direct and significant effect on the personnel of the safety department. The time it takes the safety professional to review the process, evaluate personnel hazards, evaluate the environmental impact, evaluate and recommend controls and plan the monitoring strategy can be substantial.

Materials required for the successful implementation of a safety program can be a serious cost factor if they are not planned. When considering materials, the budget for safety management should include the administrative supplies and services necessary for the smooth operation of the safety department. Materials should be budgeted for on the basis of annual use. This might seem to be a minor cost, unworthy of much effort, but in the area of documentation, the safety department must be exceedingly thorough. The cost of the reproduction paper, for example, can be staggering when the company is large and the distribution list contains forty to fifty persons. A twenty-page document becomes an eight hundred-page distribution when forty copies are needed. Sometimes the cost of reproduction may exceed the effort needed to generate the original work.

Equipment acquisition or repair must be included in the budget. The repair of a copying machine no longer covered by a maintenance agreement or repair contract may represent a greater cost than its replacement with newer and better equipment. The need for a computer in the safety department to allow for efficient database management is becoming well recognized. However, the costs of maintenance and supplies necessary for the effective use of a computer may not be so readily apparent. Many computer systems use a floppy disc drive system and have a printer attached. The cost of software, on-line services, floppy discs, computer paper, ribbons, floppy-disc storage containers, and annual maintenance must be considered in the budget.

Accident Investigation

To have an effective safety program, accidents and incidents must be investigated. For the purposes of this discussion, an incident is considered a "near miss" when it is an unplanned event that did not hurt anyone or damage any equipment but could have under different circumstances.

In any effective accident investigation, photographic equipment including a camera, a couple of lenses, a tripod, a camera bag, and the normal accessories for photography is needed. Other supplies can include tape measures, flashlights, drawing sets, and equipment cases. The cost of this equipment as well as the cost of film and film processing must be included in the budget.

Emergency Evacuation

Every effective program provides for the safe evacuation of personnel from the facility. This requires the generation of a detailed emergency plan. Such a plan should include medical, fire, and other emergency assistance. The research, development, and implementation of such a plan are commonly assigned to the safety department. When the plan is finished, it should be tested.

The evacuation plan may require the purchase of communication equipment, emergency medical equipment, special aids for the handicapped, and special tools for rescue. The plan should also be tested by conducting periodic drills. These activities must be stated in the budget.

Coordination with local agencies that provide support can be time-consuming, requiring many meetings to ensure that the agencies understand the facility, what they can support, and the hazards they may have to deal with in an emergency.

Some companies recognize that there may be no effective way to evacuate personnel in certain situations. For example, a massive earthquake, such as the 1985 earthquake in Mexico City, can paralyze all emergency services for a period of time. Many companies are considering how to provide food, a good source of water, shelter, and emergency medical support for their personnel in this type of emergency. If a company decides that it needs to provide for this situation, the planning and implementation costs can be very high.

It does not matter whether the emergency plan is detailed, general, or comprehensive, or deals with only specific problems. Time and materials are necessary to develop and place an effective plan in use. The tasks discussed herein do not represent the entire safety effort, nor was that the intent of this discussion, but they are common to almost any safety department and should be considered when developing the company's safety budget.

METHODS FOR SELLING SAFETY TO MANAGEMENT: USING ENGINEERING/FINANCIAL PRINCIPLES

Did you ever wonder why the CEO's or the comptroller's eyes glassed over when you talked incident rates and environmental management reports (EMRs)? The same reason why *your* eyes glassed over when they talked return on investments (ROIs) and an equivalent uniform annual cost (EUAC).

$$P = \frac{\left(\$400,000\left(1+0.10\right)^{3} - 1\right)}{\left(0.10\left(1+0.10\right)^{3}\right)}$$

$$P = \$300,526$$

That is, to have $400,000 in three years with a current interest rate of 10 percent, your company must be able to invest $300,526 now to be able to make the allocated expenditure for EPA modifications at your plant.

Why is it a benefit to be able to illustrate to your management that you can perform this calculation? Well, you may not have noticed but the investment calls for a $400,000 expenditure. However, the current investment required is about $100,000 less than the planned target expenditure. This makes good fiduciary sense, especially if management knows that these modifications are going to be required. Being able to make these calculations is important. However, knowing why and what the calculations mean is even more important.

Future Worth

The future worth of an asset is the current value of the asset plus the compound interest. This value is also a good check for the feasibility of a project. It can be discounted to present worth in order to compare the value of a product or project with the investment necessary to create it. If the value is less than the investment required, the project should be terminated in favor of more profitable projects. The formula for future worth is

$$F = P\left(1+i\right)^{n}$$

where

    F = future worth of a present sum of money after n interest periods, or the future worth of a series of equal payments;

    P = the sum of money at the present time;

    i = interest rate for a given interest period; and

    n = number of years.

Future worth is a function of

- initial value of investment;
- interest rates (compounding), which are composed of
  - true cost of borrowing money (3 percent)
  - risk of investment/project (junk bonds versus AAA bonds);
  - rate of inflation (currently somewhere on the order of 3–6 percent annually).

For example: Your company may need to install pollution abatement equipment in three years (1999) to bring the facility up to EPA specifications. After speaking with vendors and accountants, you estimate that at that time (1999), $2 million will be needed to make the necessary improvements. It is currently 1996, interest rates are at 10 percent and you have $750,000 to set aside and invest in anticipation of spending needs in 1999. What will the future worth of $750,000 be and how much will you have to borrow to make up the difference? Using the formula:

$$F = P(1+i)^n$$

where

F = future worth of a present sum of money after n interest periods, or the future worth of a series of equal payments (the unknown variable);

P = the sum of money at the present time ($750,000);

i = interest rate for a given interest period (10 percent); and

n = number of years (3).

Therefore,

F = $750,000 $(1+0.10)^3$

F = $998,250

If you will need $2 million for the project, and you anticipate having $998,250, you will need to borrow $1,001,750.

Why is it a benefit to be able to illustrate to your management that you can perform this calculation? You can illustrate to management that saving $750,000 in preparation for a $2 million expenditure three years from now shows forethought and vision.

Depreciation

Depreciation is the process of allocating, in a systematic and rational manner, the expense of an asset to each period benefited by the asset. The cost of the asset is divided up and spread across and charged against the accounting periods of its estimated lifetime. This allows a company to charge the expenses associated with an asset against the profits it generates during the periods in which the asset is used. There are many methods to calculate depreciation, such as sum-of-year digits, declining-balance, group and composite depreciation, and straight line. For the following example, straight-line depreciation is used. The formula for straight-line depreciation is

$$D = \frac{(P - SV)}{n}$$

where

D = annual depreciation;

P = initial cost of the asset;

SV = salvage values of the asset; and

n = expected depreciable life of the asset.

Depreciation is a function of

- cost of the asset;
- estimated lifetime of the asset;
- salvage value (if any) of the asset;
- method of depreciation used (straight line, accelerated cost recovery standard, and so on).

For straight-line depreciation, using an emergency response vehicle example, let us say that $10,000 is spent, that we plan to keep the vehicle four years, and that the salvage value is $5,000. Using the formula

$$D = \frac{(P - SV)}{n}$$

where

D = annual depreciation (unknown variable);

P = initial cost of the asset ($10,000);

SV = salvage values of the asset ($5,000); and

n = expected depreciable life of the asset (4).

Therefore,

$$D = \frac{\left(\$10,000 - \$5,000\right)}{4}$$

$$D = \$1,250 \text{ per year}$$

Why is it a benefit to be able to illustrate to your management that you can perform this calculation? You can illustrate to management that spending $10,000 on an emergency response vehicle now will be worth a $1,250 write-off every year for four years.

Rate of Return

The rate of return is a measure that allows comparison between two different alternatives. It is a function of the ratio of the present value of the net income generated over time by the asset divided by the cost of the asset, usually expressed as a percentage. In other words, the amount of money generated by two alternative projects is translated into something resembling an interest rate. In this manner, the company can choose which project will yield the highest return for its money. Many companies also have a minimum attractive rate of return, which is the lowest rate of return acceptable before a project will even be considered.

The formula for rate of return is

$$R = \frac{P_i - P_0}{P_0}$$

where

R = rate of return;

P_i = net present value of all expected inflows; and

P_0 = net present value of all expected outflows.

The rate of return is a function of

- initial cost of the asset(s);
- expected cash outflows including the cost of
  — safety program implementation (for example, confined space, lockout/tagout);

— equipment (for example, fire extinguishers, fire protection);
— training;
— new employees (for example, safety professionals, industrial hygienists);
— maintenance;
— repairs.

• Cash inflows such as cost savings or payments for goods and services and expected
— workers' compensation and insurance savings;
— decrease in overtime, productivity, turnover, training costs, and so on;
— decrease in OSHA citations.

• salvage value (if any) of the asset;

• discount rate used by the company such as
— minimum acceptable rate of return; or
— cost of capital (company's borrowing costs) or rate of inflation.

For example: Let us say we have the option between two emergency response vehicles. One vehicle will cost $10,000, it has a salvage value of $5,000, and will require about $1,000 per year in maintenance. The interest rate is 15 percent and we plan to keep the vehicle for four years. The alternative is a $20,000 vehicle with a salvage value of $9,000 that will require about $500 per year in maintenance.

For option 1

$$R_1 = \frac{P_i - P_0}{P_0}$$

where

$P_i =$ $5,000+$5,000
Salvage Value + Depreciation

$P_0 =$ $10,000+$1,000
Cost + Maintenance

$R_1 = -0.09$

For option 2

$$R_2 = \frac{P_i - P_0}{P_0}$$

where

$P_i =$ $9,000+$11,000
Salvage Value + Depreciation

$P_0 =$ $20,000+$500
Cost + Maintenance

$R_2 = -0.02$

Why is it a benefit to be able to illustrate to your management that you can perform this calculation? You can illustrate that even though R_1 is cheaper than R_2, the rate of return is not as good. In the case of the negative calculation, the value closest to zero is the best option. When considering cost inflows, salvage value and depreciation are the included values. When considering outflows, the cost of the vehicle and forecasted annual maintenance are the included values. Expected inflows that appear to cancel each other out include workers' compensation savings, insurance savings, overtime, productivity, turnover, training cost, savings, and a decrease in OSHA citations. The only expected outflow that appears to cancel is training. Maintenance and repairs do not cancel out and are included in the calculation. The other variables are not applicable to this particular problem.

Replacement Analysis

Replacement analysis provides an economic comparison of two asset choices, a defender (the current assets) versus a challenger (asset being considered for purchase). It is usually used when determining whether to replace an existing asset with a new or more efficient one, or when comparing different options of procuring equipment such as buying or leasing. The costs and expenses associated with the assets are converted into an EUAC. This determines how much expense will be associated with a given asset in one year's time, thus providing a uniform benchmark for comparison. Once the EUAC has been determined, all the company needs to do is choose the lowest cost option. For example, to determine which is more feasible—buying or leasing a particular asset—the following formulas can be used. Here the defender formula is used to perform the purchase option and the challenger formula to calculate the lease option.

The formula for the defender *(purchase option)*

The formula for the challenger *(lease option)*

$$EUAC_d = P - SV + AOC$$

$$EUAC_c = L + AOC$$

where

where

P = purchase cost of the asset;
SV = salvage value of the asset;
AOC = annual operating cost.

L = lease cost of the asset; and
AOC = annual operating cost.

Replacement analysis is a function of

- initial cost of the asset;
- salvage value (if any) of the asset;

- annual operating cost;
- lease cost.

For example: using a similar example as that shown previously, let us say we have the option between two emergency response vehicles. One vehicle will cost $10,000, it has a salvage value of $5,000, and will require about $1,000 per year in maintenance. The alternative is to lease an emergency response vehicle for $2,000 and will require about $500 per year in maintenance. Using

The formula for the defender	The formula for the challenger

$$EUAC_d = P - SV + AOC$$

where

P = $10,000
SV = $5,000
AOC = $1,000

$$EUAC_c = L + AOC$$

where

L = $2,000 x 4 years
AOC = $500

$$EUAC_d = \$10,000 - \$5,000 + \$1,000$$
$$= \$6,000$$

$$EUAC_c = \$8,000 + \$500$$
$$= \$8,500$$

Why is it a benefit to be able to illustrate to your management that you can perform this calculation? Initially, by using this formula you can illustrate the benefits of purchasing over leasing. Also, you may be able to illustrate the relatively low cost to purchase this vehicle in the first place. Using this calculation illustrates that you have examined all of the options, rather than merely indicating that the company needs to purchase an emergency response vehicle.

Retirement Analysis

Retirement analysis is the method used to find the lowest EUAC of an asset based on the number of years it will be used. This method of analysis allows the company to decide the most economical length of time to utilize an asset. With this information, the company can decide when to replace the asset, or assign a length of time during which the asset can be most economically utilized as a factor in a replacement analysis. This allows the replacement analysis to be conducted more accurately.

For example, let us look at the emergency vehicle based on a service life of four, five, six, seven, or eight years. The formula is as follows:

$$EUAC_d = P - SV + AOC$$

where

P = purchase cost of the asset;
SV = salvage value of the asset;
AOC = annual operating cost.

Retirement analysis is a function of

- initial cost of the asset;
- salvage value (if any) of the asset;
- annual operating cost.

For example: using a similar example as that shown previously, suppose that we have to determine how long we want to keep the emergency response vehicle. Given the information for each year, salvage value (SV), annual operating costs (AOC), capital recovery, and equivalent operating costs, a table can be generated for each year of use.

EUAC for n Years

YEARS n (1)	SV (2)	AOC (3)	CAPITAL RECOVERY (4)	EQUIVALENT OPERATING COSTS (5)	TOTAL (6) = (4) + (5)
1	$9,000	$2,500	$5,300	$2,500	$7,800
2	8,000	2,700	3,681	2,595	6,276
3	6,000	3,000	3,415	2,717	6,132
4	2,000	3,500	3,670	2,886	6,556
5	0	4,500	3,429	3,150	6,579

The result is that the minimum EUAC for the emergency response vehicle of $6,132 per year for three years indicates the remaining life of the asset.

Why is it a benefit to be able to illustrate to your management that you can perform this calculation? When you are trying to justify purchasing a new emergency response vehicle, this calculation could provide additional information regarding the economic value of keeping the current vehicle. It is certainly better than justifying an expense by saying, "This one sure is getting old and we need a new one." The calculations give credibility to your conclusions for needing a new emergency response vehicle.

Cost-Benefit Analysis

Cost-benefit analysis is a method used to analyze the effects of making a change in a process. Typically, cash flows of present procedures are compared against predicted cash flows incurred under the change. The advantage of using the cost-benefit analysis is the ability to monetize costs of intangibles (for example, goodwill, reputation of a company, the cost of a life, cost of future injuries, decreased turnover, and decreased training). The estimates used must be accompanied by realistic, conservative accounting assumptions. Without realistic assumptions to force the solution to the worst-case scenario, errors could occur that will invalidate the estimate basis. Cost-benefit analysis is a function of

- Decrease in legal liabilities, legal fees, settlements
- Decrease in OSHA citations
- Increase in goodwill (for example, company reputation, and union negotiating)

Cost-Benefit Analysis Techniques. Cost-benefit analysis is commonly used to assess estimated costs and projected benefits for goods and services. For example, to help justify proposed standards, the OSHA uses this method to project a proposal's compliance costs and estimate reductions in employee exposure to hazards.

Return-on-Investment (ROI) is the primary calculation used in cost-benefit analysis. It requires the following information:

- Associated labor costs (wages and benefits x hours worked)
- Proposed intervention costs (that is, equipment, installation, maintenance, employee training)
- Anticipated benefits (that is, cost savings, production improvements)

Assigning dollar values to an intervention's expected costs and benefits produces a financial ratio, which enables management to select the solutions that seem most promising. The exact ROI formula is as follows:

$$ROI = \frac{\text{Total Estimated Benefit}}{\text{Total Cost of Goods and Services}}$$

Decisionmakers often want to know how long it will take to break even on an investment. An investment's payback period can be determined via a series of calculations, as described in the following model:

1) Calculate productive hours worked and paid for by the employer.
2) Calculate wage or salary costs.
3) Determine employee turnover and training costs.
4) Project productivity shortfall (productivity losses due to absence).
5) Calculate total costs for employment and productivity shortfall.
6) Calculate the estimated health, safety, and productivity benefits.
7) Determine cost for improvements.
8) Calculate payback period.

Steps 1 through 5 determine total cost for existing conditions and are used to establish projected benefits. The payback period can be calculated in months or years.

Key Points for Success

Several steps must be taken in order to analyze the financial impact of safety-related projects. First, cost estimates must be accurate; these numbers are often overstated, which distorts the cost-benefit analysis ratio. Second, existing operating costs must be distinguished from proposed intervention costs. Existing costs can help identify an intervention's potential benefits.

Identifying Operating Costs. Operating costs can be divided into two categories: direct and indirect. *Direct costs* include associated labor and accident costs (that is, workers' compensation and medical costs). *Indirect costs* are more difficult to identify. They include costs related to reduced productivity (that is, downtime due to accidents, using less-skilled replacement workers or workers on restricted duty); costs of hiring and training temporary workers; costs of damaged employee morale (that is, increased absenteeism, tardiness, lack of cooperation or job ownership); and costs of reduced product quality.

Identifying Potential Benefits. Once operating costs are identified, it becomes easier to project potential savings and benefits. Benefits can be divided into direct and indirect categories as well.

Direct Benefits.

- Reduced labor costs (that is, eliminating excessive handling or rework, combining production steps)

- Lower accident costs (that is, reduced injury frequency and severity)
- Reduced insurance costs
- Productivity gains

Indirect Benefits.

- Quality improvements (that is, reduced scrap and rework, reduced product liability exposure and product recall expenses, improved corporate image, increased market share)
- Improved employee morale (that is, reduced turnover and absenteeism, improved teamwork, ownership)
- Reduced risk of compliance penalties

Projecting Injury Reduction Potentials

Predicting injury severity and frequency is a challenge for any safety and health professional. Convincing management of the potential frequency of injury, as well as determining how safety can prevent injury, are equally challenging tasks.

To estimate the cost of potential injuries, one can determine an "average injury cost" using accident experience from the past three to five years. When performing such an analysis, only those injuries that the budget item can control should be selected.

Next, "lost-time cost" averages for target injuries should be calculated, which involves multiplying days lost as a result of related injuries by daily labor costs of injured workers. Daily labor costs are hourly wage and benefits paid per employee (for examples, eight hours x $20.25 = $162 per employee). Next, average annual lost-time costs should be added to the average medical costs to produce a total injury cost profile. Methods such as statistical process control charting can enhance the credibility of this analysis.

Other related costs include the following:

- Overtime pay
- Temporary-worker expenses
- Reduced productivity due to use of less-skilled workers
- Lower-quality work and products
- Replacement-worker training
- Increased supervision

Nonquantifiable Losses

So far, these analyses have focused on numbers and dollar values. Although measuring qualitative issues is difficult, these issues can significantly affect financial results, productivity, and quality levels.

Employee morale and work ownership are keys to a successful operation. Unhappy workers have more injuries, are less productive, and resist management. In extreme cases, poor morale can prompt an employee to seek legal aid in settling a workers' compensation claim or to invite a union into a nonunion shop.

A survey of employee perceptions of safety can help gage their morale. Therefore, findings that could benefit from safety-related solutions should be discussed in the proposal. Repeating the survey following the interventions helps measure their success.

Identifying Costs

Costs of proposed safety-related enhancements must be accurately determined even though many such improvements are relatively inexpensive. Estimated costs should encompass initial and ongoing costs, which may involve the following:

- Equipment purchase or design
- Installation and setup
- Operator training
- Projected downtime
- Annual service/maintenance costs

With these data, first-year and subsequent-year costs can be analyzed. The illustration here does not reflect interest rates, time value of money, or actual total costs of a closed workers' compensation claim.

Cost Feasibility Process

When applying cost feasibility, safety professionals must determine which areas can be improved, then rank them according to priority. Some key considerations are as follows:

- Number of people affected
- Severity or magnitude of problem
- Ease of implementation
- Long-range outlook

By starting with a small project that will produce a high success rate, safety professionals can convince management of the value of large projects. For a project to succeed, affected employees must be involved in the problem-solving process.

Next, safety professionals must determine which projects to present to the decisionmakers. Priority should be given to those projects that have high benefit-to-cost ratios (benefits anticipated versus cost of changes) and are easy to implement.

The formal proposal should include a summary of anticipated cost and projected benefits and may also include the following:

- Rationale for improvements
- Severity or magnitude of losses
- Description of proposed solutions
- Personnel responsible for implementation
- Estimated implementation time
- Cost of proposed solution
- Projected benefits

Try to limit proposals to one page, highlighting main points with bullets. Detailed information can be provided in appendices or during formal discussions.

Financial analysis techniques, such as cost-benefit analysis, can successfully communicate the value of safety-related enhancements. Analysis of previous projects can also help justify such projects. Learning to use these techniques and to speak the language of business management will help safety professionals communicate their ideas and objectives successfully—and ultimately achieve their goals.

As an example, when Ford Motor Company performed a cost-benefit analysis to determine the benefits and cost relating to the fuel leakage associated with the static rollover test portion of the FMVSS 208 (Ford Pinto), Ford failed to make conservative accounting estimates of the worst-case scenario. In 1970, Ford used $200,000 as the cost of a life [provided by the National Highway Traffic Safety Administration (NHTSA)]; the value was based almost entirely on deferred future earnings (DFE). At the time this decision was made, there were at least three different DFE-based figures, ranging from $200,000 to $350,000, being used by as many different federal agencies. Willingness to pay (WTP) has since replaced DFE as the preferred method of assessing the value of life. Further, research has shown that various WTP studies have revealed a higher median value than the one used by Ford. On the basis of this research, the value of a life is greater than future earnings.

The Ford cost-benefit analysis presented a $137 million cost and $149 million benefit. In the formula, the number of deaths, the cost per vehicle to make the design change, and the proportion of deaths to be attributed to small, light vehicles were all subject to such dramatic change that the $2.75 cost to $1.00 benefit ratio achieved could have easily been changed so that the benefit exceeded the cost even if the value of life used was accepted and left unchanged.

The formula Ford used was as follows:

Benefits
Savings: 180 burn deaths, 180 serious burn injuries, 2,100 burned vehicles
Unit cost: $200,000 per death, $67,000 per injury, $700 per vehicle
Total benefit: [(180 x $200,000) + (180 x $67,000) + (2,100 x $700)] = $49.5 million

Costs
Sales: 11 million cars, 1.5 million light trucks
Unit cost: $11 per care, $11 per truck
Total cost: [(11,000,000 x $11) + (1,500,000 x $11)] = $137.5 million

The cost-benefit analysis performed by Ford on crash-induced fuel tank leakage and fires presents a startling example of the imprecision of cost-benefit analysis, as well as the strong possibility for manipulation of figures to achieve a desired result.

By using the high estimated death figure from the NHTSA, the benefit total would have been $161.2 million, even if everything else was held constant. By using the low estimated death figure of $5.08, the figure would have been $63 million. If only small cars had been used rather than all automobiles and light trucks, the cost figures would have been lower still. Increasing the value of life would have further skewed the results. In retrospect, the design changes would have been made based on a *conservative* accounting estimate rather than a *liberal* estimate.

What is life worth today? Various federal regulatory agencies value life differently when determining the cost-benefit of a new rule, as follows:

Consumer Product Safety Commission (CPSC)	$2 million
Environmental Protection Agency (EPA)	$8 million
Nuclear Regulatory Commission (NRC)	$5 million
Occupational Safety and Health Administration (OSHA)	$3.5 million
Office of Management and Budget (OMB)	$1 million

The total cost to implement a new rule is divided by the number of lives expected to be saved as a result. For example, if a new rule is estimated to cost $100 million to implement and it is expected to save twenty lives, the rule is too expensive for the CPSC, the OSHA, and the OMB, but acceptable for the EPA and the NRC. It seems that the federal government values life more than some workers value their own lives.

When conducting a cost-benefit analysis, one may choose to use a particular agency's numbers depending on what is being justified. For example, if you are justifying on the basis of OSHA, you would want to use $3.5 million; on the basis of the EPA, you would want to use $8 million for the cost of a life.

Justifying Safety Using Sales and Profit Margins

In times of keen competition and low profit margins, safety may contribute more to profits than an organization's best salesperson. It is necessary for the salesperson of a business to sell an additional $1,667,000 in products to pay the costs of $50,000 in annual losses from injury, illness, damage, or theft, assuming an average profit on sales of 3 percent. The amount of sales dollars required to pay for losses will vary with the profit margin. Table 15-2 shows the amount of sales dollars required to pay for different amounts of costs for accident losses; that is, if an organization's profit margin is 5 percent, it would have to make sales of $500,000 to pay for $25,000 worth of losses. With a 1 percent margin, $10,000,000 of sales would be necessary to pay for $100,000 of the costs involved with accidents. The formula is as follows:

$$\text{Sales to offset losses} = \frac{\text{Dollars of losses multiply 100}}{\text{Profit margin (\%)}}$$

Table 15-2. Sales of Dollars Required to Pay for Different Amounts of Costs for Accident Losses

YEARLY INCIDENT COSTS	PROFIT MARGIN ($)				
	1%	2%	3%	4%	5%
$1,000	100,000	50,000	33,000	25,000	20,000
5,000	500,000	250,000	167,000	125,000	100,000
10,000	1,000,000	500,000	333,000	250,000	200,000
25,000	2,500,000	1,250,000	833,000	625,000	500,000
50,000	5,000,000	2,500,000	1,667,000	1,250,000	1,000,000
100,000	10,000,000	5,000,000	3,333,000	2,500,000	2,000,000
150.000	15,000,000	7,500,000	5,000,000	3,750,000	3,000,000
200,000	20,000,000	10,000,000	6,666,000	5,000,000	4,000,000

BUSINESS CONCEPTS FOR SAFETY PROFESSIONALS WHO ARE CONSULTANTS

Not all safety professionals work in large corporations. As downsizing has taken its toll, many safety professionals work in small businesses or as consultants in their own business. Safety professionals who are consultants often fail due to a failure to manage their finances. Knowing financial management principles is imperative to survive. Therefore, safety professionals who are consultants require smart financial management in order to flourish. You do not have to be a CPA to manage your business, at the very least you must know the basics. You should be familiar with at least the following key areas.

The Financial Statement

Your company's goal is to make money, right? One way is to maintain a good financial statement. It is not unlike the company diary. The financial statement is a portfolio that contains a balance sheet, statement of cash flow, and a statement of retained earnings. It shows the way business is conducted, where profit centers are, and where potential land mines are festering. One principal mistake is thinking that you can keep track of all this in your head; you must have a consistent way of tracking activity, and it needs to be on paper.

Lenders look at the balance sheet to see how much your company is worth or how liquid it is. The balance sheet shows the assets you have versus the assets you owe. Ultimately, it shows your net worth. Both of these assets are described in terms of current and noncurrent items. Current items are those items that will be collected or are due within twelve months. Current assets include accounts receivable, inventory, and prepaid expenses. Noncurrent assets include property and equipment. Current liabilities include items such as payroll taxes, accounts payable, and deferred income. Noncurrent liabilities include multiyear loans and other debt commitments that extend past one year.

Total assets minus total liabilities calculates a company's net equity or net worth. The goal is to keep the balance sheet positive rather than negative. Negative numbers mean you owe more than you own. It is not necessarily a death knell, but it should be a concern. Ideally, your debt-to-equity ratio should be no more than 2:1.

The Income Statement

The income statement shows the company's bottom line as net income or net profit. A useful way to arrange income statements is in a year-to-year comparative format. The income statement lists annual net sales-gross sales minus returns and other allowances. For example, if you have $300,000, the statement deducts the cost of goods sold, say $160,000, leaving a *gross profit* of $140,000.

The income statement also covers net operating income. This is the sum of your selling and administrative expenses subtracted from your *gross profit*. Salaries, employee benefits, rent, payroll taxes, utilities, office supplies, and costs for marketing and advertising would be subtracted from gross profit. For this example, let us assume that your selling and administrative expenses total $97,000. When this figure ($97,000) is subtracted from gross profit ($140,000), you have net operating income, which in this case is, $43,000. Finally, subtract interest income and interest expenses to arrive at a net profit before tax. If you were operating in the red, it would show up here as negative profit, or a loss.

Cash Flow Analysis

Based on the information on your balance sheet and income statement, an accountant can prepare a statement of cash flow for your company. It shows the sources and uses of cash (for example, net borrowing under credit agreements, cash used in investing, proceeds from long-term debts, and dividends paid). Obviously, tracking the flow of money in and out of the business is fundamental to the big picture, but unless you are particularly adept with accounting terms and practices, arranging a statement of cash flow is an area best left to an accountant. However, this does not mean that you should not stay as close to the numbers as possible. Safety managers should review balance sheets once a month, and every three months or so put together a statement of changes in financial condition. This statement will tell you where funds are coming from and how they are being applied. It also lets you compare working capital from one period to the next.

These simple principles can assist you in tracking your costs. Perhaps it will give you the tools to turn your safety department from a cost center into a profit center. Perhaps you will discover a mentor who will spy your attention to detail and the bottom line. Perhaps you will change your perception from the safety guy to business person. Perhaps others in the company will notice that you are on the fast track while they are muddling through the day-to- day grind and wondering how you made it to the top. All this from merely knowing more than safety and showing how you track your costs and show fiduciary responsibility. Now that is a formula for success.

CONCLUSION

The budgeting process is vital to every program. The individual responsible for the safety department's budget must learn to describe the tasks in such a manner that they are defensive and realistic. The task definition method of budgeting provides this detail. The best data available for budgeting is historical, but in lieu of that, good judgment and use of experience must suffice. The key to successful task definition is to break down each task into some smaller portions that can be budgeted for, and to budget against a specific measurable effort. The budget, if properly used, can be the basis for effective planning year after year. Once a total budget exists, it may change very little on a year-to-year basis. If there is a major change, the method of budgeting ongoing changes in the overall company's or organization's budget must be done in a responsible manner. This also makes it much easier to prepare subsequent budgets because most of the work is done and there is a detailed example to use in correcting the estimates.

SUGGESTED FURTHER READING

Allison, W. W. (1986). *Profitable Risk Control.* Des Plaines, IL: American Society of Safety Engineers.

Brigham, E. F. (1989). *Fundamentals of Financial Management,* Chicago, IL: Dryden.

Canada, J. R., and J. A. White (1980). *Capital Investment Decision Analysis for Management and Engineering.* Englewood Cliffs, NJ: Prentice-Hall.

English, W. (1988). *Strategies for Effective Workers Compensation Cost Control.* Des Plaines, IL: American Society of Safety Engineers.

Grant, E. L., and L. F. Bell (1964). *Basic Accounting and Cost Accounting.* New York: McGraw-Hill.

Grant, E. L., W. G. Ireson and R. S. Leavenworth (1990). Chpt. 17 in *Principles of Engineering Economy.* "Aspects of Economy for Regulated Business." New York: John Wiley & Sons.

Hansen, M. D., H. W. Grotewold and R. M. Harley (June 1997). "Dollars and Sense: Using Financial Principles in the Safety Profession." *Professional Safety* 42(6):36–40.

Jones, C. A. (1989). "Standard Setting with Incomplete Enforcement Revisited." *Journal of Policy Analysis and Management* 8(1):72–87.

Lyon, B. K. (March 1997). "Ergonomic Benefit/Cost Analysis: Communicating the Value of Enhancements." *Professional Safety* 42(3):33–36.

Nikolai, L. A., and J. D. Bazley (1988). *Intermediate Accounting.* Boston: PWS-Kent.

Oxenburgh, M. S. (1997). "Cost-Benefit Analysis of Ergonomics Programs." *American Industrial Hygiene Association Journal.* 58(2):150–156.

Riggs, J. L. (1976). *Production System: Planning, Analysis and Control.* Wiley/Hamilton Series in Management and Administration. New York: John Wiley and Sons.

Slote, L. (1987). "How to Apply Economic Decision Analysis Techniques to Occupational Safety and Health." and "How to Reduce Work Injuries in a Cost-Effective Way." Chaps. 20 and 22 in *Handbook of Occupational Safety and Health.* New York: John Wiley & Sons.

Tarquin, A. J., and L. T. Blank (1976). *Engineering Economy.* New York: McGraw-Hill.

Theland, D. S. (1988). *Management of the Work Environment, Selected Safety & Health Readings, Volume III,* Project Minerva, U.S. Department of Health and Human Services.

Turner, W. C., J. H. Mize and K. E. Case (1987). *Introduction to Industrial and Systems Engineering.* Englewood Cliffs, NJ: Prentice-Hall.

Viscusi, W. K. (1986). "The Impact of Occupational Safety and Health Regulation, 1973–1983." *Rand Journal of Economics* 17(4): 567–580.

Watts, J. P., and M. L. Ruder (1992). Las Vegas Safety Workshops, Certified Safety Professional, Home Study Workbook. Las Vegas, NV: Las Vegas Safety Workshop.

Zeleny, M. (1992). *Multiple Criteria Decision Making.* New York: McGraw-Hill.

Safety in the Project Design Process

Steven F. Kane

INTRODUCTION

The evolution of a project, from design through construction and the initiation of operations, can be an intimidating process, even to most engineers. Projects are always understaffed, underfunded, and aggressively scheduled. This means that there are never enough people, money, or time to do the project correctly the first time, but that there is always enough to do the project again when it does not work. The old cliché that 90 percent of the job gets done in the last 10 percent of the allotted time lends credence to the "never enough" problem. But, somehow, projects are completed. This chapter begins with an overview of the design process. Few engineers and even fewer safety engineers know or understand the design process. The safety engineer first must understand the design process in order to understand how safety can be efficiently incorporated and reduce project cost in the near and long term.

OVERVIEW OF THE PROJECT DESIGN PROCESS

The design process begins with the first idea for a project and does not end until achieving full production rate, occupation, or product deployment. The efforts involved in this process can be divided into eight phases. Different aspects of the same project may progress through the phases at different times; hence, the design phases appear to overlap. The following are the eight phases of the design process, 5 through 8 falling under the construction phase:

Copyright ©1998 by Steven F. Kane. Reprinted by permission of author.

1) Preliminary or conceptual design phase
2) Design validation or prototype phase
3) Detailed design phase
4) Construction phase
5) Procurement phase
6) Acceptance phase
7) Final testing phase
8) Project completion and commencement of operations phase

Preliminary or Conceptual Design Phase

The preliminary or conceptual design phase begins with the first idea for a project and continues until the decision to pursue the idea. Included in this phase are definitions of design parameters, initial systems definition, preliminary layouts, and a design review that may be a part of a project review or may serve as the project review.

Defining parameters for the project continues throughout the project, but the most significant parameters, the basis for the entire project, are defined during this phase. Parameters include duration of operations, production rates, performance, and cost. All parameters ultimately result in a cost definition. Cost constrains a project or the approach to part of a project. Schedule is an important part of a project that also results in a cost. The cost is to the organization funding the project and encompasses both project cost and lost sales or increased operating expense pending project completion. Thus, schedules are shortened during negotiations, and financial penalties can be included in the contract for delayed deliveries or operations. Penalties and delayed payments for schedule delays are the usual reasons for acceleration at the end of a project.

Developing initial systems requirements begins once the project parameters are defined. Defining *how* to do something can begin only after defining *what* to do. This includes many support systems necessary for operation of the primary systems. Examples of support systems are power, lighting, sewage, and ventilation. These systems may not contribute directly to the primary purpose of the project, but the project could not function without them; hence, they are every bit as important as primary systems.

Preliminary layouts also begin with initial systems definition. These identify site, size, and shape requirements, but might not identify or link "internal" requirements. An artist's rendering of a building or product might not reflect the practicality or cost of the depicted article. Engineers are also beginning to make schematics and drawings for the systems, and those involved in the project will be meeting frequently to make sure the different schematics and systems work well together. Initial cost estimates and trade-off studies help in deciding the best approach for the project.

The efforts expended during the preliminary or conceptual design phase are summarized and reviewed at a preliminary design review. Everyone associated with a project should attend this first, important review. This is the first opportunity for project personnel and the group funding the project to identify problems and risks associated with the suggested approach to the project. Significant design direction is provided during this review. Months of work may be discarded because a particular approach is not resolving issues or is based on expectations of an unproved technology. As a minimum, a preliminary design review agenda should include the following subjects:

- Performance criteria, and deviation from originally proposed criteria
- Potential impact on other systems
- Code/standards compliance
- Cost
- Installation/maintainability
- Requirement to prototype
- Development and production schedules
- Quality assurance classification and requirements
- Outline of design
- Availability/producibility
- Trade-off issues

The preliminary design review determines the need for prototyping and technology development. The purpose of these efforts is clearly defined, so the effort is focused and cost-effective. The preliminary design review also identifies issues, features, and functions to be addressed during the detailed design phase. Of course, the outcomes are the result of a successful preliminary design review. A review that goes poorly and results in more questions than clarifications will require another preliminary design review before proceeding to the next phase. Finally, the preliminary design review is documented. Significant issues, design direction, and action items are included in a report for future reference.

Design Validation or Prototype Phase

Design features or technology development is addressed during the design validation or prototype phase. Project elements that are unproved represent a risk to project completion. The sole purpose of the design validation phase is to develop the risky, but potentially more fruitful, approaches to an element of the project. Failure during this phase will immediately conserve time and cost, and will permit a change to a less risky approach in the event that the feature or technology cannot be developed to the extent necessary for successful project completion. Specific, detailed test objectives are necessary for this phase to be productive. The test objective must be relevant to the requirements for the project element under development. Therefore, test objectives are geared toward proving that a feature or technology can be used in the operational environment. This means that test environments must be much harsher than the foreseen operational environment to compensate for the abbreviated test duration.

The prototype or test article generally does not resemble the production article. It does not have to be appealing nor does it have to be as practical or "user friendly" as the final product, but it does have to replicate the functions, features, or materials to be tested. Cost of the prototype is a significant issue because of the limited usefulness of the prototype—it is usually scrapped when testing is completed. No one wants to spend much money on something that will be "thrown away."

A prototype design review should be conducted before materials are procured or construction begins. The focus of the review is the test objective and how the prototype and proposed test will validate the approach to this project element. The prototype design review should begin with a review of the issues and actions discussed at the preliminary design review for the applicable project element. The review also needs to focus on the relevance of the prototype to the production article. A perfect prototype and a flawless test will be useless if the project element cannot be made or cannot operate like the prototype. The prototype design review agenda should address the following subjects:

- Preliminary review minutes and action item close-out
- Performance criteria, and deviation from the proposed design
- Differences between prototype and production article, with rationalization

- Summary of the test parameters and how they satisfy the need for prototyping
- Summary of the test configuration and comparison to the production environment
- Prototype development, manufacture, and testing schedule
- Layout of prototype
- Qualification testing proposal
- Availability/producibility
- Trade-off issues
- Cost
- Incoming inspection/test

The purpose of the prototype design review is to approve the design of the prototype and to certify the prototype and proposed testing that will validate the approach in question. Approval for construction of the prototype should not be given without this certification. The review is documented and a report issued for future reference.

Prototype construction is a part of the design process because constructed articles rarely comply with the initial design. Material or tool unavailability, technological or product breakthroughs, and better ways to build things crop up during any construction. These changes are either requested of engineering, or developed and implemented on the spot. These changes should always be reviewed to ensure that the changes do not invalidate the prototype design.

Written test procedures are essential, and must be written to meet the test objectives. Incremental buildup testing and initial systems testing should be completed before the qualification testing of the design element begins. This assures correction of discrepancies before the test begins, and prevents any questioning of the test results. Data collection systems should not be overlooked; "bad data" will cast doubt on the test validity and a potentially good technology. Data collection system testing should be included in the test procedures and should be validated before testing begins. The only thing worse than bad data is no data. It may not be possible to instrument all possible parameters, but the parameters that are instrumented should be reviewed for necessity. Every unnecessary data point prevents collection of other, potentially important information.

Test results and observations are detailed and documented. A test log-book with timed entries is indispensable. All extraordinary occurrences are documented and evaluated for potential impact. Finally, a test report is issued to summarize the included data and publish conclusions regarding validation of the project element.

Detailed Design Phase

The detailed design phase finalizes the project design for construction. Specifications and engineering drawings (blueprints) are completed and prospective sources for materials and services are investigated. Specifications for subcontracted items should be performance based and should not be unnecessarily restrictive unless warranted. Performance-based specifications must include test procedures. Expected testing includes first-article testing and acceptance testing. This testing verifies the specified performance for the item. Specifications should address the following:

- Parts, materials, and processes
- Design, construction, and workmanship
- Performance
- Detail requirements

Detail requirements generally compose most of the design requirements. The design portion of the specification is used to impose engineering requirements on subcontractors responsible for design and construction of a project element. Fire protection systems are an example of commonly subcontracted design elements.

Engineering drawings detail the exact form of project elements. A small project may involve hundreds or even thousands of drawings. Engineering drawings are checked for format, tolerance problems, missing elements, and special instructions. A parts list is included in engineering drawings and is used to generate a bill of materials for procurement.

Some projects include a design safety group. This group's charter is to analyze independently a design's conformance to engineering standards or specifications. Stress analysis is a common function of this group.

The final design review is the final review and approval of the design elements. These may be conducted for individual components, systems, or for the entire project. The final design review can be held in conjunction with a project review and should include the customer. This may be only the second and final time all project personnel and the customer have to identify problems and risks associated with the suggested approach to the project. The final design review should begin with a review of previous design review results, action items, and prototype test results. This will prevent covering old issues and catch outstanding action items. The final design review agenda should address the following subjects:

- Preliminary review meeting minutes and action item closeout
- Prototype design review meeting minutes and action item closeout, as applicable
- Performance criteria, and deviation from the original or contracted criteria
- Production and installation schedule, and comparison to the project schedule
- Results of qualification testing, as applicable
- Installation requirements and procedures
- Trade-off issues
- Engineering drawings
- Specifications
- Code/standard requirements
- Quality assurance requirements
- Reliability/maintainability
- Availability/producibility
- Incoming inspection/test
- Cost

The design should be reviewed carefully. Cost of the effort to this point is small when compared to the cost of the entire project. This review is the last chance to abandon the project at minimum cost. It also is the last chance to provide design direction before construction begins. This review is documented and a closed-loop system is established to ensure that action items are resolved; there are no other design reviews to catch issues and action items that may not have been addressed. The review will conclude that construction of the element, system, or project may proceed, or that another review is necessary. Approval and release of specifications and engineering drawings establish the design baseline. A strict configuration control system is necessary once the design baseline is established.

Construction Phase

The procurement process, acceptance, installation, and final test compose the construction phase.

Procurement Process. The procurement process is very much a part of the design process. Decisions regarding the type of procurement method should be based on technical requirements. Thus, the engineer is the only person qualified to make these decisions and to ensure the technical success of the project. Awarding the procurement to a sole source is the first decision. What the project loses in cost by eliminating competition may be more than compensated for by the assurance of a low-risk technology essential to the project's success. Sole source decisions may be based on compatibility with existing elements from the same source, or reduced cost by upgrading existing equipment. The project must be careful with sole source procurements if the source is unstable or unreliable. Completing the procurement after default of a sole source will jeopardize project schedule and cost, and may affect other project elements.

Competition minimizes project cost. The decision the engineer makes is the basis for the procurement award, cost, or technical superiority. Strict cost evaluation can be used for low-risk technologies. Higher-risk technologies or technologies critical to project success are heavily weighted in favor of technical competence. This requires a proposal from each prospective subcontractor and a proposal evaluation plan. The plan is prepared before proposals are requested and details the weighting for the elements of the proposal. The engineer must decide the size of the technical risk and assign commensurate weighting. The proposal evaluation plan also prescribes the evaluation team. Engineers comprise the majority of the evaluation team for a technical proposal.

Prospective subcontractors will propose an approach that is compliant with the proposal request and instructions. Often an alternate approach is proposed. This alternate approach is evaluated with the same rigor used for the final design review. The evaluation team includes engineers from all potentially affected areas for a thorough review.

After the contract is awarded, the contractual process controls all changes. Engineers must approve all technical change proposals submitted by subcontractors. A statement of work is also prepared for proposals and becomes a part of the contract on award. The statement of work should detail documentation required to show compliance with the specification(s), engineering drawings, and contractual requirements. Test reports, design reports, and witnessing of subcontractor testing are elements included in a statement of work.

Acceptance. Acceptance of procured elements and systems is a phase of the design process. Incoming inspection is based on procedures developed by the design engineer to ensure compliance with the specification and statement of work. The subcontractor usually performs functional testing before shipping. The purchaser usually repeats the tests to detect damage during shipment. This is prudent before the element is installed—it will save time spent removing and reinstalling the element if it does not work when received.

Discrepancies are inevitable. A nonconforming material review system addresses the discrepancies detected during incoming inspection and acceptance testing. This review is based on technical requirements and is documented. An element painted the wrong color is usually not detrimental, but an element lacking a specified function or incorrect interface may be unacceptable. Schedule or cost may force acceptance of the element, and the engineer must design an alternative approach to use the element. Thus, the design task continues through the acceptance phase.

Installation. In-house manufacturing procedures can be classified as specifications and dictate methods and parameters for construction and installation. These procedures are developed by the design engineer or by manufacturing engineers. Manufacturing engineers take specifications and engineering drawings and create the steps necessary to make or install an element or system. The manufacturing engineer also reviews the proposed design and may suggest changes to simplify manufacturing or decrease the cost to manufacture the element. However, only the design engineer may approve the proposed design changes.

Incremental testing is performed as elements are installed and systems are built. This testing verifies the function of the elements and systems before they become too expensive or time-consuming to remove or repair. The design engineer develops and documents these test procedures and discrepancies discovered during testing are reviewed for design problems. Discrepancies may be the result of problems with the test procedure. The design engineer must make this determination and design changes to correct problems.

Final Test. The final test is the last phase before project completion and commencement of operations. The final test requires written procedures. The design engineers develop these procedures, which may be just a compilation of the incremental test procedures. Operating procedures and manuals should be available for the final test. The final test phase is an ideal time to validate these manuals. Design engineers develop the initial draft of the procedures, and technical writers make them user friendly. A final test report may be generated to document compliance with the project objectives or contract requirements.

CHANGE CONTROL

Change control begins with the technical baseline of the project elements and continues for the life of the constructed article or facility. Change control is a management system to ensure that the same degree of engineering review applied to the original baselined design is conducted for changes and modifications. Management also uses this system to monitor cost, schedule, and performance. Changes usually are restricted to the following:

- Safety
- Making the project work
- Reducing cost
- Meeting the schedule

Making it "right" or making it "better" is not a valid reason for a change after baseline design.

Changes are proposed by the subcontractor or by the project. Subcontractor changes are processed through the contracting group, but must be approved by the design engineer. The design engineer may develop required changes to correct deficiencies discovered during the phases following technical baseline. Other project personnel may propose changes to correct deficiencies. The design engineer must review and concur with all proposed changes. A change review, similar to a design review, is conducted for the proposed change. The change review agenda should include the following subjects:

1) Requirement for the proposed change
2) Overview of the proposed change and how it satisfies the need for the change
3) Review of the change proposal for
   - Classification
   - Completeness

- Technical adequacy
- Change effectively

4) If the change is and is not implemented, the impact on
   - Form, fit, function
   - Design requirements
   - Cost and schedule for
     — tooling
     — material
     — labor
     — equipment
   - Other systems

The change review also is documented. The result may be approval or disapproval of the change proposal. Changes are made only with written authorization. These changes and their authorizations are maintained for the life of the constructed article or facility. Since the approved changes modify the originally approved baseline, the approved changes and the original drawings represent the "as-built" configuration. Engineering drawings and specifications may be updated to reflect the approved changes and simplify identification of the final configuration.

SAFETY IN THE PROJECT DESIGN PROCESS: APPLICATION

Now that you know what will happen during the design process, we can look at how safety must be involved to maximize effectiveness and minimize cost. Previously, the constraints under which the project must operate were discussed. The safety engineer should have an understanding of the problems imposed on the design engineer. Design engineers need more help, not more work. This explains the less than desirable welcome received on your last project. Safety engineers also may have arrived on the scene too late. The design is easy to modify before technical baseline. The change control process and procurement status of the element make changes difficult and costly to accommodate after the technical baseline has been established. Safety engineering must begin with the start of the conceptual or preliminary design phase.

Let us also discuss the safety philosophy that is trying to put all of us out of work. This is the philosophy that safety is a line responsibility. Engineers are not required to take any safety courses for an engineering degree. There are no safety questions or subjects addressed by professional engineering licensing examinations, except for safety engineering examinations. And, there are no additional courses an engineer must take to continue the practice of engineering. Therefore, there is no assurance that design engineers know anything about safety. So how can they be expected to fulfill the safety function? They cannot. Safety engineers must be responsible for safety, or at least impose on themselves the responsibility for safety.

I was very fortunate to be selected to function as an assistant to an engineering manager. I had no official authority or title, but I did have his confidence and trust. One day I complained about a problem and how I could not convince the design engineers to correct the problem. He told me to design it myself. I did not know what to say, so I blurted out, "But I'm not allowed to do that!" So he asked me, "Who said you aren't allowed?" The fact is, no authority ever said a safety engineer was not allowed to design. Sometimes the safety engineer is the best engineer to design some elements of a project because he or she knows how the element should *fail* as well as work. Safety systems also are best designed by safety engineers. Since that discussion with my engineering manager, I have always asked that same question of any engineer that challenges my authority to design. The irony is that, as a Licensed Professional Engineer and Certified Safety Professional, I am frequently more qualified to design than the individual challenging my qualifications.

Safety engineers need to work with the design engineers. If safety engineers work with the design engineers, and if safety engineers are seen performing engineering tasks, safety engineers will be accepted by design engineers over time and will be sought out for their help and advice.

Safety can be incorporated in the design process although the engineering culture may resist. Quality assurance is firmly established and accepted in nearly all organizations. Quality assurance is the safety engineers' greatest ally. Its priorities are based on the consequences of a failure. The dollar value of damages, injury potential, and schedule impact define failure consequences. Schedule impact can be assigned a dollar value; hence, the quality assurance priorities are based on *safety* impact. The safety engineer is the best source for failure consequences; you must make sure that the quality assurance group knows this. Design engineers and project management will usually comply with quality assurance requirements if they are reluctant to comply with safety requirements. Use this to your advantage.

Safety Analysis

Safety parameters are no less important than other parameters. In fact, safety parameters should look like performance parameters. The project design first must satisfy performance parameters; safety will be addressed from the initial design if it is part of the initial criteria. Emergency and failure conditions should be addressed in the formulation of these initial design parameters. A system configured to a safe state with loss of power is an example of a safety parameter.

Safety systems and features must be developed during initial systems definition. Security, lighting, fire protection, and environmental systems are just a few. Define as many of these systems as possible during this phase. It is much easier to delete systems than it is to add them later. Propose safety systems and define the hazards they are to address. If the design evolves to eliminate the hazard, be the first to eliminate the safety system. This shows a desire to minimize cost. Safety systems should be independent of primary control systems. The safety engineer also should define failure modes for these systems. The loss of one system should not result in an undesirable impact on another, related system.

A preliminary safety analysis should begin in earnest at the initiation of the conceptual or preliminary design phase. This analysis should be documented in a report and should be discussed at every opportunity with the customer and project management. Potential hazards usually can be placed in the following categories:

- Acceleration
- Contamination
- Corrosion
- Chemical dissociation
- Chemical replacement
- Fire
- Explosion
- Electrical
  — Shock
  — Thermal
  — Inadvertent activation
  — Power source failure
  — Electromagnetic radiation

- Heat and temperature
  — High temperature
  — Low temperature
  — Temperature Variation
- Leakage
- Pressure
  — High
  — Low
  — Rapid changes
- Radiation
  — Electromagnetic (radio frequency)
  — Ionizing
  — Ultraviolet

- Moisture
  - High humidity
  - Low humidity
- Weather/environment
- Shock

- Stress reversal
- Stress concentration
- Vibration and noise
- Structural damage or failure
- Toxicity

Safety analyses should consider all life cycle phases:

- Fabrication
- Assembly
- Test and checkout
- Handling
- Storage
- Transportation and shipping

- Maintenance (scheduled)
- Repair (unscheduled)
- Operation
- Modification
- Disposal

The safety analysis should include identification of safety design criteria detailed in applicable codes, regulations, imposed standards and specifications, and industry voluntary standards. The purpose of nearly all codes, regulations, standards, and specifications is safety. These codes may prescribe design requirements, but the purpose is to prevent accidents caused by failures. Too few engineers know of applicable design requirements because of the proliferation of these requirements. The American National Standards Institute alone lists almost 10,500 standards in its latest catalog. The SAE, ASME, IEEE, UL, NFPA, AISI, AISC, ASTM, NEMA, BOCA, CGA, ASHRAE, IES, DOT, DOE, NRC, AWS, UNS, ISO, and OSHA are just some of the organizations promulgating design standards. The safety engineer must investigate all potential sources for design requirements. Using a design standard to justify a safety requirement is much easier than using an opinion. It also gives the safety engineer an irrefutable position—his or her hands are tied by the design standards. Hence, the safety engineer is not imposing a requirement; the industry or government is. I have not failed to define safety requirements using these standards. Such standards give the safety engineer another opportunity for acceptance within the design engineering community. Design engineers will be unaware of these standards because they do not have them. Once they are "surprised" with a standard, they should start coming to you to borrow your copy of the standard or to ask about the existence of a standard. Then, the safety engineer will be a part of the team.

We have discussed the preliminary design review and agenda, but safety also needs to be addressed. Safety implications can be found in trade-off issues, quality assurance classification and requirements, code/standards compliance, potential impact on other systems, cost, and requirement to prototype. Prototyping new technologies for safety features and safety systems is essential for project success. Safety features or systems that do not function as planned can stop a project just as surely as a performance failure. This prototyping will limit unfounded claims of "inherent" safety.

Safety engineers need to participate in the review process to ensure that safety issues are adequately addressed. The purpose of the review is to provide design direction. Multiple alternatives for the same element may be proposed to the review committee to elicit direction. These proposals may not have been reviewed by the safety engineer. Problems with a potential approach must be immediately identified for consideration by the review committee. The agenda should include safety issues such as potential hazards and proposed controls, a review of previous safety analyses, a summary of the conceptual phase safety report, an overview of safety systems, and a presentation of unresolved safety issues.

Testing of Safety Systems

The design validation or prototype phase begins with the definition of test objectives. Validation of safety features and safety systems must be included in these objectives. Applicable specifications and standards must be identified for prototype design. However, the short life and limited use of the prototype does not necessarily require adherence to the same standards necessary for public or operational use. The intent of the safety requirement should be followed. Again, limited exposure and a controlled environment permit using special procedures and training to compensate for less than optimum safety in design. Additional safeguards beyond written requirements also may be necessary to ensure limited exposure and a controlled environment. The objective is a safe test, not representative production article.

The prototype design review needs to include potential safety issues and comparison to the preliminary review safety issues. The differences from the production article and safety implications must be discussed. The review committee should define and approve the safety limits of the test and the prototype. Applicable specifications and standards should be reviewed with emphasis on noncompliance and safety measures to provide equivalent safety. It is important that project managers and engineering managers know when the design does not or cannot comply with requirements. They must be willing to accept the alternatives. Safety engineers must be convinced that alternatives are viable enough to make the project managers and engineering managers comfortable. The safety engineer must accept some responsibility for the safety of prototype testing. Finally, safety test objectives and procedures should also be reviewed.

Construction of a prototype requires as strict an adherence to the design documents as the production article. Change control also is essential. Watch out for design modifications and "design-in-place" efforts. Carefully review these changes for safety impact.

The safety engineer must review the qualification test procedures. Additional cautions, warnings, and safety procedures are required by the limited exposure and controlled environment used to justify less than optimal safety features. The design engineers draft the procedures, but review the procedures with the technicians who are building and testing the prototype. The technicians know exactly how the prototype functions and can relate all the pitfalls of the procedure. They also can explain the shortcuts, which may affect safety. Qualification test procedures should define the limits of safe operation of the prototype and the limits of the testing. Emergency procedures also must be developed, which means that the safety engineer and the design engineer must review the prototype for failure modes and potentially hazardous operations. Devising emergency procedures after an emergency occurs makes it extremely difficult to prevent an accident.

The safety engineer should witness the qualification test, especially the safety tests, and review all anomalies for unanticipated safety problems. Often a problem "discovered" during operation or the final test was observed during prototype testing but was not investigated. No problems unique to the prototype will disappear without engineering intervention. Testing should follow the procedure. Deviation from the procedure must be documented. Do not permit "what-if" testing without a review of the potential consequences and a written procedure. The prototype should not be permitted to fail before all planned testing is completed. The prototype represents a significant project investment. Premature failure caused by unplanned testing may jeopardize the project schedule. These tests may result in prototype failure and personal injury because they are unreviewed operations.

Test results should be revealed during the draft stage of report writing. The safety engineer should ensure that safety tests are truthfully reported, without "artistic license." The consequences of anomalies also must be faithfully reported.

Safety Specifications in the Detailed Design Phase

Specifications are, perhaps, the single most important element in the detail design phase, yet all engineers seriously overlook them. The specification defines the performance and requirements for an element, and usually becomes part of a procurement contract. Thus, the contracting organization will be stuck with an undesirable element, with no recourse, if the specification is not correct. Prohibit use of hazardous materials defined by 29 CFR 1926 without written approval. This will make subcontractors think twice, and it will make them work to justify the use of a hazardous material that will increase your company's workload.

One of the best things a safety engineer can do is impose requirements prescribed by other standards, specifications, and regulations, as discussed for the preliminary design phase. The safety engineer will be effective in imposing these standards for several reasons. First, the requirements are not the safety engineer's idea; manufacturing and engineering organizations prescribed them. This makes the requirements "design" requirements and not really "safety" requirements. Second, these requirements are neither new nor unique to your project. The cries about "raising the cost of the job" are unfounded because any good manufacturer would not attempt to design or construct the specified project element without complying with the requirement. Finally, the cognizant engineers drafting the specification should have been familiar with these industry standards. They generally will accept the standard and, after being "caught" a few times, will come to the safety engineer to determine whether there are any other standards they should know about.

Use the specification to prescribe safety devices and features for a project. Always specify performance rather than "how-to-do" requirements. The element will not comply with the specification if it cannot meet the performance requirement. However, if the element incorporates the detailed design requirements but does not perform safely, the project, again, will have a hazardous element with no recourse. Fail-safe features are particularly important. Impose the requirement for the element to ensure that it does not fail, and make sure that the fail-safe condition or mode is specified. This is necessary to ensure that your idea of fail-safe is the same as everyone else's. Fail-safe may have different meanings for different operating modes; be sure to define these as well. Performance standards require verification by testing. The quality assurance group should address this testing. Qualification testing is particularly important and may be taken to full destruction to test fatigue and stress limitations. Use this testing to validate all safety features, especially fail-safe features. Some features are accepted for the entire production run based on these first article tests, which is acceptable because the "inherent" safety of the element has been validated. The quality assurance group also should address acceptance testing. Acceptance testing is not taken to destruction, but is used to verify element performance. This testing also verifies interlock function, warning device function, inspection for guards, and so forth. The specification should require that these tests be documented because test failures require investigation. The specification should also require failure correction and test rerun.

Safety engineers must complete safety analyses quickly during the detail design phase. Safety analyses include fault tree, fault hazard, HAZOP, what-if, failure mode effect, and sneak circuit. Design changes to correct safety problems will increase cost and will be harder for engineers and project management to accept. Design engineers generally will accommodate design changes for safety before drawings and specifications are released as official documents because change control does not become involved. As a project design matures, more element designs become fixed. Design engineers will attempt to resolve design problems in the uncompleted elements. These compromises can lead to creation of safety problems. Therefore, the safety engineer must be available when these problems are discussed so that these safety issues can be addressed. Sometimes "task teams" are formed to resolve design issues. Safety engineers should be members of the "task team," but will only be invited if they are known as problem solvers. If you go out of your way to be a part of the problem-solving group, you will always be invited to participate.

Design safety analyses, primarily stress analyses, must be completed before drawing approval and must be documented. These analyses may require certification of compliance with design standards, such as ASME and UL. The safety engineer should ensure that certifications are obtained when required and keep copies of such certifications in safety files.

A safety assessment should be completed and documented before a final design review. This will be easier to accomplish when done for each element before its individual design review. The safety assessment completes the analytical effort begun with the preliminary safety analysis during the conceptual or preliminary design phase. It shall consider all potential hazards throughout the life-cycle phases. The safety assessment must identify all unresolved hazards and recommend changes for resolution. Recommendations should be communicated to the design engineers before the final design review. This will give them the opportunity to resolve the problems before the final design review and to review and concur with your analysis. Design engineers may discover additional hazards or find hazards resolved since the issue of the design documents used by the safety engineer. This effort preserves everyone's integrity and promotes team building.

The final design review and agenda have been previously discussed. Safety implications can be found in trade-off issues, quality assurance classification and requirements, code/standards compliance, results of qualification testing, cost, reliability, maintainability, production, and installation. The agenda should include a review of the safety assessment, a review of safety issues discussed at previous design reviews for that element, and a review of the preliminary safety analysis. The safety engineer should ask many questions during the review, especially during the technical portions, to ensure a good understanding of the element under review. I guarantee that you will not be the only person at the review that does not understand. Once the ice is broken, many others will begin asking questions. Then listen to ensure that all safety issues are addressed. Watch out for "hand waving" and "smoke and mirrors." Be firm in getting an understandable answer to your questions. Insist on action items if the answer cannot be readily provided. The final design review is the last chance to make "cheap" changes, so be thorough.

Document the issues and their resolution in a report along with action items. There must be a fail-safe system established to ensure that action items are resolved before specifications and drawings receive final approval. Again, the quality assurance group is your greatest ally. It is this group's job to ensure work product quality, so capitalize on this extra "staff."

The outcome of the final design review for the safety engineer is fixing of the design. This ensures that all hazards are known and resolved to the greatest extent possible by design. Use safety procedures, personal protective equipment (PPE), and training to address hazards that are not eliminated. A final design review establishes the minimum for the operational safety program.

Safety Considerations in the Procurement Phase

Safety is addressed through the procurement phase. Safety must be considered during the decision for sole source procurement and off-the-shelf procurement. A sole source may be woefully inadequate from a safety standpoint. Any cost savings gained by using the sole source may be lost in the cost for safety modifications or increased costs for safety equipment and training. Off-the-shelf equipment can have the same shortcomings.

Safety must be addressed in the request for proposal for competitively procured elements. Include requirements to describe safety features in the definition for the proposal response. Also include safety in the list of criteria to be evaluated in the proposal. It would be best to have a separate safety description required for the proposal response, to include a weighting factor for proposal evaluation. Safety engineers should be a part of the proposal evaluation team to ensure that safety is adequately addressed by the respondents. Since many design requirements have safety implications, a poor response to safety requirements also may cause lost points for the technical design section of the proposal evaluation. Review estimates for safety efforts for rationality. I usually request safety reports with design changes. One subcontractor included a bid of $10,000 for these reports. During negotiations I asked them how many design changes they were planing to make. They had to say "none," because a contract could not be awarded to someone that built that type of cost-escalation factor into a proposed design effort. I reduced the amount from $10,000 to zero because they could not charge me for a report that they did not plan to write. In this manner the safety engineer will reduce the cost of the project in the eyes of the project management, yet ensure that safety is addressed in the subcontractor's effort.

Acceptance Testing and Reviews

The safety engineer should be involved with acceptance test failure review and nonconforming material review during the acceptance portion of the construction phase. A system should have been established to document acceptance test failures. The safety engineer must work closely with the quality assurance group to review all failures for potential hazards not previously detected or expected. An unexpected hazard that is out of compliance should have well-written specification performance requirements. The subcontractor should be obligated to correct the hazard. Minimize the cost for retrofit by quickly addressing all discovered hazards. Nonconforming material also must be reviewed. Again, the quality assurance group is your closest ally. The material review board should include a safety engineer to ensure that potential safety issues are considered in the decisions to accept, modify, or reject nonconforming material. A documentation system should have been established to document these decisions.

Installation Reviews

The installation portion of the construction phase is the most exciting yet distressful of the design phases, especially for the safety engineer. Months or years of outstanding safety effort can be destroyed in a matter of days by the feverish pace and burning desire to finish the project. Earlier delays and schedule slips are often the primary cause.

Assembly of individual elements and systems is being attempted for the first time. The installation group is discovering unexpected problems. Most are minor, but all must be processed through the change control system to ensure adequate review. Witness points are reached and incremental buildup testing is conducted.

The safety engineer must be responsive to all these demands. Quickly review minor changes for safety impact, but with the same degree of thoroughness as the detail design phase. The safety assessment must be reviewed and updated to reflect the latest changes, no matter how insignificant they may seem. Construction cannot continue until witness point inspections are adequately and successfully completed. The safety engineer must respond quickly, as the installation team will be idle until the inspections are complete. Incremental buildup testing also must be witnessed, but work is immediately redirected or stopped by failures. Check procedures for errors. True failures will be simply repaired, but incompatibilities require work-around designs. The safety engineer must review test failures for safety implications and review work-around designs for impact on safety features and systems.

As installation progresses, the schedule continues to slip. Project management will look for ways to speed up installation. Doing less is a logical approach, but not at the expense of testing, especially safety testing. A lack of failures detected during installation is not adequate justification for discontinuing a test, especially a safety test. The safety engineer must be firm. Later detection of failures will cost more in both time and money, if it also does not result in injury. Use the documentation from the design reviews to reinforce your position.

Final Test Procedures

The final test is the last portion of the construction phase and the conclusion of the design effort. The incremental buildup tests, operator training programs, safety procedures, PPE, and final test procedures must be completed before the final test can commence. This establishes the minimum for the operational safety program. Qualify final test operators to the same degree as production operators. Ensure that safety feature and safety devices are successfully tested before an operational test is initiated. Be vigilant for test and project personnel urges to continue with testing in spite of safety test failures—this can jeopardize the entire project. The violation of test procedures is most likely to occur during off-hour shifts, so ensure safety engineer coverage during these hours. Again, failures require review for safety implications, a review of the test procedures for errors, and a safety review of last-minute redesigns. Document all failures and their resolution. The changes in design and procedures must be incorporated in the production operating procedures, training, and PPE.

Change Control Agenda and Reviews

Safety engineering participation and review have been preached for changes during each of the design phases. Safety is a primary justification for changes after technical baseline. The safety engineer may propose changes for safety reasons. However, the cognizant design engineer should concur with the proposed change. This does not mean that the proposed change is dead if the cognizant design engineer should not concur, just that it will be harder to get approval. Good project managers will understand safety problems and the need for change, and will help the cognizant design engineer to concur.

The change review agenda should include a review of the safety impact regardless of whether or not the change is approved. This must include a review of the safety assessment documentation regarding the element or system to be changed, for an understanding of the safety of the originally approved design. Update the safety assessment documentation if a change is approved.

CONCLUSION

Safety in the design process requires a concentrated effort at the beginning. Safety can be more easily accommodated before pencil is put to paper (stylus to CAD machine?). Additions to designs for all reasons only become more difficult, and expensive, with time. The final phases of design may require more checking than at design and analysis stages for the safety engineer, but that checking is necessary for all engineering specialties.

Safety engineers must rise to the challenge of a design project just as other engineering specialties in order to make the project succeed. Be a part of the design team—*do* safety engineering!

Chapter 17

Safety Program Implementation: Using Audits and Safety Committees Effectively

Paul G. Specht

INTRODUCTION

Previous chapters in this book have dealt with the development of a safety and health program. This chapter covers some of the primary issues associated with the successful implementation of a comprehensive program and reviews some of the tools an effective manager will use to produce positive results. Specifically, two major topics are discussed: 1) the utilization of safety committees, and 2) the establishment of an effective audit system.

CASE STUDY: A MULTILEVELED APPROACH FOR IMPLEMENTATION

For nine years, Devon Manheim was the safety and health "program" for the York Manufacturing Company. Devon did an adequate job of maintaining OSHA and EPA records, conducting weekly inspections, enforcing safety rules, investigating all accidents, and responding to numerous safety-related emergencies. But he never seemed to have enough time to do much else. When the company decided to cut the costs of middle managers, Devon was one of the first to be let go.

Devon was replaced by Thomas Martone, an inexperienced, but enthusiastic, safety and health professional. During his first month on the job, Tom discovered that his predecessor had written a comprehensive safety and health program but that it apparently had never been fully accepted or implemented. Tom needed help to stimulate employee interest and involvement at all levels if he was to survive. He took two major steps in implementing a comprehensive program. First, he began teaching supervisors and managers about the benefits of a comprehensive program, which includes holding all employees and managers responsible for safety and health. Second, he established a company safety committee. Tom's multilevel approach included short training programs on how to recognize loss potential, how to conduct effective safety audits, and how the organization can benefit financially if everyone assumed some responsibility for safety.

ROLES AND RESPONSIBILITIES

Many organizations have impressive written safety and health programs that have never been successfully implemented. Implementing, supporting, and directing the program then becomes one of management's primary tasks. Achieving coordinated action, continuously monitoring performance, evaluating feedback and results, and revising the program to meet developing needs are only a few of the safety and health manager's tasks.

Implementing a comprehensive safety and health program requires commitment on the part of the organization at all levels if the program is expected to succeed. This includes assigning responsibility first to a program manager. The control process rests solidly on a base of effective communications: relaying information down the line, gathering necessary feedback, and receiving the information from all levels of the company. There must always be a two-way system of communication—reaching all the way from the CEO to the employees, and from the employees back to the CEO.

All persons above the worker level must know and understand their roles and responsibilities for safety and health. They must also be aware of the roles and responsibilities of other functions and persons, because these may overlap or depend on each other. Rarely are managers effective in isolation. Moreover, managers can fulfill their responsibilities only if they know and understand them. How can these roles be communicated? Among other ways, they should be

- Included in job and position descriptions
- Made a part of entry training and briefings
- Integrated into audit forms
- Published in company handbooks
- Referred to in company statements
- Specifically mentioned in job assignment

determine whether his or her mandate is clear, whether committees know their responsibilities and tasks, and whether the safety and health program works as expected.

Some Practical Considerations

A committee without an agenda is like a ship without a rudder: the agenda is a necessary tool to help the committee manage its time and to ensure that all important topics are dealt with in an orderly and efficient manner. A typical agenda outline includes the following:

1) Review minutes (of previous meeting)
2) Announcements
3) Subcommittee reports
4) Monthly statistics (including a review of all accidents and injuries)
5) Old business (issues from a previous meeting that were not completed)
6) New business

All meetings should be scheduled in advance for a convenient time. Having them at regular times (every fourth Tuesday at 9 a.m., for instance) will help avoid conflicts. Do not overschedule meetings. Allow enough time between meetings to follow up on actions. Meeting once a week is probably too often for most groups, whereas every two months is not often enough. The most frequently used schedule is once per month. This scheduling ensures that members do not spend too much time attending meetings and that actions recommended at the last meeting are well under way.

Minutes are kept and include the names of attendees, those who make motions, and those who are assigned to take action. Making this information a matter of record is essential for follow-up and helps management assess the performance of individual committee members. If there are not enough agenda items to make a meeting worthwhile, *cancel it.* Nothing destroys a good committee faster than the thought that members are wasting their time. If most committee members are absent, the meeting should also be canceled. In either case, the meeting should be rescheduled, if possible.

When to Use a Committee

The task of controlling injuries in the workplace is more than a safety and health professional, or even a team of professionals, can handle alone.

- Incorporated into job and position definition reviews
- Used as a rationale for coordinating problem-solving efforts

As new problems develop or new safety and health measures appear, this communication process must be repeated to make sure that all levels of the organization are aware of their safety and health responsibilities. In briefings, meetings, and reviews, higher management should seek to determine whether if lower management levels grasp and understand these communications, and clarify them when necessary.

SECURING PROPER INPUT AND INFORMATION

For a safety and health manager to succeed, he or she must be an effective communicator. He or she must be competent at providing and receiving the correct information. Only through this ability can a manager evaluate the effectiveness of the safety program and the environment in which it must function. The proper input and information that the safety and health manager must gather includes the following:

- Notable changes in safety and health performance (either better or worse)
- Injury and illness rates
- Average annual losses by company, division, and department
- Comparison of the company's safety and health performance with industry-wide performance
- Employee's direct involvement with safety and health issues
- Management's direct involvement with safety and health issues
- Identification of departments or functions with special risks
- Environmental risks involving the workforce and property
- Impending legislation or regulations
- Public or political pressures regarding safety and health matters
- Items having a significant or unusual cost impact

Intelligent and effective management decisions require this input. It allows the manager to compare actual performance with goals and objectives, to review resource allocations and priorities for dealing with a situation, and to consider alternative actions. Management can then accept the situation as satisfactory or take steps to remedy it.

As the previous list shows, much information and feedback is involved. It is not possible or good practice for the manager to attempt to become an expert on and deal personally with each item. Intelligent screening and allocation are necessary. Top management should separate the items and send them to the proper managers for action. Management can eliminate those items with little significance and pass the rest through the organization for routine action. Sometimes a systems specialist must furnish details before such a decision is possible. Perhaps a risk/loss analysis is needed or the manager needs more information. Constant attention will alert the manager to any trends in hazards or risks, or to any new situations that have occurred since the last analysis. Only in this way can early preventive action be taken.

Watching for trends is particularly critical, because it points to areas of change. The change may be for the better (as when a safety system is being properly managed), or it may be for the worse (showing problem areas). Changes may be detected through

1) System performance statistics
2) Field monitoring activities
3) Appraisals, audits, and inspections
4) Specific factors such as

- Personnel changes
- Hardware changes
- Management changes
- Environmental changes

The wise manager knows that he or she needs relevant facts in all these areas and will see to it that such information is furnished on a regular and timely basis—not just when a problem develops. It is not hard to merge this information system into the reporting and feedback systems that already exist on every management level. Such channels already provide information on sales, costs, quality, production, and so forth. Often, the previously listed safety and health factors can simply become part of the regular department or division reports. When special problems do occur or trends appear, the manager can secure more information as needed.

The information and feedback system is, of course, a means of furnishing the data that a manager will need. However, raw data alone are rarely useful. The manager should also be familiar with techniques for analyzing the data to discover what they mean and what actions are needed.

Many safety and health analysis techniques are available for this purpose. However, these sometimes sophisticated techniques require the services of an expert. Moreover, they may be limited because

1) Some risks cannot be detected by the worker or first-line supervisor on the site.

2) Some hazards and risks can be found only by the worker or first-line supervisor. (How does the manager get information from them?)

3) Hazards not directly associated with the worker may be detectable only by an expert or specialist (as when the hazard cannot be detected by the physical senses).

Realizing the limitations of the feedback system is necessary in order to provide for an adequate information base. Given the previous information, adequate feedback requires

1) Regular feedback from inspections or audits made by someone other than the first-line supervisor

2) Regular feedback from the worker and supervisor

3) Regular feedback from experts or specialists (such as those concerned with air-quality sampling or noise monitoring)

SAFETY AND HEALTH COMMITTEES

The safety and health committee is one of the most underutilized tools of an effective accident prevention program. However, few organizational devices generate as much debate as the committee. Great controversy exists over successfulness of a committee in dealing with any corporate function. Despite the doubt about any committee's real value, a safety and health committee can accomplish many of the previously listed objectives. Employees' direct involvement with safety and health issues can be realized as members provide management with regular feedback from audits, accident investigations, and activities that can provide long-term benefits to the organization.

If properly established and administered, the safety and health committee becomes one of the manager's most effective operational and control devices. How valuable a safety committee is depends largely on management itself. Depending on management's approach, the safety committee can be an important two-way communication device with broad powers for action or a forum for small and insignificant talk.

Advantages and Disadvantages of Safety and Health Committees

Most managers who understand the value of empowering employees know the various advantages of committees as operating tools. However, without the correct leadership, participants, and upper management support, the committee could become a waste of time and money. Consequently, there are peculiar features of safety and health committees that make a discussion of their specific advantages and disadvantages appropriate. The following sections concentrate on these advantages and disadvantages from a safety and health standpoint.

Advantages. A properly constituted and administered safety and health committee offers the organization several advantages:

- It can serve as a communication channel for safety and health policy and practices throughout every level of an organization.
- It can give management the benefit of multiple judgments and group opinions on proposed and actual safety and health practices.
- It can foster understanding and cooperation where matters cross-departmental lines.
- It can accomplish more than one expert working alone.
- It increases active participation in programs and problem solving.
- It can lead to consensus, which may be necessary to start new practices.
- It multiplies safety and health efforts and allows members from various departments to share their knowledge and expertise with each other.
- It can bring the skills of several different persons to bear on a single department's problem.
- In small operations, it is sometimes the only device needed for handling all safety and health needs.

Disadvantages. There are also many disadvantages in relying on safety and health committees. The organization should be fully aware of these disadvantages, because most of them can be easily remedied once they are recognized. The disadvantages are as follows:

- Management may try to use the committee to shift safety and health responsibility, avoid making decisions, and avoid taking action.

- It may be difficult to provide adequate leadership with the requisite safety and health expertise and to provide enough management support.

- Committee actions may not be recognized, appreciated, or properly valued in terms of their effect on safety and health.

- Committee recommendations may not, or cannot, be acted on or responded to by management.

- Membership may not be fully representative. For instance, workers or union members may be excluded, or a department with an unusual number of problems may not be represented.

- Scheduling problems may keep the committee from meeting on a timely basis; this may postpone action on crucial safety and health matters.

- Committees may not have adequate guidelines on what to do or how to proceed. Someone from a higher management level must take the time to prepare and relay such guidelines.

- Proper records of meetings may not be kept or passed on to the proper management level for action or recognition.

- The safety and health committee, if not properly organized and chaired, may waste the participant's time on trivial issues, which could be solved more effectively by one person.

- The committee could be in violation of the National Labor Relations Act if management exercises too much control and does not understand its role in an employee committee.

Most of these problems can be solved or avoided by implementing proper management action in the first place. If the committee is properly mandated, organized, recognized, and directed, and if two-way communications are open between the committee and top management, none of these problems need develop. In short, these disadvantages are all factors that management can control.

Organizing the Safety and Health Committee

There are many ways of defining and organizing safety and health committees. These committees may be authoritative, advisory, informative, or specialized, such as an ergonomics or an emergency response committee. They may also be ad hoc committees informally convened to solve one problem, or they may be permanent and formally constituted as part of the organizational structure. This chapter concentrates on the permanent committee.

Establishing the Foundation. Without the proper foundation, a workplace safety committee is doomed to fail. The organizer of the committee should be certain that several items are in place before the committee begins to deal with safety and health issues. First and foremost, the committee must have top management support. This support will need to be visible as the committee functions within the organization. Second, the long-term goals and objectives of the committee must be identified. These objectives must be reasonable, attainable, and measurable. Third, the extent of the committee's authority needs to be agreed upon.

Membership. On an advisory safety and health committee, properly selected members can be effective even without special knowledge or qualifications. Expertise in managerial decisionmaking may be helpful but is rarely available in operating committees. It is probably better that the committee members simply be thoroughly familiar with the sectors of the company they represent—knowing their operations, problems, and particular needs.

How many persons should be on the committee? Most companies tend to select representatives from all major departments and staffs so as not to slight anyone, with an equal number of employer and employee representatives. However, this often entails too many persons for effective action or good communication; the larger the group, the more difficult and ineffective it tends to become. The ideal size for a committee for information exchange and decision making is five to seven persons, allowing maximum two-way participation and input of valuable views. The question may then become, for some managers, how to represent, say, fifteen departments on a committee of seven. Here are some suggestions for dealing with this problem:

- Select someone from the department with the most potential for mishaps and injuries (he or she is going to have a problem).
- Select someone from the department with the poorest mishap/injury record (he or she already has a problem).
- Select someone with a demonstrated interest in and enthusiasm for safety and health matters (he or she will work on the problem).
- Select someone from the department with an outstanding safety and health record (he or she will have experience from which others can benefit).
- Select someone who represents the interests of the workers, such as a labor representative (he or she has a vested interest in preventing safety and health problems).

- Supplement these members, as needed, with experts or temporary subcommittee members to solve specific problems. For example, someone from the maintenance department may be an invaluable resource. Such persons need not be regular members to add to the committee's success.

Chairperson. The chairperson should be a respected leader who will keep committee meetings on track, properly direct and follow up on problem-solving actions, and handle people skillfully. The chair should be able to draw out quieter people, keep more vocal ones from dominating the meeting, and encourage all to contribute.

The regular safety and health manager *should not be the chairperson.* The chairperson may be the secretary/recorder or merely an advisor or observer. The safety and health manager can still ensure that the proper items are considered and acted on and that recommendations get action, but if the safety and health person is the chair, he or she tends to dominate the meeting. The idea is for others to take responsible roles as members of the safety and health team.

Operating Levels

Ideally, there should be a safety and health committee at every operating level. These committees interrelate on a regular basis, help solve each other's problems, and bring lower-level problems to the attention of the next higher level. Finally, the top-level committee brings senior management's attention to those things that cannot be solved at a lower level.

When this ideal arrangement is not possible, a plan must be made to disseminate the committee's findings or recommendations to all levels. A method is also needed to allow the employee at all levels to give the committee input, bring problems to its attention, or pass on information from above.

In a larger organization, top management may have its own committee, which might be called the "policy" safety and health committee. At the next lower level, there may be a management "advisory" safety and health committee. Finally, at the operating level, committees may represent the major functions, departments, or divisions. The minutes of the lower-level committees move up through the other two committees until the chief executive receives them, probably in condensed form. With this process, safety and health problems not solvable at a lower level move up until they reach a level where they can be solved. The higher-level committees can then pass recommendations down the line. Moreover, the chief executive can

The assistance of an effective safety and health committee is almost always an asset to an organization, particularly under the following situations:

- Information or input comes from several different sources.
- Full support and understanding of different departments are needed to carry out policies.
- Necessary action requires three or more management functions to be coordinated.
- Employee participation is needed and desired.

On the other hand, there are times when input from a safety and health committee is not practical, such as under these circumstances:

- Speed and time are important in making a decision.
- The problem can be handled by one or two directly concerned people.
- The problem is plainly the responsibility of a single person or function.
- The decisions required are routine or trivial.

This section has not covered all the types of safety and health committees, or how any particular organization may choose to incorporate them in its organizational structure. It has, however, given valuable guidelines for the senior manager. A little attention to properly setting up and using a safety and health committee can pay large dividends. By contrast, a carelessly organized or badly handled committee is not only a waste of time but also a liability in successful safety and health action.

ESTABLISHING STANDARDS FOR MEASUREMENT

One of management's most vital roles in the safety and health mission is to establish standards against which performance can be measured. Only then can one meaningfully assess safety performance and take corrective action. This is certainly true for the safety committee.

The law mandates some standards; industrial associations or labor unions set others; still others are dictated by the state of the art itself. The manager should be fully aware of all these standards and integrate them into his or her measurement and control systems. Certain other standards of measurement can be directly controlled and determined by management.

Once the duties and responsibilities of the committee have been established, the committee should be evaluated on how well it fulfills these duties. Although measuring the committee's success against the organization's frequency and severity rates is beyond a reasonable expectation, it is certainly reasonable to expect the committee to demonstrate positive results. These results could be demonstrated by doing the following:

1) Conducting thorough inspections on a regular basis.

2) As a result of inspections, recommending corrective actions in an effective and timely manner.

3) Conducting accident/incident investigations in an effective and timely manner.

4) Communicating input from the committee through the appropriate channels.

There is no single rule that can be applied to setting standards against which performance can be measured. Every organization differs in its definition of what is "acceptable" performance. The preceding list simply serves as a guideline to managers in measuring a committee's value to the organization. If a committee is not performing up to expectations, the prudent manager should determine how to fix the problem(s) rather than condemn the committee. A committee's failure can often be traced to management's failure to provide proper support.

Safety and Health Audits

One of the keys to program evaluation, and one of the major control devices, is the safety and health audit or inspection. Properly conducted, it can provide much of the needed input on the state of the safety and health program. Audits can be conducted on an organization's facilities, management system, and personnel. By far, the most common is the facility audit; however, a comprehensive safety and health program should conduct an audit at least annually to evaluate behaviors that affect workers within the physical environment. The safety performance of managers and supervisors is as important as the condition of the equipment and facilities.

The Audit Process. The term *audit* is used throughout this chapter, although the reader may easily substitute the word *inspection*. An audit, however, implies a more thorough examination of conditions than the ordinary "walkaround" inspection.

Management and staff at every level take part, both on the giving and on the receiving ends of audits. Although the safety committee, supervisors, and managers typically conduct audits, the audit is a managerial responsibility. It is management that has the authority and resources to ensure correction of deficiencies.

A safety and health audit is no different from any other type of audit. Managers should not wait for safety expertise before conducting an audit, nor should they wait until an outside consultant, such as an insurance company's loss control representative or an OSHA compliance officer, requests one. Outside auditors of this sort are generally looking for different things (and will advise management accordingly), such as fire safety or regulatory compliance, which may be quite different from what the manager wishes to achieve. What the top manager really wants is the most efficient operation. His or her objectives and those of the outside auditor may be related, but they are seldom identical.

Relying on safety and health personnel shifts the burden from line management, but top management should still audit the line manager. There should be an established schedule for auditing every line manager. By law, employees are not usually held responsible for safety and health matters. However, by choice, employees may be involved through the safety committee or an established union. The first-line supervisor can be held responsible only for what is specified in writing. However, line managers should, with instruction, make audits a part of their regular control system for monitoring supervisors and assessing their own performance. A problem with this system is that line managers cannot objectively grade their own performances. This problem can be resolved by making the manager part of the safety committee audit team or having other managers conduct the audits. Of these two possibilities, the idea of making the line manager part of an audit team is preferable. There are several reasons for this position:

1) The line manager is aware of day-to-day problems.

2) The line manager is in constant touch with employees, the main source of information about unsafe conditions.

3) The line manager knows the processes, tools, materials, and equipment that would be unfamiliar to an outsider. In addition, placing line managers on the audit team gives them the opportunity to appraise the performance of supervisors who report to them.

Audit Management. The audit process should be managed like any other corporate function. It should be thoroughly defined, described, and recorded. The format for written documentation of the audit varies. Numerous checklists and worksheets are available and may differ significantly based on the goals and objectives of the auditor. Audit management involves five key functions:

1) *Description.* The description should include a list of responsible parties, their duties, procedures, and expected performance.

2) *Definition.* Audit management should define what needs to be audited (that is, what will be especially looked at and examined).

3) *Designation.* Audit management should clearly specify who will make the audit. It should consider and define the respective roles of line, staff, and middle management in the audit process.

4) *Communication.* Audit management must work out reporting procedures to ensure two-way communication. All inspection forms should go to the appropriate manager, with copies to safety and health personnel.

5) *Recordkeeping.* It is necessary to make audits a matter of permanent record. The specter of future lawsuits involving cumulative trauma, delayed injuries, product liability, and so on places new record-keeping burdens on management. Also, current legislation requires that certain safety and health records to be kept; more may be required in the future.

At the end of this chapter is an example of a comprehensive audit form entitled "Safety and Health Program Assessment Worksheet," currently being used in OSHA consultation programs, of which the advantage is that, in addition to evaluating facilities, it rates items of which management should exercise control. It also contains a scoring matrix that allows a company to measure continuous improvement.

Authority to Shut Down. Should the audit team or individual inspectors have the authority to stop work when an unsafe condition is found? If the group or person represents appropriate line management, that group or person would certainly have the authority. If possible, such shutdowns should be a function of line management. Shutdown actions by staff or safety personnel should be taken only in case of imminent danger, when death or serious injury could result if operations continued. This action might then be justified, but such authority must be used with great care. The safety manager sometimes has this authority, but should use it with caution. Including the line manager on the audit team eases this problem.

Types of Audits. Audits can be divided roughly into four types: periodic, intermittent, continuous, and special. Each type has particular uses and applications.

Periodic Audits. These are conducted on a recurring basis at regular intervals. Auditors may inspect the same or different items each time. These audits are particularly valuable for inspecting processes or functions in which changes occur often.

Intermittent Audits. These are conducted at irregular intervals, without forewarning. An unannounced OSHA visit, a city sanitation inspection, or a surprise visit from top management are examples.

Continuous Audits. Some operations must undergo continuous auditing, particularly where legal standards exist. Routine quality control sampling, monitoring of air quality, radiation or toxic exposures, and temperature or humidity levels are examples.

Special Audits. These are conducted for a single purpose. An electrical or construction inspection, an insurance rating visit, or a municipal fire safety check are examples. Management may arrange for a special audit after a new system goes on line or to gather specific information on a problem.

Basic Steps in Performing an Audit. Proper planning and procedures, as the following guidelines show, are essential to a successful audit:

Plan Ahead. Prepare all details well in advance before starting. Prepare for known problems. Have all the materials needed for the audit, including records, writing materials, checklists, safety equipment, meters and other measuring devices, reference regulations and company procedures, and records from the last audit (particularly recommended corrective actions).

Communicate. Contact the proper line management on arrival. Ask for someone to accompany the audit team. Go over the findings with the supervisor and/or manager before leaving. Answer any questions or clarify any ambiguities that may arise at that time.

Follow Up. After the audit, quickly send written reports of the findings to management and the safety and health department. Complete other required reports and forms promptly and file or forward as directed. Include the paperwork and office work, and any later activity to confirm corrective actions, such as a follow-up visit or inspection.

audit is little better than none at all, and does not give top management the control it seeks or the help it needs.

Legal Issues and Audits. With all the positive information that an audit can provide, it is easy to overlook the liability that exists when you document program safety deficiencies. OSHA has the power to obtain audits and use them to identify past violations and support willful violations. In an article entitled "Protecting Your Audits from Compelled Disclosure," Dreux and White (1993) state:

> Audit reports should be written on the assumption that they will become public documents. As such, any statement in them will be construed as an admission by the company and will probably be admissible into evidence in any lawsuit. Given this potential liability, great care should be given to the identification and description of any problems, dangers, risks, or potential violations of company policy or OSHA regulations in the audit (collectively "Safety and Health Concerns"). Factors that mitigate the Safety and Health Concerns and prior corrective actions should be described. Any form or degree of exaggeration, overstatement, or oversimplification of the Safety and Health Concerns should be avoided. An audit should not be undertaken without a commitment to follow up on the Safety and Health Concerns that are identified.

CONCLUSION

This chapter has dealt with the importance of safety committees and audits as critical components of a comprehensive safety and health program. Successful implementation of such a program requires commitment on the part of the organization at all levels. The safety and health manager cannot do everything alone. A properly functioning safety committee can be a valuable asset in stimulating employees' interest in controlling unnecessary human losses and sharing the responsibility for their own safety and health.

REFERENCES AND SUGGESTED FURTHER READING

Dreux, M. S., and F. A. White (February 1993). "Protecting Your Audits from Compelled Disclosure." *Occupational Hazards*, pp. 53–55.

Findlay, J. V., and R. L. Kuhlman (1980). *Leadership in Safety.* Loganville, GA: Institute Press.

Germain, G. L., et al. (1997). *Safety, Health, and Environmental Management.* Loganville, GA: AEI and Associates, Inc.

Hartnett, J. (1996). *OSHA in the Real World—How to Maintain Workplace Safety.* Santa Monica, CA: Merritt Publishing.

Pennsylvania Workers' Compensation Bureau (1997). *Workplace Safety Committees—Technical Assistance Manual.* Harrisburg, PA: Pennsylvania Workers' Compensation Bureau.

Pope, W. C. (1973–88). *Selected Monographs.* Alexandria, VA: Safety Management Information Systems.

Chapter 18

Industrial Hygiene Management

Robert D. Soule

INTRODUCTION

Industrial hygiene is that science and art "dedicated to the anticipation, recognition, evaluation, and control of environmental factors arising in or from the workplace that may result in injury, illness, impairment, or affect the well-being of workers and members of the community." Although diseases suspected of being related to the work environment were recognized more than 2,500 years ago—for example, lead poisoning of workers engaged in mining operations—the systematic application of what have become recognized as industrial hygiene principles is a relatively recent development. The "profession" of industrial hygiene is a little more than half a century old, coming into prominence only since the 1930s.

The value of effective industrial hygiene management programs has become apparent in increasing and convincing fashion since the establishment of the profession. In all occupational settings, the sustained presence of a safe and healthful workforce is a vital contributor to overall quality, efficiency, productivity, and profitability of operations.

Of the four basic responsibilities of the industrial hygienist mentioned previously—anticipation, recognition, evaluation, and control—the first two require a fundamental knowledge of toxicology, epidemiology, and other sciences, which extends beyond the scope of this chapter. Thus, this chapter focuses on 1) the critical elements of industrial hygiene management, for example, evaluation of the magnitude of exposures, primarily of production workers or other members of the organization, and 2) the strategies, facilities, and work practices necessary to ensure adequate control of potentially hazardous materials in the work environment.

GENERAL CONSIDERATIONS IN INDUSTRIAL HYGIENE EVALUATION

The reasons for conducting industrial hygiene surveys are many and varied. They include the following:

1) Identification and quantification of specific contaminants present in the environment

2) Determination of exposures of workers in response to complaints

3) Assessment of compliance status with respect to various occupational health standards

4) Evaluation of the effectiveness of engineering controls

The reason(s) for sampling will dictate, to some extent, the sampling strategy that should be used. For example, sampling in a grid pattern throughout a plant can be used to document the environmental characteristics of the workplace. Personal breathing-zone samples can be obtained to document actual exposure conditions. The substances being evaluated will determine the type of sampling devices to be used, and the analytical requirements will specify the time and perhaps flow rate of sampling. The occupational health standard will indicate whether continuous or "grab" sampling is appropriate. In short, consideration must be given to a number of questions pertaining to the fundamental purpose of the sampling.

Many analytical methods available to industrial hygienists have been so standardized and simplified that they require relatively little experience. On the other hand, many seemingly simple tests call for a basic understanding of solubility and gas laws, partial pressures, and chemical reactions. In many instances, only qualified specialists can answer questions arising from such considerations. The ultimate methods of analysis to be used will depend on the problem at hand rather than mere application of a "standard method." Recently, the trend has been toward the development of methods that give relatively prompt results with a high degree of accuracy. The latter aspect has been given increased importance because of the legal significance given to occupational health standards, particularly as promulgated under authority of the Occupational Safety and Health Act of 1970. The National Institute for Occupational Safety and Health (NIOSH) and OSHA have developed specific methods for sampling and analyzing many atmospheric contaminants in the workplace. One criterion for these procedures is an accuracy of at least ± 25 percent with 95 percent confidence at the permissible exposure limit.

The following discussion focuses on a general strategy for sampling in the workplace. Sampling techniques for gaseous and particulate contaminants, summaries of techniques available for analyzing atmospheric samples, and a specific discussion of industrial hygiene monitoring required for purposes of compliance with regulations are presented. The role of biological monitoring as it relates to industrial hygiene management is also included. More complete discussions of detailed methods of analysis for specific contaminants, as well as other aspects of industrial hygiene evaluation, are available in other references identified at the end of this chapter.

The magnitude of environmental stresses can be evaluated in various ways. One form of evaluation is qualitative and/or subjective, using one or more of the human senses without necessarily taking any actual "measurements." This kind of inspection and evaluation of a work situation is quite beneficial, particularly when done by an experienced industrial hygienist. Another form of evaluation is quantitative, for example, involving collection and analysis of samples representative of actual conditions at the location sampled. Generally, this type of evaluation is most desirable and necessary in many cases, particularly when the purpose of the sampling is to determine compliance with occupational health standards or to form the basis for designing engineering controls.

Preliminary Survey

An experienced, professional industrial hygienist often can evaluate, quite accurately and in some detail, the magnitude of chemical and physical stresses associated with an operation without benefit of any instrumentation. In fact, the professional uses this qualitative evaluation every time a survey is performed, whether it is intended to be the total effort of the work or a preliminary inspection prior to actual sampling and analysis of potential stress. Qualitative evaluation can be applied by anyone familiar with an operation, from the worker to the professional investigator, to ascertain some of the potential problems associated with work activities.

The first step in evaluating the occupational environment is to become as familiar as possible with particular operations. The person evaluating the operation should be aware of the types of industrial process and the chemical raw materials, by-products, and contaminants encountered. This is a particularly important step for someone evaluating an operation for the first time, or for the "outside" expert brought in for the evaluation—for example, a corporate specialist or a consultant. The investigator should know what protective measures are provided, how engineering controls are being used, and how many workers are exposed to contaminants generated by specific job activities.

The number of chemical and physical agents capable of producing occupational illnesses continues to increase, even in processes that are relatively straightforward. It is important that the responsible health and safety professional establish and maintain a list of the chemical and physical agents encountered in particular areas of administration. In fact, OSHA's hazard communication standard, and many state "right-to-know" laws, make such inventorying a legal obligation. The composition of the products and by-products and as many as possible of the associated contaminants and "undesirables" should be known. This means that the industrial hygienist must obtain complete information on the composition of various commercial products. In most instances, the desired information can be obtained from descriptive material provided by the suppliers in the form of a Material Safety Data Sheet. Similarly, the labels on the containers of the material should be read carefully. Although there are explicit labeling requirements under hazard communication and right-to-know regulations, many still do not always give complete information, and further investigation of the composition of the materials is necessary.

After the inventory is obtained, it is necessary to become familiar with information on toxicity and other hazardous properties of the chemicals. Information of this type can be found in several excellent reference texts on toxicology and industrial hygiene, some of which are identified at the end of this chapter.

During a qualitative walk-through evaluation, many potentially hazardous operations can be detected visually. Operations that produce large amounts of dusts and fumes can be spotted. However, "visible" does not necessarily mean "hazardous" levels of a dust exist. In fact, airborne dust particles that *cannot* be seen by the unaided eye normally are more hazardous, since they are more likely to be inhaled into the lungs. Concentrations of dust of respirable size usually must reach extremely high levels before they become visible. Thus, the absence of a visible cloud of dust is not a guarantee that a "dust-free" atmosphere exists. Generally, activities that generate dust that can be spotted visually are likely to warrant implementation of additional controls.

In addition to sight, the sense of smell can be used to detect the presence of many vapors and gases. Trained observers are able to estimate rather accurately the concentration of various gases and solvent vapors present in the workroom air. Unfortunately, the odor threshold concentration—for example, the lowest concentration that can be detected by smell—for many substances is greater than the permissible exposure level. In these cases, if the substance can be detected by its odor, it is indicative of excessive levels. Some substances, notably hydrogen sulfide, can cause olfactory fatigue or paralysis of the olfactory nerve endings to the extent that even dangerously high concentrations cannot be detected by odor. Suffice it to say, odor cannot be relied on as a "warning property" for most materials.

Although it is usually possible to determine the presence or absence of potentially hazardous physical agents at the time of the qualitative evaluation, rarely can the potential hazard be evaluated without the aid of special instruments. As a minimum, however, the sources of physical agents such as radiant heat, abnormal temperature and humidity, excessive noise, improper or inadequate illumination, ultraviolet (UV) radiation, microwaves, and various other forms of radiation, can be noted.

An important aspect of the qualitative evaluation is an inspection of the control measures in use at a particular operation. In general, the control measures include such features as shielding from radiant or UV energy, local exhaust and general ventilation provision, respiratory protection devices, and other personal protective measures. General indices of the relative effectiveness of these controls are the presence or absence of accumulated dust on floors, ledges, and other work surfaces; the condition of ventilation ductwork (for example, whether there are holes or badly damaged sections of ductwork); whether the system appears to provide adequate control of contaminants; and the manner in which personal protective measures are accepted and used by workers.

Quantitative Surveys

Although the information obtained during a qualitative evaluation or walk-through inspection of a facility is important and always useful, only by measurement can the hygienist document the actual level of chemical or physical agent associated with a given operation. Of course, the strategy used for any given air-sampling program depends, to a great extent, on the purpose of the study. The specific objectives of any sampling program may include one or more of the following:

- To identify the sources of contaminant release
- To assist in design and/or evaluation of engineering controls
- To provide a record of environmental conditions
- To correlate disease or injury with exposure to specific stresses
- To document compliance with health and safety regulations

Regardless of the objectives of the sampling program, the investigating industrial hygienist must answer the following questions to be able to implement the correct strategy.

- Where should sampling be done?
- Who should be sampled?
- How long should the samples be taken?
- How many samples are needed?
- How should the samples be obtained?
- When should the samples be taken?

In answering these questions, the importance of adequate field notes must be emphasized. While the sample is being obtained, notes should be made of the time, duration, location, and operations underway, and all other factors pertinent to interpretation of the sample, and the condition it is intended to define. Printed forms with labeled spaces for essential data help to avoid the common failure to record needed information.

Where Should Sampling Be Done? Three general locations are used for the collection of air samples: at a specific operation, in the general workroom air, and in a worker's breathing zone. The choice of sampling location is dictated by the type of information desired; in some cases, a combination of the three types of sampling may be necessary. Probably the most frequent industrial hygiene sampling in nickel-cadmium battery operations is done to determine the levels of exposure of workers to a given contaminant, notably cadmium, throughout a workday. To obtain this type of information, it is necessary to collect samples at the worker's breathing zone as well as in the areas adjacent to his or her particular activities. On the other hand, when the purpose of the survey is to determine sources of contamination or to evaluate engineering controls, a strategic network of area sampling would be more appropriate.

Who Should Be Sampled? Logically, samples should be collected in the vicinity of workers directly exposed to contaminants generated by their own activities. In addition, samples should be taken in the breathing zones of workers in nearby areas not directly involved in the activities that generate the contaminant, as well as in those of any workers remote from the activities who either have complained or have reason to suspect that contaminants have been drawn into their work areas.

How Long Should the Samples Be Taken? In most cases, minimum sampling time is determined by the time necessary to obtain a sufficient amount of the material to allow accurate analysis. The duration of the sampling period, therefore, is based on the sensitivity of the analytical procedure and the acceptable concentration of the particular contaminant in the air. Preferably, the sampling period should represent some identifiable period of time of the worker's exposure, usually a minimum of one complete cycle of activity. This is particularly important in studying nonroutine or batch activities, which are characteristic of many industrial operations in nickel-cadmium manufacture. Exceptions are operations that are highly automated and enclosed operations in which the processing is done automatically and in which the operator's exposure is relatively uniform throughout the workday. In many cases, it is desirable to sample the worker's breathing zone for the duration of the full shift, particularly if sampling is being done to determine compliance with occupational health standards.

Evaluation of workers' daily time-weighted average exposures is best accomplished, when analytical methods permit, by allowing the person to work a full shift with a personal breathing-zone sampler attached. The concept of full-shift integrated personal sampling is much preferred to that of short-term or general area sampling if the results are to be compared to standards based on time-weighted average concentrations. The current OSHA permissible exposure limit is such a standard. When methods that permit full-shift integrated sampling are not applicable, it may be possible to calculate time-weighted average exposures from alternative short-term or general area sampling methods.

The first step in calculating the daily, time-weighted average exposure of a worker is to study the job description for the person under consideration, and to ascertain how much time during the day is spent at various tasks. Such information usually is available from the plant personnel office or foreman on the job. Investigator may need to make time studies on their own to obtain the correct information. In fact, even information obtained from plant personnel should be checked because, in many situations, job activities as observed by the investigator do not fit the official job descriptions. From this information and the results of the environmental survey, a daily, 8-hour, time-weighted average exposure can be calculated, assuming that sufficient sampling has been performed.

When sampling for the purpose of comparing results with airborne contaminants whose toxicological properties warrant short-term and ceiling limit values, it is necessary to use short-term or grab-sampling techniques to define peak concentrations and estimate peak excursion durations. For purposes of further comparison, the 8-hour, time-weighted average exposures can be calculated using the values obtained by short-term sampling.

How Many Samples Are Needed? The number of samples needed depends, to a great extent, on the purpose of sampling. Two samples may be sufficient to estimate the relative efficiency of control methods: one sample being taken while the control method is in operation and the other while it is off. On the other hand, several dozen samples may be necessary to define accurately the average daily exposure of a worker who performs a variety of tasks. The number of samples also depends, to some extent, on the concentrations encountered. If the concentration is quite high, a single sample may be sufficient to warrant further action; if the concentration is somewhat near the acceptable level, a minimum of three to five samples usually is desirable for each operation being studied. There is no set rule regarding the duration of sampling or the number of samples needed. These decisions usually can be reached quickly and reliably only after much experience in conducting such studies.

When Should the Samples Be Taken? The type of information desired and the particular operations under study will determine when sampling should be done. If the operation continues for more than one shift, it usually is desirable to collect air samples during each shift, since the airborne concentrations may be quite different for each shift. Similarly, it is desirable to obtain samples during both summer and winter months, particularly in plants located in areas where large temperature variations occur during different seasons.

How Should the Samples Be Obtained? In general, the choice of instrumentation for sampling a particular substance depends on a number of factors, including the portability and ease of use, efficiency of the instrument or method, reliability of the equipment under various conditions of field use, type of analysis or information desired, and other factors. No single, universal air-sampling instrument is available today, and it is doubtful that such an instrument will ever be developed. In fact, the present trend in the profession is toward a greater number of specialized instruments. The sampling instruments used in the field of industrial hygiene generally can be classified by type as follows:

1) Direct-reading

2) Those that remove the contaminant from a measured quantity of air

3) Those that collect a fixed volume of air.

These three methods are listed in order of their general application and preference in use today.

The industrial hygienist should consider a proposed sampling system in relation to his or her familiarity with the sampling and analytical method. As a rule, a method should never be relied on unless and until the individual has personally evaluated it under controlled conditions. One approach is to sample a synthetic atmosphere produced using a proportioning apparatus or gastight impervious chamber of sufficient size to permit creation and sampling of mixtures without introducing significant errors. Another technique to evaluate instruments is to introduce measured amounts of contaminants into a device attached to the sampling arrangement in a manner that utilizes the entire amount deposited. A third technique is to compare performance of the method of interest with a device of proven performance by sampling from a common manifold. In other words, any sampling program must be preceded by appropriate calibration of all the equipment to be used. Regardless of the sampling instrument selected for use in industrial hygiene surveys, it is imperative that the actual performance characteristics be known. This requires that appropriate calibration be done at any time that the performance of the device is questioned.

SAMPLING FOR GASES/VAPORS

For purposes of definition, a substance is considered to be a gas if at 70°F and atmospheric pressure, the normal physical state is gaseous. A vapor is the gaseous state of a substance in equilibrium with the liquid or solid state of the substance at the given environmental conditions. This equilibrium results from the vapor pressure of the substance causing volatilization or sublimation into the atmosphere. The sampling techniques discussed in this section are applicable to a substance in "gaseous" form, regardless of whether it is technically a gas or a vapor. The potential for exposure to gases and vapors exists in many occupational settings, for example, where solvents and degreasing formulations are in use.

Particulate substances can be readily scrubbed or filtered from sampled airstreams because of the larger relative physical dimension of the contaminant and the operation and interaction of agglomerative, gravitational, and inertial effects. Gases and vapors, however, form true solutions in the atmosphere, thus requiring either sampling of the total atmosphere using a gas collector or the use of a more vigorous scrubbing mechanism to separate the gas or vapor from the surrounding air. Sampling reagents can be chosen to react chemically with the contaminants in the airstream, thus enhancing the collection efficiency of the sampling procedure. In the development of an integrated sampling scheme, it is necessary to consider several basic requirements. First, the method must have an acceptably high efficiency of collecting the contaminant of interest. Second, a rate of airflow that can provide a sufficient sample for the required analytical procedure, maintain the acceptable collection efficiency, and be accomplished in a reasonable time period must be available. Third, the collected gas or vapor must be kept in the chemical form in which it exists in the atmosphere under conditions that maintain the stability of the sample before analysis. Fourth, it must be practical to submit the sample for analysis in a suitable form and medium. Fifth, a very minimal amount of analytical procedure in the field must be associated with the overall method. Finally, to the extent practicable, the use of corrosive (for example, acidic or alkaline) or relatively toxic (for example, benzene) sampling media should be avoided.

Of these general requirements, perhaps the most important is the first: knowing the collection efficiency of the sampling system chosen or anticipated. This efficiency information can be obtained either from published data describing the method or as a result of independent evaluations as an essential part of planning the industrial hygiene survey.

There are two basic methods for collecting samples of gaseous contaminants: instantaneous or "grab" sampling, and integrated or long-term sampling. The first involves the use of a gas-collecting device, such as an evacuated flask or bottle, to obtain a fixed volume of a contaminant-in-air mixture at known temperature and pressure. This is called "grab sampling" because the contaminant is collected almost instantaneously, for example, within a few seconds or minutes at most, thus being representative of atmospheric conditions at the sampling site at a given point in time. This method is commonly used when atmospheric analyses are limited to such gross contaminants as sewer gases, carbon dioxide, or carbon monoxide, or when the concentrations of contaminants likely to be found are sufficiently high to permit analysis of a relatively small sample. However, with the increased sensitivity of modern instrumental techniques such as gas chromatography and infrared (IR) spectrometry, instantaneous sampling of relatively low concentrations of atmospheric contaminants is becoming feasible.

The second method for the collection of gaseous samples involves the passage of a known volume of air through an absorbing or adsorbing medium to remove the contaminants of interest from the sample airstream. This technique makes it possible to sample the atmosphere over an extended period of time, thus "integrating" the sample. The contaminant that is removed from the airstream becomes concentrated in or on the collection medium. Therefore, it is important to establish a sampling period long enough to permit collection of a quantity of contaminant sufficient for subsequent analysis, but not so long that the capacity of the sorbent material is exceeded.

Instantaneous Sampling

Various devices are available to obtain instantaneous or grab samples: vacuum flasks or bottles, gas- or liquid-displacement collectors, metallic collectors, glass bottles, syringes, and plastic bags. The temperature and pressure at which the samples are collected must be known, to permit reporting of the analyzed components in terms of standard conditions, normally 25°C and 760 mm Hg for industrial hygiene purposes.

Grab samples are usually collected when analysis is to be performed on gross amounts of gases in air (for example, methane, carbon monoxide, oxygen, and carbon dioxide). The samplers should not be used for collecting reactive gases such as hydrogen sulfide, oxides of nitrogen, and sulfur dioxide unless the analyses can be made directly in the field.

The introduction of highly sensitive and sophisticated instrumentation has extended the applications of grab sampling to low levels of contaminants. In areas where the atmosphere remains constant, the grab sample may be representative of the average as well as the momentary concentration of the components, thus approximating an integrated equivalent. Where the atmospheric composition varies, numerous samples must be taken to determine the average concentration of a specific component. The chief advantage of grab sampling methods is that collection efficiency is essentially 100 percent, assuming no losses due to leakage or chemical reaction preceding analysis.

Since the introduction of flexible containers, evacuated flasks and displacement containers are being used less frequently for the collection of grab samples. Plastic bags can be used to collect air and breath samples containing organic and inorganic vapors and gases in concentrations ranging from parts per billion to more than 10 percent volume in air. They also are convenient for preparing known concentrations of gases and vapors for equipment calibration. The bags are available commercially in a variety of sizes, up to 9 cubic feet and can be made easily in the laboratory. These bags are manufactured from various plastic materials, most of which can be purchased in rolls or sheets cut to the desired size. Some materials, such as Mylar, may be sealed with a hot iron using a Mylar tape around the edges. Others, such as Teflon, require high temperature and controlled pressure in sealing. Certain plastics, including Mylar, may be laminated with aluminum to seal the pores and reduce the permeability of the inner walls to sample gases and the outer walls to sample moisture. Sampling ports may consist of a sampling tube molded into the fabricated bag and provided with a closing device or a clamp-on air valve.

Plastic bags have the advantages of being light, nonbreakable, and inexpensive, and they permit the entire sample to be withdrawn without the difficulty associated with dilution by replacement air, as is the case with rigid containers. However, they must be used with caution because generalization of recovery characteristics of a given plastic cannot be extended to a broad range of gases and vapors. Important factors to be considered in using these collectors are the absorption and diffusion characteristics of the plastic material, concentration of the gas or vapor, and reactive characteristics of the gas or vapor with moisture and with other constituents in the sample.

Integrated Sampling

Integrated sampling of the workroom atmosphere is necessary when the composition of the air is not uniform, when the sensitivity requirements of the method of analysis necessitate sampling over an extended period, or when compliance or noncompliance with an 8-hour, time-weighted average standard must be established. The observations and judgment of the industrial hygienist are called on in devising the strategy for the procurement of representative samples to meet the requirements of an environmental survey of the workplace.

Integrated air sampling requires a relatively constant source of suction as an air-moving device. A vacuum line, if available, may be satisfactory. However, the most practical source for prolonged periods of sampling is a pump powered by electricity. These pumps come in various sizes and types and must be chosen for the sampling devices with which they will be used. If electricity is not available or if flammable vapors present a fire hazard, aspirator bulbs, hand pumps, portable units operated by compressed gas, or battery-operated pumps are suitable for sampling at rates up to 3 liter/min. The latter have become the workhorses of the industrial hygiene profession, particularly in judging compliance with health standards.

The common practice in the field is to sample for a measured period of time at a constant, known rate of airflow. Direct measurements are made with rate meters such as rotameters and orifice or capillary flowmeters. These units are small and convenient to use, and have become quite accurate even at the very low flow rates common with modern industrial hygiene sampling, for example, 10–20 cc/min. The sampling period must be timed carefully.

Many pumps have inlet vacuum gages or outlet pressure gages attached. These gages, when properly calibrated with a wet or dry gas meter, can be used to determine the flow rate through the pump. The gage may be calibrated in terms of cubic feet per minute or liters per minute. If the sample absorber does not have enough resistance to produce a pressure drop, a simple procedure is to introduce a capillary tube or other resistance into the train behind the sampling unit.

Samplers are always used in assembly with an air-moving device (source of suction) and an air-metering unit. Frequently, however, the sampling train consists of a filter, probe, absorber (or adsorber), flowmeter, flow regulator, and air mover. The filter serves to remove any particulate matter that may interfere in the analysis. It should be ascertained that the filter does not also remove the gaseous contaminant of interest. The probe or sampling line is extended beyond the sampler to reach a desired location. It also must be checked to determine that it does not collect a portion of the sample. The meter that follows the sampler indicates the flow rate of air passing through the system. The flow regulator controls the airflow. Finally, at the end of the train the air mover provides the driving force.

Various types of absorber may be utilized for collecting gases and vapors, although they are used more in research applications than in routine exposure monitoring. The absorbers provide sufficient contact between the sampled air and the liquid surface to ensure complete absorption of the gaseous contaminant. In general, the lower the sampling rate, the more complete is the absorption.

Columns, packed with materials such as glass beads, wetted with the absorbing solution, provide a larger surface area for the collection of a sample and are especially useful when a viscous liquid is required. The rate of sampling is low, usually 0.25 to 0.5 liters of air per minute.

When collecting insoluble or nonreactive vapors is desired, an adsorption technique frequently is the method of choice; activated charcoal and silica gel are common adsorbents. Solid adsorbents require less manipulative care than do liquid absorbents; they can provide high collection efficiencies, and with improved adsorption tube design and a better definition of desorption requirements, they are becoming popular in industrial hygiene surveys. Activated charcoal is an excellent adsorbent for most vapors boiling above 0°C; it is moderately effective for low boiling gaseous substances—between (−100°C) and 0°C—particularly if the carbon bed is refrigerated, but it is a poor collector of gases having boiling points below (−150°C). Activated charcoal's retentivity for adsorbed vapor is several times that of silica gel. Because of their nonpolar characteristics, organic gases and vapors are adsorbed in preference to moisture, and sampling can be performed for long periods of time. Silica gel has been used widely as an adsorbent for gaseous contaminants in air samples. Because of its polar characteristics, silicon gel tends to attract polar or readily polarizable substances preferentially. The general order of decreasing polarizability or attraction is as follows: water, alcohols, aldehydes, ketones, esters, aromatic compounds, olefins, and paraffins. Organic solvents are relatively nonpolar in comparison to water, which is strongly adsorbed onto silica gel; such compounds will be displaced by water in the entering airstream. Consequently, the volume of air sampled under humid conditions may need to be restricted. Despite this limitation, silica gel is a very useful adsorbent.

The collection efficiency (the ratio of the amount of contaminant retained by the absorbing or adsorbing medium to that entering it) need not be 100 percent as long as it is known, constant, and reproducible. The minimum acceptable collection performance in a sampling system is usually 90 percent, but higher efficiency is certainly desirable. When the efficiency falls below the acceptable minimum, sampling may be carried out at a lower rate or, in the case of liquid absorbers, at a reduced temperature by immersing the absorber in a cold bath to reduce the volatility of both the solute and solvent. Frequently, the relative efficiency of a single absorber can be estimated by placing another in series with it. Any leakage is carried over into the second collector. The absence of any carryover is not in itself an absolute indication of the efficiency of the test absorber, since the contaminant may be stopped effectively by either absorber. Analysis of the various sections of silica gel or activated charcoal tubes used in sampling a contaminant is a useful check on the collection efficiency of the first section of the tube.

Direct-Reading Instruments

A variety of direct-reading devices has become available for use in measuring airborne concentrations of gases and vapors. These include instruments capable of direct response to airborne contaminant, various reagent kits that can be used for certain substances, colorimetric indicator (detector) tubes, and passive dosimeters.

A direct-reading instrument is "an integrated system capable of sampling a volume of air, making a quantitative analysis and displaying the result." Direct-reading instruments can be portable devices or fixed-site monitors. To the industrial hygienist, these devices are generally characterized by disadvantages that limit their application for measuring the low concentrations of significance. They are used commonly for onsite evaluations for a variety of reasons, depending primarily on the understood purpose of the survey. Direct-reading instruments are useful to find sources of emission of hazardous substances "on the spot," to determine the performance characteristics of specific operations or control devices, usually by comparing results of "before and after" surveys; as a qualitative industrial hygiene monitoring instrument to ascertain whether specific air-quality standards are being complied with; or as continuous monitoring devices, by establishment of a network of sensors at fixed locations throughout a plant. In the latter application, readout from such a system can be used to activate an alarm or an auxiliary control system in the event of process upsets, or to obtain permanent recorded documentation of concentrations of contaminants in the workroom atmosphere. The advantages of having direct-reading instruments available for industrial hygiene surveys are obvious. Such onsite evaluations of atmospheric concentrations of hazardous substances make possible the immediate assessment of unacceptable conditions and enable the industrial hygienist to initiate immediate corrective action in accordance with his or her judgement of the seriousness of the situation without causing further risk of injury to the workers. It cannot be overemphasized that great caution must be taken in the use of direct-reading instruments and, more important, in the interpretation of their results. Most of these instruments are nonspecific, responding to a property of the substance rather than the material itself. Before recommending any action based on sampling results, the industrial hygienist often must verify his or her onsite findings by conducting supplemental sampling and laboratory analyses to characterize adequately the chemical composition of the contaminants in a workroom area and to develop the supporting quantitative data with more specific methods of greater accuracy. Such precautions become mandatory if the industrial hygienist has not had extensive experience with the process in question or when a change in the process or substitution of chemical substances may have occurred.

The calibration of any direct-reading instrument is an absolute necessity if the data are to have any meaning. Considering this to be axiomatic, it must also be recognized that the frequency of calibration is dependent on the type of instrument. Certain classes of instruments, because of their design and complexity, require more frequent calibration than others. It is also recognized that quirks in an individual instrument produce greater variations in its response and general performance, thus requiring a greater amount of attention and more frequent calibration than other instruments of the same design. Direct personal experience with a given instrument serves as the best guide in this matter. Another unknown factor that can be evaluated only by experience is the environmental variability of sampling sites. For example, when locating a particular fixed-station monitor at a specific site, consideration must be given to the presence of interfering chemical substances, the corrosive nature of contaminants, vibration, voltage fluctuations, and other disturbing influences that may affect the response of the instrument. Finally, the required accuracy of the measurements must be determined initially. If an accuracy of ± 3 percent is needed, more frequent calibration must be made than if ± 25 percent accuracy is adequate in the solution of a particular problem.

Reagent Devices

Direct-reading colorimetric techniques, which utilize the chemical reaction of an atmospheric contaminant with a color-producing reagent, are available in a variety of forms. Reagent kits that allow detection of specific substances are available, although becoming less common in use than detector tubes. These colorimetric indicating tubes, containing solid reagent chemicals, provide compact direct-reading devices that are convenient to use for the detection and semiquantitative estimation of gases and vapors in atmospheric environments. Presently, there are tubes for more than two hundred atmospheric contaminants on the market.

Operating procedures for colorimetric indicator tubes are simple, rapid, and convenient. However, because of their simplicity, misleading results may be obtained with these devices unless they are used under the supervision of an adequately trained industrial hygienist who enforces a strong quality assurance program.

Colorimetric indicating tubes are filled with a solid granular material, such as silica gel or aluminum oxide, that has been impregnated with an appropriate chemical reagent. The ends of the glass tubes are sealed during manufacture. When a tube is to be used, its end tips are broken off, the tube is placed in the manufacturer's holder, and the recommended volume of air is drawn through the tube by means of the air-moving device provided by the manufacturer. This device may be one of several types, such as a positive displacement pump, a simple squeeze bulb, or a small electrically operated pump with an attached flowmeter. In most cases, a fixed volume of air is drawn through the detector tube, although in some systems varied amounts of air may be sampled. The operator compares either an absolute length of stain produced in the column of the indicator gel or a ratio of the length of stain to the total gel length against a calibration chart, to obtain an indication of the atmospheric concentration of the contaminant that reacted with the reagent. To make estimates using another type of tube, a progressive change in color intensity is compared with a chart of color tints. With a third type of detector, the volume of sampled air required to produce an immediate color change is noted; it is intended that this air volume be inversely proportional to the concentration of the atmospheric contaminant. Various resources are available and should be consulted for a full appreciation of the complex interrelationships among the factors affecting the kinetics of indicator tube reactions.

Recent years have witnessed the introduction of passive dosimetry. A passive monitor is attached to the worker in his or her breathing zone. The total exposure time is noted, and analysis results give the amount of vapor collected. These data provide an average mass collection rate, which can then be used to calculate the time-weighted average concentration. The physical parameters of the sampler design are chosen according to desired exposure time and the substance to be monitored. Corroborative testing of prototype passive monitors and comparison of results to reference methods have shown excellent agreement. With the increasing emphasis on development of specific methods for monitoring workers' exposures to contaminants, it is likely that the passive monitor concept will become a vital basis for a new generation of industrial hygiene monitoring equipment.

SAMPLING FOR PARTICULATES

In classifying airborne particulates, the term *aerosol* normally refers to any system of liquid droplets or solid particles dispersed in a stable aerial suspension. To be classified as being aerosol, the particles or droplets must remain suspended for significant periods of time. Liquid particulates usually are classified into two subgroups—mists and fogs—depending on particle size. The larger particles generally are referred to as mists, whereas small particles result in fogs. Liquid droplets normally are produced by such processes as condensation, atomization, and entrainment of liquid by gases. Solid particulates usually are subdivided into three categories: dusts, fumes, and smoke—the distinction among them resulting primarily from the processes of production. Dusts are formed from solid organic or inorganic materials by reducing their size through some mechanical process such as crushing, drilling, or grinding. Dusts vary in size from the visible to the submicroscopic, but their composition is the same as the material from which they were formed. Fumes are formed by such processes as combustion, sublimation, and condensation. The term *fume* is generally applied to the oxides of zinc, magnesium, iron, lead, and other metals, although solid organic materials such as waxes and some polymers may form fumes by the same methods. These particles are very small, ranging in size from 1 μm to as small as 0.001 μm in diameter. *Smoke* is a term generally used to refer to airborne particulates resulting from the combustion of organic materials (for example, wood, coal, or tobacco). The resulting smoke particles are all usually less then 0.5 μm in diameter.

The nature of the airborne particulate dictates, to a great extent, the manner in which sampling of the environment must be accomplished. Sampling is performed by drawing a measured volume of air through a filter, impingement device, electrostatic or thermal precipitator, cyclone, or other instrument for collecting particulates. The concentration of particulate matter in air is denoted by the weight or the number of particles collected per unit volume of sampled air. The weight of collected material is determined by direct weighing or by appropriate chemical analysis. And the number of particles collected is determined by counting the particles in a known portion or aliquot of the sample and extrapolating to the whole sample.

The general requirements discussed previously for gases and vapors apply, for the most part, to particulate contaminants as well. Several aspects of sampling apply only for particulates; however, because of the wide range of particle sizes of airborne particulates confronting the industrial hygienist in most industrial settings.

The sampling train for particulates consists of an air-inlet orifice, particle collection medium (and preselector, if classification of the total particulate is being done), flowmeter, flow rate control device, and air mover or pump. Of these, the most important, by far, is the particle collection medium, used to separate the particles from the sampled airstream. Both the efficiency of the device and its reliability must be high. The pressure drop across the medium should be low, to keep the size of the required pump to a minimum. The medium may consist of a single element, such as a filter or impinger, or there may be two or more elements in series, to classify the particulate into different size ranges.

Usually, it is important that the sampling method not alter chemical or physical characteristics of the particles collected. For example, if the material is soluble, it cannot be collected in a medium capable of dissolving it. If the particles have a tendency to agglomerate, and it is important to be able to distinguish individual particles, deep-section collection on a filter should not be used.

An additional consideration, which is of much greater concern with particulate than with gases and vapors, is the variation of concentration in space. Many cases have been reported of significant differences in concentrations being documented with sampling units placed equidistant from a source of contaminant generation. Similarly, with personal breathing-zone sampling, it is not uncommon for simultaneous samples obtained from both shoulders of a worker to indicate substantially different concentrations. The importance of these observations lies in the understanding that particulate sampling results by themselves are indicative of conditions within a short distance of the sampling unit and should be augmented with additional information, such as studies of airflow patterns within the workroom.

Collection Media

The concept of grab sampling for particulates is not as valuable as for gases or vapors. Although there are methods for collecting instantaneously a sample of airborne particles, these are primarily of historical interest and are discussed here.

Today, the most meaningful sampling for particulate is done over extended periods of time with various collection techniques, depending on the material being collected and the availability of sampling equipment. The most common collection techniques are filtration, impingement, impaction, elutriation, electrostatic precipitation, and thermal precipitation.

Undoubtedly the most common method of collecting airborne particulate is filtration. The fibrous type of filter matrices consists of irregular meshes of fibers, about 20 μm or less in diameter. Air passing through the filter changes direction around the fibers and the particles impinge against the filter, where they are retained. The largest particles (> 30 μm) deposit, to some extent, by sieving action; the smaller particles (submicrometer sizes) also deposit through their Brownian motion, which carries them into the filter material. The efficiency of collection generally increases with airstream velocity, density, and size for particles greater than 0.5 μm in diameter.

Filters are available in a wide variety of matrices including cellulose, glass, asbestos, ceramic, carbon, metallic, polystyrene, and other polymeric materials. Filters made of these fibrous materials consist of thickly matted fine fibers and are small in mass per unit face area, making them useful for gravimetric determinations. Of these, cellulose fiber filters are the least expensive, are available in a wide range of sizes, have high tensile strengths, and are relatively low in ash content. Their greatest disadvantage is their hygroscopicity, which can present problems during weighing procedures. The filters made of synthetic fibers, particularly glass and polyvinyl chloride, are more common, partly because stable tare weights can be determined easily.

Membrane filters, microporous plastic films made by precipitation of a resin under controlled conditions, are used to collect samples that are to be examined microscopically, although they can also serve for gravimetric sampling and for specific determinations using instrumentation. Thus, the cellulose ester membrane filters are the most commonly used filters for sampling for such substances as asbestos (analyzed microscopically for fiber count) and metal dusts and fumes (analyzed for specific elements by atomic absorption techniques).

Methods other than filtration that are used to collect samples of particulates include impaction, impingement, elutriation, electrostatic precipitation, thermal precipitation, and centrifugal collection. Impactors take advantage of a sudden change in direction in airflow and the momentum of the dust particles to cause the particles to impact against a flat surface. Usually impactors are constructed in several stages, to separate dust by size fractions. The particles adhere to the plate, which may be dry or coated with an adhesive, and the material on each plate is weighed or analyzed at the conclusion of sampling. Impingers also utilize inertial properties of particles to collect samples. Although the interest in, and application of, impingers is waning, they still play an important role in industrial hygiene sampling. The impinger consists of a glass nozzle or jet submerged in a liquid— frequently water. Air is drawn through the nozzle at high velocity, and the particles impinge on a flat plate, lose their velocity, are wetted by the liquid, and become trapped. Gases that are soluble in the liquid also are collected. Usually the contents of the impinger samples are analyzed microscopically, gravimetrically, or, in a few cases, by specific methods. The principles of collection for impaction and impingement are quite similar. The primary distinction between them is that with impaction, the particles are directed against a dry or coated surface, whereas a liquid collecting medium is used with impingement.

Elutriators have been essential elements in the sampling trains used to characterize dust levels in many mineral dust surveys, usually as preselectors at the front of the sampling train. Elutriators can have either a horizontal (for example, in mining applications), or a vertical orientation (for example, for cotton dust). Elutriators are quite similar to inertial separators in the theoretical basis of operation. The primary difference is that elutriators operate at normal gravitational conditions, whereas the inertial collectors induce very high momentum forces to achieve collection of particles.

Electrostatic precipitators have been used for many decades for industrial air analysis in workrooms. These systems have the advantage of negligible flow resistance, no clogging, and precipitation of the dust onto a metal cylinder or foil liner whose weight is unaffected by humidity. Very high separation efficiencies are attainable with electrostatic precipitators; they are particularly well suited for particles of submicrometer size, for example, metal fumes.

A particle in a thermal gradient in air is directed away from a high temperature source by the differential bombardment from gas molecules around it. This action is taken advantage of in the design of thermal precipitation units, another specialized type of sampling device. With this device, air is drawn past a hot wire or plate, and the dust collects on a cold glass or metal surface opposite the hot element, driven there by the thermal gradient produced.

With the increasing emphasis on personal monitoring, there has been increasing use of preselectors in conjunction with industrial hygiene sampling to document concentrations of "respirable" dust. This sampling is most commonly done with centrifugal separators, such as the cyclone. Air enters the cyclone tangentially through an opening in the side of a cylindrical or inverted cone-shaped unit. The larger particles are thrown against the side of the cyclone and fall into the base of the assembly. The smaller particles are drawn toward the center of the unit, where they swirl upward along the axis of a tube extending down from the top. The air in the cyclone rotates several times before leaving, and, consequently, the dust deposits as it would in a horizontal elutriator having an area several times that of the cyclone's outer surface. Thus, the volume of a cyclone is much smaller than a horizontal elutriator, or other inertial collector with the same flow rate and efficiency. Cyclones used to sample for respirable dust, such as in the course of determining compliance with the respirable mass standard for silica-containing dusts, should have performance characteristics meeting the specified criteria. The orientation of the cyclone is not as critical as for the elutriators, and small 10-mm diameter cyclones have become commonplace for personal breathing-zone sampling.

ANALYTICAL PROCEDURES

Continuing advances and improvements in analytical capabilities have made it possible to measure minute quantities of specific compounds, elements, or ions. The industrial hygienist can process a very small sample of air and accurately determine the presence of suspected contaminants. As discussed previously in this chapter, the field industrial hygienist should work closely with the industrial hygiene analyst, to become familiar with the limitations of the analytical equipment of interest, thus being able to plan his or her sampling strategy with maximum efficiency.

It is beyond the scope of this chapter to outline specific procedures for industrial contaminants. Instead, brief descriptions of the various analytical methods and techniques that have been applied to industrial hygiene samples are presented, with the expectation that the reader will consult more detailed sources for complete understanding of analytical requirements for particular substances of interest. Of particular note are the descriptions of analytical methods published by NIOSH and OSHA.

The most frequently utilized analytical method for industrial hygiene samples is gravimetric analysis of filters or other collection media to determine the weight gain. This requires careful handling and processing of the media before collecting the sample, as well as the conditioning of the media after collection of the sample in the exact manner as used to obtain tare weights. In so doing, any necessary correction for the "blank" can be incorporated into the analysis. Often, gravimetric analysis is done as a "gross" analysis, or general indicator of conditions, with subsequent analyses performed for specific constituents of the sample. Another type of gravimetric technique involves the formation of a precipitate by combining a sample solution with a precipitating agent, and subsequent weighing of the solid precipitate formed.

Acid-base and oxidation-reduction volumetric procedures are outstanding examples of simple but useful analytical methods still used in the analysis of industrial hygiene samples. Hydrogen chloride gas and sulfuric acid mist can be collected in an impinger or bubbler containing a standard sodium hydroxide solution and quantified by back titration with a standard acid. Ammonia and caustic particulate matter can be collected in acid solution with similar apparatus, and the airborne concentration determined by titration with a standard base. Oxidation-reduction titrations, principally iodimetric, are useful for measuring sulfur dioxide, hydrogen sulfide, and ozone. Improved volumetric methods utilize electrodes to indicate acid-base null points, and amperometric methods are available for oxidation-reduction titrations. These electrical techniques increase analytical precision and speed up the analyses but do not affect the sensitivity appreciably.

Much of the sampling for dust requires analysis by microscopic techniques. The use of light microscopy for "dust counting" is decreasing, being replaced by more specific, and more reproducible, mass sampling techniques. However, it is frequently necessary to determine the particle size distribution of airborne particulate, and optical methods offer an effective way of doing this. In addition to the more classic counting applications of microscopy, the present sampling and analysis method for asbestos is based on actual fiber counting at 400–450X magnification using phase contrast illumination. As with all optical methods, the analytical results (for example, actual counts) are somewhat analyst dependent, since much of the technique requires subjective analysis by the microscopist.

Changes of color intensity or tone have been the bases of many useful industrial hygiene analytical methods. For example, the use of a reagent in a fritted glass bubbler to determine the airborne concentrations of nitrogen dioxide is a classic application of such methods. Under controlled conditions of sampling, the concentration of nitrogen dioxide in the air is inversely proportional to the time required to produce the color change. Titrations employing acid-base and iodimetric reactions with color indicators are conducted in similar fashion. Usually such titrations of air samples lack the accuracy and precision obtainable with careful laboratory procedures, but they are adequate for most field studies and have the great advantage of giving a direct and immediate indication of the environmental concentrations. Relatively sensitive and specific analyses of many contaminants can be made, using the spectrophotometers available as both laboratory and field instruments. Colorimetric methods involve analytical reactions to produce a color in proportion to the quantity of the contaminant of interest in the sample. For example, in the determination of metals, the dithizone extraction method is able to determine selectively the various metallic elements depending on the pH of the solution.

In addition to the colorimetric procedures, which take advantage of spectrophotometry operating in the visible range, IR and UV spectrophotometers have considerable application in the industrial hygiene area. The interaction of electromagnetic radiation with matter is the basis for such analytical techniques. Principles of operation extend from the IR radiation spectra, to the UV and, in fact, to the X-ray region. The latter can be used to provide information on elemental composition (fluorescence) and crystal structure (diffraction). In most cases, the sample, whether gas, liquid, or solid, is exposed to radiation of known characteristics and specific wavelengths (fluorescence), and the fractions transmitted or scattered are determined and quantified. Color production, turbidity, and fluorescence are examples of properties determined by electromagnetic radiations that are widely used for quantifying industrial hygiene air samples.

Since the smallest trace of materials can be detected by the spectrograph, spectrographic procedures may be used for small amounts of metallic ions and elements when other procedures cannot be used. The chief limitations are the high cost and the need for a highly trained technician, having access to a rather complete spectrographic laboratory to do quantitative work. Generally, the degree of sensitivity afforded by these units is not required by the industrial hygienist, although it frequently is desirable to obtain a complete elemental analysis of a sample of unknown composition as a starting point for an elaborate analytical program. In applying emission spectroscopy, a solid sample is vaporized in a carbon arc, causing the formation of characteristic radiation, which is dispersed by a grating or a prism, and the resulting spectrum is photographed. Each metallic or metal-like element can be identified from the spectra that are formed. Elemental analyses of body tissues, dust, ash, and air samples can be qualitatively analyzed by this technique. With mass spectroscopy, gases, liquids, or solids are ionized by passage through an electron beam. The ions thus formed are projected through the analyzer by means of an electromagnetic or electrostatic field, or simply by the time necessary for the ions to travel from the gun to the collector. Each compound has a characteristic ionization pattern that can be used to identify the substance. This analytical tool, in conjunction with gas chromatography, has become a powerful technique for separating and identifying a wide range of trace contaminants in industrial hygiene and ambient air samples.

The development of chromatographic methods of analysis has given the industrial hygienist an extremely versatile means of quantifying low concentrations of airborne contaminants, particularly organic compounds. Gas chromatography utilizes the selective absorption and elution provided by appropriately chosen packings for the columns to separate mixtures of substances in an air sample or in a desorption solution. The various compounds in the air sample have different affinities for the material in the column, thus "slowing" some of the constituents more than others, with the result that as the individual compounds reach the detector associated with the chromatograph, they can be quantified by running standards of known concentration of the various substances along with the unknowns. This separation is achieved without any appreciable change in the entities; thus, the chromatograph can serve as an analytical technique in its own right by attaching an appropriate detector. Thermal conductivity, flame ionization, and electron capture detectors are commonly used for this purpose. The chromatograph can also serve to "purify" a sample by separating the constituents and selecting a narrow portion of the eluted sample. This portion than can be subjected to other more sophisticated types of analysis, such as mass spectroscopy.

With atomic absorption spectrophotometry (flame photometry), monochromatic radiation from a discharge lamp containing the vapor of a specific element, such as nickel or cadmium, passes through a flame into which the sample is aspirated. The absorption of the monochromatic radiation is measured by a double-beam method, and the concentration is determined. This technique permits rapid determination of almost all metallic elements. Solutions of the metals are aspirated into the high temperature flame, where they are reduced to free atoms. The absorption generally obeys Beer's law in the parts-per-million range, where quantitative determinations can be made. The characteristic absorption gives this technique high selectivity. Most interferences can be overcome by proper pretreatment of the samples. Atomic absorption methods have found substantial use in industrial hygiene in the determination of both major and trace metals in industrial hygiene samples, as well as in blood, urine, and other body fluids and tissues.

Many additional, specialized analytical methods are available to the industrial hygienist and analytical chemist for application to specific qualitative and quantitative needs. For detailed information on particular methods or procedures, the reader is advised to consult the references listed at the end of this chapter and, more important, to keep abreast of current developments in the industrial hygiene analytical field by subscribing to journals or routinely reviewing the wealth of new information constantly coming forth.

BIOLOGICAL MONITORING

In the final analysis, the degree of success associated with attempts to provide a safe and healthful work environment can be determined by assessing the amount of the contaminant of concern that has been actually absorbed by the workers. Regardless of the degree of sophistication applied to environmental and personnel monitoring, the extent to which workers have absorbed the contaminant should be determined by some clinical measurement on the individuals. In the industrial hygiene profession, the presence of the substance or metabolite can be derived from direct quantitative analysis of body fluids, tissue, or expired air. An indirect determination of the effect of the substance on the body can be made by measurements on the functioning of the target organ or tissue. With the possible exception of carcinogenic substances, even the most hazardous materials have some "no effect" level below which exposure can be tolerated by most workers for a working lifetime without incurring any significant physiological injury.

Perhaps the most common biological fluids analyzed in attempts to determine the extent to which individuals have been exposed to contaminants are blood, urine, and exhaled air. However, it must be emphasized that for most compounds, the specific concentrations indicating excessive buildup of contaminants or metabolites within the body are not known for most compounds with any certainty. The biological threshold limit values, for example, concentrations indicative of excessive exposure for some compounds, have been determined; biological exposure indices (BEIs), published annually by the American Conference of Governmental Industrial Hygienists, are examples of these. An example of useful application of blood analysis as an index of exposure to contaminants is the routine determination of carboxyhemoglobin in blood as an indication of exposure to carbon monoxide. Extensive studies of the concentration of carbon monoxide in the air and the consequent level of carboxyhemoglobin in the blood have been conducted.

It is likely that, with the increasing interest in providing safe work conditions for persons exposed to a tremendously wide range of contaminants, development of cause-and-effect relationships between the level of exposure to a particular contaminant and the amount of the substance remaining in the body will become a more integral part of the total occupational health monitoring procedure. Accordingly, greater emphasis will be placed on the medical aspects of the total monitoring effort in attempts to document, as an end result, that workers have not been exposed to excessive levels of contaminants. When the results of biological monitoring indicate excessive exposures, and the environmental monitoring aspects indicate that concentrations are within acceptable limits, analysis of individuals' work practices, or at least specific analysis of an individual's work activity, may be required to ascertain the cause of the elevated readings in a given individual. As such, the biological monitoring program is a viable and extremely useful supplement to the ongoing environmental and medical surveillance programs for contaminants for which such coordinated efforts are possible.

INDUSTRIAL HYGIENE CONTROLS

The industrial hygiene engineering control principles are deceptively few: substitution; isolation; and ventilation, both general and local. In a technological sense, an appropriate combination of these principles can be brought to bear on any industrial hygiene control problem to achieve satisfactory quality of the work environment. It may not be (and usually is not) necessary or appropriate to apply all these principles to any specific potential hazard. A thorough analysis of the control problem must be made to ensure that a proper choice from among these methods will produce the desired results. Proper control must be accomplished in a manner that is most compatible with the technical process, is acceptable to the workers in terms of day-to-day operation, and can be completed with optimal balance of installation and operating expenses.

With advancing technology and the tendency for acceptable exposure limits to become increasingly more stringent, the industrial hygiene engineer's function must include an ongoing analysis of installed control measures. Aside from problems associated with the operation and maintenance of these provisions, the concept of the control provisions itself might become outdated by changes in process or regulations. The engineer then must be able to develop effective control methods, and he or she must have the capability to continue to evaluate the effectiveness of these methods on a regular basis.

Substitution

Although it can be argued that substitution is not an engineering option, it frequently offers the most effective solution to an industrial hygiene problem. There is a tendency to analyze any problem from the standpoint of correcting rather than eliminating it. The first inclination in considering a vapor exposure problem in a degreasing operation is to provide or increase the ventilation rather than consider substituting a solvent having a much lower degree of hazard associated with its use. In its broadest sense, substitution includes replacing hazardous substances, changing from one type of process equipment to another, or in some cases even changing the process itself. Material, equipment, or an entire process can be substituted to provide effective control of a hazard, often at minimal expense.

There are many examples of substituting materials of lower toxicity; some are classics in the history of industrial hygiene. Most industrial hygienists are familiar with the substitution of red phosphorus for white in the manufacture of matches. Although this was done primarily in reaction to a tax law, the result was a markedly reduced potential hazard. The sequence of substitutions of degreasing solvents is an interesting one: from petroleum naphtha to carbon tetrachloride to chlorinated hydrocarbons to fluorinated hydrocarbons. Each of these substitutions alleviated one problem but resulted in a new one. This underscores a basic problem associated with using substitution as a control method in that one hazard can be replaced by another inadvertently.

In some processes, there is only limited opportunity to substitute materials; however, it might be possible to substitute, or at least modify, process equipment. This approach almost always is taken as a result of an obvious potential, usually physical, hazard. Applications to counter potential safety hazards are common: substituting safety glass for regular glass in enclosures, replacing unguarded equipment with properly guarded machines, replacing safety gloves or aprons with garments made of materials more impervious to the chemicals being handled. Since substitution of equipment frequently is done as an immediate response to an obvious problem, it may not be recognized as an engineering control, even though the end result is every bit as effective. Substituting one process for another may not be considered except in major modifications of a process. In general, a change in any process from a batch to a continuous type of operation carries with it an inherent reduction in potential hazard. This is true primarily because the frequency and duration of potential contact of workers with the process materials is reduced when the overall process approach becomes one of continuous operation. The substitution of processes can be applied on a fundamental basis. For example, the substitution of airless spray equipment for conventional spray equipment can reduce the exposure of a painter to solvent vapors. The substitution of a paint dipping operation for a paint spray operation can reduce the potential hazard even further. In any of these cases, the automation of the process can further reduce the potential hazard.

Isolation

The application of the principle of isolation frequently is envisioned as consisting of the installation of a physical barrier between a hazardous operation and the workers. Fundamentally, however, this isolation can be provided without a physical barrier by appropriate use of distance and time. Perhaps the most common example of isolation as a control measure is associated with the storage and use of flammable solvents. The large tank farms with dikes around tanks, the underground storage of some solvents, the detached solvent sheds, and the fireproof solvent storage rooms within buildings are all commonplace in American industry. Although the primary reason for the isolation of solvents is the risk of fire and explosion, the principle is no less valid as an industrial hygiene measure.

Frequently the application of the principle of isolation maximizes the benefits of additional engineering concepts such as local exhaust ventilation. For example, the charging of mixers is the most significant operation in many of the processes that use formulated ingredients. When one of the ingredients in the formulation is of relatively high toxicity, it is worthwhile to isolate the mixing operation. That is, install a mixing room, thereby confining the airborne contaminants potentially generated by the operation to a small area, rather than having them influence the larger portion of the plant. Isolation in this manner permits the application of ventilation principles to control this contaminant at the source—for example, the mixer, a much more effective option.

General Ventilation

For purposes of industrial hygiene engineering, ventilation is a method for providing control of an environment by the strategic use of airflow. The flow of air may be used to provide either heating or cooling of a work space, to remove a contaminant near its source of release into the environment, to dilute the concentration of a contaminant to acceptable levels, or to replace air exhausted from an enclosure. Ventilation is by far the most important engineering control principle available to the industrial hygienist. Applied either as general or local control, this principle has industrial significance in at least three applications: control of heat and humidity primarily for comfort reasons, prevention of fire and explosions, and, most important to the industrial hygienist, maintenance of concentrations of airborne contaminants at acceptable levels in the workplace.

Detailed discussions of the principles and application of general and local exhaust ventilation appear in many excellent reference books. Application of these principles to industrial problems, whether general or local, requires a basic understanding of the fundamentals of airflow. The scientific laws that define completely the motion of any fluid, including air, are complex, and except in the relatively simple case of laminar flow, we know relatively little about them. Nevertheless, there are fundamental relationships that must be understood and conscientiously applied by the industrial hygiene engineer. These are described in elaborate discussions of these principles in several of the general references identified at the end of this chapter.

General ventilation is the practice of supplying and exhausting large volumes of air throughout a work space. It is used typically in industry to achieve comfortable work conditions (temperature and humidity control) or to dilute the concentrations of airborne contaminants to acceptable limits throughout the work space. Properly used, general ventilation can be effective in removing large volumes of heated air or relatively low concentrations of low-toxicity contaminants from several decentralized sources. General ventilation can be provided by either natural or mechanical means; often the best overall result is obtained with a combination of mechanical and natural air supply and exhaust.

Natural ventilation can be provided either by gravitational forces, using primarily thermal forces of convection, or by forces created by differences in "wind pressure." These two natural forces operate together in most cases, resulting in the natural displacement and infiltration of air through windows, doors, walls, floors, and other openings in an industrial building. Unfortunately, the wind currents and thermal convection profiles on which natural ventilation is dependent are erratic and frequently unpredictable. Thus, it is perhaps a misnomer to refer to natural ventilation as a "control" method, since to utilize this technique requires dependence on, rather than control of, natural forces. On the other hand, there are applications for general natural ventilation. The pressure exerted on the upwind side and concurrent suction exerted on the downwind side of a building as a result of wind movement can be predicted for flat terrain fairly reliably. Thus, the wind forces exerted on an isolated building in a relatively flat area permit the prediction of natural ventilation forces. In the more common complex industrial building, however, the effects of the presence of one building on the others normally cannot be calculated; hence, the use of wind pressure models in the development of general ventilation systems is not feasible.

A modern industrial complex characterized by a low-profile building structure—for example, large floor space, low height, as well as the multistory buildings of masonry and glass construction—present ventilation problems that were not found among the older industrial plants. For the most part, the industrial facilities constructed thirty or more years ago incorporated by design many features that permitted—in fact, expected—natural ventilation of the work space. By contrast, modern buildings generally defy the exertion of natural ventilation forces, and mechanical ventilation must be relied on almost completely. Mechanical ventilation systems exhaust contaminated air by mechanical means (exhaust fans), with the concomitant use of appropriate air supply to replace the exhausted air. The best method of achieving this in modern closed buildings is to supply air through a system of ductwork, distributing the air into the work areas in a manner that will provide optimum benefit to the workers for both comfort and control contaminants.

Normally, general ventilation is applied to provide an environment that is comfortable to the workers, one that is free of harmful concentrations of airborne contaminants, or both. General ventilation for comfort, principally heat relief, includes certain aspects of what is commonly considered to be "air conditioning engineering." Here, air is treated to control temperature, humidity, and cleanliness, simultaneously, and is distributed to maximize effects in the conditioned space. In the typical residential or office building, these requirements are primarily associated with comfort for the occupants. In many industrial situations, however, comfort conditions are impractical if not impossible to maintain, and the chief function of ventilation for comfort control is to prevent acute discomfort and the accompanying adverse physiological effects. General exhaust ventilation also may be used to remove heat and humidity if a source of cooler air is available. If it is possible to enclose the heat source, as in the case of ovens or furnaces, a gravity or forced-air stack may be all that is necessary to prevent excessive heat from entering the workroom.

Dilution ventilation has, as its primary function, the maintenance of concentrations of airborne contaminants at or below acceptable exposure limits, either in terms of potential fire and explosion or from occupational health considerations. If enough clean air is mixed with contaminated air, the concentration of the contaminant can be diluted to any reasonable level. Of course, dilution ventilation for purposes of fire and explosion control should not be utilized in areas occupied by workers, since the concentrations of concern from an occupational health standpoint invariably are orders of magnitude below those of concern for explosive limits. Exposures to atmospheres controlled to concentrations below the lower explosive limit or even a fraction thereof could cause narcosis, severe illness, or even death.

Therefore, it is extremely important not to confuse dilution ventilation requirements for health hazard control with those for fire and explosion prevention. When considering whether dilution ventilation is appropriate for health hazard control, it should be remembered that dilution ventilation has four limiting factors. First, the quantity of contaminant generated must not be excessive, or the air volume necessary for dilution would be impractically large. Second, workers must be far enough away from the evolution of the contaminant, or the contaminant must be of sufficiently low concentration, that the workers will not be exposed above acceptable limits. Third, the toxicity of the contaminant must be relatively low. And fourth, the evolution or generation of the contaminant must be reasonably uniform and consistent. A review of these factors indicates clearly that dilution ventilation is not normally appropriate for the control of fumes and dust, since the high toxicity often encountered requires excessively large quantities of dilution air. Moreover, the velocity and rate of evolution usually are very high, resulting in locally high concentrations, thus rendering the dilution ventilation concept inappropriate.

In general, dilution ventilation is not as satisfactory as local exhaust ventilation for the primary control of health hazards. Occasionally, however, dilution ventilation must be used because the operation of the process prohibits local exhaust. Dilution ventilation sometimes provides an adequate amount of control more economically than a local exhaust system. However, this condition is an exception rather than the rule, and it should be kept in mind that the economical considerations of the long-term use of dilution ventilation systems often overshadow the initial cost of the system because such a system invariable exhausts large volumes of heated air from the building. This workload can easily result in huge operating costs because of the need for conditioned makeup air, and the general ventilation scheme would be much more expensive over an extended period of time.

In practice, dilution ventilation for the control of health hazards is used to best advantage in controlling the concentration of vapors from organic solvents of relatively low toxicity. To apply the principle of dilution to such a problem successfully, data must be available on the rate of vapor generation or on the rate of liquid evaporation. Usually such data can be obtained from the plant records on material consumption.

Ventilation requirements based on room volume alone have very little validity. Calculations of the required rate of air changes can be made only on the basis of a material balance for the contaminant of interest. Similar calculations can be made for the rate of concentration increase or decrease; however, they require not only the air change rate but also the rate of generation of contaminant. In the design of industrial ventilation, *x air changes* has valid application only rarely. This term is useful when applied to meeting rooms, offices, schools, and similar spaces, where the purpose of ventilation is simply the control of odor, temperature, or humidity, and the only contamination of air is the result of people-oriented activity. Dilution ventilation requirements should always be expressed in cubic feet per minute or some other absolute unit of airflow, not in "air changes per hour."

General ventilation presents some disadvantages, the most significant one being that it permits the occupied space to become, in effect, a large settling chamber for the contaminants, even though the concentration may be within acceptable limits as far as potential health hazards are concerned. In some cases, the settling or separation of contaminants from the air may represent the condensation of materials that were vaporized by high temperature processes and accumulated on surfaces after condensation. The undesirability of the settling of contamination onto surfaces within the plant has been dramatically demonstrated in plants handling highly combustible dusts, where, although the dust concentrations at any one time were not sufficient to constitute a hazard, accumulation over many months resulted in disastrous explosions. It is therefore necessary to maintain an effective ongoing good housekeeping program in conjunction with any broad-based engineering control such as general ventilation.

The design of general ventilation systems for a plant is not complete without consideration of the routes by which the air will enter and leave the work space. These routes are of critical concern not only to the occupants of the building but potentially to the neighbors and the community at large. In general, the relative locations of air inlets and outlets should be considered in the implementation of a general ventilation system.

Sufficient combined or total inlet area should be planned to accommodate the required volume of makeup air during the heating or cooling season, whichever needs the larger volume. This will prevent excessive inlet velocities that consume power, create drafts on workers, stir up dust, and interfere with the performance of local exhaust systems. The chosen inlet velocity is likely to represent an engineering compromise between low rates, which require large inlet areas, and higher rates, which facilitate rapid dilution of contaminants. If widespread hazardous operations are controlled by dilution or general ventilation, inlets should be well distributed around the building to provide uniform circulation. If contaminants are controlled by localized dilution, or by forceful diffusion into the general room-air reservoir, inlets may be purposely located to give nonuniform air supply distribution. Inlets should be located to take full advantage of any thermal or convection effects within the building. For this purpose, the designer must avoid the location of inlets (and outlets) near the "neutral zone" of inside-outside pressure differentials, which is approximately midway between the floor and roof. Inlets should be located remote from stacks or ventilators discharging contaminants from the same or neighboring structures.

Outlets should be located as far as possible from air inlets to prevent "short-circuiting." This advice holds for both natural and mechanical ventilation systems. If widely scattered operations are controlled by general dilution, outlets should be uniformly distributed around the building. Similarly, if contaminants are controlled by localized dilution, air outlets may be purposely located to short-circuit a corner of a large room, with the intent of creating a high rate of air change there without involving the atmosphere of the entire space. This method of localized space ventilation requires a higher rate of exhaust air than supply air within the area to be controlled, to prevent the spread of contaminants throughout the room. Outlets should be placed to take full advantage of thermal effects. They should be protected against the direct force of prevailing winds, which reduce the capacity of any exhaust fans or destroy the anticipated airflow route planned on the basis of thermal effects inside the building. In general, dilution air should enter the work space at approximate breathing-zone height, pass through the workers' breathing zone, then through the zone of contamination; afterward, if it is not exhausted, it should enter a space of higher relative contamination. A useful concept in the interest of air-handling economy is *progressive ventilation*. The air removed from an industrial plant frequently may be directed in a way that will provide both local and general ventilation. In fact, the air may be routed through several areas in succession, always in the direction of increasing air contamination, as long as the exhaust from one area is acceptable as the supply for the next. Progressive ventilation saves energy in the form of airborne heat and horsepower needed to move the air.

Local Exhaust Ventilation

Local exhaust ventilation incorporates the concept of controlling a contaminant by capturing it at or near the place where it is generated and removing it from the work space. Local ventilation relies more heavily on mechanical methods of controlling airflow than does general ventilation. A local exhaust system usually includes all the following components—a hood or enclosure, ductwork, an air-cleaning device (where necessary for air pollution abatement purposes), and an air-moving device (usually an exhaust fan)—to draw the contaminated air through the exhaust system and discharge it to the outside air. In general, local exhaust systems consist of more individual components than do general exhaust systems; since local exhaust systems also offer more operational arameters that must be controlled within acceptable ranges, they require more maintenance and involve higher operating expenses, as well.

When the primary purpose of the ventilation is to provide control of airborne contaminants, local exhaust systems generally are much superior to general ventilation. The advantages of the local exhaust ventilation system over general exhaust, for any particular application, will include many of the following. First, if the system is designed properly, the capture and control of a contaminant can be virtually complete. Consequently, the exposure of workers to contaminants at the sources exhausted can be prevented. With general ventilation, the contaminant is diluted when the exposure occurs, and at any given workplace this dilution may be highly variable and, therefore, inadequate at certain times. Second, the volumetric rate of required exhaust is less with local ventilation; as a result, the volume of makeup air required is less. Third, local ventilation offers savings in both capital investment and heating costs. The contaminant is contained in a smaller exhausted volume of air. Therefore, if air pollution control is needed, it is less expensive because the cost of air pollution control is approximately proportional to the volume of air handled. Fourth, local exhaust systems can be designed to capture large settleable particles or at least confine them within the hood, thus greatly reducing the effort needed for good housekeeping. Auxiliary equipment in the workroom is better protected from such deleterious effects of the contaminant as corrosion and abrasion. Finally, local exhaust systems usually require a fan of fairly high-pressure characteristics to overcome pressure losses in the system. Therefore, the performance of the fan system is not likely to be affected adversely by such influences as wind direction or velocity or inadequate makeup air. This is in contrast to general ventilation, which can be affected greatly by seasonal factors.

The four components of a simple local exhaust system are a hood, ductwork, an air-cleaning device (cleaner), and an air-moving device (fan).

Typically the system is a network of branch ducts connected to several hoods or enclosures, main ducts, air cleaner for separating the contaminants from the airstream, exhaust fan, and discharge stack to the atmosphere.

A hood is a structure designed to partially enclose an operation and to guide airflow in an efficient manner to capture the contaminant. The hood is connected to the ventilation system with ductwork that removes the contaminant from the hood. The design and location of the hood is one of the most critical aspects in the successful operation of a local exhaust system.

The ductwork in an exhaust system provides a path for flow of the contaminated air exhausted from the hood to the point of discharge. The following points are important in the design of the ductwork. In the presence of dust, the duct velocity must be high enough to prevent the dust from settling out and plugging the ductwork. In the absence of dust, the duct velocity should strike an economic balance between ductwork cost and fan, motor, and power costs. The location and construction of the ductwork must furnish sufficient protection against external damage, corrosion, and erosion to maximize the useful life of the local exhaust system.

Most exhaust systems installed for contaminant control need an air cleaner. Occasionally the collected material has some economic reuse value, but this is seldom the case. To collect and dispose of the contaminant is usually inconvenient and certainly represents an added expense. Yet, the growing concern with air pollution control, and the need to comply with legal restrictions on discharges from sources of atmospheric emissions, place new importance on the air-cleaning device within a local exhaust system.

The heart of the local exhaust system is the fan, usually of the centrifugal type. Wherever practicable a fan should be placed downstream from the air cleaner so that it will handle uncontaminated air. In such an arrangement, the fan wheel can be the backward-curved blade type, which has a relatively high efficiency and lower power cost. For equivalent air handling, the forward-curved blade impellers run at somewhat lower speeds, which may be important when noise is a factor. If chips and other particulate matter have to pass through the impeller, the straight blade or paddlewheel-type fan is best because it is least likely to clog. Fans and motors should be mounted on substantial platforms or bases and isolated by antivibration mounts. Ducts should be connected using vibration isolators, for example, sleeves or bands of very flexible material, such as rubber or fabric.

The *local exhaust hood* is the point at which air first enters the exhaust ventilation system. As such, the term can be used to apply to any opening in the exhaust system regardless of its shape or physical disposition. An open-ended section of ductwork, a canopy-type hood situated above a hot process, and a conventional laboratory booth-type hood all could be called hoods in the context of this discussion. The hood captures the contaminant generated by a particular process or operation and causes it to be carried through the ductwork to a convenient discharge point. The quantity of air required to capture and convey the air contaminants depends on the size and shape of the hood, its position relative to the point of generation of the contaminant, and the nature and quantity of the air contaminant itself. It should be emphasized that there is not necessarily a "standard hood" that is correct for all applications of a particular operation, since the methods of processing are unique to each operating plant. On the other hand, standard concepts for exhaust hooding have been developed and have been recommended for specific types of operations.

Hoods can be classified conveniently into four categories, based on the concept of contaminant capture/control: enclosures, booth-type, receiving, and exterior. Enclosure hoods normally surround the point of emission or generation of contaminant as completely as practicable. In essence, they surround the contaminant source to such a degree that all contaminant dispersal action takes place within the confines of the hood itself. Consequently, enclosure hoods generally require the lowest rate of exhaust ventilation and, therefore, are economical and quite efficient. Enclosure hoods should be used whenever possible, and they deserve particular consideration when a moderately or highly toxic contaminant is involved.

The common spray painting enclosure, and other booths, are special cases of enclosure-type hoods. These are typified further by the common laboratory hood in which one face of an otherwise complete enclosure is open for ready access. Air contamination takes place within the enclosure, and air is exhausted from it in such a way, and at such a rate, that an average velocity is induced across the face of the opening sufficient to overcome the tendency of the contaminant to escape from within the hood. The three walls of the booth greatly reduce the exhaust air requirements, although not to the extent of a complete enclosure.

The term *receiving* refers to a hood in which a stream of contaminated air from a process is exhausted by a hood located near the source of generation of the contaminants specifically for purposes of control. Two examples of this type of hood are canopies situated above hot processes and hoods attached to grinders, positioned to take advantage of centrifugal and gravitational forces to maximize control of the dust generated by the process. Canopy hoods, frequently located above hot processes, are similar to exterior hoods in that the contaminated air originates beyond the physical boundaries of the hood. The fundamental difference between receiving and exterior hoods is that in the former the hood takes advantage of the natural movement of the released contaminant, whereas in the latter, air is induced to move toward the exterior hood. In practice, receiving hoods are positioned to be in the pathway of the contaminant as normally released by the operation. If hood space is limited by the process, baffles or shields may be placed across the line of throw of the particles to remove their kinetic energy. Then the particles may be captured and carried into the hood by lower air velocities. Additional examples of receiving hoods include those associated with many hand tool operations (surface grinders, metal polishers, stone cutters, and sanding machines).

Exterior hoods must capture air contaminants generated from a point outside the hood itself, sometimes at quite a distance. Exterior hoods therefore differ from enclosure or receiving hoods in that their sphere of influence must extend beyond their own dimensions in capturing contaminants without the aid of natural forces such as natural drafts, buoyancy, and inertia. In other words, directional air currents must be established adjacent to the suction opening of the hood to provide adequate capture. Thus, exterior hoods are quite sensitive to external sources of air disturbance and may be rendered completely ineffective by even slight lateral movement of air. They also require the most air to control a given process and are the most difficult to design of the various hoods. Examples of exterior hoods include the exhaust slots on the edges of tanks or surrounding a workbench such as a welding station, exhaust grilles in the floor or workbench below a contaminated process, and the common propeller-type exhaust fans frequently mounted in walls adjacent to a source of contamination.

Local exhaust hoods perform their function in one of two ways: capture or control of air streams. With the "capture" approach, air movement is created to draw the contaminant into the hood. When the air velocity that accomplishes this objective is created at a point outside a nonenclosing hood, it is called *capture velocity*. Some exhaust hoods essentially enclose the contaminant source and create an air movement that prevents the contaminant from escaping from the enclosure. The air velocity created at the openings of such hoods is called the *control velocity*. The successful design of any exhaust hood is based on correct determination of these two quantities—control velocity and capture velocity. The air velocity, which must be developed by the exhaust hood at the point or in the area of desired control, is based on the magnitude and direction of the air motion to be overcome and is not subject to direct and exact evaluation. Many empirical ventilation standards, especially concerning dusty equipment such as screens and conveyor belt transfers, are based on parameters such as cubic feet per minute per foot of belt width. These so-called exhaust rate standards usually are based on successful experience, are easily applied, and usually give satisfactory results if not extrapolated too far. In addition, they minimize the effort and uncertainty involved in calculating the fan action of falling material, thermal forces within hoods, and external air currents. However, such standards have three major pitfalls: 1) they are not of a fundamental nature; that is, they do not follow directly from basic "laws"; 2) they presuppose a certain minimum quality of hood or enclosure design, although it may not be possible or practical to achieve the same quality of hood design in a new installation; and 3) they are valid only for circumstances similar to those that led to their development and use. It should be clear that the nature of the process generating the contaminant will be an important determinant of the required capture velocity.

The flow characteristics at a suction opening are differ greatly from the flow pattern at a supply or discharge opening. Air blown from an opening maintains its directional effect in a fashion similar to water squirting from a hose and, in fact, is so pronounced that it is often called *throw*. However, if the flow of air through the same opening is changed so that the opening operates as an exhaust or intake with the same volumetric rate of airflow, the flow becomes almost completely nondirectional and its range of influence is greatly reduced. As a first approximation, when air is blown from a small opening, the velocity 30 diameters in front of the plane of the opening is about 10 percent of the velocity at the discharge. The same reduction in velocity is achieved at a much smaller distance in the case of exhausted openings, such that the velocity equals 10 percent of the face velocity at a distance of only 1 diameter from the exhaust opening. Therefore, local exhaust hoods must not be applied for any operation that cannot be conducted in the immediate vicinity of the hood.

To provide efficient capture with a minimum expenditure of energy, the airflow across the face of a hood should be uniform throughout its cross section. For slots and lateral exhaust applications, this can be accomplished by incorporating external baffles. Another method of design is to provide a velocity of 2,000–2,500 ft/min into the slot with a low-velocity plenum or large area chamber behind it. For large, shallow hoods, such as paint-spray booths, laboratory hoods, and draft shakeout hoods, the same principle may be used. In these cases, unequal flow may occur, with resulting higher velocities near the takeoffs. Baffles provided for the hood improve the air distribution and reduce pressure drop in the hood, giving the plenum effect. When the face velocity over the entire hood is relatively high or when the hood or booth is quite deep, baffles may not be required.

The air volume for a hood can be specified by giving the hood static pressure, SP_h, and duct size. For example, the hood static pressure at a typical grinding wheel hood is 2 inches of water. This reflects a conveying velocity of 4,500 ft/min and entrance coefficient, C_e, of 0.78. For other types of machinery in which the type of exhaust hood is relatively standard, a specification of the static suction and the duct size can be found in various reference sources. Specification of the static suction without duct size is, of course, meaningless because decreased size increases velocity pressure and static suction, while actually decreasing the total flow and the degree of control. Therefore, static suction measurements for standard hoods or for systems in which the airflow has been measured previously are quite useful to estimate, in a comparative way, the quantity of air flowing through the hood.

When applying local exhaust ventilation to a specific problem, control of the contaminants is more effective if the following basic principles are followed:

1) A process to be exhausted by local ventilation should be enclosed as much as possible. This generally provides better control per unit volume of air exhausted. Nevertheless, the requirement of adequate access to the process must always be considered. An enclosed process may be costly in terms of operating efficiency or capital expenditure, but the savings gained by exhausting smaller air volumes may make the enclosure worthwhile.

2) Air velocity through all hood openings must be high enough not only to contain the contaminant but to remove the contaminant from the hood. The importance of optimum capture and control velocity was discussed in the preceding sections.

3) Exhaust hoods that do not completely enclose the process should be located as near to the point of contaminant generation as possible and should provide airflow in a direction away from the worker toward the contaminant source. This principle, drawing on the characteristics of blowing and exhausting from openings in ductwork, was considered in more detail in the preceding sections.

4) All air exhausted from a building or enclosure must be replaced to keep the building from operating under negative pressure. This applies to local as well as general exhaust systems. Additionally, the incoming air must be tempered by a makeup air system before being distributed inside the process area. Without sufficient makeup air, exhaust ventilation systems cannot work as efficiently as intended.

It is disconcerting to observe the frequency with which this last principle is violated. The beneficial effect of a well-designed local exhaust system can be offset by undesired recirculation of contaminated air back into the work area. Such recirculation can occur if the exhausted air is not discharged away from supply air inlets. The location of the exhaust stack, its height, and the type of stack weather cap all can have a significant effect on the likelihood of contaminated air reentering through nearby windows and supply air intakes.

RECENT DEVELOPMENTS IMPACTING INDUSTRIAL HYGIENE MANAGEMENT

In the past decade or so, changes have taken place, some quite subtly, that will have long-lasting effects on the practice of industrial hygiene. Perhaps the two most significant of these changes have been the infusion of computer assistance and the trend toward "prospective epidemiology."

Computers have made their presence felt, and appreciated, in almost every walk of life today. It is logical to have expected the computer to be recognized as a true ally of the industrial hygienist. Projects that only could have been talked about decades ago, because of the tremendous volume of data needed to complete the work, are now rather trivial exercises and are accomplished with minimal effort. Calculations of time-weighted average exposures, using air-sampling data and analytical results from dozens of individual samples, can be made in seconds; hours of calculations were required prior to computer assistance. More important than assisting in the more mundane activities of industrial hygiene (equipment calibration, exposure calculations, and so forth), the computer has made it possible to analyze data in a manner, and to a degree, virtually impossible without computer assistance. Indeed, many of the field instruments available today incorporate microprocessors that allow immediate manipulation and analysis of collected data. Another long-term benefit is the application of computers to *prospective epidemiology*. This term refers to the continuing maintenance of databases that document the environmental condition, as compiled by routine industrial hygiene monitoring, and the various indices of health status, derived from periodic medical surveillance. Information of this nature, referenced to individual workers to whom the data apply, is being compiled by many corporate industrial hygiene departments. Many software programs, incorporating database management, spreadsheet profiles, rapid search-and/or-match provisions, and other aids, are available to the industrial hygienist.

It is logical to expect these tools to result in a significant increase in monitoring activity. Additionally, as employees are informed periodically of the hazards of the materials with which they work, via hazard communication and or other "right-to-know" regulations, they are likely to begin asking for information regarding their own exposures.

Regardless of the motivation—for example, to add information to an ongoing database of corporate exposure/medical information or to satisfy new exposure monitoring regulations—it appears likely that the next decade or two of industrial hygiene practice will witness a significant increase in the emphasis of industrial hygiene sampling and analysis.

SUGGESTED FURTHER READING

American Conference of Governmental Industrial Hygienists. (1995). *Air Sampling Instruments for Evaluation of Atmospheric Contaminants, 8th ed.* Cincinnati, OH: ACGIH.

American Conference of Governmental Industrial Hygienists (1971). *Documentation of the Threshold Limit Values, 3rd ed.* Cincinnati, OH: ACGIH. (also see supplements).

American Conference of Governmental Industrial Hygienists (1987). *Microcomputer Applications in Occupational Health and Safety.* Chelsea, MI: Lewis Publishers.

American Conference of Governmental Industrial Hygienists (1998). *Threshold Limit Values for Chemical Substances and Physical Agents and Biological Exposure Indices.* Cincinnati, OH: ACGIH.

American Industrial Hygiene Association (1997). *The Occupational Environment—Its Evaluation and Control.* Fairfax, VA. AIHA.

Clayton, G. D., and F. E. Clayton, Eds. (1993). *Patty's Industrial Hygiene and Toxicology.* 4th rev. ed. New York: Wiley Interscience.

Confer, R. G. (1994). *Workplace Health Protection: Industrial Hygiene Program Guide.* Ann Arbor, MI: CRC/Lewis Publishers.

Eller, P. M., Ed. (1985). *NIOSH Manual of Analytical Methods. 3rd. ed.* Cincinnati, OH: U. S. Department of Health and Human Services.

Garrett, J. T., L. J. Cralley, and L. V. Cralley (1988). *Industrial Hygiene Management.* New York: Wiley-Interscience.

Harrison, L. (1995). *Environmental, Health, and Safety Auditing Handbook, 2nd Ed.* New York: McGraw-Hill.

Scott, R. M. (1995). *Introduction to Industrial Hygiene.* Ann Arbor, MI:CRC/Lewis Publishers.

Chapter 19

Ergonomic Process Planning and Occupational Stress

James P. Kohn and Mark D. Hansen

INTRODUCTION

This chapter examines the physiological and psychological factors that contribute to human performance in the occupational environment. People, workplaces, job processes, and environments are the critical components necessary for the recognition, evaluation, and control of the stressors that have an adverse impact on employee health and safety. Establishing an ergonomics process to eliminate workplace stressors is discussed in great detail. In addition, occupational stress cause and control strategies are presented.

CASE STUDY: ERGONOMIC HAZARD RECOGNITION

A midsize appliance manufacturing facility placed the responsibility for plant safety in the hands of Tonya, a young human resources manager. The facility had approximately 250 employees working two shifts. Although a corporate health and safety staff was available for assistance and direction, the day-to-day responsibilities were assigned to Tonya. Tonya thought she had a good safety program at her facility. She had tailored all of the written programs that she had received from the corporate office and had made sure that training was conducted on a regular basis. She even had the facility occupational nurse inform her of first aid trends and near misses being reported to the plant infirmary. Then, a state OSHA compliance officer arrived at the plant door.

After a two-day wall-to-wall inspection of the plant, only a few violations were discovered. In the closing conference, Tonya and the facility were congratulated for their excellent effort. However, the compliance officer indicated that the company should start an ergonomics program. This rattled the young human resources manager. Where should she start?

After discussions with the plant manager, the occupational nurse, corporate safety and health professionals, and representatives from the training branch of the state OSHA office, Tonya decided to develop an ergonomics program plan. She decided to model the program after the OSHA meatpacking guidelines and to establish an ergonomic committee. She sought and obtained the plant manager support for this project and got approval for the plan that she developed.

Tonya then asked the occupational nurse to evaluate the extent of the problem and report the results back to the committee. At the same time, she brought in a professional ergonomist to train the committee members on the basics of ergonomics. This training included the skills necessary to recognize, evaluate, and control ergonomic problems in the workplace.

Tonya gave the committee the freedom to establish short-term goals and identify the activities necessary to obtain those goals. The committee decided to train the employees at the plant in ergonomic hazard recognition and to ask them to work with the committee members to identify and correct ergonomic problems. Six months after starting the process, Tonya began to see results. Although she knew they had a lot of work ahead of them, she was confident that a good system was in place to start correcting the ergonomic problems at her plant. The plan was well established and moving forward.

ERGONOMIC PROCESS PLANNING

Overview

Safety and health planning requires that goals, activities, and resources be in place to eliminate those conditions that result in losses to the organization. Ergonomics must be considered a key component of the safety and health professionals' plan. In recent years, ergonomics has become a major topic of concern to industry due to the sharp rise in costly work-related injuries. The U.S. Bureau of Labor Statistics reports that more than half of workplace illnesses are cumulative trauma disorders. These ergonomic illnesses are estimated to cost industry more than $38 billion annually. From 1984 to 1994, the number of cumulative trauma disorders rose 800 percent (Fletcher 1996). In addition, it has been found that more than one-third of all workers' compensation costs, approximately $10 billion annually, goes toward cumulative trauma disorder claims (Kohn 1997). OSHA predicts that by the next millennium, ergonomic-related cumulative trauma disorders will make up the majority of all business medical costs. Ergonomics must be part of the safety and health management plan for the ultimate reason that a good prevention program will be more cost-efficient than reacting to problems after injuries have occurred (Hess 1996).

Ergonomics is a multidisciplinary science in which human capabilities, limitations and other characteristics are identified and applied to the design of the workplace. Elements included in this workplace redesign include tools; machines; systems tasks; jobs; and environments for safe, comfortable, and effective human use. Simply expressed, ergonomics is the study of how work affects people. Thus, it is important to bring together engineers, medical personnel, maintenance personnel, and the ergonomic management system to make any ergonomics program beneficial and effective. Organizing an effective ergonomics team can solve and eliminate ergonomic problems by taking a proactive approach before equipment or workstations are designed and used in the workplace.

The Joyce Institute states, "the companies that are concerned about productivity and quality are including ergonomics as part of what they do" (Hough 1995). Ergonomics seeks to improve worker productivity and health by making the workplace design, activities, and tools suit individual workers. In fact, ergonomics encompasses many aspects of the workplace, including furniture, lighting, air temperature, air quality, and training. Ergonomics promises that well designed work spaces increase production by increasing efficiency and reducing the risk of injury. The proactive approach to ergonomics must incorporate planning and design into all future jobs.

THE ERGONOMIC DOMAIN

When examining ergonomic problems in the workplace environment, it is important to study all aspects including employees, machinery/equipment, environment, and work processes. The human element associated with ergonomics includes four factors of concern: physiological, psychological, behavioral, and psychosocial. Machinery/equipment variables include the tools, equipment, machinery, and furniture that employees in the workplace use to accomplish the job. Environmental factors include temperature, lighting, noise, humidity, air contamination, and other factors that can have an adverse effect on people. Finally, the work process may include product manufacturing and material handling; organizational factors such as scheduling, task design, department interface and team structures; and even compensation strategies.

Let us focus on the human element of the ergonomic system to understand how physiology and psychology can impact ergonomics. *Physiology* is the branch of the biological sciences concerned with the function and process of the human body. Physiological capabilities and limitations include the structure, strength, and movement of anatomical components. These are studied in the disciplines concerned with anthropometrics and biomechanics.

Anthropometrics

Anthropometrics, the measurement and collection of the physical dimensions of the human body, is used for design criteria to improve the human fit in the workplace. These measurements can also be used to determine problems that exist between existing facilities or equipment and the employees using them.

There are two types of anthropometrical dimensions that are useful for the study of human physiology and its impact on workplace layout and design: 1) *structural* or *static* anthropometry and 2) *functional* or *dynamic* anthropometry. The body measurement and dimensions of subjects in fixed standardized positions is referred to as structural or static anthropometrics. Common structural anthropometrical measurements include the following:

- Stature (height)
- Sitting height
- Body depth
- Body breadth
- Eye-level height, sitting or standing
- Knuckle height
- Elbow height
- Elbow to fist length
- Arm reach

Functional or dynamic anthropometry is the body measurements and dimensions taken during the performance of various physical movements. These movements may be required for the performance of particular types of tasks. Some of the frequently used functional measurements include the following:

- Crawling height
- Crawling length
- Kneeling height
- Overhead reach
- Bent torso height

Range of movement for upper-body extremities is another example of dynamic anthropometric measurement.

Anthropometric measurements help designers determine furniture or workplace layout requirements based on typical human body sizes. Differences in nationalities can certainly create problems in workplace layout and equipment design. For example, the employees at a Japanese automobile parts manufacturing plant operating in the United States complained of back, shoulder, and neck pain. The machinery that these employees were using was built in Japan and designed to accommodate Japanese workers, who are usually shorter than Americans. Consequently, the employees at this plant had to bend over to perform their assigned tasks. Simple observation clearly pointed to the cause the employees' complaints.

What does the use of anthropometric data actually mean when examining the workplace? Suppose that safety professionals are interested in eliminating ergonomic problems that result from workplaces not being correctly designed for their employees. Taking anthropometric measures and readjusting the workplace could eliminate the ergonomic workplace design problems.

Biomechanics

Biomechanics, defined as the study of the mechanical operation of the human body, is the science of motion and force in living organisms. In biomechanics, the function of the body components are monitored and job requirements are modified to lower internal and external stresses. The measurement and mechanism of body movement are important to safety professionals.

These measures provide the safety professional with the ability to determine how moving body components could contribute to ergonomic injuries. For example, factors such as extension and force applied by the arm muscles while repeatedly lifting parts from a pallet might be a biomechanical area of concern. In addition, safety professionals would also study the weight of the part being lifted, the height of the pallet above the working floor surface, and other external forces that would affect employee performance.

In summary, physiological factors include anthropometric and biomechanical variables that influence an individual's ability to perform work-related tasks. Through the application of anthropometric and biomechanical principles, it is possible to reduce the physiological stressors placed on the worker's body via the redesign of tools, equipment, and facilities. This, in turn, reduces the likelihood of strain and sprain injuries prevalent in poorly designed occupational environments.

In addition to the physiological factors associated with ergonomics, psychological, behavioral, and psychosocial factors are additional worker variables that can cause ergonomic problems.

Psychological Factors

Psychology is the science that studies human behavior. Some of the psychological factors that have been found to contribute to ergonomic hazards in the occupational environment include attention, memory, fear, boredom, fatigue, job satisfaction, and occupational stress (Kohn et al., 1996). For example, an employee was going through a divorce initiated by his spouse. During that time, he was involved in two minor automobile accidents and suffered a fall-related injury in the company warehouse. Were these events coincidental or were they related to his psychological predisposition? This individual had not been involved in an automobile accident for more than eight years when the first of the two accidents occurred. Then, less than two months later, he was involved in the second fender bender. Furthermore, the fall-related incident was this individual's first worker-related accident after seventeen injury-free years of employment with the company. This example of a psychological ergonomic reaction points to health-related problems that go beyond the frequently cited sprains and strains we normally consider as typical ergonomic injuries.

Behavioral Factors

Behavioral factors refer to changes in worker activity that are observable and measurable. Reaction time, response accuracy and appropriateness, adaptation, and endurance are just a few examples represented by this category. An example of a behavioral factor might be response accuracy, reaction time, and endurance of a hazard chemical spill cleanup worker during the hot temperatures of summer. While the psychological factor of fatigue and the physiological factor of exhaustion are readily apparent, behavioral factors under these conditions must also be considered. The environment and equipment used can adversely affect the worker's ability to react to an emergency or perform a complex sequence of tasks.

Psychosocial Factors

Psychosocial factors are worker behaviors that are influenced by coworkers, supervisors, or the organization. In other words, psychosocial factors refer to worker behaviors in a group environment (Kohn et al. 1996). Concepts such as leadership style, employee motivation, organization reward systems, and attitude formation and change are just a few of the elements studied under this worker ergonomic category. An example of a psychosocial factor could be the influence of coworkers versus management on the use of personal protective equipment. If an employee wearing a hard hat at a construction site is the focus of jokes from his or her peers, will that employee follow hard-hat safety rules when management is not around? If coworker acceptance is important to the employee, you know what the worker will probably do in this example.

Equipment and Work Process

In addition to the human element in ergonomics, it is important for safety professionals to evaluate the tools, equipment, workplace environment, and work processes at their facility. For example, are there tool vibration issues at your facility that could result in vibration-induced white finger problems? Do the tools and machinery used at your facility require workers to assume awkward and unnatural postures (biomechanical hazards) for long periods of time? Is there a substantial amount of material handling necessary to carry raw and finished products from one section of the plant floor to another? These are issues that can have an impact on ergonomic injury and illness in concert with the human variable issues associated with the human element issues just discussed.

MANAGING ERGONOMICS

Reducing losses and related costs associated with ergonomic problems in an organization requires the implementation of sound workplaces principles, tools, and techniques. These principles, tools, and techniques come from many disciplines including business and management, quality control, psychology, behavioral sciences, training and development, and safety and industrial hygiene. The challenge for the occupational safety and health professional is to obtain input from as many members of the organization and as many functions (such as operations, production, purchasing, and personnel) as possible. This is necessary to ensure that when tasks are designed, all the requirements of the organization as well as the human capabilities and limitations of the workers are considered.

To organize an ergonomics program effectively the safety professional should refer to the OSHA meatpacking guidelines. These guidelines provide a structure for establishing and managing effective ergonomics programs. This structure includes two major components: ergonomic program management, and program elements. Ergonomic program management involves management commitment, development of a written program, employee involvement, and regular program review and evaluation.

Management Commitment and Employee Involvement

Commitment and involvement are complementary and essential elements of a sound safety and health program. Commitment by management provides the organizational resources and motivational forces necessary to approach ergonomic hazards effectively (See ErgoWeb [on-line]).

An effective program should be in writing, use a team approach with top management as the leader, and should include the following:

1) Management's involvement demonstrated through personal concern for employee safety and health by the priority placed on eliminating the ergonomic hazards.

2) A policy that places safety and health on the same level of importance as production. The responsible implementation of this policy requires management to integrate production processes and safety and health protection to assure that this protection is part of the daily production activity within each facility.

3) Employer commitment to assign and communicate the responsibility for the various aspects of the ergonomics program so that all managers, supervisors, and employees involved know what is expected of them.

4) Employer commitment to provide adequate authority and resources to all responsible parties, so that assigned responsibilities can be met.

5) Employer commitment to ensure that each manager, supervisor, and employee responsible for the ergonomics program in the workplace is accountable for carrying out those responsibilities.

Written Program

A written program is another important step for implementing an effective program. Endorsement by upper management is a must, as these plans outline the employer's goals and activities that will be implemented to address ergonomic problems. The written program should be suitable for the size and complexity of the workplace operations, and should permit the guidelines to be applied to the specific situations of each plant. It should be communicated to all personnel, as it encompasses the total workplace Make sure to include everyone regardless of the number of workers employed or the number of work shifts. Establish clear goals, clear objectives to meet those goals, and ensure that everyone knows his or her role and responsibilities in the program.

Employee Involvement

Companies should provide an opportunity for employees to participate and, at the same time, encourage employee involvement in the ergonomics program. This involvement should include the following (see ErgoWeb [on-line]):

1) An employee complaint or suggestion procedure that allows workers to take their concerns to management and provide a mechanism for program activity feedback without fear of reprisal

2) A procedure that encourages prompt and accurate reporting of ergonomic symptoms by employees so that problems can be evaluated and, if appropriate, treated

3) Ergonomic teams with the required skills to identify, analyze, and recommend solutions for jobs in which ergonomic problems are observed

Regular Program Review and Evaluation

Since results are the bottom line, procedures and mechanisms should be developed to evaluate the progress and accomplishments of the program. Regular reviews, by upper management, should be conducted for evaluating the progress being made to meet ergonomic program goals and objectives. The following are some examples of evaluation technique methods:

- Analysis of trends in injury/illness rates
- Employee surveys
- Before and after surveys/evaluations of job/work site changes

- Review of results of plant evaluations
- Up-to-date records or logs of job improvements tried or implemented

Ergonomic Management Program Elements

The meatpacking guidelines recommend that safety professionals address ergonomic hazards in the industrial environment by including in the ergonomic management program the four major program elements of work site analysis, hazard prevention and control, medical management, and training and education.

Work Site Analysis. Work site analysis identifies existing hazards and conditions, operations that create hazards, and areas where hazards may develop. This type of analysis is referred to as *active surveillance* and may include NIOSH lifting analysis, job hazard analysis, or biomechanical techniques. Work site analysis also includes close scrutiny and tracking of injury and illness records to identify patterns of traumas or strains. This analysis is referred to as *passive surveillance* since it tends to study records accumulated by the safety function and/or workers' compensation insurance carrier. Remember that by utilizing active and passive surveillance methods, the objective of work site analysis is to recognize and identify ergonomic hazards.

Hazard Prevention and Control. Once ergonomic hazards have been identified through systematic work site analysis, the next step is to design measures to prevent or control ergonomic hazards. Thus, a system for hazard prevention and control is the second major program element for an effective ergonomics program.

Ergonomic hazards are prevented primarily by effectively designing the workstation, tools, and job. To be effective, an employer's program should use appropriate engineering and work practice controls, personal protective equipment (PPE), and administrative controls to correct or control ergonomic hazards. PPE is the least desirable method for controlling worker exposure to ergonomic hazards. Thus, modification of workplace layouts and selection of ergonomically designed tools should be a high priority. Process changes such as the elimination of manual material handling is another example of a high-priority approach for the elimination of ergonomic hazards.

Medical Management. Implementation of a medical management system is the third major element in the employer's ergonomics program. Proper medical management is necessary to both eliminate and substantially reduce the risk associated with ergonomic problems through early identification and treatment. It can also prevent future problems through the development of information sources. In an effective program, health care providers will be part of the ergonomic team, interacting and exchanging information routinely in order to prevent and properly treat ergonomic health problems.

Training and Education. The fourth major program element for an effective ergonomics program is training and education. The purpose of training and education is to ensure that employees are aware of the ergonomic hazards to which they may be exposed. Training and education are critical components of an ergonomics program for employees potentially exposed to ergonomic hazards. Training allows managers, supervisors, and employees to understand the ergonomic hazards associated with a job or production process, prevent and control ergonomic hazard exposures, and comprehend the medical consequences of ergonomic exposures.

OBTAINING ORGANIZATIONAL SUPPORT
Management Support

The primary reason that safety professionals implement ergonomic programs is to reduce the incidence of injury and illness in the workplace. Management has a totally different view. Management is driven by the costs of implementing or not implementing something, whether it is ergonomics or any other requirement or perceived requirement. As a result, management will ask the question, "Why should ergonomics be implemented?" It is incumbent on the safety professional to cogently provide a compelling response to this question. Therefore, gaining management support to implement ergonomics programs requires articulating several approaches to convince management to implement ergonomics. These approaches include regulatory, business, and legal.

Regulatory Requirements. There are several ways to achieve management support to implement ergonomics. The strongest impetus to implement on ergonomics program is, of course, regulatory. OSHA's proposed ergonomics standard is stealthily closing in on publication at the time of this writing. Long live ergonomics! We know how political the proposed ergonomics standard was, as it was being presented to an antiregulation Republican majority house and senate. Most of us are now asking, *What happened?*

The mechanics of how a proposed ergonomic standard came to be, at this point, are immaterial. The fact is, look out, here it comes! Unfortunately, the bad news is that, yet another regulation is forthcoming. The good news is that an ergonomics standard is long overdue. There is ample information to substantiate an ergonomics standard on workers' compensation claims alone, even after you filter out all the erroneous data and false positives. Many ergonomists and safety and health professionals agree, not because of potential business opportunities or to provide another job duty in these downsizing times, but because our primary concern is to keep people from getting hurt in the first place. Design it correctly in the first place and the hazard will not exist. Ergo, people will not get hurt. Period.

Ergonomics Is Good Business. How can doing ergonomics be good business? Well, why do you think companies hire ergonomists and safety and health professionals in the first place? Is it because they are altruistic? Is it because they are safety minded? Is it because they like you? The answer to these questions is, unequivocally, *no*. Companies hire ergonomists and safety and health professionals for one reason, and one reason only—*it makes good business sense.* Beware of anyone telling you differently. The bottom line is that companies have ergonomists and safety and health professionals on staff *to save money.* Whether it is to prevent OSHA citations, lawsuits, uncontrolled workers' compensation claims, or improve productivity, the bottom line is to save money.

Ergonomics, like safety, pays for itself in many ways. Quantifiable productivity increases of from 5 percent to 50 percent have been seen. Decreased absenteeism, increased worker satisfaction, reduced error rates, and decreased workers' compensation are just a few of the benefits and cost savings of implementing ergonomics. In this day and age of competitive advantage, ergonomics is playing a stronger role. It just makes good business sense to implement ergonomic workstations that are easily adjustable. Businesses are realizing that if they provide an ergonomic work environment, people can work longer, more productively, without injury, and enjoy work more.

Ergonomics Saves Money. By training employees on the basic ergonomic principles of how to recognize early indicators of potential ergonomic injuries, employers can save money and time, and prevent injuries. By catching ergonomic injuries early rather than later in the injury continuum, employers can save the costs of incident/accident investigations; lost production, restricted duty; workers' compensation costs such as therapy and surgery;

and worker-job satisfaction. In 1994 the average ergonomic injury was about $4,000 per case [(carpal tunnel syndrome (CTS): $4,674 per case; back injuries: $3,700 per case; and upper-extremity cumulative trauma disorders: $3,350 per case)]. When the other factors like workers' compensation, insurance, and administrative costs are included, the cost jumps to $12,000 per injury. Most ergonomists and safety and health professionals can understand that the further down the time continuum the injury is discovered and treated, the higher the workers' compensation costs will be. Some estimates top $33,000 per ergonomic injury at this end of the continuum.

Ergonomics Saves Time. Employers save time by training employees to recognize symptoms early, line supervisors to be vigilant of working conditions and postures, and management to understand the cost-benefit of doing ergonomics. The result is that employees report injuries earlier, thus reducing workers' compensation claims, increasing productivity, improving their health, and having fewer lost workdays. Furthermore, in 1994, the BLS reported that (CTS) was the leading cause of lost workdays, exceeding amputations, fractures, sprains and strains, and heat burns.

Ergonomics Saves Where "Time Equals Money." When all is said and done, implementing ergonomics saves money, and in these fast-paced times, ergonomics also saves time. Today, many professionals will agree that time equals money. Treating money as we treat time, especially in the case of ergonomics, makes perfect sense. When combining the fact that ergonomics saves time and money, the bottom line is that fewer employees will be injured in a proactive environment in which the ergonomic injury is not allowed to progress to a debilitating state.

Ergonomic-related injuries are a rapidly growing concern in today's workplace. The cost and frequency of ergonomic-related injuries easily justify the benefits of incorporating ergonomics into the workplace. But there is more.

There is substantial social and legal justification for incorporating ergonomics in today's workplace. The central social trend involves accommodating workers with diverse physical capabilities and disabilities. The result has meant that denying persons of a certain age, gender, race, or apparent disability of fair opportunity to perform a job are no longer valid (Miner and Miner 1978). It has also been realized in recent years that a person's physical attributes, such as strength, flexibility, and endurance vary greatly within gender and age groups, and that these variances must be accounted for in the work environment. For instance, an allowable weight-lifting standard recommended in the mid-1960s.by the International Standards Organization is no longer valid in many countries because it discriminates against women and persons of advanced age by stating limits for these groups that are far below those for a younger male.

For these reasons, it has become necessary to know, with as much precision as possible, the functional capacity of an individual relative to a job's performance requirements to ensure that the person is not unjustly denied a job and that the job is fit for the worker. This is also good medical practice and, indeed, is required by some occupational health and safety statutes. The use of functional capacity tests can assist in ensuring that a job applicant will not be overstressed when placed on a new job.

For instance, recent studies indicated that individuals not capable of demonstrating specified levels of isometric strength had a threefold increase in both the incidence and severity of musculoskeletal injuries as compared to their stronger peers. The development and validation of appropriate functional capacity tests of both job applicants and persons who are seeking to return to work after suffering a musculoskeletal injury is a major occupational biomechanics effort.

Legal Trends. Numerous OSHA fines and individual lawsuits have been filed against employers. Some examples of OSHA fines are as follows:

- Samsonite—fined $1.6 million for the lack of ergonomics in the workplace.
- Cargill—fined $400,000 for the lack of ergonomics in the workplace.

The following are examples of lawsuits:

- *Dennis v. Communication Machinery Corporation.* Five workers seeking $40 million each in compensatory damages and $20 million in punitive damages for ergonomic-related injuries.
- *Aikman v. Electronic Pre-Press Systems, Inc.* Each of the plaintiffs are seeking $10 million for negligence and another $10 million for strict liability for ergonomic-related injuries. Their spouses are seeking $2 million each for the loss of consortium.
- *Taylor v. System Integrators, Inc.* The plaintiffs are seeking $10 million in punitive damages and attorney's fees and costs.

Taking lawsuits to the next level, workers are joining together to file class action suits against their employers. The following is a list of companies and the number of class action suits against them:

- IBM Corp. (67)
- AT&T (45)
- Computer Consoles (43)
- ATEX Inc. (36)
- NCR Corp. (21)
- Unisys Corp. (18)
- Northern Telecom (13)
- Systems Integrators, Inc. (12)
- Memorex Corp. (11)
- WANG Laboratories, Inc. (9)

These examples of fines and lawsuits should be sufficient to convince any company that implementing an ergonomic program is both good business sense as well as good for the workforce.

Corporate Policy

Once management has agreed to implement ergonomics, it is recommended that a policy and a program be developed and enforced. The following is a sample ergonomics policy/procedure.

**Safety, Health and Environmental
Policy and Procedures
XYZ Corporation**

Date: Revision 10/96   Number: xyz   Title: ERGONOMICS

A. Purpose:
   The following guidelines have been established to prevent the occurrence of work-related musculoskeletal disorders by controlling exposure to the ergonomic workplace risk factors for all personnel working at XYZ.

B. Scope:
   This policy shall include all personnel at XYZ whose normal duties require them to perform the same motion or motion pattern every few seconds for more than a total of 3-4 hours, work in awkward postures, perform repetitive motions, use vibrating or impact tools, and/or use forceful manual material handling.

   The provisions of this policy and procedure are intended to be slightly more stringent than the current proposed ergonomics standard of the OSHAct as set forth in the 29 CFR 1910.

C. Requirements:

1. Ergonomic equipment is provided by XYZ at no expense to the employees. This includes but is not limited to workstations and or work tables adjusted to the employee, chairs adjusted to the correct seating height required by the task for employees, and hand tools that eliminate or minimize the exposure to ergonomic stressors.

2. XYZ will provide health & safety personnel to perform routine workstation analyses and training for all employees. These activities include all workstation analyses for manufacturing, aftermarket and service operations, as well as office situations and video display terminals (VDTs).

3. All personnel are required to be trained in ergonomics which includes, at a minimum: an overview of ergonomics, information about musculoskeletal disorders (MSDS), risk factors associated with MSDS, signs and symptoms of MSDS, how to report MSDS, ways to reduce exposure to MSDS, the benefits of implementing an ergonomics program, how to recognize symptoms of ergonomics injuries, the function of the ergonomics committee, and the basic elements of the XYZ ergonomics program.

4. XYZ will establish an ergonomics committee, which will conduct regularly scheduled meetings to address ergonomics issues. This committee will establish a charter, which includes at a minimum: management commitment, scope, functions, activities, management involvement, employee involvement, and selection of committee officers.

5. XYZ has established an ergonomics program that will comply with the proposed ergonomics standard issued through the occupational safety and health administration, as well as increase productivity and reduce ergonomics-related injuries and illnesses.

6. Annual physicals are not required by the OSHA standard. However, on a periodic basis, it is recommended that "at risk" employees be examined by a company physician to identify any physical indications of ergonomics-related injuries. The employees shall ensure they report for their physical exam at the place, time and date as specified.

D. Responsibilities of the employee:
    The employee shall be responsible for performing manual material handling correctly, using ergonomic-related equipment as prescribed, and correctly performing other related ergonomics procedures, once trained.
    The employees shall be responsible for reporting any ergonomic symptoms relating to potential ergonomic injuries and assisting in implementing ergonomics interventions into their workplace.

E. Responsibilities of the supervisor:

It shall be the responsibility of the supervisor to ensure that all employees in his work area are familiar with this policy and are in compliance at all times. The supervisor shall further assist the ergonomics committee in implementation aspects of ergonomics at each site.

Each employee must be advised of the OSHAct, which clearly states that each worker will perform his work in a safe and healthful manner, as required by the employer. Progressive disciplinary action is indicated for obvious violations of this requirement.

F. Responsibilities of the Health & Safety Department:

H&S personnel will conduct workstation and job analyses, symptoms surveys, administer checklists, and implement engineering interventions. They will identify, through primary and secondary surveillance, areas requiring the application of ergonomics design guidelines and principles.

H&S shall determine the effectiveness of the ergonomics program and assist in the implementation of engineering and administrative controls to mitigate ergonomics hazards to an acceptable level of risk.

H&S personnel will monitor the use of ergonomics equipment during periodic audits and inspections and notify appropriate line and staff management of nonconformances with the requirements.

G. Responsibilities of engineering:

The proposed ergonomics standard by OSHA places primary emphasis on engineering and administrative controls to minimize the incident of ergonomics injury and illnesses.

1. Ergonomic engineering control problems requiring ergonomic engineering expertise beyond the scope of XYZ in-house capabilities will be referred to an approved local ergonomics consulting firm. This consulting firm should have on their staff, an Ergonomist with the following basic requirements:

    a. Certification as a certified professional Ergonomist (CPE) issued by the board of certification on professional ergonomics.

    b. Education which includes, at a minimum, a master's degree in an ergonomics-related field, such as industrial engineering, or safety engineering, from an accredited college or university.

    c. Experience that includes a minimum of seven years of applied experience in a related industry.

2. Engineering control feasibility determinations:

    a. The first step in determining a course of action is to determine the technical feasibility for each ergonomic engineering control based on a trade study of acceptable alternatives. Trade studies will be used to assist in assessing the level of risk, incident rate and severity, and the cost of ergonomics engineering controls for feasibility of implementation. The feasibility trade study of ergonomics engineering controls must be carefully documented to ensure that ergonomics issues are properly addressed.

   b. The use of administrative controls by job rotation, job alternation, work/rest cycles or reassignment of jobs to other areas may not appear feasible in many cases at XYZ.

   c. Purchasing of new equipment: establish a purchase control program through XYZ's purchasing director to ensure that equipment is specified to meet ergonomics design criteria.

   d. Appropriate H&S personnel shall review and sign off on all XYZ authorization for expenditures (afes) which involve the design of new equipment, machine tools, or facilities, the purchase of new equipment, machine tools or facilities, or modification of existing equipment, machine tools or facilities.

3. Whenever an ergonomics issue is discovered through the initial workplace risk factor evaluation or passive and active surveillance (using the signal risk factors and the workplace risk factor checklist), necessary steps to reduce the ergonomics hazard exposure shall be ascertained. A detailed action plan, with projected completion dates for individual steps shall be prepared. When the compliance plan involves long term engineering projects, it may be revised from time to time as conditions change. The orderly completion of schedule, of the various phases of the action plan, together with other components of the ergonomics plan, will be considered compliance with the proper regulations.

Joe Bossman
Vice President
Finance and Administration

Supervisory Support

Once management support is garnered, the battle is not over. If management is not supportive, convincing supervisors can be difficult at best. Having corporate management support the implementation of ergonomics certainly helps. However, in order to implement ergonomics successfully, supervisors must also feel as though they are part of the process. Otherwise, ergonomics simply becomes another element that will pass anyhow and therefore deserves only lip service. Many of the same reasons used to convince management can be used to convince supervisors.

Supervisors must see the benefit. Decreasing workers' compensation is not enough to convince supervisors because they rarely see those funds anyway. Illustrating the productivity increases along with communicating the potential for being fined or sued are the strongest elements that can be used to gain support from supervisors. If there is a safety committee, employees can sometimes provide testimonials that clearly illustrate productivity increases along with a decrease in ergonomically related injury and illness.

Keeping supervisors apprised of ergonomics-related activities is also a must. Supervisors must not feel that employees are plotting behind their backs. Include them in the review just to let them know what is going on. This will help supervisors know if capital investments for ergonomics are in the works, which can prevent unnecessary confrontation.

All these activities can draw supervisors into the solution process rather than casting them as part of the problem. Remember, supervisors can thwart an ergonomics program to the point of frustration. The result is a lose-lose situation for everyone. When supervisors are included in the process and convinced of the value of ergonomics, everyone wins.

Worker Support

The primary reason that ergonomic programs should be implemented is that it benefits the employees. If the employees do not believe in implementing ergonomics, it does not matter whether supervisors and management are supportive. Since employees are doing the jobs being analyzed, listen closely for their concerns, both concrete and abstract. Concrete concerns include specific ergonomic criteria for workstation and work design. Abstract concerns include factors that affect how work is being done. The corporate culture must be assessed, and the plan of action must reflect that culture.

Employees are often skeptical that the company really intends to implement ergonomics. They feel that the company will not spend money to implement ergonomics. Once you illustrate that implementing ergonomics is not a free ride or a blank check, but that veritable concerns and problems are addressed and corrected, employees will become more receptive. Building on successes allows ergonomics to be successfully implemented.

Ergonomics Committee

Assembling an ergonomics committee is yet another activity that must be addressed. There are several strategies: one is to include ergonomics in the safety committee, which is perfectly legitimate; another is to set up an independent ergonomics committee. Regardless of the strategy, choose the members carefully. The ergonomics committee should not be a place for malcontents to voice their dissatisfaction with the company. Pick pro-company employees to start with so that the goals and mission of the committee stay focused. At the onset, choose simple problems that can be fixed with minimal cost to the company. Build on success so that when significant potential costs are required the justification is simple based on past successes. Furthermore, when significant costs are required, assist members of the ergonomics committee in generating a cost-benefit study of options and costs. This will help them see the potential expenditures clearly, and potentially see the costs as their money, rather than only the company's money.

Roles and Responsibilities of the Ergonomics Committee. The ergonomics committee should be responsible for overseeing, coordinating, supporting, and reviewing the ergonomics process at the company. Its responsibilities include:

- Establishing performance expectations and benchmarks
- Coordinating problem resolution in areas that affect more than one business unit
- Providing assistance and advice on technical matters when necessary
- Reviewing the ergonomics program and making recommendations to the site safety professional(s).

Ergonomics committees bear responsibility for operating the ergonomics program within the business unit. Ergonomics committees are specifically responsible for:

- Performing passive and active surveillance, analysis, and design of jobs
- Identifying specific training needs
- Coordinating its activities with facility engineering, the site safety professionals, procurement, labor organizations (where applicable), and appropriate company management
- Reporting findings and activities to management

Preparation for Implementing Ergonomics

Preparing to implement ergonomics requires two primary efforts: developing an ergonomics program, and training all employees on ergonomics. Once the ergonomics program is developed with the coordination of all related disciplines, employees must be trained on ergonomics. This training includes what ergonomic injuries consist of, what to do when reporting an ergonomic injury, and how the company addresses ergonomic injuries. Training of ergonomics should be treated just like any other safety discipline in training and should require annual refresher training.

A Final Word on Organizational Support

Obtaining organizational support requires buy-in from all levels, management, supervision, and employees. Missing any of these groups could cause the implementation of the ergonomics program to fail. Attending to each of these groups ensures that everyone is on board and believes in the ergonomics program throughout the life of the program.

OCCUPATIONAL STRESS

Stress may best be defined as physiological or behavioral worker reactions to any environmental demand. The demand (also known as a *stressor*) may be pleasant or unpleasant and may evoke conscious or unconscious reactions in the body. Detrimental worker reactions are the result of physical, chemical, biological, or interpersonal causative agents or stressors found in the workplace.

In general, the body's adaptive mechanisms are flexible enough to meet and resist the stresses imposed by the environment. Indeed, most of us are even stimulated by the challenge of meeting each day's many needs and changes. For the most part, we may even be unaware of our responses to stress. Automatic changes continuously occur in the nervous system and glands, adjusting the body's function to meet these demands. Rarely do these adjustments disturb the normal stability and balance of bodily systems.

However, this is not always the case; there are times when stress does interfere with normal functioning. This often occurs when the stress represents some kind of a threat. It may be the threat of losing a job, or frustration resulting from an excessive workload, or the actual danger of physical harm. These and other factors may exist in both work and nonwork environments.

In either setting, not everyone will show signs of stress-induced illness or disease. Usually several factors must be present to reduce the ability to cope with any given situation. It is essential that the safety professional understand these factors, particularly their interaction, and learn to recognize the early signs of stress or maladaptation. Only then can effective prevention and modification techniques be incorporated into the overall safety and health program.

Body Response

To function normally, the body must operate with very narrow biological and chemical limits. The normal operation of the cardiovascular, respiratory, gastrointestinal, nervous, and urinary systems depends on automatic adjustments in the glandular and nervous systems in response to stimuli that arise within the body itself. These changes may affect us in any of several ways:

1) Psychologically or emotionally
2) Functionally
3) Biochemically
4) Pathologically

Psychologically, when we feel excessive stress, we may develop depression, difficulty in concentrating, and loss of appetite, insomnia, fatigue, and withdrawal. Functionally, we may experience headaches, muscle aches, nausea, constipation, diarrhea, irregular heartbeat, and frequent urination. Biochemically, there is an increase in the output of adrenals, or stress glands. This is accompanied by an increase in blood fat levels, as well as changes in blood sugar and mineral levels.

Chronic tension may result in increased use or even dependence on tobacco, alcohol, and drugs. This type of abuse depletes the normal body constituents and interferes with its normal defense mechanisms. Various studies have examined occupational stress factors and deteriorated employee performance. Alcohol abuse and drug dependence are just one example of a growing body of evidence supporting the relationship between interpersonal or psychosocial workplace factors and disease (Crum et al 1995).

Eventually, chronic stress can lead to the destruction of body tissues, resulting in disease. Pathological disorders such as migraine, coronary heart disease, high blood pressure, bronchial asthma, peptic ulcers, colitis, neurosis, some skin diseases, and possibly arthritis are frequently associated with chronic stress.

Predisposing Factors

Obviously some persons are better at handling stress than others. Individuals vary widely in their ability to tolerate stress, either on a short-term or a prolonged basis. Although there is no handy formula for predicting the exact degree to which a prospective worker will be able to handle stress, we do know that numerous factors predispose some persons to react more strongly than others. The major predisposing factors are heredity, environment, and the general state of health or illness.

The exact influence of heredity in determining stress tolerance is uncertain. However, we do know that congenital defects in the body systems involved in the adaptive process can reduce tolerance. Defects of the brain, autonomic nervous system, and glandular system are the most critical, although an abnormality in any major body organ may be significant. Even in cases in which no physical abnormality exists, the psychological environment of the growing child can have an enormous influence on his or her later ability to handle stressful situations. Psychological factors include the relationship with parents and siblings, role in the family unit, interactions with peers, the influence of teachers or social institutions, and the opportunities provided for maturation and growth. Physical illnesses during the growth period (especially those involving the adaptive body systems) may greatly reduce the individual's later ability to combat threatening stressors.

Work Environment

Job Demands. A certain amount of stress is involved in every work situation: the need to perform, to meet deadlines, to cope with normal changes or unexpected events. The universality of stress may obscure its importance as a factor in safe, efficient operations, and the importance of identifying and mitigating it as much as possible.

Work stress may be measured by worker's reaction to any situation that is consciously or unconsciously perceived as threatening. Such threatening situations are almost invariably associated with change. For example, the new employee wonders whether he or she will succeed on the job. How will the work environment affect him or her? Will he or she fit in socially with fellow workers? Will the job require certain new skills or physical abilities applied in a unique environment? There may be a strong demand for concentration, attention, sensory alertness, perception, and precision. Unless the worker is already knowledgeable and experienced in the job, the fear of failure, loss of self-esteem, and possible job loss can pose a real threat. In the absence of strong support from the manager and fellow workers, this threat may easily become exaggerated.

The nature and amount of work itself may also create stress, and understimulation and overload may be just as equally stressful. Some workers relish deadlines as challenges and perform best under pressure. For others, the very thought of having to meet a deadline creates enormous anxiety. Some workers view menial tasks or monotonous assembly line work as meaningless and unfulfilling. They develop feelings of frustration and hostility toward themselves (for their inability to control their lives) and toward the company (for its apparent lack of interest). Other workers perform the same repetitious jobs without experiencing any apparent conflict. They find security in knowing exactly how to perform their work without having to make difficult choices, decisions, or readjustments. Most workers probably fall somewhere in between these two extremes, able to tolerate a certain amount of dull, routine work while enjoying some challenges or changes in routine. When all workers, however, are faced with excessive workloads, adverse health and performance effects will occur. For example, excessive workloads have been linked to general health complaints including headaches and fatigue (Barnett et al 1991).

The effects of the physical demands of a job are often difficult to assess, since much depends on the physical condition of the workers. Work requiring unusual postures, heavy or repeated lifting, repetitive assembly line tasks, or frequent travel, as well as work considered dangerous, all pose greater threats to the average worker than safer, more routine sedentary work. Hence, these jobs may be perceived as more stressful. Unless the worker can choose his or her hours, mismatching may also occur with respect to personal desires, home or school demands, and body clock. A small number of people are "night workers," at their best on late shifts; others are more alert early in the day. It is well known that frequent shift changes or rapid travel through many time zones can disrupt the biological rhythms of the body. Young persons are able to adapt rapidly to shift changes without bodily injury or significant changes in work performance. This is not necessarily true for older workers, for those with marginal health, or for those taking certain types of medication. Even when there is no physiological problem, personal or domestic reasons may create difficulties when the worker must deal with late shifts or sudden or repeated shift changes.

Environmental Stress. Many studies have been conducted on the effects of certain environmental stressors. We have known for years that noise can cause permanent damage to hearing unless earplugs are worn. Noise also interferes with voice communication, which can cause great annoyance, especially to those exposed for the first time. Most workers can quickly make the psychological adaptation, but a very small group cannot adapt and must be removed from further noise exposure. This group may experience irritability, insomnia, headaches, and tension, which disappear once they have been removed from the noise. More recently, it has been alleged that excessive noise exposure can also cause vascular problems, including high blood pressure. Most investigators believe that noise may be a nonspecific stressor to those who are not accustomed to it. They state that an initial increase in blood pressure may occur but will subside in time.

Moderate background noise is not necessarily a stressor, nor is it necessarily counterproductive. In some occupations, large numbers of employees work in close proximity. In areas where production and safety are not sound dependent, production may actually increase. It has also been noted that workers doing repetitive small-assembly bench work may perform better and with less fatigue if allowed to use radios with individual earphones.

Ultrasonic acoustic energy, generated by many machines currently used in industry, does not seem to constitute a hazard to workers. However, vibrations below 20 Hz are very annoying, as they may cause resonance in various body organs and interfere with muscular action, vision, speed, and coordination. Muscular aches and fatigue may persist after exposure. However, with the exception of a few occupations and exposures in which workers have sustained permanent vascular and neuromuscular injuries, vibrations should have no permanent effects.

Lighting may also be a significant stress factor. Illumination of the proper intensity and direction, without glare, is essential to avoid eyestrain. Both the muscles and the lens of the eye are involved in making the necessary adjustments for different light levels. Under adverse conditions, they may have to work continuously, resulting in eyestrain. Headaches, aching eyes, fatigue, and irritability are the common complaints associated with eyestrain. Generally, eyestrain does not mean permanent eye damage, but can seriously impair worker comfort and productivity.

Proper temperature is also essential to controlling stress. With the exception of noise, there are more worker complaints about workplace temperature than about any other factor. The first hot days on the job are difficult to tolerate, especially if the humidity is also high. This environment may pose health and comfort problems. Depending on the level of physical work performed, temperatures in the 68°F to 75°F range, with relative humidity between 25 and 70 percent, are well tolerated. One or two weeks may be necessary to acclimatize workers to greater heat, after which they feel less discomfort. In most factory and office environments, airflow and temperature are primarily comfort and performance considerations. However, they can also be stress considerations for some workers.

Potential hazards abound in most factory operations. Chemicals, radioactive substances, electrical and electronic equipment, and new materials and processes all represent potential physical and health hazards to workers. The recent OSHA emphasis on work-induced diseases, especially cancer, has made workers more aware of workplace hazards and may add to their anxiety on the job. To a certain extent, this anxiety can be allayed by presenting adequate information concerning the exact nature and extent of the hazards, the measures in use to prevent harmful exposure, and the availability of immediate assistance in the event of a harmful exposure.

Morale and Stress. Properly motivated and contented workers are the most productive and have fewer complaints. Any alteration in the work environment can precipitate a stressful response, whether the change is a promotion or demotion, possible layoff, transfer, loss of friends or coworkers, a new supervisor or manager, a change in work activity or process, net shift, impending company difficulties, or even rumors about such changes. Anticipating such changes, preparing for them, and providing both peer and company support for the worker may help minimize the impact of the change. Many companies today are not considering the consequences of the stresses that they are creating. Future career ambiguity, commonly referred to today as "downsizing," has been reported to be a serious stressor. As a consequence, employee stress reactions to these layoffs have been reported by the insurance industry to have resulted in a substantial rise in workers' compensation and health insurance claims (Koco 1996).

Contributing Factors

Personal Health Habits. The ability to tolerate stress depends both on the intensity of the immediate stimulus and on the susceptibility of the body at that time. So far, we have spoken of some of the stress factors that may have been found in the workplace. However, most of us spend only a small portion of our time at work. In addition, we live in a 24-hour environment, in which many other factors produce stress. Some of these physical and emotional stresses may be carried over into the workplace.

Some of the events and circumstances of our daily lives are beyond our control; others are self-induced and can be modified. Our resilience and state of well-being at any moment can be greatly influenced by what we eat, how much we drink or smoke, whether we take any drugs, how much sleep or rest we get, what exercise we take, and what pleasurable pursuits we enjoy.

Much has been written about nutrition and its role in the causation of heart and blood vessel disease, high blood pressure, and diabetes. The role of high blood fat levels and obesity in causing these diseases is still under investigation, but there is little doubt that they do constitute risk factors. Anxiety, fear, and increased tension appear to increase both circulating blood fats and, initially, blood sugar. In recent times, much emphasis has been placed on the condition known as hypoglycemia, or low blood sugar, which may cause light-headedness, weakness, or even unconsciousness. Both diet and workplace stress can contribute to a hypoglycemic reaction. The stress of stimulation can cause the system to produce too much insulin, which, in turn, causes the blood sugar level to drop too low. This sudden drop in blood sugar may cause the worker to become faint, panicky, or unconscious. Fasting, missing meals, or eating refined sugars, which leave the blood very rapidly, may lay the groundwork for a hypoglycemic reaction when paired with a sudden stressful stimulus to the body.

The consumption of alcohol, particularly if done regularly or often, can also have a radical effect on the ability to deal with stress both on and off the job. Alcohol should be thought of as a depressant drug with the same addictive potential as any other dangerous drug. It is a water-soluble drug that diffuses into all fluid compartments of the body and can interfere with the normal oxygenation of body cells. This produces its most sought-after effect—relaxation, often accompanied by a release of inhibitions or anxieties. Alcohol, therefore, interferes with judgment, self-evaluation, and reaction time, any one of which could be crucial in the work environment.

There is no evidence that small quantities of alcohol, even if consumed daily, are detrimental to health or longevity in a healthy person. However, some individuals can become addicted to alcohol, developing increased tolerance, and consuming progressively greater amounts of it. In addition, the alcoholic experiences toxic symptoms of withdrawal (such as tremors), and will drink even more to avoid these symptoms. If alcoholism continues unchecked, it eventually will lead to early death, whether due to accidents, infectious disease, or liver or heart problems.

Drugs are used for many of the same reasons as alcohol. Other than simple analgesics and cold preparations, the drugs most frequently consumed by workers are tranquilizers, sedatives, and antidepressants. All act on the nervous system, either depressing electrical activity or blocking transmission of certain impulses. In addition to their sedative effect, many of these drugs are addictive and may have unpleasant side effects, including withdrawal symptoms. A person's habitual use of these drugs indicates an underlying emotional problem, which in itself tends to increase the person's sensitivity to other stressors. Increased dependency on these drugs is an index of emotional turmoil and fear of losing control without them.

Increased smoking is another sign of inadequate response to stress. The role of smoking in the development of respiratory cancer has been well publicized, especially among workers exposed to various chemicals. Less well publicized has been the increase in coronary heart disease among smokers. Nevertheless, many workers continue to smoke, indicating a disregard for their own health or addiction to nicotine.

It is impossible to feel well when the body is not functioning at maximum efficiency. Maximum efficiency, including increased tolerance for fatigue, rests on an appropriate program of sleep, rest relaxation, and exercise. Sleep is an essential physiological process during which waste products are removed from the body and all organs are rested and reenergized. The average person requires seven to nine hours of sleep with slightly more for young persons and less for the elderly. Both the quality and quantity of sleep are important. Sleep cannot be stored up, and it may take two or three nights of normal sleep to make up for one night's lack of sleep. Insomnia is a common complaint of persons under tension, indicating an inability to relax at bedtime. The insomnia, in turn, contributes to both mental and physical fatigue, lowering resistance to other stimuli. This sets up a vicious cycle of reduced tolerance to stressors, higher tension, and increased sleep difficulty, which further reduces tolerance to stress and continues the cycle.

Lack of adequate rest and the inability to relax during and after work contribute to fatigue. Fatigue may be either mental or physical, slowing both mental processes and physical responses. In addition, fatigue is characterized by irritability, anxiety, poor concentration, and withdrawal. It renders the worker more susceptible to stimuli that he or she would normally be able to cope with, or ignore.

Lack of adequate exercise also decreases the body's efficiency, hastening the onset of fatigue. Many factory jobs involve simple, repetitive body movements but little whole-body activity. Boredom and stagnation due to lack of activity are common contributors to fatigue. Consequently, many workers find that a short, brisk walk during break periods is more invigorating than a snack or a smoke.

Domestic and Social Stresses. Domestic and economic problems may engender stresses that are equal to or greater than those associated with the workplace. The death of a close family member, a divorce or separation, a serious family illness or injury, sexual problems, and the assumption of a mortgage all represent significant changes. The inability to cope with such changes results in uncertainty, severe anxiety, and reduced resistance to additional stress.

We are also influenced by world, national, and local events, often to an extent we do not recognize. Situations that threaten our security, whether threats of war, neighborhood crime, pollution, inflation, lack of transportation, housing inadequacies, or energy shortages, can affect us emotionally and physically. When these insecurities are severe, we feel unable to control them or our environment, and, therefore, we develop a feeling of helplessness. If severe or persistent, this feeling can ultimately degrade our coping ability and render us susceptible to other stresses.

Most of the factors and conditions discussed so far are not work related, and managers may have little knowledge or control over them. They cannot, after all, see to it that every worker gets enough sleep, cuts down on smoking or drinking, or solves his or her domestic problems. Managers should, however, realize that these factors can create serious problems in the workplace and may explain why a worker's previously satisfactory performance suddenly deteriorates. Lecturing the worker may simply increase the worker's level of stress and further erode his or her performance. By contrast, an effort to pinpoint the difficulty and take preventive or ameliorative action may improve his or her performance.

Preventive Measures

Proper Placement and Motivation. The new employee brings certain talents and capabilities to the job, as well as certain hopes, aspirations, desires, needs, and fears. Generally, during the interview, the employer will try to match the worker to a job that is compatible with his or her skills, training, experience, and aims. The worker may then be required to undergo a physical examination to determine his or her state of health and establish the physical limitations. The extent of pre-employment physical examinations vary with the company, but they are usually quite superficial. Rarely are they designed to detect emotional problems. In addition, much of what is gleaned from the examination is information volunteered by the subject; it is unreasonable to expect job applicants to volunteer information that they feel may jeopardize their chance of employment. Hence, very important facts may be suppressed or even falsified, only to surface later on the job.

With increasing government emphasis on health and safety in the workplace and the steady liberalization of workers' compensation laws, the employer has an urgent need not only to conduct thorough pre-employment examinations but also to expand their scope. Matching a worker's physical abilities to a job is much easier and often less crucial then emotional matching. The latter problem, in fact, cannot be resolved using most of the present methods of job matching. A skilled employment interviewer is often better at assessing the applicant's personality, dependency needs, desires, and hopes than are most health care personnel. A private meeting, during which the interviewer expresses a genuine interest in the applicant, can do more to allay fear and anxiety than other, more sophisticated testing.

Many relatively simple things can be done to reduce or eliminate workplace stress. Many of these, you will notice, are simply sound management practices. Assuming that the worker is placed in the proper job, it is still important to provide proper motivation. Workers must feel that they are being supported by both their coworkers and their supervisor. Job requirements and objectives must be clearly detailed, yet workers must be given some personal control over their activities. If possible, they should be allowed to make decisions relevant to their work. Above all, they must be made to feel that they are important to the effort, that they are making a significant contribution to achieving certain goals. Meaningful rewards should be given frequently. (We must emphasize that rewards that enhance the worker's self-esteem are often more reinforcing and motivating than mere cash awards, though these too have their place.) Major changes in assignments, procedures, personnel, or benefits should be announced well in advance to allow adequate preparation for readjustment; few things produce greater insecurity and stress than the unexpected. Workers with greater potential should be encouraged to develop their skills and be given more responsibility where possible. Promotions should never be arbitrary, especially if they involve giving one worker responsibility over others.

Early Detection of Stress. Managers are not doctors and cannot be expected to act as such. They should, however, learn to detect the early signs of stress in subordinates and know the actions they can reasonably take to alleviate it. Most work-related stress reactions, if detected early enough, can be resolved at the manager-worker level before they create serious problems. Frequently, this simply means listening sympathetically to a worker's complaints or worries. On other occasions, relatively minor changes in work detail or work site can resolve a worker's frustration. Unless information is volunteered by the worker, however, the manager should make no attempt to delve into the details of the worker's personal life. When the primary stresses are related to nonwork conditions, the manager can take very little direct action to alleviate them. He should, however, know what company or community resources are available to the worker, and should offer both the information about such help and the encouragement to seek it.

Early stress reactions are most readily noticed in changes in the employee's work performance, behavior, or appearance. Unusual delays in completing work, increased frequency of errors, increased accidents or sick time, lower standards of performance, and general apathy on the job may be the first signs of stress. Similarly, the unexplained appearance of poor personal hygiene, tremors, sudden weight change, argumentativeness or hostility, chronic confusion or hesitation, suspiciousness, and increased smoking may signal personal problems that need attention.

The manager should not take on the responsibilities of a qualified medical diagnostician or skilled psychological counselor, which could have serious repercussions for both the manager and the company. If personal counseling services are available in or through the company, the manager should take advantage of this resource. Suspected medical problems should be referred to the company nurse or doctor. Most corporations have clear policy guidelines and legal constraints on what the manager can and cannot do in these cases. Other company personnel may be available to help in the areas of economics, career planning, and additional training, if indicated. A list of community agencies in the area should be maintained—if not by the manager, then by the personnel counseling and medical staff. Ideally, the company should establish a preventive medical program including the following:

1) Education in recognizing and avoiding stresses
2) Use of tension-reducing techniques, such as exercise, progressive muscle relaxation, biofeedback, and meditation
3) Screening programs to detect high blood pressure and diabetes
4) Alcohol and smoking avoidance programs
5) Periodic examinations of key personnel
6) In-plant counseling services

Keep in mind that stress is disruptive to the workplace. Most people are happy to cooperate when offered some means of reducing or coping with uncomfortable stress.

Stress Prevention. Stress prevention in managers requires that the company understand their dependency needs, desires, and fears, as well as their psychological and physiological limitations. The need for good communication, responsible and clear delegation of authority, and sensitivity to work overload is clear. The company should also consider the stress to the manager's family when sudden or drastic changes in the manager's duties or location are contemplated. Excessive pressure or problems in the family can seriously affect the manager's performance on the job. The company should provide full support, using all of its resources to ensure maximum job satisfaction and minimum extraneous conflict. Managers, after all, are among the company's most valuable resources. The company can certainly justify the considerable expenditure in time and funds required to keep key persons functioning at their best.

Managers should be encouraged to have periodic health evaluations, performed by a physician who understands workplace stress in general and (ideally) the particular situation in which the manager works. The company should pay for the examination, and the executive should have the absolute assurance that the results will be released only to the company's physician and the executive's own doctor. These periodic examinations may show early changes or trends that can be reversed before they become damaging to the manager's health or performance.

Personal counseling services can also play a vital role in an executive stress management program. Counseling services should be available for discussing personal as well as job-related problems. No stigma should ever be attached to these visits. In fact, top management should actively encourage managers to make use of them. Some companies have found that group sessions for managers and industrial psychologists, seminars in handling stress, role playing, and relaxation techniques are equally valuable. Wherever possible, the company should make every effort to deemphasize the crisis aspects of these counseling services and activities in favor of "personal enhancement" or "self-actualization."

Managers should be encouraged to maintain maximum physical health as well. Exercise facilities, including jogging tracks, handball courts, swimming pools, and gyms, can pay handsome benefits in executive health and satisfaction. Relaxation techniques, including whole-body exercise, progressive muscular relaxation techniques, biofeedback training, and meditation, are being taught and used in many companies, with good results. Some of these techniques can be practiced several times a day without disrupting daily work activities, effectively reducing stress in both managers and their subordinates.

CONCLUSION

The complete avoidance of stress is neither possible nor desirable: a certain amount of stress is inherent in any business or work activity, whether making a decision involving the corporation's direction or simply making one's quota in the machine shop; such stress is natural and normal. However, when allowed to escalate, stress can produce substantial problems for the individual and the company, ranging from errors of judgment to increased sick time, higher accident rates on the job, and the costly waste of valuable resources. A safety and health management program that deals only with physical factors, ignoring the crucial emotional and psychological aspects of safety and health associated with workplace stress, is only half a program. It may, in fact, be ignoring safety and health problems that will be far more costly in the long run, eroding efficiency and profits without a visible accident or injury to bring to management's attention.

REFERENCES AND SUGGESTED FURTHER READING

Barnett, R. C., et al. (1991). "Physical Symptoms and the Interplay of Work and Family Roles." *Health Psychology* 10:94–101.

Caplan, R., et al. (1975). *Job Demands and Worker Health.* NIOSH Publication No. 75-160. Cincinnati, OH: National Institute for Occupational Safety and Health.

Crum, R. M., et al. (1995). "Occupational Stress and the Risk of Alcohol Abuse and Dependence." *Alcohol Clinical Experimental Research* 19:647–655.

ErgoWeb. [on-line]. "Ergonomics Program Management Guidelines for Meatpacking Plants." (http://www.ergoweb.com/Pub/Info/Std/mpg1.html).

Fletcher, M. (May 1996). "Labor Defends OSHA's Funding." *Business Insurance* 61(1).

Friedman, M., and R. H. Rosenman (1975). *Type A Behavior and Your Heart.* Greenwich, CT: Fawcett.

Hess, C. F. (November 1996). "Building a Better Ergonomics Program." *Facilities Design and Management* 15(11):28.

Hough, J. R. (1995). "Making a Business Case: Selling Ergonomics" [on-line]. http://204.71.249.80/ishn/9510/ergofeat.htm.

Jacobson, E. (1963). *Tension Control for Businessmen.* New York: McGraw-Hill.

The Joyce Institute (1996). "Design for Product Design End User" [on-line]. http:www.arthurdlittle. com/joyce/joyce_workergs.html.

Koco, L. (1996). "Downsizing Spurs Disability Claims, Survey Finds." *National Underwriter* 100(52):27–30.

Kohn, J. P. (1997). *The Ergonomic Casebook: Real World Solutions.* Boca Raton, FL: CRC Lewis.

Kohn, J. P. et al. (1996). *Fundamentals of Occupational Safety and Health,* Rockville, MD: Government Institutes, Inc.

Margolis, B., et al. (1974). "Job Stress: An Unlisted Occupational Hazard" *Journal of Occupational Medicine.*

Poulton, E. C. (1970). *Environment and Human Efficiency,* Springfield, IL: Charles C Thomas.

Rahe, R. H. (1967). *Life Crisis and Health Change.* Report No. 67-4. San Diego: Naval Medical Neuropsychiatric Research Unit.

Repetti, R. L. (1993). "Short-Term Effects of Occupational Stressors on Daily Mood and Health Complaints." *Health Psychology* 12:125–131.

Selye, H. (1974). *Stress without Distress.* New York: Lippincott.

U.S. Department of Labor, Occupational Safety and Health Administration (1990). *Ergonomics Program Management Guidelines for Meatpacking Plants.* OSHA Publication No. 3108. Washington, DC: OSHA.

Chapter 20

Fire Prevention and Protection Management

James P. Kohn

INTRODUCTION

The previous chapters of this book have examined the development of safety and health programs along with various critical areas of OSHA compliance areas. This chapter examines some of the vital issues associated with the successful implementation of a comprehensive fire prevention and protection management program. Following the case study, this chapter addresses OSHA fire prevention and protection requirements. Issues associated with the basic concepts of fire science are then reviewed. Next, elements of the fire management programs are presented. Finally, coordination of in-house and external agency resources is examined.

CASE STUDY: THE DANGERS OF A FIRE EMERGENCY

The following item was printed in a Massachusetts newspaper dated December 1996.

> METHUEN, MA—A boiler explosion rocked the Malden Mills textile complex in Methuen, MA leaving at least 30 people injured, 13 critically. Fire engulfed and destroyed most of the complex as a result of the explosion.

The Malden Mills complex, located near the New Hampshire border, was a major producer of clothing and upholstery fabrics. The complex, which employed 2,500 workers, occupied more than one million square feet of space in nine separate buildings.

The explosion occurred in the Manomac building, which housed three boilers used in the manufacturing process. There were some fifty workers in the Manomac building at about 8:00 p.m. when one of the boilers exploded. One witness who was in the building at the time of the explosion said, "The building started shaking and smoking and then there was a boom." The fire, fanned by a 40-mph wind and several exploding propane tanks, quickly spread to nearby buildings on and off the complex.

The workers in the Manomac building, some badly burned, helped each other escape from the fire as the building collapsed around them. Others, more severely injured, remained in the inferno for up to fifteen minutes, waiting to be rescued.

Propane tanks, used to power forklifts, exploded from the intense heat and flames. The entire complex was equipped with a sprinkler system, but one of the many major explosions severely damaged the system. The sprinkler systems in all of the buildings were interconnected, and after the blast, the effectiveness of the sprinkler system was severely reduced. The situation worsened later as collapsing floors further damaged the sprinklers.

Firefighters were hindered in their battle by low water pressure and collapsing walls. A turbulent and bone-chilling 40-mph wind aided the spread of the flames and hampered fire- fighting efforts.

As the fire consumed the buildings, the sound of heavy machinery crashing through burned-out wooden floors could be heard over the roar of the fire. At one point, the fire became so intense that flames were blasting out of windows 50 ft horizontally and 150 ft into the sky. Smoke rose high in the sky and could be seen from as far away as 10 miles.

Hazardous materials became a problem as the fire progressed. Although hundreds of chemicals were used at the complex, less than fifteen were considered toxic. Most were in small quantities and were consumed in the fire. However, a tank containing 3,000 gallons of sodium hydroxide ruptured when the building's roof collapsed. The sodium hydroxide flowed onto a loading dock and mixed with water and debris. State officials set up air quality monitors the following day and called in a private hazardous waste removal company to contain and clean up the chemical. The fire raged through the night, and at daybreak the next morning, firefighters were still on the scene suppressing spot fires and watching for flare-ups.

The emergency response to the fire was massive. Mutual aid fire companies came from thirty-five communities including nearby New Hampshire. Fire and rescue equipment from southern New Hampshire was dispatched from the Salem Fire Department. Nearly 300 firefighters fought the blaze. The Northern Massachusetts Law Enforcement Council Tactical Patrol was called out to join local and state police to control the watching crowd and to assist in evacuations. Alcohol, Tobacco, and Fire-arms agents as well as the Massachusetts Department of Environmental Protection also responded.

Fire officials initially set up a command post at a Malden Mills guard shack. As the fire raged, the command post was moved farther away to a second guard post and, as the fire continued to spread, was eventually moved still farther away, to another building. Several nearby homes were evacuated.

Area hospitals were put on red alert status. Of the twenty-seven persons injured, at least thirteen were admitted with severe burns over much of their bodies. OSHA is investigating the cause of the incident.

FIRE PREVENTION MANAGEMENT

The last event that companies want to experience is a fire as catastrophic as the one reported in this case study. Unfortunately, fires adversely effect hundreds of companies every year. Fire losses in the United States are estimated to be in excess of $8 billion annually. Nearly $2 billion of this annual loss has been attributed to incendiary and suspicious "structural" fires. Fire-caused property losses are far in excess of those caused by all classes of crime, and rival those produced by hurricanes and earthquakes (Building Research Establishment Ltd. 1997). Only injuries and fatalities due to traffic accidents exceed the loss in human lives and injuries due to fire. More than 6,000 fire-related fatalities and more than 13,000 nonfatal injuries occur each year.

Most workplaces will usually have some form of a fire protection and fire prevention program or emergency action plan. These programs or plans may be in a written form, or they may exist as verbally communicated procedures of what to do in the event of an emergency. The primary goal of fire prevention programs and emergency action plans is to protect employees and the facility from potential injuries and property damage. Programs and plans, however, are only as good as the time and effort put into them. In addition, management support is absolutely necessary for successful program implementation. Before examining the elements of an effective fire program, let us first examine fire prevention and protection requirements found in the OSHA general industry regulations and then review the basic principles of fire science.

29 CFR 1910 FIRE PREVENTION/FIRE PROTECTION

OSHA general industry standards address fire prevention and fire protection issues in several locations. The following subparts contain standards related to fire prevention and protection program requirements.

- Subpart E—Means of Egress: general requirements, means of egress, general and employee emergency plans, and fire prevention plans
- Subpart H—Hazardous Materials: compressed gases and flammable and combustible liquids; spray finishing; dip tanks; explosives and blasting agents; and storage and handling of liquefied petroleum gases
- Subpart J—General environmental controls
- Subpart L—Fire Protection
- Subpart N—Materials Handling and Storage, Housekeeping: requirements for industrial trucks for specific hazardous locations
- Subpart Q—Welding, Cutting, and Brazing; requirements for fire extinguishers, fire prevention, hot-work permits
- Subpart R—Special Industries
- Subpart S—Electrical
- Subpart T—Commercial Diving Operations

Topics covered by Subpart L, Fire Protection, include the following:

- 1910.156: Fire brigades
- 1910.157: Portable fire extinguishers
- 1910.158: Standpipe and hose systems
- 1910.159: Automatic sprinkler systems
- 1910.160: Fixed fire extinguisher systems, general
- 1910.161: Fixed fire extinguisher systems, dry chemical
- 1910.162: Fixed fire extinguisher systems, gaseous agent
- 1910.163: Fixed fire extinguisher systems, water and foam
- 1910.164: Fire detection systems
- 1910.165: Employee alarm systems

To ensure that an organization is in compliance with OSHA fire prevention and protection requirements, these subparts should be thoroughly studied. It is especially important that the health and safety professional review Subparts E and L to determine the specific activities and written programs that should be developed to ensure compliance with safety standards.

THE CHEMISTRY OF COMBUSTION

Despite its importance, most people know little about fire. Three things are needed to make fire: a fuel, air, and an ignition source. Let us consider the combustion of wood by the following chemical formula:

Wood + Air + Heat →   Ash + Smoke + Even More Heat

The following list includes the basic principles of fire:

- Fire is a chemical reaction known as combustion.
- It is the rapid chemical union of fuel with oxygen that releases heat and light.
- For years the fire triangle was used to explain combustion and how to extinguish it.

The Fire Triangle

For years the basic theory of combustion was referred to as the fire triangle. The fire triangle required three elements to sustain combustion: fuel, an ignition source, and oxygen.

1) *Fuel*—Physical state
   - Gases: natural gas, propane, butane, hydrogen, acetylene, and so on.
   - Liquids: gasoline, kerosene, turpentine, alcohol, paint, varnish, lacquer, and so on.
   - Solids: coal, wood, paper, cloth, wax, grease, plastic, grain, and so on.
2) *Ignition source*—To reach ignition temperature: open flame, hot surfaces, sparks and arcs, friction, chemical action, electrical energy, compression of gases.
3) *Oxygen*—Approximately 16 percent required. (Normal air contains 20.9 percent oxygen. Some fuel materials contain sufficient oxygen within their makeup to support burning.)

The Fire Pyramid

W. M. Haessler developed the fire tetrahedron or pyramid theory (see Figure 20-1). It differs from the fire triangle by including the concept of the chemical chain reaction. This new concept helps explain the action of some agents such as dry chemicals and halon.

- *Fuel*—In both theories fuel is any solid, liquid, or gas that can combine with oxygen in the chemical reaction known as oxidation.

- *Ignition source*—Combustion requires a level of energy that will cause an increase in molecular activity within a substance's chemical structure. Temperature is the measure of the molecular activity within a substance.

- *Oxidizing agent*—Under most conditions, the agent will be oxygen in the air, but some materials release their own oxygen during combustion, such as sodium nitrate and potassium chlorate.

- *Chain reaction*—Certain agents will put out fires because they combine with the substances before the substances can combine with oxygen.

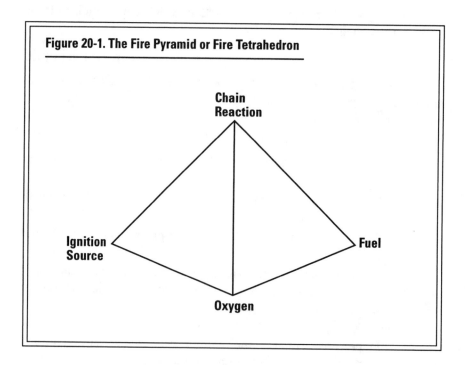

Figure 20-1. The Fire Pyramid or Fire Tetrahedron

Definition and Classification of Fire

Fire is the rapid oxidation of a substance often with the evolution of heat and light in varying degrees of intensities. A common misconception is that fire burns the actual solid or liquid fuel (for example, wood or organic solvent). However, it is the vapor given off by an object that burns. Heat causes objects to give off these flammable vapors (or gases). When the gases reach their ignition temperature, you see the light given off during the oxidation, known as fire. Fire itself generates more heat to the object, and thus an endless cycle begins until all of the gases have been exhausted from an object. All that remains of the object are particles or ash.

Fires may be classified into two groups: simple and complex. A simple fire results in complete combustion and produces no soot, no free carbon, and no appreciable amounts of corrosive gases, fumes, or smoke. A relatively pure fuel, such as natural gas, gasoline, or a high- quality fuel oil, consists of many compounds of carbon, hydrogen, and oxygen. However, if any of those relatively pure fuels are burned efficiently and completely, the products of combustion would be essentially carbon dioxide (CO_2) and water. Only trace impurities would be present.

Most fires are classified as complex—the result of incomplete combustion—and are fueled by synthetic materials. Incomplete combustion occurs when there is insufficient oxygen present to react with the carbon and hydrogen in the fuels. The products of incomplete reactions could include carbon monoxide, unburned free carbon, and a variety of complicated hydrocarbon products. Materials acting as synthetic fuels form acid gases and corrosives such as hydrochloric, hydrofluoric, sulfuric, and nitric acids. Some of the synthetic fuels are foams, films, polyethylene, polypropylene, melamine, acrilan, saran, synthetic rubbers, Teflon, polyurethane, polyvinyl chlorides, and fluorides. Objects made from these materials include carpets, flooring tiles, sheet goods, furniture, clothing, shoes, appliances, plumbing, dishes, and restroom equipment. Even wood fires have been analyzed and found to produce over a dozen different organic acids.

The following definitions are for flammability and combustibility based on consensus standards as given in 29 CFR 1910.106. Additional definitions describe some of the critical physical and chemical characteristics of flammable and combustible liquids.

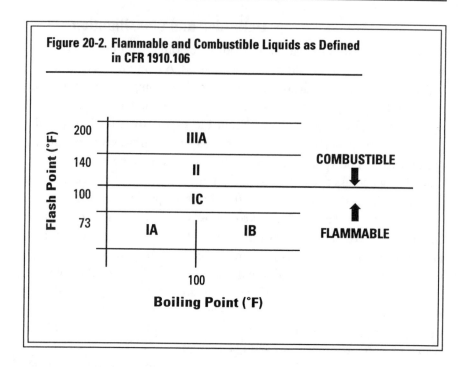

Figure 20-2. Flammable and Combustible Liquids as Defined in CFR 1910.106

Boiling Point—shall mean the boiling point of a liquid at a pressure of 14.7 pounds per square inch absolute (psia). This pressure is equivalent to 760 millimeters of mercury (760 mm Hg).

[(Note: The boiling point is the temperature of the liquid at which its vapor pressure equals atmospheric pressure. Above this temperature, the pressure of the atmosphere can no longer hold the liquid in a liquid state. The lower the boiling point of a liquid, the greater the vapor pressure, and, consequently, the greater the fire hazard (NFPA 1994)].

Flash Point—means the minimum temperature at which a liquid gives off vapor within a test vessel in sufficient concentration to form an ignitable mixture with air near the surface of the liquid. The flash point is normally an indication of susceptibility to ignition.

[(Note: Liquids with higher flash points are a lesser hazard because of a decreased chance of ignition and a decreased potential for vapor spread (NFPA 1994)].

Combustible Liquid—means any liquid having a flash point at or above 100°F (37.8°C). Combustible liquids shall be divided into two classes as follows:

Class II Liquids shall include those with flash points at or above 100°F (37.8°C) and below 140°F (60°C), except any mixture having components with flash points of 200°F (93.3°C) or higher, the volume of which make up 99 percent or more of the total volume of the mixture.

Class III Liquids shall include those with flash points at or above 140°F (60°C). Class III liquids are subdivided into two subclasses:

Class IIIA Liquids shall include those with flash points at or above 140°F (60°C) and below 200°F (93.3°C), except any mixture having components with flash points of 200°F (93.3°C) or higher, the total volume of which make up 99 percent or more of the total volume of the mixture.

Class IIIB Liquids shall include those with flash points at or above 200°F (93.3°C). This section does not regulate Class IIIB liquids. Where the term "Class III Liquids" is used in this section, it shall mean only Class IIIA Liquids.

When a combustible liquid is heated to within 30°F (16.7°C) of its flash point, it shall be handled in accordance with the requirements for the next lower class of liquids.

Flammable Liquid—means any liquid having a flash point below 100°F (37.8°C), except any mixture having components with flash points of 100°F (37.8°C) or higher, the total volume of which make up 99 percent or more of the total volume of the mixture. Flammable Liquids shall be known as Class I liquids. Class I liquids are divided into three classes as follows:

Class IA shall include liquids having flash points below 73°F (22.8°C) and having a boiling point below 100°F (37.8°C).

Class IB shall include liquids having flash points below 100°F (37.8°C) and having a boiling point at or above 100°F (37.8°C).

Class IC shall include liquids having flash points at or above 100°F (37.8°C) and having a boiling point below 100°F (37.8°C).

[(Note: Flash point was selected as the basis for classification of flammable and combustible liquids because it is directly related to a liquid's ability to generate vapor for combustion. The volatility of a liquid, a liquid's ability to produce vapors, determines the likelihood of a fire since it is the vapors and not the liquid itself that burns. The relationship of "low flash-high hazard" is an important one to remember.) (See Figure 20-2 for a presentation of these classes.)]

Vapor Pressure—shall mean the pressure, measured in pounds per square inch (absolute) exerted by a volatile liquid as determined by the "Standard Method of Test for Vapor Pressure of Petroleum Products (Reid Method)," American Society for Testing and Materials ASTM D323-68.

[(Note: Vapor pressure is the liquid's ability to evaporate. The higher the vapor pressure, the more volatile the liquid, and the more readily the liquid gives off vapors.)]

Other related definitions include the following:

Explosive/Flammability Limits—The vapor to air ratio capable of supporting combustion. (See Figure 20-3 for a presentation of these limits.)

Lower Flammability Limits—The minimum amount of vapor in air necessary to support combustion. Below this limit the mixture is "too lean" to support combustion.

Upper Flammability Limits—The maximum amount of vapor in air necessary to support combustion. Above this limit the mixture is "too rich" to support combustion.

Vapor Density—The weight of a flammable vapor compared to air. Air = 1. Vapors with a high density are dangerous because they tend to collect near the floor or other low spots. Vapors with a low density collect near ceilings.

Ignition Temperature—The ignition temperature is that to which a substance must be raised to ignite by itself. The ignition temperature of a liquid is generally several hundred degrees above its flash point temperature. (NFPA 1994).

Upper and Lower Flammability Range

Lower flammable limit, also known as the *lower explosive limit,* is the lowest concentration of gas or vapor (percentage volume in air) that burns or explodes if an ignition source is present. The *upper flammable limit* or *upper explosive limit* is the highest concentration of a gas or vapor that burns or explodes if an ignition source is present. A mixture can have too little of the concentration (and be too lean) or too much of the concentration (and be too rich) to support combustion.

The hazard associated with flammable and combustible liquids is the *flash point.* This is the temperature at which the substance gives off enough vapors to sustain combustion. The source of ignition might be a flame, a hot object, or even a spark from a tool or static electricity.

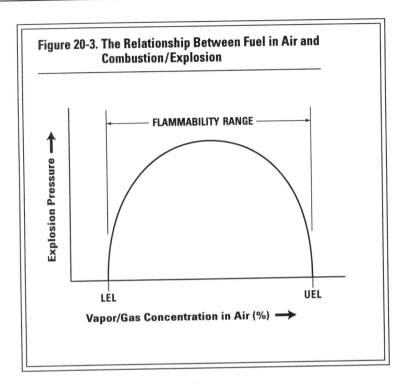

Figure 20-3. The Relationship Between Fuel in Air and Combustion/Explosion

Another factor that becomes important is where fuels tend to collect in the workplace. *Vapor density* becomes a key issue in determining where a vapor might be found in the atmosphere. If the vapor has a low density (below 1.0), it will float in the air and collect along ceilings. If the vapor has a high density (above 1.0), it will tend to travel to the floor and collect in low-lying locations. An ignition source below fuels that are heavier than air may ignite and cause the fire to spread from the point of ignition to the source of the spill. Refer to the list of definitions from 29 CFR 1910.106 for more details on specific terms of importance when studying fire science.

Phases of Burning

The methods used to extinguish a fire depend largely on which phase the fire is in. Elements that must be considered depend on such factors as

- Time the fire has burned
- Ventilation characteristics of the structure
- Amount and type of combustibles

The Three Progressive Stages of Burning

There are three progressive stages of burning: the incipient or the beginning phase, the free-burning phase, and the smoldering phase:

1) *Incipient (beginning) phase:* The air is still full of oxygen and the fire is producing water vapor, carbon dioxide, CO_2, and perhaps sulfur dioxide and other gases. It is also generating some heat, which will increase as the fire grows.

2) *Free-burning phase:* The flames draw in air as heated gases rise and carry the heat to the top of the confinement area. Heated gases spread out from the top downward, forcing cooler air to lower levels and eventually igniting all the combustible material in the upper levels of the room. The area is "fully involved." Temperatures in the upper regions can exceed 1,300°F. As the fire progresses through the latter stages of this phase, it continues to consume free oxygen until there is nothing left to react with the liberated fuel gases. The fire can burn or explode if more oxygen becomes available, that is "backdraft."

3) *Smoldering phase:* Flames may die and leave only glowing embers if the area is airtight during this phase. The area fills with dense smoke and fire gases at enough pressure to seep from the cracks of the building. The temperature will rise to more than 1,000°F, and the heat will vaporize lighter fuel components, such as hydrogen and methane, increasing the hazard.

Flashover occurs when a room or other area is heated enough that flames sweep over the entire surface. Combustible gases released early in fire development collect at the ceiling and mix with air. As the fire burns, it gradually heats all the contents of ignition temperatures, which burst into flames and involve the area.

Products of Combustion

When a fuel burns, there will always be four categories of products of combustion:

1) *Fire gases:* These are gases that remain when other products of combustion cool to normal temperature. Common combustible materials contain carbon, which forms CO_2 and carbon monoxide when burned. Other fire gases include hydrogen sulfide, sulfur dioxide, ammonia, hydrogen cyanide, nitrous and nitric oxide, phosgene,

and hydrogen chloride. The gases that form depend on the chemical composition of the fuel, the percent of oxygen present, and the temperature of the fire. The complete combustion of fuels containing carbon will produce CO_2 but seldom is there complete combustion. Hydrogen sulfide is a fire gas formed during fires involving organic material containing sulfur such as rubber, hair, wood, hides, and so forth. Hydrogen cyanide is a toxic fire gas found in oxygen-starved fires involving nitrogen-containing materials such as wool, silk, urethane, polymides, and acrylics. Hydrogen chloride is produced in fires involving chloride-containing plastics found in furnishings to electrical insulation, conduit, and piping.

2) *Flame:* Flame is the luminous body of a burning gas that gets hotter and less luminous when mixed with more oxygen. It is considered a byproduct of incomplete combustion.

3) *Heat:* A form of energy measured in degrees of temperature is the product of combustion that spreads the fire.

4) *Smoke:* This is a visible product of incomplete combustion and is usually a mixture of oxygen, nitrogen, CO_2, carbon monoxide, finely divided particles of soot and carbon, and a miscellaneous assortment of products released from the burning material.

GENERAL METHODS OF EXTINGUISHMENT

Whether discussing portable or fixed suppression equipment, the general methods of fire extinguishment are basically the same:

- Fuel removal
- Cooling
- Limiting oxygen
- Inhibiting the chemical reaction

Portable Fire Extinguishers

When selecting a portable fire extinguisher, the most important consideration is the nature of the area to be protected. National Fire Protection Association (NFPA) 10, *Portable Fire Extinguishers,* classifies fires according to the fuel involved as Class A, B, C, or D. Extinguishers are classified for use on one or more of these types of fires.

Class A-rated extinguishers are the most often chosen for ordinary building protection. They are used for fires of ordinary "combustible materials such as paper, wood, cloth, and some rubber and plastics" [(29 CFR 1910.155 (c)(8)] that require the heat-absorbing, cooling effects of water or water solutions, the coating effects of certain dry chemicals that retard combustion, or the interruption of the combustion chain reaction by dry chemical or halogenated agents. Extinguishing agents used for Class A extinguishers are water, loaded-stream aqueous film-forming foam (AFFF), multipurpose dry chemical (ammonium-phosphate-base), and halogenated types. Halogenated extinguishing agents are being discouraged due to resulting environmental problems (Demers 1991).

Class B-rated extinguishers are for use on fires "involving flammable or combustible liquids, flammable gases, greases and similar materials, and some rubber and plastic materials" [(29 CFR 1910.155 (c)(9)] that must be put out by excluding air (oxygen), by inhibiting the release of combustible vapors or by interrupting the combustion chain reaction. For Class B fires, there are four basic types of extinguishers that can be used: CO_2, dry chemical, halogenated agent, and AFFF. The three general types of flammable liquid fires (Demers 1991; Peterson 1991) are:

1) Fires in liquids of appreciable depth (greater than ¼ in.)
2) Spill fires or running fires in liquids of no real depth (¼ in. or less)
3) Pressurized flammable liquid or gas fires from damaged vessels or product lines.

Class C-rated extinguishers are for use on fires "involving energized electrical equipment where safety to the employee requires the use of electrically nonconductive extinguishing media" [(29 CFR 1910.155 (c)(10)]. They should be chosen according to how the electrical equipment is constructed, the degree of agent contamination that can be tolerated, and the nature of other combustibles in the area. The agents classified for Class C use include CO_2, dry chemical, and halogenated compounds (Demers, 1991; Peterson 1991).

Class D-rated extinguishers are for fires " involving combustible metals such as magnesium, titanium, zirconium, sodium, lithium and potassium" [(29 CFR 1910.155 (c)(11)] that require a heat-absorbing extinguishing medium that does not react with the burning metals. Generally, these extinguishing agents are dry powders. (Demers 1991).

Some portable extinguishers will extinguish only one class of fire, whereas others are suitable for two or three. However, none is suitable for all four. NFPA extinguisher labels, as required by ANSI standards, are provided so that users may quickly identify the class of fire for which extinguishers may be used. NFPA 10 *(Portable Fire Extinguishers),* contains this classification as well as applicable symbols and color codes (Demers 1991; Peterson 1991).

Rating numerals are also used on the labels of both Class A- and Class B-rated extinguishers. This rating is an indication of the size of fire that labeled extinguishers can handle. It is expressed as a number from 1 to 40 for Class A and 1 to 640 for Class B fire extinguishers. The rating numeral gives the relative effectiveness of the extinguisher. The higher the number, the greater is the extinguisher's effectiveness. Class C fire extinguishers do not have this numeral rating since electrical equipment has either ordinary combustibles or flammable liquids, or both, as part of its construction. Extinguishers for Class C fires should be chosen according to the nature of the combustibles in the immediate area. Extinguishers for Class D fires contain different dry powders that are effective on fires in different kinds of combustible metals. These extinguishers also lack the numeral rating and instead have a nameplate detailing the type of metal that the particular agent will extinguish ("Extinguishers" 1997).

Since there are several kinds of extinguishers and since differences exist among them, it is imperative that persons be trained to use them properly. Currently, based on the extinguishing medium each contains, fire extinguishers are classified into five major groups. These are water based, carbon dioxide, halogenated agent, dry chemical, and foam. Class D fire extinguishers, which use dry powder, underwent special testing and are listed based on their effectiveness on specific combustible metals.

Extinguishing Agents

Water-Based Extinguishers. Water-based extinguishing agents include water, antifreeze, loaded stream, wetting agent, soda-acid, and foam. All except AFFF foam are to be used solely on Class A fires. Antifreeze, loaded stream, AFFF, and wetting agent all use water as a base to which chemicals are added to improve the extinguisher's performance. Both antifreeze and loaded-stream agents are specially treated to withstand low temperatures. In wetting agents, a chemical is added to reduce the surface tension of the water so that it will spread and penetrate better (Demers 1991).

If ordinary or nonfoam water-based extinguishers are used on flammable liquids or electrical fires, they may spread the fire and/or injure the operator. After activating the extinguisher, the operator should point the stream at the base of the flames and work from side to side or around the fire. When flames are high, the range of the stream should be used to the operator's best advantage. As the flames diminish, it is possible to move closer and change the solid stream to a spray by holding a fingertip over the end of the nozzle. A spray stream is more effective on burning embers. Deep-seated smoldering or glowing areas should be wetted thoroughly. If necessary, a hand tool can be used to poke apart burning material.

Carbon Dioxide Extinguishers. CO_2 is a compressed gas agent. Though primarily intended for use on Class B and C fires, it may be used on Class A fires until water or some other Class A-rated agent can be obtained.

CO_2 prevents combustion by displacing the oxygen in the air surrounding a fire and by cooling the fuel. Its principal advantage is that it does not leave a residue, a consideration that may be important in laboratories and in areas in which food is prepared. The rapid expansion and pressure reduction of the pressurized liquid CO_2 in the extinguisher's cylinder result in a cooling and refrigeration effect. When most of the CO_2 is discharged, rapid expansion from liquid to gas converts about 30 percent of the liquid into a very cold "snow" or "dry ice." This snow or dry ice then sublimes into a gas (Demers 1991).

Because CO_2 extinguishers have limited range and are affected by draft or wind, they should be applied as close to the base of a fire as possible. Agents should be applied even after the flames have been extinguished to allow time for cooling and to prevent reflash. Due to the relatively short discharge range, CO_2 extinguishers are of limited effectiveness on large Class B fires. For fires involving electrical equipment, the discharge should be directed at the source of the flames. It is important to deenergize the equipment as soon as possible to prevent possible reignition.

Halogenated Agent Extinguishers. In general, two agents are widely used: Halon 1211 (bromochlorodifluoromethane) and Halon 1301 (bromotrifluoromethane). Years ago, Halon 104 (carbon tetrachloride) was widely used. However, due to undesirable byproducts, it was phased out and is currently illegal as an extinguishing agent. Halon 1211 and 1301 are similar to CO_2 in that they are "clean agents." Though intended for use mostly on Class B and C fires, Halon 1211 is also effective on Class A fires. In fact, extinguishers with a size capacity of nine pounds and greater have Class A ratings (Demers 1991).

Halon 1211 leaves no residue, is virtually noncorrosive and nonabrasive, and when compared on a weight-for-weight basis with CO_2, it has about twice the range and is at least twice as effective on Class B fires. And, unlike CO_2, it needs no cold weather protection. Although Halon 1211 does not have the cooling effect of CO_2, it is not as severely affected by wind as CO_2, but strong air currents may disperse the agent too rapidly. When Halon 1211 is used on a fire, the decomposition products include hydrogen chloride, hydrogen fluoride, hydrogen bromide, traces of free halogens, and various chlorofluorocarbons (CFCs), which have implications with ozone depletion, the greenhouse effect, and other environmental problems. Normally, only small quantities of these chemicals are formed, and, as a warning of their presence, they give off an acrid odor. Because Halon 1211 is retained under pressure in a liquid state, a booster charge of nitrogen is added to ensure proper operation.

Since the late 1960s, several manufacturers have produced Halon 1301 extinguishers in 2.5-lb size containers. These extinguishers are rated only for small Class B and C fires and are significantly less toxic than Halon 1211. Although Halon 1301 produces a smaller array of decomposition products than its chemical cousin, Halon 1211, it still produces halogenated and CFC by-products.

Acute health hazards are known for both extinguishants—Halon 1211 is acutely toxic at concentrations greater than 4 percent by volume and Halon 1301 at concentrations greater than 10 percent by volume. However, chronic health effects are not known. Furthermore, due to the vapor pressure, the discharge stream of both Halon formulations is principally a gas, which may complicate effective application. In addition, due to the vapor density, both formulations tend to collect in low-lying areas. Halogenated agent extinguishes are also available containing a blend of Halon 1211 and Halon 1301. Finally, halogenated agent extinguishers should be operated and applied to fires in the same general way as CO_2 extinguishers (Demers 1991).

Alternatives to Halogenated Agent Extinguishers. With current environmental concerns and the "recent ban of Halon gasses, fire extinguisher sales and service companies are looking for alternatives to traditional extinguishers" (First 1996). A product known as HAWK™ HR-95 is attempting to fill this gap. According to the manufacturer, the extinguishing agent in HR-95 is UL listed, biodegradable, effective for all fire classifications, and its effectiveness is not diminished by rain or windy weather. Instead of harmful gases, it utilizes an organic propellant that contains no CFCs and thus eliminates the chemical "cloud" projected into the atmosphere with dry chemical extinguisher use. It is the first nonhazardous, environmentally safe portable fire extinguisher with an EPA listing that is currently available for purchase ("First" 1996).

Also attempting to get a piece of the action is a product from Buckeye Fire Equipment Company known as Halotron I. The manufacturer claims that it has been approved by both the FAA and the EPA, is noncorrosive, leaves no residue, and is designed to replace Halon 1211. This agent is listed as HCFC Blend B (hydrochlorofluorocarbon).

Dry Chemical Extinguishers. There are two basic kinds of dry chemical agents: ordinary dry chemicals, which may be used on Class B and C fires and include sodium bicarbonate, potassium bicarbonate (Purple K), urea potassium bicarbonate (Monnex), and potassium chloride–based agents; and multipurpose dry chemicals, which may be used on Class A, B, and C fires. An ammonium phosphate-based agent is the only multipurpose dry chemical currently manufactured. There are also two basic designs of dry chemical extinguishers: one uses a separate pressurized cartridge to expel the agent, and the other pressurizes the agent chamber for the same purpose. The stored-pressure or rechargeable type is the most widely used. It is the best suited to areas where infrequent use is anticipated and where skilled personnel with professional recharge equipment are available. By contrast, the cartridge-operated type can be refilled quickly in remote locations without special equipment (Demers 1991).

Dry Powder Extinguishers. Dry powder extinguishers are intended for use on Class D fires. The agent, extinguisher, and method of application should be chosen according to the manufacturer's recommendations. Depending on the type of both agent and metal, the agent may be applied to the fire from an extinguisher, with a scoop, or with a shovel. In any case, the agent should be applied so that it covers the fire and provides a smothering blanket. More agent may be necessary on hot spots. Care should be taken not to scatter the burning material, which should be left undisturbed until cooled (Demers 1991).

The two most common agents are sodium chloride and G-1 powder. The latter consists of graded, granular graphite to which compounds containing phosphorous have been added. The sodium chloride can be used in an extinguisher or applied by hand. The G-1 must be applied by hand. When G-1 is applied to a metal fire, the heat of the fire causes the phosphorous compounds to generate vapors that blanket the fire and thus prevent air rich with oxygen from reaching the burning metal. The graphite, being a good conductor of heat, cools the metal.

Aqueous Film-Forming Foam. AFFF, a surfactant, when added to water, forms a solution that creates a foam when discharged through an aspirating nozzle. On Class A fires, the agent both cools and penetrates to reduce temperatures below the ignition point. On Class B fires, it acts as a barrier to exclude oxygen from the surface of the fuel.

Currently, AFFF is available in a 2.5 gal stored-pressure model rated at 3-A and 20-B. It discharges its contents in approximately 50 s and has a range of 20–25 ft. It should only be installed in areas not subject to temperatures below 40°F. A 33-gal unit with a 20A:160-B rating is also available (Demers 1991).

Distribution of Extinguishers

Regardless of how carefully the correct fire extinguisher is chosen, it will do no good unless it is readily available when an emergency arises. The proper distribution of extinguishers is imperative in order to respond to fires appropriately.

Determining the minimum number and rating of extinguishers for *Class A* fires in a particular area involves several considerations. According to NFPA 10 *(Portable Fire Extinguishers),* the first step is to determine whether an occupancy load is light hazard, ordinary hazard (groups 1–3), or extra hazard (groups 1 and 2). This occupancy load classification system is a measure of how much combustible load a building has (for example, light applies to churches and residences; ordinary group 2 applies to wood products, cereal mills, and so forth). Second, the extinguisher rating should be matched with the occupancy load hazard to determine the maximum area that an extinguisher can protect. Third, since maximum travel distance to Class A fire extinguishers is 75 ft [(29 CFR 1910.157 (d)(2)], one method widely used to determine the correct distribution is the square-foot rule. The maximum area that a Class A extinguisher can protect is determined by taking the largest square inside a circle with a 75-ft radius (106 ft on a side) and then squaring it: 106 ft x 106 ft = 11,250 ft^2. (Demers 1991; Cherry 1996; Gundersen 1997).

The maximum distance to Class B fire extinguishers is 50 ft [(29 CFR 1910.157 (d)(4)]. Since flammable liquid fires reach their maximum intensity almost immediately, the extinguisher must be closer at hand. With lower-rated extinguishers, the travel distance is reduced to 30 ft. When flammable liquids reach an appreciable depth (greater than $\frac{1}{4}$ in.), the extinguisher's rating number, except for foam types, should be at least twice the number of square feet of surface area of the largest tank in the area, assuming that other requirements are met. The same travel distances required for flammable liquids that are not of appreciable depth would also be used to locate extinguishers for protection of spot hazards. Sometimes one extinguisher can be installed to provide protection against several different hazards, provided that individual travel distances are not exceeded. When hazards are widely separated and travel distances are exceeded, individual protection should be installed according to the square-foot requirements previously described (Demers 1991; Cherry 1996; Gundersen 1997).

The maximum travel distance to Class C fire extinguishers can vary. Class C extinguisher use is determined by whether or not the electrical equipment is energized. Once the power to live electrical equipment is cut off, the fire becomes a Class A, Class B, or Class A:B fire, depending on the nature of the burning electrical equipment and the burning material in the vicinity. Extinguishers for Class C fires should be selected according to the size of the electrical equipment, the configuration of the electrical equipment (particularly the enclosure of units that influence agent distribution), and the range of the extinguisher's stream. At large installations of electrical equipment where continuity of power is critical, fixed fire protection is desirable. However, even when fixed fire protection is present, it is recommended that some Class C extinguishers be provided to handle incipient fires (Demers 1991; Cherry 1996; Gundersen 1997).

The maximum travel distance to all extinguishers for Class D fires is 75 ft. It is particularly important that the proper extinguishers be available for Class D fires. Since the properties of combustible metals differ, an agent for Class D fires may be hazardous if used on the wrong metal. Agents should be chosen carefully and according to the manufacturer's recommendations. The amount of agent needed is normally calculated according to the surface area of the metal, plus the shape and form of the metal, which could contribute to the severity of the fire. (Demers 1991; Cherry 1996; Gundersen 1997).

Inspection and Maintenance of Extinguishers

Once a fire extinguisher has been purchased, it becomes the responsibility of the purchaser or an assigned agent to maintain the device. A fire equipment servicing agency is usually the most reliable way for the general public to maintain extinguishers, but large industries often train employees to handle this maintenance in-house. Adequate maintenance consists of periodically inspecting, recharging following discharge, and performing hydrostatic tests as needed for each extinguisher (Demers 1991; Cherry 1996; Gundersen 1997).

An *inspection* is a quick check that visually determines whether the fire extinguisher is in the correct location and will operate. Its purpose is to give reasonable assurance that the extinguisher is fully charged and will function effectively if needed. To be effective, inspections must be conducted frequently, regularly, and thoroughly. An individual evaluation must be made of each property to determine how frequently inspections should be conducted; however, they should be conducted at least monthly [(29 CFR 1910.157 (e)(2)]. It is also important to consider the nature of the hazards present, which can influence the potential use of the equipment.

Maintenance means a complete and thorough examination of each extinguisher. A maintenance check involves examining all the extinguisher's parts; cleaning and replacing any defective parts; and reassembling, recharging, and, when appropriate, pressurizing the extinguisher.

Maintenance checks sometimes reveal the need for special testing of extinguisher shells or other components. Maintenance should be performed periodically, but at the least once every year [(29 CFR 1910.157 (e)(3)], after each use, or when an inspection shows that the need is obvious.

FIRE PROTECTION/PREVENTION MANAGEMENT

Fire protection/prevention programs start with commitment from top management. Company management must motivate employees to participate in the fire protection program. Active support from top management must include assigning an individual to oversee the fire program. The duties and responsibilities should be clearly identified for this individual along with the appropriate authority to complete these duties.

One aspect of management support is adequate program funding. Financial resources are necessary for the purchase, installation, and maintenance of manual and automatic fire suppression systems. Management commitment also requires that individuals involved in the fire program are released from normal responsibilities to obtain the training and participate in the drills necessary to develop and maintain necessary knowledge and skills.

Components of the fire program should include

- Planning for fire and other emergencies
- Acquiring and maintaining fire protection systems and equipment
- Practicing fire prevention and conducting inspections
- Organizing and training fire brigades (if appropriate)
- Developing cooperation and coordination with outside agencies
- Meeting insurance carrier and regulatory agency requirements

In addition, the program should ensure compliance with OSHA regulations including

- Provision of emergency escape routes
- Employee emergency action and fire prevention plans
- Control of hazardous materials and processes
- Organization and training of fire brigade
- Installation and maintenance of fire protection systems
- Installation and maintenance of fire detection and alarm systems

Under 29 CFR 1910.38 (a), OSHA requires written plans for responding to fires. This is part of the Emergency Action Plan, which includes

- Organization statement
- Emergency escape procedures and escape routes (map of the plant)
- Procedures to be followed by employees who remain to operate and/or shut down critical plant operations before they evacuate
- Procedures for accounting for all employees after evacuation
- Duties and responsibilities for personnel performing rescue and medical work (including organizational chart)
- Procedures for reporting fires and other emergencies
- Description of expected emergencies, their hazards, and the plan of action to combat such emergencies

- Procedures that describe the actions to be taken in situations involving special hazards such as flammable liquids, toxic chemicals, and radioactive materials

Another item that should be included in the plan is the alarm system to be used to notify affected employees. In addition, resource lists of personnel and equipment should also be maintained for planning purposes.

Under Fire Protection Plan Development [29 CFR 1910.38 (b)], OSHA requires the formulation of specific components of the plan. These components include

- Means for notifying employees of emergencies
- Evacuation procedures
- Pre-fire planning

In general, fire prevention plan requirements under OSHA [found under Subpart E 29 CFR 1910.38 (b)(2)] are as follows:

1) A plan is required, even if employers choose not to fight fires in their facilities.

2) Plans must include, but are not limited to
   - A list of major workplace fire hazards
   - Proper storage and handling procedures for fuel source hazards
   - Potential ignition sources
   - Types of fire protection equipment and systems protecting hazards and ignition sources
   - Names of personnel responsible for the maintenance of fire protection equipment and systems
   - Procedures for controlling accumulations of flammable and combustible waste and housekeeping
   - Maintenance procedures for process safety devices used on heat-producing equipment to prevent accidental ignition of combustible materials

3) The employer must provide training to recognize fire hazard properties of materials and processes to which employees may be exposed.

4) The plan must be reviewed with each employee when he or she is first hired.

5) The plan must be kept in the workplace and made available to all employees.

Goals of the Fire Protection/Prevention Management Program

Goals must be established to set the direction for the fire safety program. The number one goal should be to prevent fires in the workplace by developing the fire safety awareness of employees. They must have the necessary knowledge to recognize hazardous conditions and take appropriate action to prevent these conditions from developing into a fire.

Organizing the Staff. Someone must be responsible for the overall fire safety in the plant or facility (Hedrick 1994). The plant safety officer, or safety director is usually the person assigned this task and responsibility. Safety or fire prevention committees are an important way of involving employees in the process, but if an individual or committee is not assigned the responsibility of carrying out plans or recommendations, then little, if anything, will be accomplished.

The first responsibility or goal of this individual or committee is to make sure that there is a written fire prevention plan and an emergency action plan in place at the facility. These plans are required by OSHA standards to be in place at some facilities.

Establishing Protection Required for Particular Occupancies. The responsible individual or committee will have to determine how the facility is classified. NFPA 101, *Life Safety Code,* should be reviewed to determine the type of facility that the operation is classified as, and what requirements apply regarding arrangement of means of egress, fire-rated doors, marking of exits, emergency lighting, and fire protection (alarm systems, sprinkler systems, and portable fire extinguishers). Workplace facilities are classified into the following categories (NFPA 1994):

- *Mercantile*—occupancies that are used for the display and sale of merchandise. These include shopping centers, department stores, drugstores, supermarkets, and auction rooms.

- *Business*—occupancies that have large numbers of occupants during normal business hours and very few occupants during non-working hours. These include general offices, doctors' offices, government offices, city halls, municipal office buildings, courthouses, college and university classrooms, and instructional laboratories.

- *Industrial*—any building, portion of a building, or group of buildings used for the manufacture, assembly, service, mixing, packaging, finishing, repair, treatment, or other processing of goods or commodities by a variety of operations or processes. These occupancies include, but are not limited to, the following:

— Chemical plants
— Factories of all kinds
— Food processing plants
— Furniture manufacturers
— Metal working
— Power plants
— Refineries
— Woodworking

- *Storage occupancies*—buildings or structures used to store or shelter goods, merchandise, products, vehicles, or animals. Examples are warehouses, freight terminals, parking garages, aircraft storage hangars, grain elevators, barns, and stables.

- *Occupancies in special structures*—these include open structures, towers, underground structures, vehicles, vessels, water-surrounded structures, and windowless buildings.

Types of Protection

Automatic Sprinkler Systems. The responsible individual or committee should check to determine if the OSHA standards; NFPA 101, *Life Safety Code;* or insurance carrier requires protection by some type of automatic extinguishing system. If so, then NFPA 25 and NFPA 13 should be consulted for proper installation, inspection, testing, and maintenance of sprinkler systems. Someone should be assigned to inspect, maintain, and keep the system in proper working condition.

Portable Fire Extinguishers. Portable fire extinguishers are installed in many occupancies to give the building occupants a means of fighting a fire manually. Not all occupancies are required to have portable fire extinguishers. Normally, building codes; NFPA 101, *Life Safety Code;* or insurance company standards will have a provision that states: "Portable fire extinguishers shall be installed in accordance with NFPA 10, *Standard for Portable Fire Extinguishers.*" Once it has been established that portable fire extinguishers are required, they must be properly selected, mounted, and serviced (NFPA 1994). Some individual or committee should be assigned the responsibility of inspecting, maintaining, and servicing the extinguishers.

Standpipe and Hose Systems. Standpipe and hose systems provide a means of manually applying water to fires in buildings. However, they do not take the place of automatic fire protection systems. They usually are required where automatic protection is not provided and in areas where hose lines cannot easily reach from outside hydrants (NFPA 1994).

Refer to NFPA 25 for the inspection, maintenance, and testing of standpipes. Also, assign some individual or committee to be responsible for maintaining and servicing these systems.

Fire Brigades. If your company decides to organize a fire brigade, it must meet all the requirements of the 29 CFR 1910.156 OSHA standards. You should also refer to NFPA 600, *Industrial Fire Brigades,* for more information covering this topic. The Industrial Fire and Emergency Management Training Association has produced an excellent guide, *Implementing NFPA 600, Private Fire Brigades: A Nuts and Bolts Approach* (Hedrick 1994). This guide contains a training needs assessment, as well as performance standards for fire brigade members. Facilities that handle a large quantity of hazardous chemicals and utilize extensive chemical processes, and/or are located in rural areas, where the response of a local fire department takes more time than desirable, may decide to organize a fire brigade.

Inspections of Fire Protection Equipment

To be fully protected by the fire protection equipment at your facility, it must be inspected and serviced on a regular basis. Depending on the size of the facility and the amount of fire protecting equipment installed, the responsible individual or committee will have to determine how to set up an equipment inspection program. For example, one person could be responsible for inspecting and maintaining the sprinkler systems, another could be in charge of the portable fire extinguishers and standpipe systems, and another for maintaining the fire alarms and automatic fire detectors (smoke and heat detectors). If the facility is relatively small, a small number of individuals could be responsible for all of the fire protection equipment.

Automatic Sprinkler Systems. Depending on the type of system you have, inspections, testing, and maintenance of the system will be performed on a weekly, monthly, quarterly, annually, five-year, and ten-year schedule, according to tables 12-1 and 12-2 of the *NFPA Inspection Manual* (NFPA 1994).

Standpipe and Hose Systems. Inspections, tests, and maintenance should be performed on a monthly, quarterly, annual, five-year, and ten-year cycle, according to tables 12-3 and 12-4 of the *NFPA Inspection Manual* (NFPA 1994).

Portable Fire Extinguishers. If fire extinguishers are provided in the workplace, and intended for employee use, they must be inspected, maintained, and tested on a regular basis. Refer to OSHA standard 1910.157; NFPA 101, *Life Safety Code;* and NFPA 10, *Standard for Portable Fire Extinguishers,* for specifics in this area. In general, fire extinguishers should be visually inspected monthly, maintenance-inspected annually (conducted by a private organization certified to perform this inspection), and hydrostatically tested at five and twelve years (also performed by personnel specifically trained to conduct this type of inspection).

The monthly visual inspection is a quick check intended to ensure that the extinguisher is available, fully charged, and operable. Maintenance, on the other hand, is a thorough check of an extinguisher intended to give maximum assurance that it will operate effectively and safely. Maintenance includes a thorough examination and any necessary repair or replacement. It normally will reveal the need for hydrostatic testing (NFPA 1994). Refer to the checklist in the *NFPA Inspection Manual,* which may be useful for conducting your monthly visual inspections (NFPA 1994).

The purpose of hydrostatically testing fire extinguishers is to protect against the unexpected failure of the cylinder. See table 14-4 of *the NFPA Inspection Manual* for the test intervals for fire extinguishers (NFPA 1994).

Fire Alarm and Fire Detection Systems. Fire alarm systems perform several functions vital to limiting life and property losses during fires. They can provide fire detection and early warning for evacuation, and response by in-house fire brigades or the local public fire department (NFPA 1994). In general, industrial occupancies with a total occupancy of fewer than one hundred persons, or fewer than twenty-five persons, which is normally above or below street level, the fire alarm system may be waived (NFPA 1990). OSHA standards 1910.164 and 1910.165 do not require these systems to be installed unless they are required by another OSHA standard, or by NFPA 101 *(Life Safety Code).*

NFPA 72, chapter 7 of the *National Fire Alarm Code,* covers the testing and maintenance of fire alarm systems, manual fire alarm boxes, and automatic fire detectors (smoke and heat detectors). NFPA 101 requires that a manual fire alarm box be located near each required exit (NFPA 1994).

Education and Training

Hazard reduction or fire prevention is often the least expensive and most effective way of fighting fire hazards. Fire prevention is most effective through education, engineering, and enforcement (NFPA 1990).

All employees should be made aware of fire hazards and instructed on how to identify and report fire hazards, respond to a fire emergency, and implement emergency control procedures. A properly trained plant emergency team can help other employees learn to react promptly and effectively to a fire emergency (NFPA 1994).

Effective education and training also applies to the operation and maintenance of machinery and processes. Special considerations should be given to maintaining the reliability of fire hazard control equipment and procedures (NFPA 1994).

Fire Extinguisher Training. If portable fire extinguishers are provided at the workplace, and employees are expected or allowed to use them, they must be provided with initial training covering proper use of the extinguisher, and annually thereafter.

Emergency Action Plan and Fire Prevention Plan. Another key to an effective occupational fire prevention program is the organization of a well-written emergency action plan and fire prevention plan. These plans are located in the OSHA standard 29 CFR 1910.38. All facilities are not required to have these plans in place at their work site, unless it is required by a particular OSHA standard, or NFPA 101. However, it would be prudent to have these plans in place, because they cover all the aspects of a solid fire prevention program.

Fire Drills and Emergency Evacuation. Employee response within the first few minutes of an emergency can make the difference between a minor incident and a disaster. During a fire or emergency in an industrial facility, two problems can arise: 1) the need to move essential personnel, such as the plant emergency team, to locations where they are needed; and 2) the need to evacuate occupants quickly and efficiently (NFPA 1994).

Providing means of egress is not enough to ensure the safety of personnel. Exit drills are essential to ensure effective use of exits in an emergency. They are required and commonly practiced in educational, health care, business, and some residential occupancies. Drills should be conducted in industrial occupancies in which life hazard from fire is significant (NFPA 1990). The frequency of drills will be determined by the applicable requirements or individual/committee discretion, usually quarterly or semiannually.

All employees should be trained in the recognition of the evacuation signal and the exit route they are to follow. When employees are assembled outside the building, managers or supervisors should account for all employees under their supervision, especially the whereabouts of any handicapped persons in the area. Assign mobile persons along the exit route to accompany these persons. If one or more persons are not accounted for, report this to the safety officer so that search and rescue efforts can be initiated. Only trained search and rescue personnel should be permitted to reenter an evacuated area (NFPA 1994). After each exit drill, a meeting of the responsible managers should be held to evaluate the success of the drill and discuss items that may have been faulty or misunderstood (NFPA 1990).

All facilities should delegate the responsibility of daily inspections for the proper functioning of emergency exits, and unobstructed evacuation routes, to an employee in each area or department. Preventive maintenance programs should be in place for all fire protection equipment, including doors, panic hardware, and exit lights. The equipment should be given high priority to ensure that all identified deficiencies are corrected without delay (NFPA 1994).

COLLABORATION WITH OUTSIDE RESOURCES

A chronological examination of emergency response coordination activities will ensure adequate preparation and response of all available resources. The chronological examination breaks down community resource and in-house response team integration into three phases: 1) preincident planning, 2) incident coordination, and 3) post-incident follow-up (Kohn 1990). The pre-incident planning phase includes all preparatory activities required to implement an effective emergency response plan. These activities include everything from hazard recognition to coordinated predisaster preparation and evaluation. The next phase, incident coordination, refers to activities that maximize response efficiency and effectiveness during emergencies. The last phase, postincident follow-up, includes activities required to evaluate and improve emergency response effectiveness.

There are several components of phase one (preincident planning). The first component of phase one is to identify and analyze the hazards in the workplace. This may encompass determining the hazardous properties of all materials found at the facility, examining physical operation hazards, and analyzing operator practices and training. The second component is performing a risk assessment to determine the degree of risk and potential loss.

The third component of phase one consists of determining in-house resources, capabilities, and commitment. Considerations for internal organizational resources for emergencies include part or all of the following:

- Onsite emergency coordinator
- Evacuation procedures
- Security
- Fire brigade
- Search and rescue team
- Medical and first aid services
- Emergency transportation

Additional resources to be assessed are as follows:

- Communications
- Shelter
- Supplies and materials
- Public information officer
- Counseling services
- HAZMAT team
- Spill containment
- Personal protective equipment (PPE)
- Fire suppression equipment

Assessing in-house emergency response capabilities is an important part of the preincident planning phase. The extent of the emergency response team's training and expertise should be determined. Can these groups handle incipient stage or structural fire fighting emergencies? If a disaster were to occur, could the organization stabilize and properly transport the injured? How knowledgeable are personnel regarding hazardous chemical release response protocol?

As with all safety processes, commitment on the part of management and employees to emergency and disaster planning is essential for its success. Once the commitment has been made, it is important that all participants, including employees, visitors, and contractors, understand that the commitment is a true reflection of the spirit of the organization. Ultimately, management's commitment will determine how comprehensive the facility's emergency preparedness plan should be.

The fourth component of the preincident planning phase is the development of the in-house emergency action plan. This plan helps minimize loss of property and human suffering. Planning must involve the integration of in-house emergency resources and community emergency resources. To accomplish this integration, a core group should be named to seek input and formulate how emergency resources are to be used. Members of this group should represent the safety department, in-house emergency response groups, and outside response groups, and they should take the lead in developing the emergency plan.

A larger advisory group should be formed that will provide information from a wide variety of perspectives. The advisory group should include supervisors and employees from the various departments of the facility, and from maintenance personnel, finance and accounting, human resources, safety committee members, and other groups that may have useful planning information.

A company should begin the process by coordinating ideas between departments and divisions inside the facility. Identify who is going to be the contact person in each department and division inside the facility. Then determine which employees will participate in emergency response. Not only will they be involved in the planning process, but they will also assist in accounting for employees after an evacuation or fire event. Determine what skills and abilities the contact person or response team members have that can be used during the operation and recovery process.

Methods for notifying employees include two-way radio, building alarm systems, horns, sirens, strobe lights, and paging systems. You must ensure that the system will operate in the event of a power or system failure, or have an alternate backup to effectively notify employees, visitors, and contractors of the emergency. Do not make assumptions that everyone will understand the signal or the appropriate response.

The company's resources and those of public response organizations in the area should also be surveyed to provide information on current response capabilities and to identify potential gaps and resources. The first draft of the written plan is prepared as the next step. The committee and advisory group then reviews and revises the draft until a consensus is reached that the plan is adequate.

Determining community resources, capabilities, and commitment is the fifth component of the preincident planning phase. Community resources to be evaluated include the following:

- Police
- Fire departments
- Rescue and emergency teams (including HAZMAT)
- Hospitals and emergency clinics
- Ambulance services (including air transport)
- The Red Cross
- Industry emergency response teams (local agencies and businesses that share assets and expertise)
- Heavy equipment providers (suppliers, construction contractors, National Guard, and so on)
- Communications media

Assess how committed the community resources are to your facility. Are they only committed to fight certain fires? For example, the municipal fire department may make a commitment to fight only structural fires.

The sixth component of the preincident planning phase is the development of the community emergency action plan. It is important that community officials are involved in the company's emergency planning process and that the company is involved in local planning. The facility's plan is more likely to include the correct procedures for notifying and coordinating with government agencies during emergencies if these local agencies are involved. Community involvement during the planning stage will also allow local officials to assess their own capabilities to respond to an emergency at the local facility. Finally, these officials will have a better understanding of the facility's processes and will gain additional confidence in the facility's ability to control these operations.

Community emergency action planning factors consist of all the local conditions that make an emergency plan necessary:

- Identifying and describing the facilities in the district that possess extremely hazardous substances and the transportation routes along which substances may move within the district

- Identifying additional facilities in the plan because of their vulnerability to releases of extremely hazardous substances from facilities.
- Identifying other facilities included in the plan by virtue of their potential for releases of hazardous substances, including methods for determining the occurrence of a release, and the area of population likely to be affected by such release

Once an in-house emergency action plan and a community action plan are completed, they need to be integrated. Coordination of the two groups will maximize emergency response effectiveness and minimize the interference that often takes place during an emergency. At the minimum, the following five items need to be properly coordinated between the in-house and community resources:

1) Notification and warning
2) Communications
3) Protective actions
4) Training
5) Public awareness

Once the in-house and community plans have been integrated, the next step is to initiate the integrated response training activities. This will involve inviting community resource agencies to the facility. Community agencies such as fire, police, and other courtesy groups may be able to identify emergency response problems, which might not have been originally considered during the planning process. By conducting mock drills and emergency training activities using the raw and finished products found at the facility, all participants with emergency response assignments should be better prepared to cope with whatever problems arise.

During the second phase, incident coordination, involved personnel will examine the plan's effectiveness during the incident itself. The plan commences with initiation of the communication system and implementation of the incident command system. The actual emergency is assessed and responsibilities are assigned based on predetermined emergency action plans. The greatest advantage to using an incident command system is the modular development of an organization. A modular organization illustrates the chain of command, and the roles and responsibilities of each individual or group. The major functional areas of the incident command system are command operations, planning, logistics, and finance. The expression "one person, one boss" refers to the concept of unity of command. This means that a worker or sector leader gives and gets orders to/from one person (unless there is a dire emergency). Due to unity on command, the lines of communication are clear and direct; thus, orders and questions are less likely to be garbled.

In addition, emergency response data/information and details should be recorded. This allows assessment of the plans strengths and weaknesses under actual conditions. Successful implementation of incident coordination phase activities will ultimately be determined by how thorough pre-emergency plans were developed during the preincident planning phase.

Finally, during post-incident follow-up, one should analyze the incident. At this time, a critique of in-house and community group performance can help determine emergency action deficiencies. Modification of in-house and community action plans should be in line with initiation of response training activities in order to practice the identified changes.

The degree and effectiveness of preparedness often spell out the differences between emergency and disaster. Fire is the most common and most costly emergency faced by industries today, and the most valuable defense during the first few minutes of a fire is a well-trained and disciplined fire brigade. OSHA does not require manufacturers to maintain fire brigades, and neither do most insurance companies. However, if you form one you have to meet OSHA's regulations and minimum standards. OSHA regulations include the requirements for the organization, training, and PPE of fire brigades whenever an employer establishes them. The NFPA's document NFPA 600 builds on the fire brigade regulations issued by OSHA. Although NFPA standards are not always legally binding, loss-control experts highly recommend them to reduce the risk of fire-related losses.

CONCLUSION

Planning activities involved in the in-house fire program can reap a substantial number of rewards for the safety and health practitioner. Fire prevention is a topic that most managers and employees can understand and recognize. Using extinguishers during training sessions is an activity that most employees enjoy. Most employees also recognize the importance of fire detectors and fire extinguishers at home. By involving everyone at the facility in various fire program activities, the safety and health practitioner will do more than just improve the quality of the fire management plan at the plant. The fire program activities can serve as a rallying point and jumping board for all safety, health, and environmental activities initiated at the facility. They may serve as a rallying point for off-the-job health and safety as well.

REFERENCES AND SUGGESTED FURTHER READING

Bruegman, R.R. (1996). *Hazardous Material Handbook.* New York: Van Nostrand Reinhold, pp. 231-282.

Building Research Establishment Ltd. (1997). *Fire Research Station—Fire Investigations.* [on-line]. http://www.bre.co.uk/bre/frs/Fire_investigations.htm.

Cherry, D.T. (1986). *Total Facility Control.* Boston: Butterworth Publishers.

Cote, A. E. (1991). "Building and Fire Codes and Standards." *Fire Protection Handbook. 17th Ed.* Quincy, MA: National Fire Protection Agency.

Croft, H. (1989). "Fire Management: Strength through Diversity." *Fire Management Notes* 50:2-4.

Davis, J.L. (1996). *Hazardous Material Responder Training.* North Carolina: J.L. Davis and Associates.

Davisson, J. (January 1994). "Gauging the Need for a Fire Brigade." *Occupational Hazards* 56 (1):97-100. , Cleveland, OH: Penton Publishers.

Demers, D.P. (1991). "Selection, Operation, Distribution, Inspection, and Maintenance of Fire Extinguishers." *Fire Protection Handbook.* Boston, MA: R.R. Donnelly & Sons.

"Extinguishers Have Limits." (1997). http://www.ci/sanmateo.ca.us/dept/fire/uexting.html.

Fagel, M. (October 1997). "Creating a Disaster Plan," *Occupational Hazards.*

Federal Emergency Management Agency (1996). *Guide for All-Hazard Emergency Operations Planning.* Government Document SLG 101.

"First Non-Hazardous, Environmentally Safe Fire Extinguisher Available" (1998). http://www.gamani.com/hawk/press.htm.

Gundersen, B. (1997). "Using the Correct Fire Extinguisher." http:/www.ci.la.us/department/LAFD/extinfo.htm.

Hedrick, D. (December 1994). "Elements of an Industrial Fire Safety Program." *Voice* 23:36–37.

Herman, P.L. (September 1991). "Industrial Fire Brigades." *Plant Engineering* 45(17): 64-67.

Hopkins, R. and E. Blair. (March 1993). "Industrial Emergency Management Requires More than Lucky Breaks." *Occupational Health and Safety* 62(3):38-42.

James, D. (1986). *Fire Prevention Handbook,* Cambridge, Great Britain: Butterworth & Co., Ltd..

Kelly, R.B. (1989). *Industrial Emergency Preparedness.* New York: Van Nostrand Reinhold.

Kohn, J. (November 1990). "Integrating Community Resources and the In-House Emergency Response Team." *Professional Safety* 35(11):32-35.

Ladwig, T.H. (1991). *Industrial Fire Prevention and Protection.* New York: Van Nostrand Reinhold.

Laford, R. (February 1997). "The Missing Cog in Fire Safety." *Occupational Health and Safety* 66(2):24-27.

Laing, P.M. (1992). *Accident Prevention Manual for Business and Industry.* Chicago: National Safety Council.

National Fire Protection Association, (Fire/NFPA/standards/factshts.html#5)gov//fema/house.html.

National Fire Protection Association. (1990). *Industrial Fire Hazards Handbook.* Quincy, MA: NFPA.

National Fire Protection Association (1994). *NFPA Inspection Manual.* Quincy, MA: NFPA.

North Carolina Occupational Safety and Health Standards. (1994). *1910.106—Flammable and Combustible Liquids.* CCH Business Law Staff Publication.

Peterson, M. E. (1991). "The Role of Extinguishers in Fire Protection" *Fire Protection Handbook.* Boston, MA: R.R. Donnelly & Sons.

Schroll, C. (February 1992). "Anatomy of an Emergency Action Plan," *Occupational Hazards* 54,(2):51-56. Cleveland, OH: Penton Publications.

Self Inspection Checklist. (April 1997). *OSHA Handbook for Small Business* [on-line] www.saftek.com/osha/oshachek2.html.

Sennett, A. R. (December 1988). "Occupancy Fire-Safety Training." *Industrial Fire World* 3:6–7.

Technical Report No. 11 (1998). "Alternative Fire Extinguishing Agents: Non-Volatile Precursors to Olefinic Bromoflourocarbons: http://pprc.pnl.gov/pprc/rpd/statefnd/turi/alternat.html.

"The National Model Fire Prevention Code" [on-line]. http://www.wpi.edu/Acade.FPA_journal/nmfpc.html#2.

United States Department of the Army. (1993). *Repairs and Utilities Management of Fire Prevention and Protection Program.* Washington, DC: Headquarters, Department of the Army.

United States Occupational Safety and Health Administration. (1993). *Workplace Fire Safety.* Washington, DC: U.S. Department of Labor, OSHA.

Personal Protective Equipment

Jeffrey O. Stull

INTRODUCTION

The purposes of this chapter are to

- Provide a comprehensive approach that can be used to assist end users in selecting appropriate personal protective equipment (PPE)
- Describe different types of PPE in terms of design configurations, features, and sizing
- Discuss different performance properties that can be applied to specifying appropriate PPE
- Indicate other factors that end users should consider in selecting PPE
- List relevant regulations, standards, and guidelines that can be used in selecting PPE.

PPE AND ITS ROLE IN PROVIDING WORKER PROTECTION

PPE comprises items of clothing and equipment that are used alone or in combination with other protective clothing and equipment to isolate the individual wearer from a particular hazard or hazards. PPE can also be used to protect the environment from the individual, such as in the case of cleanroom apparel and medical devices for infection control.

PPE is considered the "last line of defense" against particular hazards when it is not possible to prevent worker exposure by using engineering or administrative controls.

- Engineering controls should be first used to eliminate a hazard from the workplace by modifying the work environment or process to prevent any contact of workers with the hazard. An example of an engineering control is the replacement of a manual task involving potential hazard exposure with an automated process.

- In the absence of engineering controls, administrative controls should be used to prevent worker contact with the hazard. An example of an administrative control is to establish a procedure dictating that workers be out of an area when hazards are present.

- Finally, when neither engineering nor administrative controls are possible, PPE should be used.

However, workers should not rely on PPE exclusively for protection against hazards, but should use PPE in conjunction with guards, engineering controls, and sound manufacturing practices in the workplace setting.

Although PPE is designed for the protection of personnel against various hazards, *PPE cannot provide protection to the wearer against all hazards under all conditions.* The use of PPE itself may create additional hazards for the wearer including heat stress, reduced mobility, dexterity, and tactility; and impaired vision or hearing. Therefore, PPE must be selected not only for the level of protection it provides, but also for the specific wearer hazards it creates, or its impact on worker productivity.

PPE covers a broad range of different clothing and equipment items, including

- Full and partial body protective garments
- Protective gloves and other handwear
- Protective footwear
- Protective headwear
- Protective face and eyewear
- Respirators
- Hearing protectors or hearing protection devices

The selection and use of PPE is relevant to several different industries, where different hazards exist requiring the following:

- Physical protection
- Environmental protection
- Chemical protection
- Biological protection
- Thermal protection
- Electrical protection
- Radiation protection

Because of the broad range of PPE types and applications, this chapter focuses on those areas in which there is less information available to the industry for selection of "appropriate" PPE.

INDUSTRY REGULATIONS GOVERNING PPE SELECTION, USE, AND TRAINING

Sources of Regulations and General Regulations on PPE

In the United States, OSHA promulgates the majority of PPE-related regulations. OSHA provides regulations that address the selection and use of PPE in Title 29, Code of Federal Regulations (CFR) Subpart I, Sections 1910.132–1910.140:

- Section 1910.132: General requirements for PPE
- Section 1910.133: Eye and face protection
- Section 1910.134: Respiratory protection
- Section 1910.135: Head protection
- Section 1910.136: Foot protection
- Section 1910.137: Electrical protective equipment
- Section 1910.138: Hand protection
- Section 1910.139: Sources of standards on which Sections 1910.132–1910.137 are based
- Section 1910.140: List of standards organizations relative to PPE.

Examples of other OSHA requirements for PPE are contained in the following:

- Section 1910.95, paragraphs (i) and (j): Occupational noise exposure and hearing protectors
- Section 1910.120, paragraphs (g)(3)–(g)(5): Use of PPE in hazardous waste operations and emergency response
- Section 1910.156, paragraphs (e) and (f): Specific requirements for PPE worn in fire fighting
- Sections 1910.1000—1910.1050: Specific occupational toxic and hazardous substances and PPE use relative to these substances
- Section 1910.1030 (c)(3): PPE used for protection against bloodborne pathogens
- Various sections of Part 1926: Use of PPE for the construction industry

- Various sections of Parts 1915, 1917, and 1918: Use of PPE for the maritime industry

States with their own OSHA plans must meet or exceed these requirements.

General Employer Selection Responsibilities under OSHA 29 CFR 1910.132

The employer is responsible for conducting a hazard assessment of the workplace to determine whether hazards requiring PPE are present or are likely to be present. If hazards are present, the employer must:

- Select and have each affected employee use the types of PPE that will protect the affected employee from the hazards identified in the hazard assessment.
- Communicate selection decisions to each affected employee.
- Select PPE that properly fits each affected employee.

The employer also must verify that the required workplace hazard assessment has been performed through a written certification that

- Identifies the workplace evaluated
- Identifies the person certifying that the evaluation has been performed
- Identifies the date(s) of the hazard assessment
- Is itself clearly identified as the documentation of certification of the hazard assessment

For selection of PPE for protection against respiratory and electrical hazards, the employer should refer to Section 1910.134 and Section 1910.137, respectively.

General Employer Training Responsibilities under OSHA 29 CFR 1910.132

The employer must provide training to each employee who is required to use PPE with each employee instructed in the following:

- When PPE is necessary
- What PPE is necessary
- How to don, doff, adjust, and wear PPE properly
- The limitations of PPE
- The proper care, maintenance, useful life, and disposal of PPE

The employer also must have each affected employee demonstrate an understanding of the required training and the ability to use PPE properly before being allowed to perform work requiring the use of PPE. If the employer has reason to believe that any affected employee who has already been trained does not have the required understanding, the employer must retrain such an employee. The employer must conduct retraining under circumstances that include, but are not limited to, situations in which:

- Changes in the workplace render previous training obsolete
- Changes in the types of PPE to be used render previous training obsolete
- Inadequacies in an affected employee's knowledge or use of assigned PPE indicate that the employee has not retained the requisite understanding or skill

In addition, the employer must verify that each affected employee has received and understood the required training through a written certification that

- Lists the name of each employee trained
- Indicates the date(s) of training
- Identifies the subject of the certification

For selection of PPE for protection against respiratory and electrical hazards, the employer should refer to Section 1910.134 and Section 1910.137, respectively.

Limitations and Shortcomings of Regulations

Regulations specifying the selection and use of PPE are often general in scope, limited to specific applications, or do not provide specific guidance. Only in a few areas are regulations specific in recommending particular types of PPE in terms of design, performance properties, and service life. The majority of OSHA and other governmental regulations simply specify the use of general types of PPE (for example, "use protective clothing") without indicating a particular configuration and required performance. The principle exceptions to generic requirements in regulations exist when national standards exist, such as in the case of

- Protective footwear (ANSI Z41.1)
- Protective face and eyewear (ANSI Z87.1)

- Protective headwear (ANSI Z89.1)
- Respirators (ANSI Z88.2 and 42 CFR Part 84)

There are no national standards for protective garments or gloves except for very specific applications (for example, emergency response). Therefore, employers and end users are faced with a variety of choices for PPE selection for meeting regulatory requirements and must decide on appropriate PPE design, performance, and service life through a risk assessment and determination of protection needs.

THE RISK ASSESSMENT-BASED APPROACH TO SELECTING PPE

OSHA 29 CFR 1910.132 requires that employers use a hazard assessment to determine the need for and then to select PPE. Appendix B to Subpart I of OSHA 29 CFR provides nonmandatory guidelines for conducting hazard assessment and for selecting PPE but is general in providing an overall approach.

Recommended Risk Assessment Approach

The risk assessment-based approach for selecting PPE in this text uses an approach for conducting a hazard assessment of the workplace; determining the risk of exposure and rank protection needs; evaluating available PPE designs, performance, and applications against protection needs; and specifying appropriate PPE. The specific steps of this process include the following:

Define Each Workplace and Tasks to Be Evaluated. The workplace should include the area that encompasses the range of hazards that may be encountered. Examples of a workplace can be the specific work locations for a particular employee, a laboratory, or a part of a production process. Work tasks should be defined as those worker activities that involve unique hazards and are accomplished by a single individual or group of individuals within a given period of time.

Identify the Hazards Associated with Each Work Task. General hazard categories include physical, environmental, chemical, biological, thermal, electrical, radiation, person-position, and person-equipment.

Determine Each Affected Body Area or Body System. For each hazard, determine which portion of the body can be affected by the hazard. General body areas and body systems typically affected by workplace hazards, include the following:

- Head
- Eyes and face
- Hands
- Arms
- Feet

- Legs
- Trunk or torso
- Entire body
- Respiratory system
- Hearing

Estimate the Likelihood of Employee Exposure to Identified Hazards. For every identified hazard affecting a specific portion of the body (or the whole body), indicate the likelihood of exposure. For this course, a proposed rating scale of 0 to 5 is based on both the risk and the frequency of exposure:

- 0: Exposure impossible
- 1: Exposure very unlikely
- 2: Exposure possible, but unlikely
- 3: Exposure likely
- 4: Multiple exposures likely
- 5: Continuous exposure likely

Estimate the Possible Consequences of Exposure to Identified Hazards. For every identified hazard affecting a specific portion of the body (or the whole body), indicate the consequences of exposure. For this course, a proposed rating scale of 0 to 5 is based on the "worst case" effects on the worker, if exposed:

- 0: No effect
- 1: Temporary effect on employee (such as discomfort) with no long-term consequences
- 2: Exposure resulting in temporary, treatable injury
- 3: Exposure resulting in serious injury with loss of work time
- 4: Exposure resulting in permanent debilitating injury
- 5: Exposure resulting in likely death

The principal source of information for conducting a hazard assessment comes from an inspection of the workplace with actual observation of specific tasks being carried out. Additional information can be obtained by interviewing the affected employee and asking about

- Types and frequency of hazards encountered in the task
- Specific instances in which hazards have been encountered in the past
- Past effectiveness or ineffectiveness of any PPE used in the task.

Another source of information is a review of the log and summary of all occupational illnesses and injuries at the workplace. This may be accomplished by examining OSHA No. 200 or equivalent forms. The specific hazard and nature of any accidents or exposures should be evaluated to determine the possible preventative role of using PPE, or improving PPE if involved. In some cases, it will be necessary to measure hazard levels using special instrumentation such as portable sampling devices, to measure airborne concentrations of chemicals, or noise-monitoring equipment. A completed hazard assessment will provide a list of hazards, which parts of the worker body may be affected, how likely exposure will occur, and what the probable consequences of exposure might be.

Determining Relative Risk and Establishing Protection Needs

The risk of exposure is determined by the hazard assessment. Protection needs are ranked by the relative risk. For determining relative risk, establish a risk assessment form for listing

- Hazards identified for the specific workplace/task
- Body areas or body systems affected by the respective hazard
- Rating associated with the likelihood or frequency of the respective hazard
- Rating associated with the severity or consequences of the respective hazard

The risk assessment form should be filled in using each row to indicate a separate hazard with the associated information and ratings. For each identified hazard, multiply the rating for the likelihood (or frequency) of the hazard by the rating for the severity (or consequences) of the hazard to determine the risk rating. Risk ratings of 0–6, 8–10, 12, 15, 16, 20, and 25 are possible. Using risk determinations from the risk assessment form, rank all hazards associated with the workplace/task. Those hazards with the highest amount of risk should be assigned higher priority for prevention or minimization. Those hazards with 0 risk or low risk should be assigned lower priority for prevention or minimization.

Examine possible engineering or administrative controls for those hazards with the highest risk. Engineering controls can encompass changes in the task or process or use of protective shields or other designed measures that eliminate or reduce possible exposure to hazards. Administrative controls can include changes in tasks or work practices to eliminate or limit employee exposure time to a hazard. If engineering or administrative controls are not possible, examine different types of PPE for elimination or reduction of exposure to hazards.

From information about affected body area or body systems, decide which type of PPE can be used to eliminate or minimize the hazard. In many cases, the type of PPE to be used will be obvious and limited to a single general type (for example, inhalation hazards can be eliminated by using a respirator). In other cases, there may be several types of PPE that can be used to provide the necessary protection.

EVALUATING PPE DESIGNS, DESIGN FEATURES, PERFORMANCE, AND APPLICATIONS

The evaluation of PPE designs, design features, performance, and applications encompasses understanding and choosing the types of PPE available for protection, as well as understanding and choosing relevant performance properties of PPE to consider during PPE selection.

The types of PPE available in the marketplace, and thus the choices available to the end user, are rapidly increasing. PPE exists in a variety of designs, materials, and methods of construction, each having advantages and disadvantages for specific protection applications. End users should have an understanding of the different types of PPE and their features in order to make appropriate selections. It is important to realize that similarly designed PPE may offer different levels of performance. Thus, PPE performance must be carefully scrutinized in addition to design and features. Furthermore, PPE must be properly sized to provide adequate protection. Improperly sized or ill-fitting PPE may reduce or eliminate the protective qualities of PPE.

PPE may be classified by

- Design
- Performance
- Intended service life

Classification of PPE by its design usually reflects how the item is configured or the part of the body area or body systems that it protects. For example, footwear by design provides protection to the wearer's feet.

The types of PPE can be generally categorized as follows:

- Full-body garments
- Partial-body garments
- Gloves
- Footwear
- Head protection
- Face and eye protection
- Hearing protection
- Respirators

Classification of PPE by design may also provide an indication of specific design features that differentiate PPE items of the same type. For example, closed-circuit self-contained breathing apparatus (SCBA) are configured with significant design differences when compared to open-circuit self-contained breathing apparatus. Some PPE designs may offer varying protection against hazards in different parts of the PPE item. For example, the material in the palm of a glove may provide a better grip surface than the material on the back of the glove. Nevertheless, it is important to realize that PPE coverage of a specific body area, in and of itself, does not guarantee protection of that body area.

Classification of PPE by performance indicates the actual level of performance to be provided by the PPE item. This may include a general area of performance or a more specific area of performance. For example, although two items of PPE might be considered to be chemical protective clothing, one item may provide an effective barrier to liquids but not to vapors whereas the other item may provide an effective barrier to both liquids and vapors. Classification of PPE by performance is best demonstrated by actual testing or evaluations of PPE with a standard test that relates to the type of desired protection. These tests can then be used as a demonstration of protection against the anticipated hazards and often become the basis of claims by the manufacturer for its products. However, intended or manufacturer-claimed performance does not always match actual performance.

Classification of PPE by expected service life is based on the useful life of the PPE item. PPE may be designed to be

- Reusable
- Used a limited number of times (limited use)
- Disposable after a single use

Classification of PPE by expected service life is based on its durability, life cycle cost, and ease of reservicing. Durability is determined by evaluating how the item of PPE maintains its original performance properties following the number of expected uses. The life cycle cost of PPE is the total cost for using an item of PPE and is usually represented as the cost per use for a PPE item. The following costs should be considered in determining the life cycle cost:

- Purchase
- Labor for selection/procurement of PPE
- Labor for inspecting PPE
- Labor and facility for storing PPE
- Labor and materials for cleaning, decontaminating, maintaining, and repairing PPE
- Labor and fees for retirement and disposal of used PPE

The total life cycle cost is determined by adding the separate costs involved in the PPE life cycle and dividing by the number of PPE items and number of uses per item. If an item of PPE cannot be reserviced to bring it to an acceptable level of performance, it cannot be reused.

PPE PERFORMANCE PROPERTIES

Performance properties describe specific characteristics of PPE or PPE materials or components. Performance properties are determined by evaluating or testing PPE. Performance testing may be conducted on specific whole items of PPE, or PPE materials or components. Categories of performance properties covered in this chapter are as follows:

- Physical properties
- Overall product integrity
- Environmental resistance
- Chemical resistance
- Biological resistance
- Heat and flame resistance

- Electrical properties
- Human factors
- Respirator properties

Some performance properties apply to several types of PPE whereas others may be applied to only one PPE category.

Physical Properties

Physical properties assess or determine the weight and thickness of PPE and materials, strength of PPE, resistance to specific physical hazards, and product durability. Physical performance properties include the following:

- Weight—measured for whole PPE items, materials, or components as an indication of burden on the wearer.
- Thickness and hardness—measured as an indication of bulk, or the general hardness of a PPE material surface.
- Breaking strength (or tensile) strength tests—measure the force required to break PPE materials or components when items are pulled along one direction; also used to measure the strength of seams and closures.
- Burst strength—measures the force or pressure required to rupture PPE materials or components when a force is directed perpendicular to the item; may be related to ability of PPE materials to prevent items from protruding through garments.
- Impact and compression resistance—evaluates the ability of PPE to resist the forces of impact from a falling heavy object, which either deforms, compresses, or fractures parts of the PPE.
- Projectile and ballistic resistance—measures the ability of PPE items, materials, or components to deflect or prevent the damage to PPE or penetration of specific types of projectiles through PPE shot at relatively high velocities.
- Tear or snag resistance—measures the force required to continue a tear in a PPE material once initiated or the resistance of a material in preventing a tear or snag from occurring; for snagging, evaluates when an individual yarn of a textile is pulled away from the material.
- Abrasion or scratch resistance—measures the ability of PPE surfaces or materials to resist wearing away or being damaged when rubbed against other surfaces.

- Cut resistance—measures the ability of PPE items or materials to resist cutting through by a sharp-edged object or machinery.

- Puncture resistance - measures the ability of PPE items or materials to resist penetration by a slow-moving, pointed object.

- Flex fatigue resistance—measures the ability of PPE items or materials to resist wear or other damage when repeatedly flexed.

- Slip resistance—measures the ability of PPE items to maintain grip or traction with a surface

- Shock absorption—measures the ability of PPE items to provide cushioning against vibration or repeated light impact.

Overall Product Integrity

Overall product integrity performance testing provides a determination of how well PPE prevents substances from entering (or leaving) the PPE through the material, seams, closures, or any other parts of the PPE that are evaluated. Overall integrity tests are usually used in conjunction with barrier tests for demonstrating resistance to environmental, chemical, or biological penetration hazards. The following are three types of integrity testing include:

- Particulate-tight integrity—determines whether particles enter or leave whole items of PPE.

- Liquid-tight integrity—determines whether liquid enters the interior side of the PPE or onto wearer underclothing.

- Gas-tight integrity—determines whether gas can penetrate PPE.

Environmental Resistance

Environmental resistance testing evaluates PPE for conditions arriving from the extremes of weather, light, or noise in occupational settings. Performance properties include the following:

- Water repellency and absorption resistance—measures the ability of PPE to resist surface wetting when contacted with water from spray or impingement, or ability of PPE to gain weight when contacted with water from spray or impingement.

- Water penetration resistance—measures the ability of PPE materials to resist the flow of liquid water onto the interior surface of

PPE material surfaces when contacted with water from spray or impingement.

- Salt spray and corrosion resistance—evaluates the ability of PPE items to resist corrosion when exposed to salt spray.

- Resistance to cold temperatures—evaluates changes in PPE, materials, and components when exposed to cold temperatures.

- Ultraviolet (UV) light resistance—evaluates changes in PPE, materials, and components when exposed to UV or visible light.

- Ozone resistance—evaluates changes in PPE, materials, and components when exposed to elevated concentrations of ozone.

- Excess light attenuation—measures the effectiveness of lenses or filters used in PPE for reducing the radiation produced by harmful light to acceptable levels.

- Excess noise attenuation—measured to determine the effectiveness of hearing protectors in preventing unsafe levels of sound from reaching the worker's ear level (provided in decibels) .

- Visibility—evaluated to determine how well the PPE item can be seen as enhanced by colors or special materials during either daytime or nighttime conditions.

Chemical Resistance

Chemical resistance performance properties include the following:

- Chemical degradation resistance—measures the deleterious changes in one or more physical properties of protective clothing material as the result of chemical exposure.

- Particulate penetration resistance—measures the flow of chemical particles through closures, porous materials, seams, and pinholes or other openings in protective clothing.

- Liquid penetration resistance—measures the flow of a liquid chemical on a nonmolecular level through closures, porous materials, seams, and pinholes or other openings in protective clothing.

- Vapor penetration resistance—measures the flow of a chemical vapor on a nonmolecular level through closures, porous materials, seams, and pinholes or other openings in protective clothing.

- Chemical permeation resistance—measures the rate at which chemicals move through a protective clothing material on a molecular level.

Biological Resistance

Biological resistance properties include the following:

- Biological resistance properties include microorganism filtration efficiency—evaluates the ability of PPE items and fabrics to prevent the passage of airborne microorganisms.
- Biological fluid resistance—discriminates barrier characteristics of different fabrics used in apparel for preventing blood or other body fluid strike-through (fabrics with fluid resistance may still allow fluid penetration under some use conditions).
- Biological fluid penetration resistance—evaluates the ability of fabrics to prevent penetration of biological fluids into the PPE (biological fluid penetration resistance testing differs from biological fluid resistance testing in that it provides a "proof" type determination).
- Viral penetration resistance—evaluates the ability of materials to prevent the passage of virus or related microorganisms.
- Antimicrobial performance—evaluates PPE and material resistance to microbial growth (such as fungus and mildew) and inactivation of microorganisms when contacted by the PPE or material.
- Insect resistance—evaluates the effectiveness of material treatments for repelling insects.

Heat and Flame Resistance

Heat and flame resistance test methods evaluate PPE, materials, and components for the effects of either heat or flame exposure, or the relative protection offered when exposed to either heat or flame. Heat and flame resistance performance properties include the following:

- Convective heat resistance—measures the effects of convective heat on different PPE items, materials, or components.
- Blocking resistance—refers to prevention of material adhering to itself under high heat conditions.
- Conductive heat resistance—measures the effects of conductive heat on PPE items or material and the amount of insulation provided by these items or materials when in contact with hot surfaces.
- Radiant heat resistance—measures the effects of radiant heat on PPE items or material and the amount of insulation provided by these items or materials when exposed to different levels of radiant heat.

- Thermal protective performance—measures the insulative and barrier properties of PPE materials when exposed to heat (usually convective or convective and radiant heat).
- Flame resistance—measures the effects of flame contact on an item of PPE or PPE.
- Molten metal contact resistance—evaluates the effects of and heat transfer through PPE contacted by molten metal.

Electrical Properties

Electrical performance properties are as follows:

- Electrical insulative performance—evaluates PPE for their ability to reduce contact with electrically energized parts.
- Conductivity—evaluates PPE or material for protection against hazards of static charge buildup and for the equalization of electrical potential of personnel and energized high-voltage sources.
- Static charge accumulation resistance—evaluates the static charge generated and the rate of its discharge.
- Electrical arc protective performance—measures the heat transfer from exposure to high-energy electrical arcs.

Human Factors

Human factors describe how PPE affects the wearer in terms of functionality, fit, comfort, and overall well being. Most human factor properties represent tradeoffs with protection, for example, barriers to chemicals versus thermal comfort. The following are human factors:

- Material biocompatibility—evaluates the potential for skin irritation or adverse reactions due to contact with certain substances that may be present in or on PPE.
- Thermal insulation and breathability tests—evaluate the ability of protective clothing materials to allow the passage of air, moisture, and the heat associated with body evaporative cooling and environmental conditions.
- Tests for PPE mobility and range of motion—evaluate the effects of PPE on wearer function in performing work tasks.
- Hand function tests—assess the effects of handwear on specific functions of the hand used to perform tasks or manipulations, such as dexterity, tactility, and grip.

- Ankle support testing—evaluates the ability of footwear to maintain support for the ankle under conditions of use.

- Back support testing—evaluates the ability of garment PPE and accessories to provide support for the lower back during lifting and other strenuous activity under conditions of use.

- Clarity testing—evaluates the ability of an individual to see through a visor or a faceshield (field of vision testing evaluates peripheral vision for an individual wearing the visor or faceshield).

- Communications tests—evaluate the ability of PPE to allow intelligible (understood) communications of the wearer to other persons.

- Sizing—determines how well PPE fits the individual wearer.

- Donning and doffing tests—evaluate how easily or how quickly individuals can don or doff PPE.

Overview of Durability and Serviceability

Durability refers to how PPE maintains its performance with use. Serviceability refers to the user's ability to care for, maintain, and repair PPE so that it remains functional for further use. Conventional approaches for measuring durability include measurement of PPE performance following different conditioning techniques intended to simulate PPE use, or expected effects on PPE.

Respirator Testing Standards

Evaluation of respirator design and testing of respirator performance properties is performed in accordance with Title 42 Code of Federal Regulations Part 84, *Approval of Respiratory Protective Devices* (42 CFR Part 84). All approvals and testing are carried out by the National Institute for Occupational Safety and Health (NIOSH) in accordance with 42 CFR Part 84. Specific tests for open-circuit, SCBA for the fire service are not covered by 42 CFR Part 84, but are included in NFPA 1981, Standard on *Open-Circuit, Self-Contained Breathing Apparatus*. Specific tests for supplied-air suits are not covered in 42 CFR Part 84 but have been established by the Department of Energy, and the certification program for these respirator/protective clothing products is administered by Los Alamos National Laboratory.

PPE DESIGN, FEATURES, AND SELECTION FACTORS

Protective Garments

Full-body garments are designed to provide protection to the wearer's upper and lower torso, arms, and legs. Full-body garments may also provide protection to the wearer's hands, feet, and head when auxiliary PPE is integrated with the garment to form a suit. Partial-body garments provide protection to only a limited area of the wearer's body, including the upper and lower torso, arms, legs, neck, and head. The level of protection varies with garment design; many designs do not provide uniform protection for all areas of the body covered by the garment.

Full-body garments may be single or multipiece clothing items:

- Full-body suits
- Jacket and trouser combinations
- Jacket and overall combinations
- Coveralls

Partial-body protective garments vary significantly with the application. Examples of partial body protective garments are as follows:

- Hoods, head covers, and bouffants for head/or face protection
- Aprons, gowns, smocks, lab/shop coat, and vests for front or upper torso protection
- Sleeve protectors for arm protection
- Chaps, leggings, or spats for leg protection

Features affecting the design of garments include the following:

- Type and location of seams
- Type, length, and location of the closure system(s)
- Amount of overlap for multipiece garments
- Design of interface areas with other PPE
- Type, size, and location of pockets
- Types, function, and location of hardware

Full-body garments are constructed of materials appropriate for the specific application; typical materials include various types of textile materials, leather, unsupported rubber or rubber materials, rubber- or plastic-coated fabrics, or aluminized fabrics. Full-body garments may be constructed of several layers to provide insulation or other properties related to comfort. Linings may enhance comfort by wicking away body perspiration. Barrier materials may prevent the penetration of liquids. And, batting materials may provide insulation from severe heat or cold.

There are few uniform sizing practices for the design of full-body garments. The availability of sizing often depends on the design of the garment and the relative volume of garments sold by the manufacturer. Sizing may be based on individual measurements for custom sizing, numerical sizing for wearing apparel, or alphabetic sizing (for example, small, medium, and large). Sizing systems for garments usually use two or more wearer dimensions, such as height and weight, or height and chest circumference. Unfortunately, sizing between manufacturers is often inconsistent. In addition, the sizing of protective garments often does not address the needs of special worker populations.

OSHA 29 CFR Subpart I offers no specific guidance for selection of protective garments and does not cite specific standards for compliance of protective garments. The basic approach for selecting garments encompasses the following:

1. Decide whether full-body or partial-body protection is needed.
   - Full-body protection is warranted when the nature of the hazards or use environment are severe or cannot be totally anticipated.
   - Partial-body protection is warranted when the nature of the hazards or use environment are limited to a specific area of the body (torso, arms, or legs) and the means of exposure can be anticipated.

   The selected garment should provide protection to those areas of the body that can be or are likely to be exposed. The garment design should provide uniform minimum protection for all areas covered by the body, or should be designed such that those areas of the garment offering protection are clear and distinct from those areas that do not offer protection, or offer relatively limited protection.

2. Before choosing design features and material systems, determine the intended service life of the PPE since this decision will affect both the garment design and the materials.

3. Choose design features as required for the application and intended performance, considering the following:

- Seams
- Closures
- Interface areas
- Accessories

4. Decide on the overall integrity that the garment should have, if any, in preventing intrusion of
   - Particulates
   - Liquids
   - Vapors

5. Choose the material or material system, considering the different types of performance that are required:
   - Use the hazard category or categories that apply as based on the hazard assessment; match performance tests and criteria with these hazards.
   - Establish protection priorities because, in some cases, all types of protection may not be available in a single product.
   - When stronger materials are desired, choose materials that show high tensile strength, bursting resistance, or tear resistance.

6. Consider other use factors:
   - Are garments being used in areas where wearer daytime or night-time visibility is important? If so, consider using high-visibility colors or materials on garments.
   - Is work being conducted next to water? If so, ensure that the worker also has a personal floatation device.
   - Is work being conducted on an elevated platform? If so, ensure that the worker also has adequate fall protection.

7. Consider potential hazards from selected garments (for example, do garment materials irritate or sensitize the wearer's skin? or do garments significantly limit wearer mobility or range of motion?)

Gloves and Other Handwear

Handwear provides protection to the wearer's hands and wrists or portions of the wearer's hands and wrists. Some handwear may have gauntlets to protect the lower arm. Hand protection may be provided to the wearer in several forms:

- Five-fingered gloves
- Two-fingered gloves
- Mittens
- Partial gloves
- Fingerless gloves
- Finger guards
- Finger cots
- Hand pads

Gloves and other handwear are constructed of materials appropriate for the specific application. Typical materials include leather, cotton and other natural textile knit or woven materials, synthetic fiber knit or woven materials, aluminized fabrics, rubber-coated or impregnated fabrics, rubber, plastic, and metal mesh. Gloves fabricated from materials using rubber-coated fabrics are known as supported gloves; rubber gloves without a supporting fabric are known as unsupported gloves.

Principal glove design features include the type of glove construction, cuff designs, and grip designs. Gloves and other handwear may be constructed in several layers to provide insulation or other properties related to comfort. Linings may enhance comfort by wicking away hand perspiration. Barrier materials may prevent the penetration of liquids. And, batting materials may provide insulation from severe heat or cold.

Glove sizing is based on either numerical hand sizes or qualitative size ratings, such as small, which may be based on hand size. Glove length varies from gloves that end at the wearer wrist crease to those that extend to the upper arm of the wearer. The availability for glove sizes depends on the glove style and relative sales volume.

OSHA 29 CFR 1910.138 requires selection of handwear that provides appropriate protection against specific hazards but does not specify compliance with a particular standard or set of standards. The basic approach for selecting gloves and handwear encompasses the following:

1. Select the handwear design.
    - When full hand and wrist protection and full hand function are needed, select full, five-fingered gloves.
    - When protective performance requirements dictate high levels of protection (particularly for thermal and ambient cold protection) and some hand function can be sacrificed, select two-fingered gloves or mittens.
    - When hazards affect only the palm or portions of the hand, select partial gloves, fingerless gloves, or finger guards.
    - When short-term finger end protection is needed, select finger cots.
    - When temporary hand protection is needed and hand function is not needed, hand pads may be selected in hot work that is well characterized.

2. Choose handwear design features, materials, and service life. The handwear design will usually be a function of the material and the

intended application (or industry). Choose design features and material or material system, considering the different types of performance that are required:

- Use the hazard category or categories that apply as based on the hazard assessment.
- Establish protection priorities because, in some cases, all types of protection may not be available in a single product.

Common types of gloves and handwear include:

- Disposable two-dimensional plastic gloves
- Disposable lightweight rubber gloves and handwear
- Unsupported rubber gloves
- Fabric gloves and handwear
- Knit fabric gloves and handwear
- Leather gloves and handwear
- Rubber- or plastic-coated fabric gloves

Specialized types of gloves and handwear include

- Insulated gloves and handwear
- Multilayer gloves and handwear

Many types of gloves are used only for certain applications; for example, two-dimensional, unsupported rubber, and coated work gloves are generally used for chemical resistance. Disposable gloves are generally used in applications involving various forms of contamination (for example, protection against bloodborne pathogens).

3. Consider potential hazards from selected handwear.

Footwear

Footwear provides protection to the wearer's feet or portions of the wearer's feet. Depending on the footwear height, additional protection may be afforded to the wearer's ankles and lower and upper legs. Foot protection may be provided to the wearer in several forms:

- Shoes
- Boots
- Overshoes or overboots
- Shoe or boot covers
- Metatarsal footwear
- Toe protectors or caps

Primary design features for footwear include the type of sole, type of closure (if present), linings, interior supports (metatarsal or arch supports, insoles), and protective hardware (toe caps, metatarsal plates, puncture-resistant devices or midsole plates, ladder shank). The majority of footwear is constructed primarily of leather. Other footwear materials include rubber- or plastic-coated fabric, unsupported rubber (for overshoes and overboots), and textiles (for shoe or boot covers). Leather or textile footwear is usually constructed on a last or foot form for creating the desired footwear item size. Rubber or plastic footwear is typically fabricated by injection molding in one or more stages. Boot linings may consist of woven or nonwoven textiles, coated barriers, or foam materials.

The majority of protective footwear uses the same footwear sizing system as utilized for standard footwear; this system is based on the Brannock measuring scale, for which foot length and width are the two key dimensions used for choosing footwear, and requires the individual to measure his or her foot and then to select the corresponding labeled size (for example, 7D or 9EE). It also involves the designation of sizes by the manufacturer for their products so that appropriate fit with the corresponding dimensions is provided.

Selection of many footwear types is governed by ANSI Z41-1991, *American National Standard for Personal Protection—Protective Footwear.* OSHA 29 CFR 1910.136 requires selection of footwear that complies with the relevant edition of ANSI Z41. The basic approach for selecting footwear encompasses the following:

1. Select the footwear design.
    - Select protective shoes when the primary concern is physical hazards to the foot from impact.
    - Select protective boots for most applications in which several different hazards exist for the wearer's feet.
    - Select overshoes or overboots for additional barrier protection; overshoes or overboots may also be used to supplement physical protection for the wearer's feet.
    - Select shoe or boot covers only when a disposable shield for preventing contamination of the wearer's primary footwear is needed.
    - Select toe caps only as a temporary measure for providing impact resistance to normal wearer footwear.
2. Select footwear materials, design features, and service life. Footwear design will usually be a function of the material and the intended application (or industry).

- Choose footwear height.
- Choose footwear sole.
- For leather footwear and some rubber footwear, choose closure system.
- Choose lining system depending on the application.

3. Decide on needed footwear performance. ANSI Z41-1991 defines five types of protective footwear:

- Metatarsal
- Conductive
- Electrical hazard
- Sole puncture-resistant
- Static dissipative

Requirements for other areas of performance must be made by matching performance properties with hazards identified in the risk assessment.

4. Consider potential hazards from selected footwear.

Headwear

Protective headwear provides protection to the wearer's head or portions of the wearer's head. Headwear may or may not provide protection to the wearer's face. Industrial headwear is typically designed for physical and electrical protection. ANSI Z89.1-1997, *Protective Headwear for Industrial Workers—Requirements,* defines two impact types of helmets:

- **Type I**—helmets intended to reduce the force of impact resulting from a blow only to the top of the head
- **Type II**—helmets intended to reduce the force of impact resulting from a blow which may be received off center or to the top of the head

ANSI Z89.1-1997 defines three electrical classes of helmets:

- **Class G (General)**—helmets that are intended to reduce the danger of contact exposure to low-voltage conductors
- **Class E (Electrical)**—helmets that are intended to reduce the danger of exposure to high-voltage conductors
- **Class C (Conductive)**—helmets that are not intended to provide protection against contact with electrical conductors

The basic form of head protection for industry is the helmet. Helmets consist of a shell, suspension (absorbs energy within the shell), and harness (secures the helmet to the wearer). Helmets may also have various accessories, including brackets for lights, faceshield or brackets for other types of face and eye protective wear, and high visibility materials. Helmets generally come in a single-sized shell; headband adjustments provide individual fit.

Helmet shells are constructed of hard plastics and composites, typically including nylon, high-density polyethylene, fiberglass, vulcanized fiber, polycarbonate, or aluminum. Helmet visors may be polycarbonate, nylon, Lexan, or steel mesh. Materials may include gold coating for protection against radiant heat. Most helmet components in the suspension are nylon, vinyl, or sponge foam for cushioning elements. Depending on the intended application of the helmet, some materials used in helmets may be heat and flame resistant.

Selection of many headwear types is governed by ANSI Z89.1-1997, *American National Standard for Industrial Head Protection.* OSHA 29 CFR 1910.135 requires selection of headwear that complies with the earlier editions of ANSI Z89. The basic approach for selecting headwear encompasses the following:

1. Select the headwear type; headwear types include

    • Bump caps
    • Helmets for impact and electrical protection
    • Helmets for specialized protection

    Head covers and hoods are considered partial-body garments.

2. Select the headwear design features, materials, and performance. Many design features and materials will depend on the intended performance for the headwear. Selection of the type of headwear for performance should be governed by ANSI Z89.1-1997 for the two impact types and three electrical classes of protective headwear. For protecting the head from environmental, chemical, biological, thermal, and radiation hazards, select hoods that conform to the same guidelines as those provided for garments (for example, for protecting the head from thermal hazards, choose helmets that are constructed of high heat- and flame-resistance materials).

3. Consider potential hazards from selected headwear.

Protective Face and Eyewear

Protective face and eyewear provides protection to the wearer's face and eyes or portions of the wearer's face and eyes. ANSI Z87.1, *Practice for Occupational and Educational Eye and Face Protection,* defines a number of general types of eye and face protectors, including:

- Spectacles (safety glasses)
- Faceshields
- Goggles
- Welding helmets
- Hand shields

Safety glasses, also known as spectacles, are protective devices intended to shield the wearer's eyes from certain hazards depending on the type of spectacles. Safety glasses are commonly used to provide protection from impact and optical radiation. They consist of a frame, type of front lens or lenses (Plano, prescription, or shaded), temples (which secure the device on the head in front of the eyes), bridges (which secure the device on the nose of the wearer), and sideshields (which provide side protection to the eyes). Faceshields are protective devices intended to shield the wearer's face or portions of the face in addition to the eyes. ANSI Z87.1 requires that faceshields be used in conjunction with spectacles or goggles (faceshields are not primary protectors). Faceshields consist of headgear that supports a window curved to surround and cover the wearer's face. Goggles are face protection devices that are intended to fit the face surrounding the eyes in order to shield the eyes from certain hazards. Features of goggles include the style (eyecup or cover), frame type (rigid or flexible), frame material, type of ventilation, lens type, and type of band. Welding helmets and handshields are a special type of eye/face protection usually intended to protect the entire face, eyes, ears, and front of the neck from optical radiation and weld splatter. Welding helmets include some form of headgear with sizing and fit constraints similar to those for faceshields. Handshields are simply held in front of the face by the wearer during use and have few sizing constraints except that these devices must be large enough to cover the faces of the general worker population.

Selection of most face and eyewear types is governed by ANSI Z87.1-1989, *American National Standard Practice for Occupational and Educational Eye and Face Protection.* OSHA 29 CFR 1910.133 requires selection of face and eyewear that complies with the relevant edition of ANSI Z87.1. ANSI Z87.1 provides a detailed selection chart for selecting face and eyewear based on specific categories of hazards. In addition, other choices can be made in terms of face and eyewear design features and materials based on the organization's needs and user preferences. As with other PPE, the potential from hazard in using selected face and eyewear must be considered.

Respirators

Respirators protect the wearer from inhalation of harmful dusts, chemicals, and other respirable substances. Respirators provide protection to the wearer by 1) removing contaminants from the air (air purifying) or 2) supplying an independent source of respirable air (atmosphere supplying). NIOSH defines respiratory designs and requirements in 42 CFR Part 84, Respiratory Protective Devices, Tests for Permissibility. Respirators are qualified by their purpose for either entry and escape, or for escape only. Respirators are further differentiated by the types of environments in which they can be used. For example, some respirators are not used in oxygen-deficient atmospheres (atmospheres containing less than 19.5 percent oxygen). Other respirators are not for immediately dangerous to life and health (IDLH) atmospheres (hazardous atmospheres that may produce physical discomfort immediately, chronic poisoning after exposure, or acute physiological symptoms after prolonged exposure). Respirators that rely on finite air supplies or filtering capabilities are also classified by their service time, ranging from three minutes to four hours as defined in 42 CFR Part 84 (approval of respiratory protective devices).

All respirators are equipped with respiratory inlet covers or facepieces to provide a barrier from the hazardous atmosphere and for "connecting" the wearer's respiratory system with the respirator. There are two types of respiratory inlet covers: tight fitting (facepieces), such as quarter masks, half masks, and full facepieces, or loose fitting (helmets, hoods, blouses, or suits).

General respirator designs include the following:

- Air-purifying respirators (APRs)
    - Disposable respirators (Many disposable respirators do not provide an adequate seal on the user's face to prevent inward penetration of atmospheric contaminants and may not easily be evaluated by fit testing)

— Particulate filter respirators
— Cartridge or canister respirators (gas mask)
— Cartridge or canister respirators (gas mask) with particulate filter

• Powered air-purifying respirators (PAPRs)
• Supplied-air respirators (SARs)
— Demand SARs
— Continuous flow SARs
— Pressure-demand SARs

• Combination supplied-air/APRs
— Continuous SAR/APR
— Pressure-demand SAR/APR

• Self-contained breathing apparatus (SCBA)
— Demand SCBA
— Continuous flow SCBA
— Pressure-demand SCBA

• Combination SARs with auxiliary self-contained air supply (SCBA/ SAR)
— Demand SCBA/SAR
— Continuous SCBA/SAR
— Pressure-demand SCBA/SAR

Respirators may be either negative-pressure or positive-pressure respirators. All nonpowered APRs are negative-pressure respirators. Other negative-pressure respirators include demand SARs, demand SCBA, and combination continuous or pressure-demand SAR/APR. Positive-pressure respirators include PAPRs, continuous flow SARs, pressure-demand SARs, continuous flow SCBA, pressure-demand SCBA, and combination pressure-demand SAR with auxiliary SCBA.

In the United States, respirators must be certified by NIOSH to the respective requirements in 42 CFR Part 84. OSHA 29 CFR 1910.134 (January 8, 1998) specifies selection of general respirator types. These regulations update previous selection practices specified by the regulation from ANSI Z88.2-1992, *American National Standard for Respiratory Protection.* OSHA 29 CFR Part 1910 provides for specific selection of respirators for protection against several substances. NIOSH also has "recommended practices" documents for several chemicals, which provide specific respirator selection guidelines. In addition, NIOSH has developed specific guidelines for respirator selection contained in *The Guide to Industrial Respiratory Protection* (DHHS/NIOSH Publication No. 87-116, 1987). Figure 21-2 provides a decision logic for selecting respirators that is based on the requirements in OSHA 29 CFR 1910.134 (January 8, 1998).

Information needed to conduct the specific respiratory hazard assessment includes the following:

- Identification of atmospheric contaminant(s).
- Measurement of concentration for specific contaminant(s).
- Determination of IDLH concentrations for contaminant(s); IDLH means an atmosphere that poses an immediate threat to life, would cause irreversible adverse health effects, or would impair an individual's ability to escape from a dangerous atmosphere.
- Measurement of oxygen concentration in atmosphere; oxygen deficiency exists when an atmosphere has an oxygen content below 19.5 percent by volume.
- Determination of whether respirator use is for work or escape.
- Determination of chemical and physical state of contaminant(s):
  — Gas or vapor
  — Particulates (aerosols, mists, fumes)

The general respirator risk assessment provides specific types of respirators for IDLH environments and only general respirator types for selection in non-IDLH environments. A more detailed analysis is required to allow decisions between different specific respirator types and features. This analysis consists of determining specific exposure limits and characteristics of the contaminants, evaluating workplace factors that affect respirator selection, and reviewing respirator features related to protection.

Specific respirators must be selected for special environments. For fire fighting, a full facepiece pressure-demand SCBA that meets the requirements of National Fire Protection Association (NFPA) 1981 standard must be selected. For chemical emergency response or hazardous waste site cleanup requiring Level A or B protection, either a full facepiece pressure-demand SCBA or a combination full facepiece pressure-demand SAR with auxiliary self-contained air supply (SCBA/SAR) must be selected. For air-line or air-supplied suits (without internal respiratory inlet covering), suits must be selected that have been approved by the requirements specified by the U.S. Department of Energy. Operations involving abrasive blasting require respirators approved for abrasive blasting such as powered APRs and Type AE, BE, or CE SARs. Last, protection against biological airborne pathogens (such as *Mycobacterium tuberculosis*) should be achieved by choosing particulate filter facepiece APRs equipped with high-efficiency particulate air (HEPA) filters. These respirators must be worn when employees enter rooms housing individuals with suspected or confirmed infectious tuberculosis (TB) diseases, employees perform high hazard procedures on individuals who have suspected or confirmed TB diseases, or when emergency medical response personnel or others must transport, in a closed vehicle, an individual with suspected or confirmed TB diseases.

Hearing Protectors

Hearing protectors are used to reduce occupational noise levels to those below OSHA permissible exposure limits or acceptable levels. There are three general types of ear protectors:

- Ear plugs that fit directly into the ear.
- Ear canal caps that cover the external part of the ear canal opening.
- Ear muffs that fit over the ear.

A new class of hearing protectors includes active hearing-protection devices.

Ear plugs are available in three types: formable, custom molded, and premolded. Ear canal caps are a relatively new type of ear protection device. Ear canal caps consist of ear plug-like caps (which are larger since they do not fit into the ear canal) and a headband that goes over or behind the head. Ear muffs consist of two cup or dome-shaped devices that fit over the entire external ear, including the lobe. The muffs seal against the head with a suitable cushion or pad. Generally, the cups are made of molded, rigid plastic and are then lined with an open cell foam material. The shape, size, and degree of noise attenuation of earmuffs vary from one manufacturer to another. Active hearing-protection devices are used to overcome communications difficulties when hearing protection is required. Active hearing-protection devices resemble ear muffs but have exterior mounted microphones with electronic amplifiers for picking up ambient sounds and delivering the sounds to the wearer by earphones in the over-the-ear muff cups. The amplifiers in active hearing-protection devices can be designed with electronic processors to filter or modify selectively the level of sound within the frequency distribution reaching the hearing mechanism.

OSHA 29 CFR 1910.95 establishes requirements for occupational noise exposure. Employers must provide hearing protectors to employees who are exposed to noise exposures at or above the action level of an 8-hour time-weighted average (TWA) of 85 dB—at no cost to the employees. Employers must ensure that hearing protectors are worn by all employees who are exposed to noise levels that exceed the levels established in 29 CFR 1910.95, or are exposed to noise at the action level or greater and either have not had a baseline audiogram established or have experienced a standard threshold shift (a change in the hearing threshold of 10 decibels or more at 2000, 3000, or 4000 Hz in either ear). Employees must provide a variety of suitable hearing protectors so that each employee can choose one that fits his or her needs. Employers also must provide training in the use and care of all hearing protectors provided to employees. In addition, employers must ensure proper initial fitting and supervise the correct use of all hearing protectors.

OSHA 29 CFR 1910.95 requires employers to evaluate the hearing protector attenuation of the specific noise environments in which the protector will be used by specific methods. Selected hearing protectors must attenuate employee exposures at least to an eight-hour TWA of 90 dB. For employees who have experienced a standard threshold shift, hearing protectors must attenuate employee exposures at least to an eight-hour TWA of 85 dB. The adequacy of hearing protector attenuation must be reevaluated whenever employee noise exposure increases to the extent that the hearing protectors provided may no longer provide adequate attenuation; employers must then provide more effective hearing protectors when necessary.

Product Labeling and Information

An important part of PPE selection is ensuring that appropriate labels and product information are provided. PPE is labeled to provide different kinds of information to the user about the specific product. The purposes of PPE product labels are to provide identification of the product, describe the uses of the product, warn about product limitations, provide care information and recommendations, and indicate certification to specific standards. Labels are also important in providing product traceability, particularly in the case of a product recall or identifying potential defects.

Instructions for the PPE should address all aspects of the product's life cycle and should be written so that the anticipated user groups can easily understand the use of the product. Instructions may contain several warning or caution statements that point out areas of concern for using, caring for, or maintaining the product. Many instructions may be supplemented by diagrams, photographs, or videotapes to provide this information in an easier-to-understand format.

DETAILED PPE SELECTION

Selection of appropriate PPE should be undertaken using a comprehensive, multistep approach.

Step 1: Conduct the Risk Assessment

From the risk assessment described previously, the following information will be provided:

- Type of hazards present
- Body areas or body system that will be affected
- Level of risk associated with each hazard

First, PPE must be selected based on a risk assessment that is specific to the task, job, or application. Second, PPE must be fit to the task, job, or application; the task, job, or application should not be fit to the available PPE unless the necessary PPE is not offered by the PPE industry.

Step 2: Use Information from the Risk Assessment to Make Selection

The PPE selection process uses input from the risk assessment by answering the following questions:

- What types of hazards are present?
- Which body areas and body systems will be affected by those hazards?
- What is the risk of exposure for each hazard and body area/system?

The risk of exposure for each hazard and body area/system is used to determine whether protection is needed and, if so, how much protection is needed. Based on its risk assessment of a task, job, or application, each organization should determine the level of risk or "action level" at which PPE is warranted. The action level can be determined by using the numerical ratings from the risk assessment and assigning a maximum acceptable risk level, above which PPE should be used, or by making individual subjective decisions as to which risks are acceptable with or without PPE (these decisions must be made by the authority who is responsible for PPE decisions within the organization). At risk levels higher than the action level, PPE that offers greater protection as demonstrated by PPE design and performance properties must be selected.

By virtue of the affected body areas and body systems, the type of PPE can be chosen. For example, partial-body protective clothing or headwear can be chosen to protect the wearer's head. In some cases, certain types of PPE may protect multiple parts of the body, such as a full facepiece respirator that provides respiratory system protection concurrent with eye and face protection.

The types of hazards present indicate which performance properties are relevant for qualifying PPE items or their materials and components as providing protection against those hazards. Performance properties may apply to the overall PPE item, certain features or components of the item, or materials used in the construction of the item.

Step 3: Choose the PPE Design, Design Features, and Minimum Performance

The selection of appropriate PPE encompasses specifying

- The specific PPE design or type
- Design features associated with the specific PPE design
- Minimum performance for the specific PPE item

The combination of specifying PPE designs, design features, and minimum performance results in the development of minimum design and performance criteria. In selecting PPE, particular designs can be specified or performance criteria can be used. Performance criteria are preferred over design criteria because different manufacturer designs may offer the same level of performance (performance criteria permit a wide range of designs to be considered). Design criteria are justified when performance tests do not capture the specific properties desired for the PPE, or when the organization has experience with specific designs that are compatible with other equipment or use processes that warrant specifying a particular design or product. Usually, it makes the best sense to specify some design criteria in combination with performance criteria. The design criteria specify the type of design that the organization has experience with and wants to use, whereas the performance criteria specify the level of protection that the product must be capable of providing.

For each general type of PPE, several potential designs exist that provide a range of coverage and performance. These designs or styles are generally intended for specific types of hazards. The level of performance for different designs of the PPE item can vary significantly. Most PPE is marketed for a specific application or a limited number of applications and may not account for all hazards for a specific situation. Each industry has its owns practices related to the selection and use of PPE; these practices are usually not consistent among various industries and applications.

The specific minimum design and performance criteria must be selected based on the severity of the hazards and the level of risk established in the risk assessment. The challenge in specifying PPE is setting minimum levels of performance for acceptable protection. In some cases, some performance area criteria will be in conflict (for example, increased thermal insulation from heat is likely to decrease thermal comfort, which can result in heat stress). Some performance properties do not have easily identifiable end points for judging acceptable levels of protection.

Step 4: Prepare the PPE Specification

Specifications may be developed in three different ways:

1) Using an established standard or specification
2) Modifying or supplementing an existing standard or specification
3) Creating a new specification

Established specifications come from several consensus standards from ASTM, NFPA, CEN, ISO and other organizations, trade organizations or labor unions (as model specifications), PPE manufacturers, and other end-user organizations. Standard specifications are of two types: 1) minimum criteria, or 2) classification. Minimum criteria-based standards set specific design features in terms of the PPE configuration, design tolerances (dimensions), and PPE sizing; specify certain test methods for each performance property; and set a minimum or maximum requirement based on test results. Qualification of PPE to minimum criteria-based standards is either "pass" or "fail." Failure to meet any one requirement may disqualify a product from meeting the standard; however, products may exceed minimum requirements. By contrast, classification standards use PPE design features to type-classify PPE types and specify certain test methods but classify product performance by different levels according to the test results. These types of specifications may designate the minimum acceptable level of the PPE to be that with the lowest defined level of performance or the PPE item having specific design features.

In using existing PPE standards, the organization must check to determine whether the selected standard applies to their application. Existing standards may be modified or supplemented to meet all the protection requirements of the user organization. Performance levels may be lowered or raised based on the risk assessment; however, lowering the minimum performance defined in a consensus standard may expose the user organization to additional liability. Additional criteria for both design and performance may be added when the specification or standard does not cover all protection needs.

When no guidance is available from PPE standards, or standards do not exist for the type of PPE to be specified, minimum criteria must be developed. It is best to avoid specifying a brand product or equivalent, unless a list of salient characteristics can be established that allows the purchaser to determine whether other products are truly equivalent. A recommended practice for setting criteria includes

- Selecting both acceptable and unacceptable products from experience or end-user surveys
- Making measurements of the relevant performance properties for both acceptable and unacceptable products
- Determining the cutoff value between acceptable and unacceptable products to establish the minimum acceptable performance

For the recommended practice to work, the selected test methods for measuring relevant performance properties must discriminate product performance consistent with field observations.

Product specifications should have a scope, a purpose, a list of definitions, and a list of referenced standards that lists all documents cited within the specification. The primary sections of a specification include the design criteria (which provide the minimum configuration of PPE item and any design features that are considered mandatory), performance criteria (based on a specific test method with a minimum or maximum acceptable value), documentation requirements (information to be provided by the manufacturer), labeling and packaging requirements, and acceptance criteria.

Step 5: Evaluate Candidate PPE Products

A critical step of the selection process is to ensure that candidate products meet the established specification. Conformance of products to the organization's specification may be accomplished by reviewing manufacturer product information and technical data, using a third-party testing laboratory or certification organization to verify product design, performance, and documentation, or conducting a field trial to determine whether the candidate product will meet the organization's needs. Reviewing manufacturer product information and technical data or using a third-party laboratory or certification organization will only determine whether the submitted or candidate product meets the specification; these on-the-job evaluation processes will determine whether a product meeting the specifications will provide adequate performance for the organization. Field trials are the most effective means for determining whether a specified product meets the organization's needs; feedback from the field trial can also be used to modify a specification. Methods used to demonstrate acceptable products should be the same as those included as part of the acceptance criteria within the product specification.

Step 6: Establish a PPE Program

Selected PPE must be incorporated into a new or existing PPE program. A subsequent section describes the necessary elements of a PPE program.

Step 7: Periodically Review PPE Selections

To ensure that selected PPE continues to meet the organization's needs, part of the PPE program must include a periodic review of PPE decisions. This review should be conducted by the individual or committee responsible for the PPE decisions and should include the following:

- Review of injury data for workers
- Specific complaints about selected PPE
- Examination of new technology available in the marketplace

No change in the incidence of injuries may indicate that PPE is not working or that appropriate work practices are not in place; a determination should be made as to whether the PPE is providing protection as intended. Specific complaints about PPE should be investigated to determine whether the selected PPE poses a risk to the employee or fails to perform as expected;

these include complaints for worker productivity and comfort. With continued development of PPE technology, new products are becoming available all the time; these new products should be examined to determine whether protection or cost benefits can be achieved by their acceptance. If the responsible individual or committee determines that the selected PPE is adequate and performing properly, this decision should be documented.

The majority of PPE represents an ensemble of individual items of protective clothing and equipment that must function together to afford the intended protection against identified workplace hazards. PPE ensembles must be selected such that

- The PPE items fit together to provide uniform protection over the areas of the body or body systems that need to be protected

- Individual PPE items do not degrade or interfere with the performance of another item

- The wearer can perform necessary tasks without substantial impacts on required productivity

- The overall ensemble does not cause extraneous bulk or weight that creates stress on the wearer

Many integration problems occur at interfaces between PPE items; example interface areas requiring attention include the following:

- Upper torso garment to lower torso garment

- Upper torso garment neck area to hood

- Lower torso garment trouser end to footwear

Ensemble integration guidelines include the following:

1) Determine whether protection must be uniform over the body or specific to certain areas or body systems as determined by risk assessment.

2) Identify interface areas for which exposure to hazards may occur.

3) Determine the appropriate underclothing to be worn by the worker or the types of clothing over which PPE must be worn.

4) Determine whether multiple layers of PPE (especially garments, gloves, and footwear) will be required to achieve desired protection.

5) Determine the type of protection that must be provided by each interface area.

6) Decide which item of PPE should be responsible for providing protection to the interface area.

7) For a given performance property, use information from the risk assessment to determine whether the type of overall protection can be the same or should be different for items of PPE that interface in a particular area.

8) Determine what type of integrity, if any, is needed for the overall ensemble or those parts of the ensemble that must provide protection. Types of integrity include

   • Particulate-tight integrity

   • Liquid-tight integrity

   • Vapor (gas)-tight integrity

9) Apply performance criteria to interface areas as appropriate.

   • Choose items of PPE that are designed to work together:
     — Match characteristics from interfacing items.
     — Specify integrated systems provided by a single manufacturer.

10) Choose PPE that collectively integrates two or more PPE items. Examples include

   • Full-body garments (combining upper and lower torso garments)

   • Garments with integrated hoods

   • Headwear that incorporates face and eyewear (or face and eyewear that directly attaches to headwear)

   • Headwear that incorporates hoods, face/eyewear, and/or respirators

   • Full facepiece respirators (combining face/eyewear with respiratory protection)

11) Ensure that combination PPE meets specifications for each separate item of PPE.

Wearing trials of specified ensemble using simulated work tasks should be conducted to determine the potential effects from integrated items.

ESTABLISHING A PPE PROGRAM

The establishment of a written PPE program provides a method for documenting all aspects of PPE selection, use, care, and maintenance. The advantages of a PPE program for an organization include the following:

- Documenting organization PPE procedures
- Establishing uniform and effective PPE usage guidelines
- Controlling PPE costs
- Creating user acceptance
- Meeting OSHA regulations for selecting and providing appropriate PPE

Elements of the PPE Program

The minimum elements of a PPE program are as follows:

- Risk assessment methodology
- Procedures for the evaluation of other control options
- PPE selection criteria and procedures for determining the optimum choice
- PPE purchasing specifications
- User training procedures
- PPE usage criteria
- PPE care and maintenance procedures
- Validation plan for PPE selection coupled with medical surveillance
- Auditing plan to ensure that the PPE program is properly implemented

The risk assessment methodology should provide a systematic approach for identifying workplace hazards, assessing risk associated with hazards, and documenting the risk assessment. Typically, some methods of hazard identification are outlined in the facility's hazard communication program. This part of the program should determine the persons who will identify the hazards and evaluate risks (safety engineer, industrial hygienist, and other trained individuals), the types of equipment that should used for identification or evaluation, and the methods and frequencies of evaluation.

While the primary objective of the PPE program is the proper selection, use, care, and maintenance of PPE, part of the PPE program should address the evaluation of other control options for eliminating or reducing worker exposure. In response to the results of the hazard assessment, the individual or group within the organization responsible for making the assessment should first consider engineering or administrative controls. Whenever possible, other control options should be implemented in lieu of PPE, particularly for routine and repetitive operations that require PPE.

The PPE program should contain detailed procedures for selecting PPE once the need for PPE has been identified through the risk assessment. PPE selection procedures should include the following:

- Developing PPE design and performance criteria
- Preparing PPE specifications
- Identifying candidate PPE
- Evaluating and choosing PPE
- Purchasing PPE
- Inspecting PPE on receipt

Several other elements of the PPE program require specific attention. Purchase specifications for each item of PPE used by the organization, no matter how simple, should be included in the PPE program documentation. Adequate education and training for users of PPE is essential. OSHA 29 CFR Subpart I requires that employers provide training to employees in the use and care of PPE and ensure that employees understand the training. Training and education should include the following:

- Nature and extent of workplace hazard(s)
- When PPE should be worn
- Which PPE is necessary
- Understanding of basic principles for how the PPE provides protection
- Use limitations of the PPE assigned
- How to inspect, don, doff, adjust, and wear PPE properly
- How to select the appropriate size of PPE
- How to recognize signs of heat stress and other ailments that may be associated with wearing PPE
- Decontamination or sterilization procedures (if needed)
- Signs of PPE wear, overexposure, or failure
- How to report PPE failure if it occurs
- Use of PPE under emergency conditions
- Proper storage, service life, care, and disposal of PPE

The PPE program should establish responsibilities for conducting training, or at least specify the individual or group that is responsible for ensuring that employees receive training.

Proper care and maintenance of PPE is important to ensure that selected PPE continues to provide the intended protection to workers over its intended service life. OSHA 29 CFR Subpart I requires employers to maintain PPE for employees even if PPE is not provided by the employer. PPE care includes procedures for cleaning, decontamination or sterilization, and storage. Cleaning is the process of removing nonhazardous soiling or surface contamination such as dirt, dust, grease, and body oils. Decontamination is the physical and/or chemical process of reducing and preventing the spread of contamination of PPE. Sterilization is the physical and/or chemical inactivation and removal of biological contamination on PPE. Storage encompasses practices and conditions for properly storing PPE.

PPE maintenance includes inspection and testing, repair, removal from service, and disposal. Inspection involves practices for routinely examining PPE for signs of wear, damage, or failure. Some types of PPE must be periodically tested for specific performance properties to ensure adequate protection. Repair encompasses manufacturer-approved practices for bringing PPE back into service. (Note, however, that some PPE cannot be repaired). PPE removal from service is warranted when certain retirement criteria are met or when, in the estimation of the worker or authority designated within the organization, the PPE performance might have deteriorated. Some PPE, particularly PPE that has been contaminated, must be properly disposed of.

Procedures for care and maintenance of PPE should be in accordance with product manufacturer instructions. Furthermore, responsibilities for PPE care and maintenance must be established within the organization.

The PPE program should establish requirements for validating PPE selection decisions and ensuring that the proper PPE is used and used correctly. Responsibilities should be established throughout the organization for use of PPE. The individual or group responsible for PPE selections should periodically evaluate how selected PPE is providing protection and allowing workers to perform their required tasks. Means for workers to provide feedback to the selection committee or individual should also be made available. Injury reports should be reviewed as another means for acquiring feedback on the effectiveness of selected PPE. Some types of PPE will require medical surveillance to determine its effectiveness. Medical surveillance programs should address the frequency and type of medical examinations or testing. Medical surveillance results must be handled in a confidential manner.

Provisions for periodically auditing the PPE program must be included to ensure the long-term adequacy of PPE selections and proper use of PPE. The audit process should review:

- New or revised standards requiring protection
- Injury statistics
- Worker complaints
- Changes in tasks
- Availability of alternative controls
- Availability of new PPE technology

In addition, the audit should be conducted at least annually.

Implementation of the PPE Program

Employers have the ultimate responsibility and duty for the execution of the PPE program for ensuring the safety of their employees and compliance with applicable regulations. Employees have the responsibility and duty for using all PPE that is provided to them in accordance with the instructions and training that they have received. The PPE program must be prepared in writing for documenting organizational procedures with respect to all elements of the PPE program. Administration of the PPE program should be made the responsibility of one person with the assistance of a PPE team. One of the essential aspects to a successful program is the formation of a PPE team for establishing the responsibilities for carrying out the plan. In some smaller organizations, the administration of the PPE program can be a responsibility of the organization's safety committee or similar group. The team members should be individuals who represent a cross section of the facility's operational and organization units. The following persons are vital to provide a working core:

- Safety/industrial hygienist
- Operational supervisor (that is, manager, superintendent)
- Purchaser (key suppliers or vendors would be helpful)
- Operational personnel

The roles and responsibilities of these individuals should be clearly spelled out prior to the initiation of the facility's PPE program. Examples of their activities include the following:

- Approving the results of the risk assessment
- Recommending acceptable control options as an alternative to PPE
- Establishing a formal new equipment approval process
- Reviewing PPE evaluation plans and results of any PPE evaluation process
- Documenting decisions for allowing or removing PPE from the organization
- Maintaining the list of approved PPE for the organization.

Significant PPE problems should be brought to the attention of the individual within the organization responsible for administrating the PPE program.

CONCLUSION

Personal protective equipment should be viewed as the last defense against workplace hazards when engineering or administrative controls cannot provide sufficient isolation from hazards. OSHA regulations in 29 CFR 1910 Subpart 1 dictate the use of PPE when hazards exist and require employers to select PPE based on a risk assessment and to provide proper training in the use of selected PPE. The key to the selection of appropriate PPE is the proper conduct of a risk assessment involving the identification of potential hazards, weighing the risks of those hazards and then choosing commensurate clothing and equipment design and performance properties. Different design features and performance properties are associated with each type of PPE. Further consideration must be given to integration of items such that all of the selected PPE works together. PPE selections must be reviewed periodically with attention given to both the effectiveness and acceptance of PPE for providing the needed protection.

REFERENCES

National Fire Protection Agency (1981). *Standard on Open-Circuit, Self-Contained Breathing Apparatus.* Quincy, MA: NFPA.
National Institute for Occupational Safety and Health (1987). *Guide to Industrial Respiratory Protection* (DHHS/NIOSH Publication No. 87-116). Morgantown, WV: NIOSH.

Chapter 22

Safety and Health Training

Jerry L. Burk

INTRODUCTION

Employees within organizations must be prepared to perform their jobs productively, attend to quality standards, and practice safe behavior. Performance requirements related to safety and health practices can be achieved only through systematically focused attention to training. Consequently, we can expect the need for safety and health training to increase with the multitude of fundamental changes that accompany paradigm shifts in business and operations.

Those performing the work within organizations must be trained to manage the hazards within their workplaces so that the risks encountered are kept at acceptable levels. Safety and health training must specifically focus on solving problems, or potential problems, in which the "learning" is practically and immediately applied on the job. The objectives of this chapter are to teach you to

- Illustrate the role of organization strategy in the development of safety and health training
- Document the impact of safety management on safety and health training processes
- Describe design and development problems that undermine safety and health training efficiency and effectiveness
- Specify requirements for improving the design and development of safety and health training
- Identify methods for effectively evaluating the safety and health training impact upon organization goals

- Explain how documented success justifies the existence of safety and health training as an investment in organization performance.

This chapter applies processes from organization "reengineering" in a systems-focused examination of safety and health training. Guidelines for answering questions about *what* should be included in effective safety and health training—design, development, and evaluation—are addressed. Additionally, recommendations for answering questions on *how* to integrate safety and health training into the overall organization strategy are included.

The following brief case study illustrates the need for integration of safety and health issues within organization values and culture to provide a meaningful context for safety and health training.

This case features two conditions that have important impacts on the ability to conduct effective safety and health training: 1) safety and health training must be perceived as a value-added function by management within an organization; and 2) any training that is conducted must demonstrate a return-on-investment (ROI) to the organization if support is to continue in the future.

Clearly there is a need to be effective and to demonstrate the value-added contribution of safety and health training. That need cannot be achieved, however, without the application of systematic training methods and an evaluation of their impacts. Safety and health professionals can achieve this objective only by the application of training methods that consider the organizational context, acknowledge the unique role of safety and health training, and take adult learning principles into account.

CASE STUDY: SAFETY TRAINING CHALLENGE

Recently you were hired as a safety manager and directed to establish a "proactive" safety department within an expanding facility. Corporate management feels that it must have such a department to manage requirements from government regulations in a time when the business is expanding and when competing organizations have such departments.

It is becoming apparent that one of your most difficult tasks will be to educate facility management and employees concerning the benefits that may be derived from a "proactive" safety function. Facility management casually discourages you from taking too much company time for safety training because production is affected when employees are in training.

Prior to your arrival, safety training was confined to posters on the bulletin board. The severity and frequency rates of accidents have been low recently, due to luck and the small size of the facility. You are convinced that, with the facility expansion and hiring of new workers, things will be different in the future.

— Adapted from McLarney and Berliner 1970

APPROACH TO SAFETY AND HEALTH TRAINING

Adult learning principles must be emphasized so that participants are actively involved in safety and health training. Involvement includes solicitation of life and work experiences, immediate application on the job, trainee participation in how the training content will be delivered, and demonstration of "benefits" from safety and health training.

THE ROLE OF SAFETY AND HEALTH TRAINING

Principles and applications can increase the impact of safety and health training within an organization. Instructional design skills are processes for making planning and preparation efforts more effective. By systematically aligning the training content with day-to-day performance, it is possible to develop trainee skills, knowledge, and attitudes/abilities that can be practically and consistently applied.

Effective safety and health training must use adult learning principles that promote practical application. Adult trainees ask themselves: "Why do I need training?" "What's in it for me?" and "How will this make me more successful?" Without answers to these questions, the impact of safety and health training is likely to be diminished.

Effective safety and health training methods that develop safe behavior practices within organizations do not *just happen*. Proven instructional design methods and training evaluation techniques will substantively increase the impact of safety and health training by

- Developing objectives that are clear and meaningful
- Designing lesson plans that meet the objectives
- Measuring training effectiveness through systematic evaluation

Training content, application, and evaluation are fundamental considerations.

DEFINITION OF SAFETY

The following definition of safety will be used to guide the development of safety and health management and training processes.

Safety — *the control of hazards to keep risk at an acceptable level.*

This operational definition reinforces the "what" and "how" of safety and health training and clearly indicates that *all hazards* cannot be eliminated and that there is an *acceptable level of risk*. Risk is not casually regarded within the workplace, but it is a reality of industrial and organizational life that can be managed. Four key issues are emphasized in this definition:

1) Control (management)
2) Hazards
3) Risk
4) Acceptable level

Paying attention to the "what" and "how" of these issues guides the development of safety and health management processes, as well as the focus of effective safety and health training. Nevertheless, effective safety and health practices will be compromised if they are not integrated into the overall organization strategy. Failure to integrate these processes will reduce the safety and health professional's ability to systematically, comprehensively, and consistently control hazards that keep risk at an acceptable level.

SAFETY MANAGEMENT AND PARADIGM SHIFTS IN TRAINING

Safety management and safety and health paradigm shifts influence training. These phenomena are interrelated, particularly because each demonstrates a priority—or emphasizes what is important—for organizations, management, and individual contributors. Petersen's (1988) book *Safety Management* provides historical insight into eight safety management eras. (See Table 22-1.) Those eight eras are also in his revision of the original work,

Table 22-1: Safety Management Eras

ERA	NAME OF ERA	DATE OF ERA
Era I	Inspection Era	1911-1931
Era II	Unsafe Act and Condition Era	1930s–1940s
Era III	Industrial Hygiene Era	1930s–1940s
Era IV	Noise Era	1951–Present
Era V	Safety Management Era	1961–1970
Era VI	OSHA Era	1970s–Present
Era VII	Accountability Era	Late 1970s–Present
Era VIII	Human Era	Early 1980s–Present

with mention of a "Psychology of Safety Management Era," which Petersen claims was delayed by the OSHA era.

The eight eras are not exclusive of one another, including the increased psychological emphasis that we see emerging presently. Nevertheless, they illustrate how different emphases in the field of safety and health might affect training. Each of the eight safety management eras reflects a different emphasis on safety and health, safe performance, environmental conditions, and/or worker behavior. In Eras II and III (the 1930s and 1940s), worker training and behavior became priorities.

Eras II and III included four dimensions:

1) Education and training
2) Attention to the behavior of workers
3) Emphasis on physical working conditions
4) Attention to environmental conditions

W. H. Heinrich's 1931 classic text *Industrial Accident Prevention* influenced the focus of Eras II and III. It was written at the conclusion of Era I—Inspection—and influenced the safety and health profession for the next six decades. But despite the attention given to education, training, and worker behavior, neither era included approaches that were systematic, comprehensive, or consistently applied.

Petersen (1991) continued his historical review of safety and health in an article entitled "Safety's Paradigm Shift." He indicated that a paradigm shift in safety and health required a critical reexamination of Heinrich's fundamental safety principles, which had guided this field for so long. Peterson wrote, "Those axioms from his [Heinrichs'] book, *Industrial Accident Prevention,* have been the foundation of safety programs for 60 years. They are not only the foundation for safety programs, but also for safety legislation. Maybe we need to examine the reality of those axioms" (Peterson 1991). That conclusion, after examination of the reality of Heinrich's axioms, produced Peterson's "Ten New Principles of Safety Management" (see Figure 22-1). Predictably—as the principles of safety management change the requirements for safety and health training change as well.

STATE OF SAFETY AND HEALTH TRAINING

An assessment of modern safety and health training lead Fred A. Manuele, in his book, *On the Practice of Safety,* to the following conclusion:

> Unfortunately, safety training is often much talked and written about but poorly done. . . . In the best situations, training needs are anticipated along with plans for new or altered facilities or the planned use of new materials, and consider the changing aspects of the work force and the continuing stream of regulatory requirements.

The responsiveness of safety and health training to paradigm changes in industry and the new *realities* of safety management will determine whether the results are successful or unsuccessful. Success may be defined as the demonstrated ability to improve the impact of safety and health training so that prescribed behaviors are applied at the moment of risk to avoid hazards that place employees in jeopardy.

Safety and health training that is poorly done directly impacts employees' lives and health more than any other training and development intervention. In fact, organizations that develop effective safety and health training interventions are making investments that support *quality and productivity as well as the safety of employees.* Contrary to popular thinking of the past, attention to safety and accompanying safety and health training is good for business performance.

Figure 22-1. Ten New Principles of Safety Management

1. Safety issues—unsafe acts, unsafe conditions, and accidents—are symptoms of management system failures.

2. Severe injuries can be predicted from certain organization settings that are identifiable and controllable:
    - Unusual, nonroutine activities
    - Nonproductive activities
    - High-energy sources
    - Particular construction situations

3. Safety is a management function that lends itself to the application of management tools.

4. Management procedures must establish accountability to develop effective safety performance.

5. Safety functions should locate and define operations that contribute to incidents:
    - Why?—Search for "root causes" of incidents.
    - How?—Examine the effectiveness of controls used.

6. Unsafe behaviors and conditions can be identified, classified, and controlled—overloading, traps, and decisionmaking.

7. Management must change the organization environment that contributes to unsafe behavior.

8. An effective safety system is composed of three major components:
    - Physical
    - Managerial
    - Behavioral

9. Safety systems must "fit" the organization culture.

10. There is no "one way" to implement organization safety systems. Certain components must include:
    - Top management visible from commitment
    - Involvement of all management levels
    - Supervisory accountability
    - Flexibility
    - Being perceived as "positive"

adapted from Petersen (1991)

SAFETY AND HEALTH PRACTICES ARE GOOD BUSINESS PRACTICES

Two complementary principles will improve acceptance of safety and health training by management and develop support for initiatives taken:

1) Safety and health training must reinforce sound business practices that are emphasized within an organization's strategic plans.

2) Safety and health training methods should be appropriate at every level of an enterprise to change organization behavior.

First, safety and health training must receive support as a component of the organization's strategic planning system. This need will increase in the emerging "information age" economies of the twenty-first century. The future will be more demanding and will rob organizations of three dimensions accompanying the industrial age paradigm. Think of them as the "Three Cs":

1) Comfort
2) Clarity
3) Certainty

A fundamental change of the type described here will be uncomfortable, in part, because it separates organizations from what is familiar to us.

Perhaps one of the most troublesome dimensions of this unfamiliarity will be a lack of clarity concerning specific directions of movement—organizations are in new, uncharted territory. This discomfort may be expected to increase further because continuation of practices that produced success, even a short time ago, will produce failure in the future.

Though practices that guarantee success will not be clear or certain, organizations can be assured that *practices of the past industrial age will be prescriptions for failure in the emerging information age.*

Organizations, like living organisms, must adapt, change, and renew themselves . . . or die. Hammer and Stanton (1995) described this need as "organization reengineering," which they defined as "the fundamental rethinking and radical redesign of business processes to bring about dramatic improvements in performance." Reengineering of organizations in general, and of safety and health training in particular, requires four strategic actions to remain viable in the 21st century:

1) Management must develop a strategic and operational vision.
2) Management must clearly assess where they are now.
3) A clear plan must be initiated to "close the gap" between "where we are now" and "where we want to be" in the future.
4) Organizations must develop means for monitoring the environment to manage changes that are undertaken and to remain responsive to future needs for change.

Safety and health training, as a component of organization strategy, can be a powerful tool for helping close the "gap" and contribute to the development and maintenance of an organization's viability, vitality, and competitiveness.

A vital component associated with the concept of organization reengineering is the need for leadership. But leadership cannot exist in an environment where productive training is not supported. This relationship is not a new development. John F. Kennedy said, "Leadership and learning are indispensable to each other." If an enterprise is to compete successfully in the information age marketplace, it must have both enlightened leadership and an informed, competent, and safe workforce.

Second, safety and health training interventions must be specifically designed to meet the needs of an enterprise in order to change organization behavior. The impact of safety and health training interventions is illustrated by six steps that were identified as the "Attenuation of Effect," (National Committee 1989) a model of decreasing of training intervention impact. An inverted pyramid (Figure 22-2) was used to illustrate the loss of desired impact on one-third of a population at each successive step when developing more demanding performance expectations following training interventions.

Effective safety and health training requires observable and measurable behavior change to satisfy the definition of safety used in this chapter. Without systematic attention to methods that promote effectiveness, application

---

Figure 22-2. Applied Behavior and Training

Total target population:	100%
Those exposed to educational messages:	66%
Those comprehending messages:	44%
Those changing behaviors:	29%
Behaviors persisting over time:	19%
Behaviors applied at moment of risk:	13%

of desired behavior will occur in only 13 percent of a population "at the moment of risk"—an unacceptable result.

COMMON SAFETY AND HEALTH TRAINING BARRIERS

Merely identifying a training failure is not especially helpful if safety and health training methods are to be systematically improved. Lynton and Pareek (1978) identified four barriers to training effectiveness that can be adapted to improve the impact of safety and health training interventions on participants and organizations.

Safety and health training must have a proactive impact on participants initially and, ultimately, on the organization itself. This will be possible only when the training content, training processes, organization values, and organization culture are in alignment. Failure to align these dimensions will not only reduce the effectiveness of safety and health training, but also reduce overall organization performance and morale. This is not a system to be impacted by safety and health training. Organization culture, for

Figure 22-3: Impact of Safety and Health Training on Individuals and Organizations

SAFETY AND HEALTH TRAINING FOCUS	IMPACT	
	On Participants	On an Organization
Input overload	• Low clarity • Discouragement • Failure to apply	• Diffused focus • Reduced confidence • No clear direction
Low practicality	• Unclear relevance • Disbelief • Failure to comply	• Confused focus • Reduced relevance • No behavior change
Value conflict	• Mixed message • Dissonance • Failure to internalize	• Conflicting focus • Reduced morale • No respect
Culture violation	• Rejection of the message • Disintegration • Failure to trust	• Contradictory focus • Reduced long-range effect • No clear leadership

example, is critical when dealing with organization behavior change and the persistence of desired performance.

This approach to safety and health, and related training, is a departure from more traditional means and methods. Effective safety and health training in the future must include the following three dimensions in order to achieve optimal results: 1) a systems view, 2) a process approach, and 3) motivation through facilitation.

Systems View

Safety and health management and training must not be viewed as separate from the strategies, goals, and objectives of the organization as a whole. Training in support of safety and health management should remain an open system, meaning it must remain flexible and responsive to the organization's needs. The characteristics of an "open system" may be most clearly understood and contrasted with those of a "closed system." Kast and Rosenzweig (1972) presented a model contrasting open and closed systems in a special issue of the *Academy of Management Journal* dedicated to General Systems Theory. They contrast the differences between "open" and "closed" systems are contrasted based on 1) environmental relationships and 2) goals and values.

Safety and health training must remain flexible, open, focused on solving organization problems, and dedicated to satisfying multiple organization goals. There is a seductive simplicity when operating in a "closed" system, but the challenges of an "open" system more accurately define the requirements of safety and health training in the 21st century.

A Process Approach

The requirements associated with operating within an open system lead to the development of a process approach to safety and health training. The need for a process approach is best contrasted using Kurt Lewin's concept of change even though it is anchored in the industrial age mind-set. Lewin described change management in three steps:

1) Unfreeze the present state
2) Change during this more flexible, "unfrozen," less rigid state
3) Refreeze to institutionalize the changed state

The very idea of reintroducing the rigidity of a frozen state is unthinkable within the mind-set that characterizes the information age. Lewin's

Figure 22-4: Organizational Characteristics and Organizational System Types

ORGANIZATION CHARACTERISTICS	CONTINUUM OF ORGANIZATION SYSTEM TYPES	
	Closed Organization Systems	Open Organization Systems
Environmental relationships		
General nature	Placid	Turbulent
Predictability	Certain, determinate	Uncertain, indeterminate
Boundary relationships	Limited participation; Fixed and well-defined	Multiple participants not clearly defined
Goals and values		
Organization goals in general	Efficient performance Stability, maintenance	Effective problem-solving, innovation, growth
Goal Set	Single, clear-cut	Multiple, determined by necessity to satisfy a set of constraints
Stability	Stable	Unstable

model should now read: Unfreeze the present state and continually change . . . change . . . change. Safety and health training must keep pace with change and remain responsive to the environment, goals, and values of the organization.

Motivation through Facilitation

Increasingly, safety and health trainers will find themselves selling the need for safe performance and safe behavior first to management and then to the participants. An important consideration will be the increased involvement of participants in their own training. A study examining the enhancement of human performance concluded that trainer input should be minimized in favor of contributions from trainees. Rather than trying to motivate trainees, it will be the job of trainers to establish an environment that elicits motivation.

That approach requires developing specific tasks to be learned, practicing those tasks, and making mistakes during the learning process. Identification of "mistakes" would come from participants themselves as they answer their own questions and provide feedback to one another.

A truly "learner-oriented" environment compounds the trainer's challenge because the direction provided by trainees is unpredictable. Trainers employing learner-oriented methods are likely, moreover, to perform as training facilitators rather than as traditional trainers.

Learner-oriented training sessions are typically noisy and might seem somewhat chaotic compared to the traditional trainer-oriented environment. But this is the way adults learn: learn, practice, make mistakes, correct mistakes, and apply to "real work" related situations. Rekus (1993) reinforced the need for a facilitative training environment. According to him, "Adults learn best in an atmosphere that permits and encourages use of their existing knowledge to question, debate and discuss the relevance of any material presented in class." Effectiveness, impact, and ultimately the value of safety and health training will be enhanced by motivation from a training environment that promotes application of learned skills on the job.

SAFETY AND HEALTH TRAINING: INVESTMENT, NOT EXPENSE

When the decision is made to invest an organization's financial resources in safety and health training, it is often top management's way of saying, "We see this training contributing value to employees and value to the organization," or "We can justify safety and health training activities because they contribute to the performance of this organization." It is an investment in the future of the organization and its capabilities—a strategic decision.

Attention to safety and health training, and other training interventions, can enhance an organization's capabilities to execute strategic initiatives. Waterman, Jr., coauthor of *In Search of Excellence,* continued his focus on organization performance in *The Renewal Factor* (Waterman 1987), where he concluded that "skill building" can be an important strategic consideration. He wrote that skill building, or training processes, make important strategic contributions to success both in the short run and long run. Waterman stated that, "The notion of skill building is what starts to put meat back on the bones of some strategy and organization ideas that got bleached white by the last decade's glare of economic reality." Training is not an "expensive luxury" in the scheme of an organization's strategy, but a powerful catalyst ensuring that capabilities are available to execute strategic initiatives.

Future-focused organizations must promote an approach to safety and health training that emphasizes a ROI that documents the contribution to overall organization strategy and performance.

TOP MANAGEMENT RESPONSIBILITIES FOR PRODUCTIVE TRAINING

Many individuals in an organization will find it easy to indict the safety and health training function for its omissions, and there will always be opportunities to document them. But meaningful remedies will come only through the development of systems-focused training processes that take into account multiple causes and multiple factors. Top management, for example, must bear some of the burden, together with those directly responsible for safety and health training. Two common omissions by top management that directly influence the effectiveness of safety and health training are 1) lack of support for safety and health processes in overall organization strategic planning; and 2) weak, or nonexistent, integration of safety and health processes into clearly stated organization mission and values.

Lack of Overall Organization Strategic Planning

Top management must think strategically, maintaining a constant and concentrated focus that includes the role of occupational safety and health within the organization. An *effective and comprehensive organization strategy* must, therefore, include leadership that supports safety and health processes.

The ability of safety and health professionals to provide input into the development of an organization's business strategy is a compelling reason to ensure membership of safety and health among top management ranks. Safety and health must assume a new mission, according to Anthony Veltri, Professor of Safety Studies at Oregon State University, by embracing a mission to *prepare, protect, and preserve* organization resources. He states that the safety profession of the past justified its existence by bureaucratic practices that kept government agencies off organizations' backs (personal communication, August 6, 1991). The profession did not identify itself with the leadership requirements of the information age.

When the safety and health profession becomes concerned with promoting the cost-effective use of organization resources, it is more likely to be empowered through membership in the ranks of top management. Veltri's research indicated that involvement in the business strategic focus will result from evolution away from the industrial age mind-set. He predicted:

> . . . it will no longer be acceptable for the safety function to simply control hazardous exposures and comply with mandates from governmental agencies and insurance carriers. Rather, the safety function will be required to create added strategic value and operating leverage to the firms business performance.

Thus, safety and health would be perceived as a value-added component of an organization rather than being exclusively perceived as "overhead" or "interference."

Participation among top management ranks must not be viewed as a "right"—it must be earned through responsible performance. Lahey (1988) described the changing workforce, nature of work performed, and the challenges confronting the safety and health professional. He identified three types of illiteracy that must be faced: functional, cultural, and computer. Truly these are significant challenges concerning the composition of tomorrow's workplace and prospective safety and health trainees.

But safety and health professionals who assume a place among the ranks of top management must not be guilty of another type of illiteracy— business illiteracy. Warren Bennis, Distinguished Professor of Business at the University of Southern California, describing business literacy as it applies to top managers, stated: "They have to be in planning, in staffing, in computers, in legal and research. They have to really experience and taste and be subculture hoppers. They have to be all around the place before they get to the top" (Bennis 1991). There is a "rite of passage" that can prepare individuals to assume the role, perform effectively, and prepare the way for those who follow.

How can safety and health training perform as an effective mechanism for change when it does not have strategic input? How can the organization expect to achieve any goal if there is no specific time frame guiding the organization planning process? The net result is that, without active integration in overall organization strategy, safety and health training is often relegated to a "fire fighting" role. This signals an organization in which safety and health issues are not truly strategic priorities; rather these issues are reactions that are likely to generate skepticism among employees.

Weak Integration of Safety and Health Processes with the Organization Mission and Values

Integration with organization mission and values can help establish safety and health's higher rank among the overall enterprise strategic values. Two related questions can help in this determination: 1) What is the focus of an organization's mission and values concerning safety and health? and 2) What is being done to attain and reinforce the mission and values of the organization relative to improved safety and health? This line of questioning is likely to illuminate the mission and values, verify the commitment of an organization to safety and health, and establish opportunities for integrating the safety and health mission and values within an organization.

Answers to these questions can also help identify the need for developing coherent safety and health values within the overall organization mission and values. These values must be represented within the overall organization strategy just as productivity and quality are fundamental organization values. When safety and health issues are not part of organization-wide values, their "disintegrated" status is a concern.

SAFETY AND HEALTH TRAINERS' SEVEN KEY RESPONSIBILITIES

Key Responsibility #1: Minimize Resistance

Safety and health trainers must first establish credibility within the ranks of management and then with training participants. Without credibility the ability to change participant's behavior is compromised. Five key factors (McGee and Burk 1991) often prevent safety and health trainers from being perceived as vital organization resources:

1) Lack of professional training expertise
2) Unclear purpose and expectations of safety and health training
3) Fuzzy safety and health needs analysis
4) Inadequate or inappropriate presentation and instructional tactics
5) Lack of evaluation

Let us look at these five key safety and health training factors individually.

Lack of Professional Training Expertise. When talking with trainers, I often ask how they got into safety and health training. Frequently the reply is, "Oh, I just kind of fell into it and learned as I went along." One of the saddest instances came from a supervisor who moved an ineffective performer into a safety training role. It seems that the individual had been with the company for a long time and would be retiring shortly. Reportedly, this individual could not do any harm there in the interim. But that was not the case.

Some safety and health trainers explain that they were high performers in their previous assignments, until someone offered them the opportunity to move into training. Still others have been assigned to safety and health training as part of their professional development as line managers. Each of these cases illustrates how safety and health "training specialists" can assume their roles with minimal professional preparation. Small wonder that their impact on the organization is questionable.

Once assigned safety and health training responsibilities, these well-meaning individuals predictably fall back on what they know about education. And experiences as students in high school and college frequently become models for their performance. Regrettably, those models and methods are ill-suited to the needs of adult learners when the goal is effective safety and health training.

This is not to say that all safety and health trainers lack expertise or an aptitude for training. Many specialists seek professional development as trainers. Although advanced programs in other organization processes are readily available, there are too few opportunities to study specifically the methods of safety and health training. This need seems to be recognized by some, but too little has been done to provide development in this sorely neglected area of formal study. Most trainers must rely on a patchwork of offerings from professional associations, and commercially sponsored workshops and seminars, or pursue a less formal, and sometimes less systematic, course of self-directed professional development. Unfortunately, there remains a cadre of safety and health trainers who continue to refine what they know about training "on the job."

Unclear Purpose and Expectations of Safety and Health Training. When asked why a particular training process exists or the reason for its development, trainers often respond, "Upper management requested it. They said we needed such a course." One might conclude from such a comment that the training activity is being provided with a vote of confidence by management for addressing organization issues and hazards. But, unfortunately, most organization issues and hazards cannot be mitigated by training alone. Trainers must not perpetuate the myth that every problem can be effectively treated by training or additional training.

Evidence of this myth can be found when training is delivered regardless of whether it is an appropriate response to an issue, incident, or hazard within an organization. Top management often uses safety and health training to address organization problems because it is quick, visible, and puts management in a positive light. But it is imperative that those responsible for safety and health training ensure that the training does, in fact, address the "real" problem.

Safety and health training specialists can wittingly or unwittingly perpetuate the myth of training as a cure-all. For example, I knew a specialist with a marvelous command of vocabulary and fine wordsmithing skills who embraced this myth. He headed the development function for a Fortune 100 organization and applied his skills in renaming and reframing his programs—old wine in new bottles. Unfortunately, participants in such programs are in the same position as the late Telly Savalas's racehorse, Telly's Pop.

After winning many races, several people suggested that the horse should be retired and put out to stud. Mr. Savalas's trainer confided that the horse could not be put out to stud because he was a gelding. Indignantly, Mr. Savalas demanded to know what difference that made in passing on the powerful traits of his racehorse. He found, however, that his celebrity and money could not overcome the cruelest cut of all. Without sound preparation in adult training skills and techniques that are required for overall organization success, safety and health training specialists are reduced to little more than geldings put out to stud.

Lack of clear expectations—learning results, outcomes, and/or impacts—can be as serious an obstacle as an unclear safety and health training purpose. Managers and safety and health training specialists alike must unite by identifying the required organization benefits expected from training. They must examine what participants will be able to do—behavior applied at the moment of risk—after participating in safety and health training. Finally, performance objectives must be aligned with the trainee's needs, or the training will not result in behavior modification or improved performance.

Fuzzy Safety and Health Needs Analysis. A lack of focus frequently occurs because many training processes suffer from inappropriate instructional design. This inappropriate design may be due to a trainer's failure to understand the need for training. *Needs assessment,* a necessary step in the development of a training process, may often be bypassed in favor of speedy development or to reduce development costs. It is true that training content can be built around what a trainer knows about the subject, what can be "borrowed" from colleagues, or from previously developed training programs, books, and other media. But, without a systematic needs analysis that is specific to the organization setting, coupled with the involvement of those who have a "stake" in the process, a safety and health training specialist is, at best, ineffective and, at worst, a safety hazard.

Among the most critical issues in the development of training are: 1) What are the reasonable expectations of the trainees? and 2) What should be expected of the trainees as a result of a training intervention? By asking these questions, we illustrate how the industrial age business paradigm has been revolutionized. In the industrial age, the employee was "fit" to the job, whereas now, in the information age, the reverse is true.

Ouchi (1981) explained that management in the information age would approach its job with the involvement of workers to increase performance. Similarly, active involvement of employees in safety and health training increases the probability of safe behavior. The employees' needs must be met, but determining the nature of those needs is the role of needs analysis. Ouchi (1981) put it this way:

What remains, therefore, is for organizations to change their internal social structure in a manner which simultaneously satisfies competitive needs for a new, more fully integrated form, and the needs of individual employees for the satisfaction of their individual self interest.

No other action by organizations can serve the self-interests of employees more than attention to their safety and health needs. The checklist in Figure 22-5 provides a sound starting point for conducting safety and health needs analyses.

Inadequate or Inappropriate Presentation and Instructional Tactics. Unfortunately, the most common instructional tactics utilized by safety and health trainers have been lecture and visual/video presentations. Following the lecture model, trainers assume a position at the front of a group and hold forth for the benefit of those in attendance. This is largely one-way communication. When observing most trainers "at work," it is easy to believe that, except for the introduction of video technologies, training methods have remained unchanged for two thousand years.

Well-meaning training specialists sometimes relish the opportunity to be the "star," "expert," or "center of attention." Though training presentations may fulfill the trainer's ego needs, they are unlikely to serve the needs

---

Figure 22-5. Guidelines for Analyzing Training Needs

Assess the Need: "Who Will Be Trained?"
- How many trainees will be involved?
- What will be the mix of trainees?

Men – Women	*Supervisors – Nonsupervisors*
Organization Levels	*Homogeneous – Heterogeneous*
Age Range	*Education Levels*
Cultural Diversity	*Disabilities – Handicapped*

What is their level of responsibility in the organization?

Does the course content interest particular members of the organization?

What is their knowledge and skill level?
- How does this group of participants learn best?
- What do they already know about the training content?
- What is their skill level in the course content area?
- What can they handle in terms of experiential training?

How familiar are they with each other?
- Do they know each other?
- Do you know any of them?
- Are there "friends" or "foes" in the audience?

What are their attitudes?
- What are their attitudes toward the topic?
- Were they "sent" or did they come "voluntarily?"
- Are they hostile? Friendly? Indifferent?
- Are they willing to learn? Are they daring you to try to teach them?
- Are there words or jargon that will offend or distract?
- Are there issues that will push their buttons?
- Are they interested?
- Are they intimidated or fearful of the topic?

Analyze the Training Need: "Why Should 'They' Be Trained?"

What are the trainee's expectations?
- What are the job/performance expectations?
- What are the gaps between reality and expectations (that is, what they will learn versus what they want to learn)?

What are your expectations? Theirs? Their supervisors'? The organizations'?
- Who put me up to this? What do they want the trainees to learn?
- Is it a requirement that they know this information?
- What does this group of trainees need?

of trainees or an enterprise. Organizations need safety and health trainers who can manage learning environments, not manage training media, or satisfy their egos. Recognizing the limitation of the following training media should help avoid abuses.

Overhead Transparencies. A well-known aircraft and aerospace organization has employees who sarcastically refer to existing training practices as "death by overhead." Instances abound in which up to sixty overhead transparencies may be used in twenty-minute presentations. This is tantamount to the old notion of "spray and pray": expose trainees to a spray of content and pray that some of it is retained. The "Attenuation of Effect" indicates that it is highly unlikely that behavior change will result from this training tactic.

Video Cassette Recorders (VCRs). Videotaped training resources may be the most abused of training media. They are simple to use, relatively inexpensive, available on a wide range of topics, and trainers are not responsible for their content. As a result, they have become popular training crutches that are substituted for training design and development.

Video-Based Training. Technological developments in the area of interactive video, however, should not be confused with VCR abuse. The software of these powerful learning environments can be programmed with systems, structures, and active involvement that should be models for developing effective learner-centered education. But, these powerful training resources are not appropriate for every safety and health training need; they do not fit every adult's learning style and can also be subject to abuse when used inappropriately.

The availability of instructional technologies within organizations can run the gamut from nonexistent to an embarrassing surplus. Some facilities have little or no equipment beyond a flip chart and markers, whereas others have thousands of dollars worth of advanced video production equipment, and still others have computer-based instructional systems. Many organizations have computer equipment at which trainers invest countless hours developing materials that might be delivered just as effectively by thoughtfully developed handouts.

Casual investigation of training facilities can yield photographic darkrooms being used as storage areas, video cameras gathering dust and rusting from corroded batteries, and other equipment suffering from similar neglect. And, of course, there are the computers. This array of unused equipment often represents attempts by trainers to solve design, development, and delivery challenges by applying media.

Lack of Evaluation. Training, in general, and safety and health training, in particular, is frequently evaluated using "smile sheet probes"—those questionnaires that are written to guarantee positive responses For example, "Did we keep you entertained?" "How did you like the workbooks?" "Did you like the meals and breaks?" "Enough selections? Healthy selections?" "Did you have fun?" These examples of pseudoevaluations are the bane of effective training.

I once observed a trainer using smile sheet evaluation probes with one of my clients. Assuming that this was being done naively, I suggested that there might be other types of questions that should be asked for feedback to guide improvements in the training process. The trainer responded, "I would never use those types of questions in a course evaluation. I might not get a good evaluation and good evaluations are very important, you know." The quality of the training process, trainee performance, and application on the job were not that trainer's concern.

It is impossible to validate the contributions and benefits of safety and health training without an effective evaluation strategy to measure the results. Socrates advised us that "the unexamined life is not worth living." This thought can be applied to the role and function of safety and health training within organizations. Training specialists who do not maintain high standards of performance evaluation diminish their abilities to be successful, both in the present and especially in the future.

Key Responsibility #2: Keep Pace with Industry and the Business Climate

One reason for training evaluation concerns a clear understanding of the impact from changing operations, changing economic and environmental issues, emerging social and governmental pressures, and higher expectations from both management and employees. Change is the operative word. Safety and health trainers need to adapt to a changing business environment. This requires the following two steps.

First, safety and health training specialists must understand the business, its processes, and attendant hazards so that the safety and health function is aligned with the enterprise's operating environment. This is similar to the advice that was given to organizations in Peters' and Waterman's classic book *In Search of Excellence:* "Stick to your knitting." Businesses must remain focused on what they do best. That need to stick to your knitting is good advice for safety and health trainers, but, first, trainers must "know the knitting."

Second, those who do the work must be prepared to perform that job productively, attend to quality standards, and practice safe behavior. These comprehensive employee performance requirements, especially behavior related to safety and health practices, are not achieved by themselves. Consequently, we can expect the need for safety and health training to increase with the multitude of fundamental changes that accompany paradigm shifts in business and operations. Employees who do the work must receive formal training so that they can manage the hazards within their workplace if organizations are going to keep risk at an acceptable level.

"Performance," must be emphasized, which requires training rather than education. Training is a specialized application of education processes (Rekus 1993).

- **Education:** A process through which learners gain new understanding, acquire new skills, or change their attitudes or behaviors.
- **Training:** A specialized form of education focusing on developing skills. The goal of training is for learners to do something new or better than before.

Safety and health training specifically focuses on solving problems or potential problems. It requires participants to apply the "learning" from the classroom practically and immediately upon their return to the job.

Key Responsibility #3: Employ Behavioral Science Methods

A behavioral science-based approach is required to guide safety practices so that safety and health specialists are focused on the fundamentals of human performance. They can no longer rely on past practices, or work from intuition alone. Scott Geller, in *The Psychology of Safety,* reinforced the need for a behavioral science-based approach by dedicating that book to Drs. B. F. Skinner and W. Edwards Deming. Their influences no doubt contributed to his conclusion that ". . . behavioral science principles provide the basic tools and procedures for building an improved safety system" (Geller 1998). A safety and health trainer's attention should be clearly focused on human factors that influence productivity, quality, and safety performance.

Moreover, employees must be guided by practical behavioral science that is embodied in, and reinforced by, the values of an organization. Geller (1996) went on to emphasize that the fundamental need to "act people into thinking differently" rather than the reverse.

- **The goal:** To change relationships between behaviors and their consequences within the organizational and community setting.
- **The method:** To apply the three new Es—Ergonomics, Empowerment, and Evaluation—when building the safety culture.

The traditional "3 Es" of safety—Engineering, Education, and Enforcement—do not lend themselves to a behaviorally focused approach to safety within organizations.

Management and safety and health professionals must be challenged to take the complexity of human behavior into account. The future will require them to follow different principles, to develop new procedures, and to strategically implement a safety culture. That need is underscored by increased numbers of potentially hazardous substances and processes that are part of day-to-day operations. The burden upon safety and health training will increase as specialists are called on to train employees using safety principles that help manage—prepare, preserve, and protect—the organization's resources.

Key Responsibility #4: Focus on Behavior Modification

Safety and health training can achieve behavior modification, and management of organization resources, only by employing methods that promote safe, high-quality performance. Four effective safety and health training practices are capable of achieving this goal:

1) Hazards and associated risks encountered on the job should be communicated in advance of exposure during new employee orientations.
2) All employees should receive training in safe job practices and receive follow-up training for reinforcement as well.
3) Supervisors should receive special safety training.
4) Multiple training methods should be utilized to match the variety of learning styles of trainees.

These documented "best practices" specify fundamental guidelines for more effective safety and health training.

Organizations appear unable to apply systematically these straightforward, proven practices. Despite the fact that training is delivered, unsafe behaviors may persist because employees fail to receive and/or apply it for several reasons. When there are training reception problems, trainers do not

- Understand that tasks are being performed incorrectly
- Understand instructions for performing tasks correctly
- Know specific instructions for performing the task
- Remember how to perform a task correctly

When there are training application problems, trainees do not

- Find instructions for performing tasks easy to apply
- Consider instructions for performing tasks important
- Use instructions despite training provided
- Forget unsafe work habits

Increasing the amount of safety and health training will not reduce unsafe behaviors that can result from these reception and application problems. The issue is not a matter of training quantity, but *quality*.

Key Responsibility #5: Focus on Documented Needs

Safety and health needs analyses can help trainers understand how organizations function as systems as well as help specify the role of safety and health. Needs analyses may have been conducted in isolation from other organization operations and functions in the past, but that approach is no longer acceptable. A priority in the development of safety and health training must be its integration with organization systems. Embracing a systems approach, for example, would mean that analyses are likely to have a character that is different from traditional "training and development" approaches.

Evaluations of organization needs must begin with broad, organization-wide issues to integrate safety and health training effectively into the organization's value system. The following checklist identifies some critical issues that should be considered to achieve integration:

1) What are the focus and priorities of organization management and leadership?

2) What are the key safety and health training issues that exist within the organization?

- Management?
- Employees?
- Supervision?

3) What levels and types of assessment are required?
   - Are different types of needs analysis required at different levels of the organization?
   - What do people know?
   - What do people *need* to know?

4) What organization data are available?
   - Develop a comprehensive list of questions.
   - Address safety issues:
     — Hazards/risks?
     — Regulations?
     — Safety culture?
   - Address "organization" issues:
     — Structure?
     — Procedures?
     — Attitudes?

5) How can organization roles and responsibilities be identified and assessed?
   - Assess employee capabilities to perform in assigned roles.
   - Assess employee capabilities to execute assigned responsibilities.

6) Is there a need to assess the *effectiveness* of present training and education?
   - Is it appropriately focused?
   - What results are being achieved?

7) Is there a need to assess the *retention* of present training and education?
   - Short term retention?
   - Long term retention?

Answers to each question are likely to generate new questions.

Key Responsibility #6:
Focus on the Safety Culture

An organization's safety culture contributes to the development of a context for developing training objectives, as well as guiding planning, preparation, and presentation. Additionally, an understanding of the organization's safety culture can provide insight into how management, supervision, job incumbents, and other stakeholders perceive safety and health performance. Consider the following organizational questions when seeking an understanding of the safety culture:

- Do stakeholders see the same problems?

- What similarities or differences exist in their perceptions?

- Is there agreement on approaches to solving performance problems?

Probing for data concerning these organization issues can provide important benefits when designing and developing safety and health training. Training effectiveness, support, and receptivity will improve with the following:

1) Involvement of key individuals in advance of training and/or development:
   - Defining the problem(s)
   - Assessing solutions
   - Implementing changes

2) Involvement of managers and supervisors:
   - Inviting participation
   - Seeking input
   - Gaining support

3) Educate management:
   - Employing high-quality probes/questions
   - Modeling desired clarity
   - Modeling desired specificity
   - Responding with purpose:
     — Practically
     — Workably
     — Desirably

A systematic understanding of the organization safety culture provides an important starting point for integrating safety into the organization's value system.

Key Responsibility #7:
Focus on Safety as an Organization Value

The attention and support of upper management are critical considerations when focusing on safety and health as values within the overall organization culture. "Value" and "culture" are not abstractions that lack practical application. Each directs and determines what people pay attention to, how people prioritize actions that are taken, and whether actions will be taken at all. The attention of upper management can be obtained by identifying with their priorities and by utilizing words that confirm their priorities.

Attention to two areas can ensure upper management attention: 1) organization performance issues—quality, efficiency, profitability; 2) anticipated resistance to change—organization systems, tasks, and technology. These can be reduced to performance concerns for what can be done to improve the organization and how that will be achieved.

Upper managers want to know: "What can you do," "What will it cost me," and "What assurances do I have that those promised results will be achieved?" Safety and health trainers must, therefore, think of themselves as change agents and assume accountability for actions taken. The five strategic performance factors identified in the following checklist must be considered.

1) What would *optimal* performance produce?
   - Results?
   - Outcomes?
   - Impacts?

2) What does *actual* organization performance produce?
   - Results?
   - Outcomes?
   - Impacts?

3) What factors are responsible for the GAP between "optimal" and "actual" organization performance?

4) What are the barriers to achieving optimal performance?
   - Structure?
   - Procedures?
   - Attitudes?

5) What organization system solutions would contribute to optimal organization performance?

The *gap* between optimal and actual performance may be created by a variety of sources, such as the following:

- Business environment
- Deficient skills and/or knowledge
- Unclear expectations

- Inappropriate or contradictory incentive(s)
- Unclear sense of plan and/or purpose
- Lack of confidence

CONCLUSION

There is no "silver bullet" when it comes to integrating safety and health training as an organization priority. Probing for data concerning organization issues can, however, provide important benefits when designing and developing safety and health training. Attention must be given to cultivating management support, integrating the values of safety and health with the organization's culture, and clearly communicating the value-added benefit of safety and health training to the organization. Correlating the benefits of safety and health training with business goals will both improve the application of practices presented and their acceptance as value-added activities.

Consider the following steps to help integrate safety and health training within the overall organization, to improve effectiveness, and to improve receptivity:

1) Develop probes/questions to understand the organization's culture:
   - Employ "sensing interview" process.
   - Seek articulation of the "future/present states" and GAP.
   - Begin with key managers.
   - Seek referrals concerning who else to interview.
   - Follow-up referrals.

2) Identify key organization contacts:
   - Who knows about the problem(s)?
   - Who cares about the problem(s)?
   - Who should know/care about the problem(s)?
   - Who will be directly affected by expected changes?
   - Who can provide information on problem(s)?
     - Customers/clients?
     - Management/supervision?
     - Sales/marketing?
     - Engineering/technical?
     - Operations/production?

3) Systematically assess incidents within the organization:

- Examine an unsuccessfully managed incident within the organization:
  - — What was the incident?
  - — What happened?
  - — What did you do?
  - — What did you learn?
  - — What would you do differently next time?
- Examine a successfully managed incident within the organization:
  - — What was the incident?
  - — What happened?
  - — What did you do?
  - — What did you learn?
  - — What would you do differently next time?

Just as there is no "silver bullet," there is no "one shot" approach that will achieve the objectives that are outlined in this chapter. Safety and health training must be understood as an ongoing need within every organization.

Finally, effective safety and health training and effective safety management practices cannot be separated. Members of the organization's management must support overall organization goals. This will require measurable safety and health training objectives, documentation of benefits—results, outcomes, and/or impacts—as well as regular status reports. When these benefits are realized, the safety and health function and associated training will be performing as partners whose alliance supports the organization's strategic goals.

REFERENCES AND FURTHER READING

Bennis, W. (May 1991). "Future Organizations Need an 'ACE'." *United News-Journal* [Special Edition], pp. 1 and 12.

Burk, J. L. (December 1996). "Nine Things You Should Know about Safety Training." *Safety + Health* 154(6):66-68.

Burk, J. L. (December 1991). "Strategies for the Future of Safety and Health." *Safety and Health* 144(6):46-48.

De Geus, A. P. (March-April 1988). "Planning as Learning." *Harvard Business Review,* pp. 70-74.

Geller, E. S. (1996). *The Psychology of Safety.* Radnor, PA: Chilton Book Company.

Hammer, M. and S. A. Stanton (1995). *The Reengineering Revolution.* New York: Harper Business.

_____ (1989). "Program Design and Evaluation." Injury Prevention: Meeting the Challenge. New York: Oxford University Press, pp. 63-88. Published by Oxford University Press as a supplement to the American Journal of Preventive Medicine 5(3).

Kast, F. E. and J. E. Rosenzweig (1972). "General Systems Theory: Applications for Organization and Management." *Academy of Management Journal* 15(4):447-465.

Lahey, J. W. (June 1988). "Tomorrow's Workplace." *Safety and Health,* pp. 46-49.

Lynton, R. P. and U. Pareek (1978). *Training for Development.* West Hartford, CT: Kumarian Press.

Manuele, F. (1993). *On the Practice of Safety.* New York: Van Nostrand Reinhold.

McGee, P. and J. L. Burk (Summer 1991). "When It Comes to Training Are You Throwing Away a Strategic Weapon in Your Corporate Arsenal?" *Goodyear Progressions in Training.* pp. 1-11.

McLarney, W. J. and W. M. Berliner (1970). *Management Training: Cases and Principles.* Homewood, IL: Richard D. Irwin.

Ouchi, W. G. (1981). *Theory Z.* Reading, MA: Addison-Wesley Publishing Company.

Peters, T. J. and R. Waterman, Jr. (1982). *In Search of Excellence.* New York: Harper & Row.

Petersen, D. (1988). *Safety Management: A Human Approach.* Goshen, NY: Aloray, Inc.

Petersen, D. (August 1991). "Safety's Paradigm Shift." *Professional Safety*, pp. 47-49.

Rekus, J. (April 1993). "Training Should Treat Participants as Adults Eager to Exchange Ideas." *Occupational Health and Safety*, pp. 42-84.

Saari, J. (1990). "Safety Strategies for the 21st Century: From Accident Prevention to Safety Promotion." *Proceedings of the Annual International Ergonomics and Safety II Conference,* edited by Biman Das, pp. 975-982.

Stata, R. (Spring 1989). "Organizational Learning—The Key to Management Innovation." *Sloan Management Review*, pp. 63-74.

Veltri, A. (August 1991). Associate Professor, Safety Studies, Department of Public Health, Oregon State University. [Personal communication].

Veltri, A. (1991). "Transforming Safety Strategy and Structure: A Research Study." [Unpublished research paper].

Wack, Pierre. (September-October 1985). "Scenarios: uncharted waters ahead." *Harvard Business Review,* pp. 73-89.

Waterman, R. H. (1987). *The Renewal Factor.* New York: Bantam Books.

Wick, J. A. (October 1989). "Roles of the Safety Professional Will Change, Expand in the Next Decade." *Occupational Health and Safety Review*, pp. 48 and 125.

Chapter 23

Medical Services

Ronald F. Tiechman

INTRODUCTION

With spiraling costs associated with workers' compensation and medical expenses, the safety professional must assemble the best team possible to minimize injuries and illnesses in the workplace. When injuries and illnesses do occur, systems must be in place to minimize their adverse effects. Health care professionals must be considered vital members of the safety and health team and the prevention/minimization system. This chapter examines the members of the health care profession and the roles they play in the medical services system.

OVERVIEW

In the turbulent business world of the late twentieth century, many companies have been forced to undergo significant change just to stay solvent. It is against this background of changing missions, product lines, alliances, and markets that many employers have come to appreciate fully their employees as their most valuable resource. The second half of the century saw American industry adopt the concept of the cost-effectiveness of preventive maintenance for machines and equipment. But, only since the 1980s have many employers realized that money spent on maintaining good operating conditions and preventing problems within their workforces is also money well spent.

Company medical departments, which grew out of "stitch and patch stations" and then evolved into industrial medicine clinics, have become much more comprehensive in their focus. They now typically provide services on a wide variety of issues within the broad field of occupational health and safety. The primary objective of an occupational health services department is to preserve and protect and, in the case of injured or ill employees, to restore the health of employees. Most enlightened employers give their occupational health services a broad mandate. Each company establishing an occupational health service should first decide which objectives it wishes the service to focus on. Much of this decision will be driven by

- The nature of the business (for example, high-hazard industry versus office-based service provider)
- The size of the business (for example, multiple thousands of employees versus a handful of relatives)
- The geographic distribution of the business (for example, all employees at one site versus multiple sites in several states and countries)
- The organization of the business (for example, health, safety, benefits, regulatory compliance, workers' compensation administration, and so forth all report to one staff head versus each of these reporting to different executives, such as personnel, finance, operations, facilities and legal)

The objectives of a company's comprehensive occupational health service may include the following:

- Ensuring that workers have the ability to perform their assigned job functions in a safe manner, without presenting a threat to themselves or others
- Preventing and minimizing the negative health impacts, on workers, of all manner of workplace exposures
- Striving to establish and maintain a drug-free workplace
- Ensuring the provision of appropriate treatment and rehabilitation for injuries and illnesses that are either work related or that can impact the workers' ability to work
- Minimizing the impact of injuries and illnesses on employees' attendance and productivity and on the company's bottom line
- Ensuring compliance with health and safety-related federal, state, and local laws
- Striving to enhance the overall health and well-being of employees, retirees, and the dependents of both of these groups

A partial listing of the occupational health and safety-related federal, state, and local laws with which a provider (individual or department) of occupational medicine should be familiar includes the following:

- Occupational Safety and Health Act General Duty Clause
- OSHA recordkeeping requirements for work-related injuries and illnesses
- OSHA Hazardous Communication Standard
- OSHA Hazardous Material Handling Standards
- OSHA Bloodborne Pathogens Standard
- OSHA Hearing Conservation Standard
- Americans with Disabilities Act (ADA)
- Family and Medical Leave Act
- Federal Department of Transportation (DOT) drug-testing requirements
- DOT commercial drivers' license medical clearance examinations requirements
- Federal Aviation Administration (FAA) aviation medical examination requirements
- OSHA first aid requirements
- OSHA medical surveillance requirements for workers exposed to various substances, for example, lead and asbestos
- OSHA Violence in the Workplace Guidelines
- State specific OSHA regulations
- State workers' compensation statutes
- State hazardous material handling regulations
- Local health and safety regulations

The patchwork of complementary and sometimes conflicting regulations continues to expand, as does the range and complexity of some of the regulations. No single listing can cover all of the applicable statutes, which is why the choice of provider and provider setting is so important to most companies.

As the list of objectives for a company's occupational health service has become longer and the mechanisms for meeting these objectives has grown, so too has the number and types of occupational health professionals, the different occupational health Services available from these professionals, and the different settings for the provision of occupational health services (Table 23-1). The remainder of this chapter discusses these three areas.

OCCUPATIONAL HEALTH PROFESSIONALS

Occupational Medicine Physicians

Occupational medicine physicians are medical doctors who have devoted their careers to protecting and preserving the health, safety, and well being of employed individuals and entire workforces. Occupational medicine is a specialty area of the field of preventive medicine, and thus, the focus of these physicians is much more on preventive practices, to reduce the need for curative and rehabilitative medicine. These physicians are often well trained clinically to evaluate and treat work-related injuries and illnesses, while minimizing the impact of the illness or injury on the life of the patient and on the operations of the employer—in that order of importance. Preventive medicine is a population-based specialty, rather than an individual patient-based one, which means that occupational medicine physicians are oriented to developing programs and interpreting health hazards to protect the health of groups of people, for example, entire workforces, neighborhoods, and plant departments.

Occupational medicine physicians receive special training, often leading to a Master's degree in occupational health. This training typically includes courses in epidemiology (the study of disease and risk factors in population), toxicology, industrial hygiene, public health law, risk assessment and risk communication, occupational diseases, business management, and ergonomics. These physicians are very knowledgeable in employee-related federal and state laws, such as the ADA, the Family and Medical Leave Act, OSHA medical surveillance requirements, and DOT medical requirements for commercial drivers' licenses. Occupational medicine physicians frequently are skilled at helping companies develop employee drug-testing policies and then reviewing the results of drug screens [(functions of the Medical Review Officer (MRO)].

Since the 1950s, there has been a board-certifying examination for physicians in occupational medicine. The American College of Occupational

Table 23-1. Health Professionals, Occupational Health Services and Settings for the Provision of Occupational Health Services

Occupational Health Professionals

Occupational medicine physicians
Physician assistants/nurse practitioners
Occupational health nurses
Practical nurses/medical assistants
First responders
Industrial hygienists
Ergonomists
Wellness specialists/health educators

Occupational Health Services

Work-related injury and illness care
Disability management programs
Early return-to-work programs
Preplacement medical examinations
Periodic medical surveillance examinations
Return-to-work and fitness-for-duty examinations
Respirator medical clearance examinations
Independent medical and second opinion examinations
Commercial drivers' license medical examinations
Drug-testing programs
Employee assistance programs (EAP)
International travel evaluations and vaccinations
First aid and cpr training and programs
Back schools and injury prevention programs
Health promotion, health screening, and wellness programs
Violence-in-the-workplace training programs

Settings for the Provision of Occupational Health Services

In-house medical department
On-site medical department with a contract physician
Hospital-based occupational health center
Industrial medical clinic
Urgent care centers
Multispecialty group practices
Mobile testing companies

and Environmental Medicine is the world's largest organization of physicians dedicated to the preservation of worker health, with some 7,000 member physicians. Many excellent occupational medicine physicians are not board certified, and do not have a graduate degree in occupational health. In fact, there is a serious shortage of board certified occupational medicine physicians, essentially all practicing board-certified occupational medicine physicians worked for large industry, the government, or academia. Now, their services are available to small-and medium-sized businesses via consulting companies, hospital-based occupational health programs, freestanding occupational health clinics, medical group practices, health maintenance organizations, insurance companies, and so on.

Physician Assistants/Nurse Practitioners

Physician assistants and nurse practitioners are usually required to obtain a minimum of two years of formal training, often at the post-baccalaureate level. Their scope of practice will vary from state to state, but usually the two are treated fairly comparably within a state. These providers can usually examine, evaluate, and treat patients on their own, as long as supervision by a physician is available. Many nurse practitioners and physician assistants specialize in occupational health. They can obtain additional formal training in many of the same areas as occupational medicine physicians. Physician assistants and nurse practitioners typically practice within an in-house medical department, or in a freestanding or hospital-based occupational health clinic.

Occupational Health Nurses

Occupational Health Nurses are typically registered nurses with particular experience and supplemental education in occupational health. They provide screening, first-aid, wellness, and health education services. They can also provide occupational health program supervision and case management/disability management services. Occupational health nurses, many of whom are members of the American Association of Occupational Health Nurses, may receive specific training, achieve experience requirements, and then be eligible to sit for the examination to obtain certification from the American Board for Occupational Health Nurses. Occupational health nurses usually provide services within an in-house or on-site medical clinic, although a growing number are providing expert services as consultants.

Practical Nurses/Medical Assistants

Practical nurses and medical assistants may be licensed by a state nursing board or certified by passing a state examination, but to perform many of the more basic functions of both of these providers often requires only minimal basic training. These levels of providers must practice under the supervision of a physician or registered nurse. Practical nurses and medical assistants provide direct patient care and administrative support services in the full array of occupational health settings.

First Responders

First responders are typically employees who have an interest in helping coworkers in trouble. For many workplaces that are too small (or too low hazard) to have full-time occupational health professionals (physicians or nurses) and for the second and/or third shifts in workplaces that have these professionals during the day, medical emergency response teams have been created. Often, the company will pay for these individuals to receive formal CPR and first aid training. The availability of paramedics and emergency medical technicians within the workforce (who may volunteer with their local rescue squad) can greatly enhance this team's ability to respond to onsite emergencies. Medical emergency response teams should have their duties and level of responsibility comprehensively spelled out—in writing. An assessment needs to be performed to determine whether these duties will fall under the bloodborne pathogen standard. A mechanism needs to be developed to ensure that records are maintained and that injuries and illnesses are properly recorded on the OSHA 200 Log of Injuries and Illnesses. Frequently, the purpose of creating these teams is to comply with the OSHA requirement to have first aid services available within reasonable proximity to the workplace, but these teams provide outstanding benefits in employee morale and team building as well.

Industrial Hygienists

Although industrial hygienists are basically trained in engineering, physics, chemistry, or biology, this individual has acquired, by study and experience, a knowledge of the effects on health of chemical and physical agents under various levels of exposure. The industrial hygienist is involved in the monitoring and analytical methods required to detect the extent of exposure, and the engineering and other methods used for hazard control.

Ergonomists

Ergonomists are typically engineers, industrial hygienists, safety professionals, or occupational health professionals who have decided to focus on preventing injuries resulting from the worker-workstation interface. Ergonomics is not a new science, and the benefits of proper job design have been known for many years, but recently these issues have received greatly increased visibility due at least in part to OSHA's intent to promulgate an ergonomics standard.

Wellness Specialists/Health Educators

Wellness specialists and health educators are either bachelor's or master's level trained professionals who work with employees and sometimes with their dependents to reduce their health-risk behaviors. There is now sufficient evidence that well-designed and executed wellness and health promotion programs at the work site result in several fold savings for employers. In the current business environment in which profit margins for many companies are getting smaller, and benefit and health care costs are getting larger, many employers are realizing the fiscal (as well as the less tangible) wisdom of using wellness specialists and health educators to help their employees and other dependents of the company's benefit plan stay healthy and reduce their lifestyle risk factors.

OCCUPATIONAL HEALTH SERVICES

Work-Related Injury and Illness Care

The most common reason for an employer initially to seek out an occupational medicine physician is for the treatment of work-related injuries and illnesses. A physician with the proper orientation and training will be able to treat injured employees, without unneeded testing and without taking them out of the workplace unless necessary. In the United States, every state has workers' compensation laws, but they can vary dramatically across state lines. Some conditions are recognized as being compensable under the laws in one state but are not in another. Some states allow the employer to designate which physicians will treat their injured employees whereas others preclude the employer from directing this care. Some states hold a very narrow view of the definition of injury whereas other states hold a broader one. Some states have a published list of fees that can be charged for these medical services whereas others allow physicians to charge their customary fee. The physician utilized by an employer for the treatment of work-related injuries and illnesses should be knowledgeable in these laws, as well as in the specific treatment of occupational diseases. Even in a state that does not allow for employer-directed care of injured workers, an occupational medicine physician can still be used to review cases, ensure appropriateness of care, and interface with the treating physician regarding work restrictions and return-to-work issues. As preventive medicine specialists, occupational medicine physicians also focus on trying to understand what caused the injury and/or illness in an effort to prevent recurrences. An occupational health nurse can perform many of these same functions, but in most states cannot be an independent provider of treatment.

Disability Management Programs

Disability management programs are a form of secondary and sometimes tertiary prevention. They are useful for employers who wish to minimize the impact of injuries and illnesses on their employees and on their workforce as a whole. The most effective of these programs are in place well before an accident occurs and function as soon as an employee is injured. The most important concept of these programs is that the employer's most valuable resource is his or her employees and that the employer needs to be made aware of this fact. Many employers have discovered that by taking an active role in managing their employees' disability, they dramatically reduce lost work time, remarkably cut the number of cases that go to litigation, markedly drop the costs of their injuries, and, perhaps most important, increase employee loyalty. Occupational medicine physicians and occupational health nurses can help establish these programs.

Early Return-to-Work Programs

Early return-to-work programs enable employees with medical restrictions to remain in, or return to, the active workforce even before fully recovering from an injury or illness. There are myriad reasons that this generally is a an excellent business decision:

1) Receiving partial work from an employee for full pay is better than no work for two-thirds pay (the usual salary under workers' compensation).

2) There is little dispute that the longer an employee is out of work for an injury the harder it is ever to return them to the workplace;

thus, the natural corollary is that the sooner they are returned the better.

3) Employees who feel that their employers are willing to accommodate their medical restrictions are more apt to return to full duty sooner and less likely to seek an attorney.

4) Knowledgeable, experienced employees can frequently be utilized for tasks other than their regular jobs during a period of restricted duty.

5) If none of these reasons matter, remember that the workers' compensation experience modifier (the calculation that determines the premium) is, in part, based on the number of lost workdays, not the number of restricted workdays worked, and thus, the same lost workdays may cost the employer money for several years consecutively.

A word of caution: I no longer have the term *light duty* in my vocabulary. I use *modified duty* or *restricted duty* instead. When injured workers are told that they will be sent back to work in a light-duty capacity, they believe this means that they will be put on a different job than they usually do. When these same employees are told that they are being given restrictions or modifications, then if their usual job meets these parameters, there is no reason that they cannot go back to their regular job.

Preplacement Medical Examinations

Preplacement medical examinations or, more precisely, postoffer, preplacement medical examinations to ensure compliance with the ADA, are medical evaluations conducted on new employees after an offer of employment has been made, but before they actual begin working. It is extremely important that the examining physician(s) has access to the job descriptions for the positions for which individuals are being hired; this allows recommendations from the physician(s) to be much more valuable. These evaluations can be very broad ranging and comprehensive, and need not only ask questions that are specific to the job for which the examinee is being hired, as long as all new employees in the category undergo the same evaluation. These examinations can be extremely valuable, both to the employer and to the new employee. They can be used to

- Detect active medical or health problems that would prevent the person from being able to perform the essential functions of the job

- Identify whether the individual meets the medical or physical requirements for the position
- Identify health problems that would present a direct threat to the individual or others if placed into the job
- Identify personal health problems that the individual should be made aware of and counseled to follow up on with his or her personal physician
- Establish a data base in the event of subsequent injury or illness during the course of employment
- Determine the presence of communicable disease, which can present a hazard to coworkers and the public.

Periodic Medical Surveillance Examinations

Periodic medical surveillance examinations are examinations conducted to monitor for health effects of exposure to any one of a wide variety of substances that are frequently found in workplaces. These can range from silicosis monitoring in miners, to lead monitoring in painters, to general monitoring in hazardous waste workers, to tuberculosis screening in health care and child care workers. Many of these programs are dictated by federal law (OSHA, MSHA) or by state or local law (health care workers, food service workers). Often, the law requires periodic medical surveillance examinations, but leaves the details up to the examining physicians. In these cases, it is especially critical that the employer seek the advice of an occupational medicine physician who understands the nature of the hazards and is aware of the latest scientific literature and the latest regulations.

Return-to-Work and Fitness-for-Duty Examinations

Return-to-work examinations are conducted on employees on their return to work from medical leave of absence, usually after they have been out of work more than a preset number of days. Fitness-for-duty examinations are requested when an employer has a concern about a worker's ability to perform his or her job safely; this is usually based on objective, documented evidence that a problem exists that may be health related. Both of these types of evaluations are designed to determine whether an existing employee can safely perform or return to his or her job duties.

Respirator Medical Clearance Evaluations

Respirator medical clearance evaluations are conducted on employees that are going to be required to wear respirator protection due to exposure to airborne toxins at the work site. The existence of a respirator protection program and policy is mandated by OSHA as are the program's components. The components of the medical clearance evaluation are usually not well defined in statute, with some exceptions, such as with asbestos exposure. Most occupational medicine physicians require pulmonary function testing and an actual medical examination prior to clearing a worker to wear a respirator in a hazardous environment.

Independent Medical Examinations

Independent medical examinations are generally used in workers' compensation cases in which there is some disagreement or question about the treatment or prognosis, or to identify answers to issues such as causation, maximal medical improvement, and permanent impairment rating. These examinations should be conducted by a physician other than the one who has been treating the injured worker, and he or she should be a specialist in the appropriate specialty, but need not be the same specialty as the treating physician. There are formal educational programs and certification examinations for physicians, which indicate a certain level of competency in performing these evaluations. There is not, however, a medical specialty examination in the independent medical examinations.

Commercial Drivers' License Medical Examinations

Commercial drivers' license medical examinations are required by the federal government (DOT) on holders of the licenses. Any physician, as well as several other levels of health care providers, can conduct these examinations. It is critically important to identify whether the professional conducting the examination is familiar with the regulations that dictate how they are conducted and what medical conditions require disqualifying a driver, for example, insulin-dependent diabetes, active coronary artery disease, seizure disorders, or blood pressure above 180/104. Medical certification cannot exceed two years, but can and should be for a shorter interval if the examiner believes that health conditions exist that place the driver, or the public, at risk if not followed more often.

Drug-Testing Programs

An ever-increasing number of employers are establishing drug-testing programs for their employees. The benefits of these programs include reduced injuries, absenteeism, tardiness, health care costs, scrap costs, and employee turnover; and increased morale, productivity, safety, and wellness. Any testing should be conducted only after a written policy is developed, adopted, and communicated to the entire workforce. The reason for testing may include pre-employment, post-accident, cause, reasonable suspicion, and random. A few of the issues that should be thoroughly discussed in developing the drug-testing policy are as follows:

1) What drugs are tested for?
2) Who will collect the sample?
3) What bodily substance will be collected (urine, breath, hair, blood, and so on)?
4) What are the consequences of a positive test?
5) What constitutes a positive test?
6) What are the consequences of refusing to submit to testing?
7) What are the grounds for reasonable suspicion testing?

Occupational medicine physicians who obtain additional training to become MROs are ideally suited to help employers develop these policies; however, other medical professionals, such as occupational health nurses, are frequently highly knowledgeable in this area.

Employee Assistance Programs

Employee assistance programs (EAPs) are, typically, programs paid for by employers to assist employees with a variety of personal problems that can affect their ability to perform their jobs. These programs can be very narrowly focused on issues of substance abuse, or they can offer help on a wide range of issues including

- Substance abuse
- Gambling
- Psychological problems
- Parenting issues
- Stress management
- Issues regarding caring for elderly or infirm parents

- Self-esteem issues
- Continuing educational programs
- Literacy projects
- Behavior modification programs

Employers frequently contract with EAPs on an employee-per-month fee basis, with a maximum number of encounters per member included in that fee. The services are usually available to the employee as well as dependents and retirees. The information gathered by the EAP is strictly confidential and often is covered by specific federal laws with more safeguards than a medical record. Part of the basic philosophy of offering EAP services is that if an employee is focused on his or her personal problems while at work, he or she is not working properly or efficiently, and often not working safely. Many studies have been conducted that support the fact that employers who offer a high-quality EAP to their employees see a several fold return on their investment within a few years of program initiation.

International Travel Evaluations and Vaccinations

As more companies compete in the global marketplace, more employees are having their health and safety put at risk by traveling to all parts of the globe without the correct orientation, protections, and immunization. There are two separate components to this problem, the more obvious one being that travel to foreign countries, and different parts of foreign countries, may pose certain health risks from air-, water-, food-, or insect-borne infectious diseases. Which countries are on the itinerary, which regions within those countries, whether or not the traveler will be staying on the beaten tourist path, or whether the traveler will be taking side pleasure excursions are all important considerations when deciding which, if any, prophylactic immunizations and medications the traveler needs to take prior to departure. The other major issue is determining whether the individual should even undertake the trip, and if so, determining what type of orientation he or she should have to minimize the risk of adverse outcome. Examples of potential problems include an employee with unstable insulin-dependent diabetes going on an extended trip to a location where eating at regular intervals may be difficult, or an employee with asthma going to Mexico City, Mexico or Sao Paulo, Brazil, both of which have high altitudes (making the air thinner) and serious air pollution problems. All these factors need to be considered prior to sending anyone on a trip for business.

First Aid and CPR Training Programs

An increasing number of employers are offering first aid and CPR training to their employees. Sometimes this training is driven by the fact that there may be no access to outside professional emergency services within minutes of the workplace; thus, employers are attempting to comply with the regulatory requirement to make these services available to workforces. At other times, this training is offered simply because the employer wants to have a core of people on-site who can respond to medical emergencies until professional help can arrive. Whatever the reason, the training must be professionally conducted, and decisions must be made regarding the role of these newly skilled employees, the hierarchy of control at the scene of a medical emergency, and the anticipated level of exposure to bloodborne pathogens to which these employees will be subjected. This latter issue may require a change in the company's exposure control plan.

Back Schools and Injury Prevention Programs

Back schools and injury prevention programs are frequently offered by local medical, rehabilitation, or physical therapy groups. They are an attempt to educate employees about how to avoid disabling injuries. There are many well-known and well-reputed back school programs, all of which focus on giving attendees a basic understanding of the anatomy of the back and teaching them how to accomplish whatever they need to while minimizing the stress they place on the structures of the back. Most of the programs also discuss regular stretching and warm-up exercises to avoid back strains, and what individuals can do, on their own, if they develop mild to moderate lower-back pain. Many of the same principles can be applied on a broader basis to prevent strains and sprains in other body parts. Thus, back schools can provide a solid foundation for a general injury prevention training program. These schools and programs usually include a primer on the causation of accidents and focus on the many steps in the life of an accident, with the ability to intervene at any of these points and prevent the adverse outcome from even occurring. Back schools and injury prevention programs can be very cost-effective, both in strictly financial terms, and in terms of avoiding unnecessary pain and suffering. Often, these skills and the newly gained knowledge are taken home to the employee's family and can provide accident prevention twenty-four hours a day, dramatically reducing an employer's group health utilization, as well as the experience with workers' compensation issues.

Health Promotion, Health Screening, and Wellness Programs

For most of the first decade of their widespread use, health promotion, health screening, and workplace wellness programs were provided by employers simply because they sounded like a good idea. Companies offered these programs for several years before there was any objective research to justify the expense of the programs. The data are now plentiful and they are consistent; funds spent on identifying and reducing risk factors and health risk behaviors in employed populations are funds that provide a several fold return on investment to the employer within a few years of implementing the program (not, as was initially thought, only over a ten-year payback period). The general consensus in the field (and the literature concurs), is that a single point-in-time health risk evaluation, without any ongoing attempt to modify the risks identified (for example, a one-day health fair with no follow-up programming), does not significantly reduce the level of health risk within the workforce. Ideally, the data obtained from screening a large segment of the workforce should be used to develop a series of behavior modification, health education, and risk reduction programs lasting one to two years, with a repeat of the overall screening at the conclusion of the series. This follow-up screening can be used to demonstrate the progress made in risk reduction during the course of the program, and it can also be used to define the issues that still need attention in the next round of programming.

Violence in the Workplace Training Programs

Violence in the workplace training programs are being offered by an increasing number of employers, not just quasi-government agencies or public assistance type offices. The problem has reached the point where it affects all manner of businesses, of all sizes, in all locations. As of the time of this writing, there are federal guidelines for the prevention of workplace violence in two industry sectors. These guidelines do not have the weight of law, but many of the recommendations made are extremely practical and reasonable. Nationally, there is a growing number of organizations offering to develop and deliver violence prevention training custom designed for your workforce and work environment. These programs tend to have two separate but equally important components: handling potential violent persons and situations, and ensuring that the work environment is as safe as possible to prevent a situation in which your employees can easily become victims. I am unaware of any objective studies that attest to the preventive value of these training programs, but they make sense and if they can help avert a workplace disaster, they are well worth the expense.

SETTINGS FOR THE PROVISION OF OCCUPATIONAL HEALTH SERVICES

In-House Medical Department

In-house medical departments are much less common now than they were a few decades ago. Many companies have cut back on their medical departments, sometimes to the point of outsourcing the entire department. Those companies that retain a company medical department usually have an occupational health physician as its head, and tend to have the department involved in most of the previously listed activities and often many others (for example, benefit package design, executive health programs, and policy development). This chapter does not devote further attention to this area, as the company medical director should serve as a valuable resource for and ally of the safety director. Therefore, the reader is strongly encouraged to seek him or her to discuss the range of services available from the department.

Onsite Medical Department with a Contract Physician

A large number of companies have their own in-house medical departments, but utilize an external contract physician for certain services, either on a full-time or, more commonly, a part-time basis. This allows small- and medium-companies, without the hazards or medical workload that would require an in-house medical director, to have access to a knowledgeable physician on a regular basis to address their occupational health needs. Sometimes these physicians are trained in occupational medicine, but more often they are not. The reduction in the number of full-time, in-house company physicians has resulted in a large number of physicians with many years of valuable experience being made available to provide their expertise to employers who may not need a full-time physician. This type of arrangement—having access to a knowledgeable physician on a regular basis—can be an invaluable enhancement to the health and well-being of a workforce. Ideally, this physician should be involved in the entire health and safety program and not utilized simply to do physicals and injury care. Periodically, the arrangement needs to be audited and evaluated to ensure that all of the appropriate services are being provided. The number of physicals and injuries, the full scope of the health services delivered, and the desired level of involvement of the physician are all determinants of how many hours of service for which the physician should be contracted.

Hospital-Based Occupational Health Center

For much of the past two decades, hospital-based occupational health centers have been opening around the United States. These programs run the gamut from very simple ones operated out of the emergency department, which utilize emergency department physicians and personnel, to full-service occupational medicine programs in freestanding facilities, with radiology and physical therapy services. These programs are sometimes operated by the parent hospital as a relationship builder with the local business community, with profitability not being the most important consideration. Often, these programs have easy, priority access to the hospital's network of specialists and diagnostic testing capabilities. A growing trend is for hospitals to hire well-trained, experienced occupational medicine physicians as the medical directors of these programs. These programs can serve as contract medical departments for employers, often offering on-site nursing staffing as one of the services. Numerous academic medical centers across the country also have occupational medicine programs. These are often associated with occupational medicine residency training programs and can provide access to some of the leaders in the field.

Industrial Medical Clinics

Industrial medical clinics have existed in the United States for more than a century and have proliferated rapidly in the latter part of the twentieth century. As company medical departments and in-house medical clinics diminish, these freestanding industrial medical clinics are filling the void. Initially, they were independently owned and operated, but in the late 1990s, there are several networks, or "chains," of industrial medicine clinics nationwide. As the market for externally purchased occupational health and safety services has developed, these facilities have often evolved from simply industrial medical clinics to comprehensive occupational health and safety service providers, including consultants, trainers, safety professionals, and even fully qualified and certified occupational medicine physicians and occupational health nurses.

Urgent Care Centers

Urgent care centers are fairly ubiquitous on the American health care scene in urban and suburban areas. Typically, these facilities offer extended hours and are designed to be able to accommodate unscheduled patients. They usually will see all manner of patients, with all manner of problems, from an infant with diarrhea, to an adolescent needing a school physical, to an employment applicant needing a commercial drivers' license medical examination, to a worker with a laceration, to a senior citizen needing treatment for several chronic medical problems. It is my experience that most of these facilities market themselves as providing industrial medical services, but have no staff members with specialized training or knowledge of these services.

Multispecialty Group Practices

Multispecialty group medical practices are increasingly including occupational medicine physicians among the specialists in their group. This trend began in California and has allowed these groups to provide a full spectrum of physicians "under one roof." This often allows for easy referrals to specialists when needed, and the opportunity for the occupational medicine physician to obtain informal consultations from a wide range of specialties. These group practices tend not to employ safety professionals, industrial hygienists, ergonomists, trainers, or the like.

Mobile Testing Companies

Mobile testing companies typically utilize customized tractor trailers to go from worksite to worksite to conduct large numbers of examinations. These can be anything from full physical examinations, complete with audiogram, x-ray, physical examination, blood work and the like, to simply audiograms or drug screen collections. The trucks of several large national companies crisscross the country daily. Mobile testing companies are usually cost-effective only when a large quantity of examinations is needed within a very short time period. The companies frequently know "to the minute" how long it takes to conduct fifty examinations. They can also provide administrative help in getting these examinations scheduled and processed. Mobile testing companies are niche providers of occupational health services, but obviously cannot provide injury care or any intermittently needed, small-volume service.

CONCLUSION

This chapter has provided a substantial, though not all inclusive, list of categories of occupational health professionals, occupational health services, and settings for the provision of these services. The challenging questions facing individuals in a safety role without much formal background in any of the occupational health and safety disciplines are as follows:

- How is the process begun of deciding which regulations are applicable?
- Which services are necessary?
- Which setting is optimal for the delivery of these services?
- Which professional should deliver the service?

Only the largest companies will need assistance with the full range of issues, but the possible combinations are myriad. It is worth stating here that the fewer different providers and settings are utilized the better. Identifying a single provider or organization that can satisfy all, or most, of your needs allows it to become more than just a vendor to your company, but allows it to serve as a partner in your efforts to provide health and safety services to your employees. The better this vendor knows your operations the greater the level of service you should be able to expect and receive from it.

In planning for the management and delivery of medical services, it is important to assess the expected needs of the workplace and the workforce that you have the responsibility of protecting. However, it is also critically important to attempt to assess the unexpected possibilities and have a plan in place for handling these. This may be as simple as maintaining a list of phone numbers of specialists and organizations that can be called on in an emergency, or it may require establishing a relationship with a professional on a retainer basis. The latter will allow for easier access to high-level assistance, some preexisting familiarity with your operations, and will not require you to go shopping in the phone book when you have a crisis and need fully qualified help.

ADDITIONAL RESOURCES

U.S. Department of Labor
Occupational Safety and Health
  Administration
200 Constitution Avenue, NW
Washington, DC 20210

National Institute for Occupational Safety
  and Health
4676 Columbia Parkway
Cincinnati, OH 45226-1998
  800-35-NIOSH

American College of Occupational and
  Environmental Medicine
55 West Seegers Road
Arlington Heights, IL 60005
  847-228-6850

American Association of Occupational
  Health Nurses
50 Lenox Pointe
Atlanta, GA 30324
  800-241-8014.

National Safety Council
1121 Spring Lake Drive
Itasca, IL 60143
  800-621-7619.

Institute for a Drug-Free Workplace
1301 K Street, NW
East Tower, Suite 1010
Washington, DC 20005
  202-842-7400

REFERENCES AND SUGGESTED FURTHER READING

American College of Occupational and Environmental Medicine (1992). "Scope of Occupational and Environmental Health Programs and Practices." *Journal of Occupational Medicine* 34(4):436–440.

Brandt-Rauf, P. W. and R. F. Teichman (1988). "Current and Future Needs for Occupational Medicine Physicians in Nonindustrial Settings: A Survey of Multispecialty Group Medical Practices and Health Maintenance Organizations." *Journal of Occupational Medicine* 30(12):928–933.

DiBenedetto, D. V., J. S. Harris, and R. J. McCunney, ed. (1998). *The OEM Occupational Health and Safety Manual,* version 2.1. Beverly Farms, MA: OEM.

Ducatman, A. M., S. Forman, R. F. Teichman, and R. Gleason (1991). "Occupational Physicians Staffing in Large U.S. Corporations." *Journal of Occupational Medicine* 33(5): 613–618.

McCunney, R. J., ed. (1996). *A Manager's Guide to Occupational Health Services.* Beverly Farms, MA: OEM.

Menzel, N. N. (1998). *Workers' Comp Management from A to Z., 2nd ed.* Beverly Farms, MA: OEM.

Moser, R. (1992). *Effective Management of Occupational and Environmental Health and Safety Programs: A Practical Guide.* Beverly Farms, MA: OEM.

Teichman, R. F., and P. W. Brandt-Rauf (1989). "Occupational Health Services in Non-industrial Settings in the U.S.: A Review." *Public Health Review* 90(17):51–68.

Teichman, R. F., and P. W. Brandt-Rauf (1990). "The Need for Occupational Health Nurses in Nonindustrial Settings: Results of a National Survey." *AAOHN Journal.* 38(2):67–72.

U. S. Department of Labor (1995). *The Occupational Health Professional's Services and Qualifications: Questions and Answers.* Washington, DC: Occupational Safety and Health Administration.

U. S. Department of Labor (1992). *"OSHA Handbook for Small Business."* Washington, DC: Occupational Safety and Health Administration.

Chapter 24

Systems Safety Engineering

Henry Walters

INTRODUCTION

An operator is putting a pump on-line after a repair. A drain valve located down stream behind a wall is overlooked and left open, resulting in a spill of acid to the ground. This would be an environmental reportable and could have easily resulted in an injury. Even though there are tagging systems that should prevent this from happening, it could also be a design problem. Do we need the drain line at that point and/or is there a system problem with the tagging procedure? If either of these two items is driving the at-risk behavior of "failing to secure the drain line" then there is a system barrier that is a root cause of at-risk behavior. (Brown and Hodges 1997).

The preceding example highlights the need for occupational safety professionals to become familiar with and use the principles of *systems safety engineering*. In the truest sense, systems safety engineering is a technical field, requiring specialized skills and training that pertains primarily to the design of safe equipment. As such, professionals in occupational safety and health are often reluctant to embrace and practice its concepts in the workplace. Therefore, many safety professionals have little or no formal training in the fundamentals of systems safety and are afraid to routinely use the many valuable tools available through its practice. The release of the Process Safety Management Standard in 1992 has forced many industries, such as the chemical industry, to incorporate systems safety principles (Roughton 1993; Goyal 1996; Roughton and Buchalter 1997). Another area closely related to principles of systems safety is *total quality management* (Esposito 1993; Weinstein 1996; Manzella 1997). In fact, if one examines the basics of systems safety engineering, it becomes obvious that in a way many of

Copyright ©1998 by Behavioral Science Technology, Inc. Reprinted by permission of author.

the principles are practiced, but that companies often fall short of formalizing the process. It is time that all companies begin formally adopting the principles of systems safety engineering.

To define systems safety engineering, let's break the term into more manageable pieces. Definitions of a "system" include "a group of elements forming a unity." "Safety" can be defined as "a state of functioning under predetermined conditions with acceptable minimum of accidental loss." A definition of "engineering" is "the act of maneuvering or managing skillfully." Putting the various elements together, one could define systems safety engineering as "the act of skillfully managing a set of interacting elements with the goal of establishing an acceptable minimum level of accidental loss." Doesn't this sound like something all managers should strive to achieve?

One of the primary goals of systems safety engineering is to be proactive rather than the traditional reactive (Brown 1993). This is accomplished by critically examining all aspects of a system in order to identify and control the hazards present before the worker is exposed to the hazard (Rogan 1994). In other words, systems safety engineering attempts to address all possible hazards before they become an incident. To accomplish this goal, one must understand that in the life cycle of any system there are possibly different hazards introduced at each stage and that these hazards may react differently based on the system interface.

The purpose of this chapter is to identify some of the systems present in the workplace that must be evaluated and to provide an introduction of many systems safety concepts and tools available for use in occupational safety. It is impossible to teach adequately all the techniques of systems safety engineering in a single chapter. For additional information on systems safety engineering, consult the references at the end of this chapter.

OCCUPATIONAL SAFETY SYSTEMS

In its most complex form, the overall system in occupational safety is the entire workplace (Fitzgerald 1997). The workplace consists of various parts or subsystems. The four major subsystems are as follows:

1) Physical plant
2) Tools and equipment
3) People
4) Managerial systems that link the other subsystems together

Each of these, in turn, consists of subsystems. Let us examine the major subsystems in more detail. The physical plant can be divided into subsystems in a myriad of ways. One way is to think of the plant as consisting of the following eight subsystems (Ferry 1990):

1) Basic structures and facilities
2) Internal storage and transport
3) Heating and cooling systems
4) Electrical power
5) Cleaning and waste
6) Utilities
7) Fire protection
8) Security and exclusion from the area

Each of these subsystems can be further subdivided. For example, fire protection consists of the fire-retardant material used in construction, the sprinkler system, fire doors, fire extinguishers, and so forth. How one determines the level of safety desired in each is addressed later.

Tools and equipment is a system consisting of the items employees use to do their job. It includes such subsystems as these:

- Workstation (for example, seating, tables, lighting)
- Tools used to do the job (for example, hammers, drills, wrenches, keyboards)
- Equipment used in processes (for example, fork trucks, presses, punches, extruders, copiers)

Even the workforce consists of subsystems. These might include the following:

- Management
- Foremen
- Hourly employees
- Teams or cells
- Committees
- Social systems such as culture

Managerial systems (Ferry 1990) include the following:

- Written procedures
- Placards, signs, notification systems, and alarms

- Training programs
- Supervisory personnel

LIFE CYCLE

All systems have a life cycle. This means that a system goes through a specific series of steps on its way from inception to the time it no longer serves a useful purpose. Life cycles can be described in various ways. In the sense of occupational safety, referring to the workplace as the system, the phases of the life cycle may be described as follows:

1) Concept
2) Definition
3) Construction
4) Operation
5) Termination

Concept Phase

The concept phase consists of the initial plans for a new company facility. Safety considerations in this phase may include initial hazard analyses for

- Industrial safety (for example, working surfaces, area access/egress, material handling)
- Fire protection
- Industrial hygiene (lighting, ventilation, noise, sanitation, heat, toxic agents)
- Human factors
- Vehicular safety (for example, loading zones, parking, traffic flow)
- Emergency preparedness
- Radiation protection
- Environmental protection

Someone must use the principles of systems safety in this phase to begin identifying potential hazards before plans proceed too far. It is much easier and less costly to implement safety during this phase than to wait and redesign a facility. Because it is difficult at this stage to identify how various subsystems may interface, the safety professional will conduct primarily what is called *preliminary hazard analyses* (PHAs). These analyses identify the obvious hazards that are encountered in almost every facility.

Systems safety engineering is proactive by identifying and eliminating all hazards before anyone is exposed to them. This means that, ideally, all hazards could and would be identified in the concept phase. However, the interaction of subsystems may generate hazards that might not have been identified in the analysis of the independent subsystem. For example, in the design of an aircraft's landing gear, the hydraulic system is designed to an acceptable level of risk. If there is a minor leak of hydraulic fluid, it is determined that there is no safety hazard involved within the hydraulic system. The brake system has also been designed to an acceptable risk. However, when the two subsystems are integrated, the heat generated by the brake system could ignite the leaking hydraulic fluid, creating a safety hazard. The same concept applies within the occupational system. Therefore, the principles of systems safety must be practiced throughout the life cycle.

Definition Phase

The definition phase begins when the initial concept has been solidified and it is time to begin intricate design of the facility. It is at this point that the interaction of subsystems may become obvious and more detailed. As the concept becomes better defined, safety professionals should conduct systems and subsystems hazard analyses (SHAs and SSHAs). Additional analyses include fault hazard analysis (FHA) and failure mode and effects analysis (FMEA). For example, it is important to analyze such things as the role of personnel in fire protection, or what happens to the fire protection system if the electrical system shuts down. The definition phase should be completed before construction of a new installation begins, which means that safety professionals should make a safety related "go/no go" decision prior to the beginning of construction.

Construction Phase

The construction phase is when the facility is built. During this phase, the safety professional's primary role is monitoring and quality control, ensuring that the plans developed in the previous phases are being followed. Training of the workforce often begins during this phase. Their introduction to the overall process requires verification of previous analyses and the initiation of operating and support hazard analyses (O&SHAs).

Operation Phase

By the time the life cycle reaches the operation phase, safety professionals should have worked themselves out of a job. All potential hazards should have been identified previously and either eliminated or their effects controlled. However, since the interaction of subsystems often produces new, sometimes unanticipated hazards, new analyses may be required in the operation phase. For example, management control systems are not usually fully introduced until this phase, and, therefore, new hazards may become apparent. However, this phase usually sees a greater use of some of the other systems safety engineering tools, to be discussed later.

Since the objective of systems safety engineering is to identify and remove hazards before workers are exposed to them, it is during the operation phase that such innovative approaches as behavioral-based safety (BBS) can play a vital role. BBS allows a systematic method of checks and balances for various subsystems. Research has found that in 85 to 95 percent of all accidents, the last common factor is a behavior. Since employees are not usually performing at-risk behaviors intentionally, BBS allows for the identification of behaviors involved in accidents, a method of collecting data to determine what is occurring, feedback to reinforce safe behaviors and identify barriers leading to at-risk behavior, and methods of problem solving to remove the barriers. Thus, BBS is another proactive tool to be used in systems safety engineering during the operation phase of the life cycle. The example given in the Introduction highlights the importance of considering behavior in systems safety approaches.

Termination Phase

As stated previously, the goal of systems safety engineering is to identify all hazards before anyone is ever exposed to such hazards. Therefore, the safety professional should be considering the residual hazards after a facility is vacated (termination phase) even as the system is being developed and operated. This is an area that has historically been overlooked. The EPA and specific acts such as the Clean Water Act, the Toxic Substances Control Act, and the Resource Conservation and Recovery Act, among others, play a key role in determining factors that should be considered during the termination phase. Otherwise, companies may be held liable for injuries and illnesses occurring years after they have vacated a property. As with the other phases, planning for the termination phase should be conducted as early as possible during the previous phases. Responsible companies consider the termination phase during concept formulation.

SYSTEM SAFETY TOOLS

Now that the life cycle has been discussed, it is time to discuss some of the various systems safety tools used in each of the phases.

System Safety Program Plan

The old saying that "people don't plan to fail, they simply fail to plan" is true in the safety field. Therefore, one of the most critical elements in any system safety effort is the system safety program plan (SSPP). The SSPP is the road map of the entire system safety effort. As such, it

- Identifies all activities and steps to be completed as part of the system safety effort.
- Establishes which analyses will be conducted and when they will occur throughout the system life cycle.
- Outlines the management review process and controls.

The success of the system safety effort is dependent on a well-written SSPP. The SSPP should denote which analyses will be conducted, the schedule for the analyses, and criteria (milestones) for management review prior to movement into the next phase. An important part of the SSPP is to establish the methods of risk assessment and determine how to test the system to ensure that acceptable risk has been obtained. The training program is also identified in the SSPP.

As a final check of the effort, a method of auditing is established. Auditing is important to ensure that

- The elements of the SSPP are being followed and that there are no problems preventing completion of the plan.
- The planned analyses and other activities are sufficient.
- The SSPP is still viable.

When properly developed, the SSPP provides a checklist that can be used to measure progress toward real system safety. Once the SSPP is written, the various hazard analyses specified by it become the major focus of the system safety effort.

Hazard Analysis

A hazard can be defined simply as the potential to do harm (Manuele 1997). Since the ultimate goal of all system safety efforts is to create a workplace with a minimal acceptable level of exposure to harm, hazard analysis is the heart of systems safety (Roland and Moriarty 1990). Each of the hazard analyses described subsequently contains at least three elements. These are a description of hazard severity, hazard likelihood, and a plan for hazard control.

Hazard Severity. Hazard severity can be defined in many ways. In its simplest form, hazard severity could be based on the extent of the injury and use standard categories such as fatality, lost-time injury, medical treatment, and first aid. Table 24-1 is a system of classifying hazard severity as described in MIL-STD-882B.

The problem is that the definition of each category often leaves much to be desired. For example, it is not unusual for companies to lower their lost-time injury rate by implementing limited duty (reduced responsibilities) for the injured person. Does this make the hazard less hazardous? Therefore, it is probably best for a company to develop its own definitions for severity.

Hazard likelihood can also be defined in many ways. If possible, one should use an actual probability based on *a priori* (before the fact) or *a posteriori* (result of test or experience) probabilities. Methods to determine these are presented subsequently. If actual probabilities can not be established, simpler methods to determine categories of hazard likelihood have been developed. They include the MIL-STD-882B example found in Table 24-2.

By combining Tables 24-1 and 24-2, it is possible to develop a hazard risk matrix that allows a semi-quantitative basis of resource allocation. Table 24-3 is such a combination and is similar to one used in marine safety (Wang and Ruxton 1997).

The resulting hazard risk index (HRI) is Table 24-4 (Vincoli 1993).

The HRI might then be used so that management would require that all HRI I items receive immediate attention, all HRI II items be scheduled to be addressed, all HRI III items be monitored, and HRI IV items be reviewed infrequently. A more precise method that would allow for prioritization within an HRI would be a formal risk analysis.

Risk Analysis. Risk analysis uses a factual probability and a sophisticated estimate of the loss as the criteria for a quantifiable risk assessment. Risk is

Table 24-1. Hazard Severity Classification

Category	Name	Characteristics
I	Catastrophic	Death
II	Critical	Severe injury
III	Marginal	Minor injury
IV	Negligible	No injury

Table 24-2. Classification of Hazard Likelihood

Description	Level	Specific definition
Frequent	A	Likely to occur frequently
Probable	B	Will occur several times in life of exposure
Occasional	C	Likely to occur sometime in life of exposure
Remote	D	Unlikely but possible to occur
Improbable	E	So unlikely assumed may not be experienced

Table 24-3. Hazard Risk Matrix

Frequency of Occurrence	Hazard Category			
	I	II	III	IV
A. Frequent	1A	2A	3A	4A
B. Probable	1B	2B	3B	4B
C. Occasional	1C	2C	3C	4C
D. Remote	1D	2D	3D	4D
E. Improbable	1E	2E	3E	4E

simply the probability of an event occurring times the most reliable estimate of loss. Thus, an occurrence with a probability of .001 and a resulting loss of $1,000 (risk of 1) is a greater risk than an occurrence with a probability of

Table 24-4. Hazard Risk Index (HRI)

HRI	Classifications	Suggested Criteria
I	1A, 1B, 1C, 2A, 2B, 3A	Unacceptable
II	1D, 2C, 2D, 3B, 3C	Undesirable
III	1E, 2E, 3E, 4A, 4B	Acceptable with review
IV	4C, 4D, 4E	Acceptable without review

.000001 but an estimated loss of $500,000 (risk of .5). The problem with quantifying loss is how does one account for intangibles such as personal anguish and lost production due to morale after the event? It is therefore advisable that the use of quantifiable methods be a basis of decision making, but it should not be the sole factor considered.

Hazard Control. The third part of all hazard analyses is the development of a method of hazard control. First, the development of hazard control begins with the hazard reduction precedence. The best solution to any hazard is to eliminate it totally through engineering. For example, if workers are exposed to a toxic substance in the workplace, is it possible to use a nontoxic substance in place of the toxic substance or use a method to avoid toxicity, such as holding one's breath and immediately vacating the area? If not, go to the second step, which is to initiate controls for the hazard through engineering: for the example, design a ventilation system that removes the toxic fumes before the worker is exposed to them.

If the hazard is still not lowered to an acceptable risk, the third step is to provide safety devices: for example, an appropriate respirator may be sufficient. The need to conduct FMEA becomes apparent in this case. The respirator may sufficiently reduce the risk of injury, but what happens if the respirator fails? The fourth step in hazard reduction is to provide warning devices. Thus, the worker would be warned that the toxic substance is present and that the respirator is not functioning properly. Should the risk still not be acceptable, the fifth step is to provide special procedures or training. In this example, it may be possible to counteract the effects of the toxic substance through some special method. Finally, if the system is still not at an acceptable level of risk, it is time to determine whether to accept the level of risk or dispose of (eliminate) the system altogether.

With these three parts—hazard severity, hazard likelihood, and hazard control—in mind, let us examine some of the types of hazard analyses more closely. Although there are numerous types of hazard analyses that should be conducted, the following ones are some of those most often used in occupational systems safety efforts.

Preliminary Hazard Analysis

The PHA is one of the first activities described in an SSPP. This analysis is initiated during the concept phase and is very cursory. During a PHA, the safety professional compares the original concept for the system to existing systems and uses historical data to determine what hazards may be encountered in the new system. Specific items considered in the PHA include the following:

- Examination of basic energy systems such as fuels, power sources, and pressure systems
- Identification of all legal regulations pertaining to the system
- Identification of initial subsystem interface hazards that may be generated
- Identification of environmental hazards such as possible natural disasters and radon exposure

In simple terms, the PHA is a basic evaluation of existing, known hazards that may be present in the system.

Subsystem Hazard Analysis

Subsystem hazard analysis (SSHA) focuses on subsystems, or elements within the system, that may be systems in and of themselves (Utley1994). Thus, an SSHA is basically a SHA conducted on a subcomponent of the system. The SSHA identifies all the hazards generated by the operation of the subsystem's components. It identifies the hazards that the subsystem generates in both its normal and fault occurrence modes. SSHAs should be conducted on every major identifiable subsystem.

System Hazard Analysis

The SHA is an in-depth investigation of the entire system to determine its level of safety. It should consist of numerous parts. A detailed description of how the system operates identifies the hazards present and provides information on the hazard severity and likelihood. Each hazard is then investigated to determine the causes. Since the hazard may be a result of the interface of subsystems, it is necessary to conduct an extensive review of the impact of the interface on the various subsystems. An evaluation of the acceptability of the risk is followed by the development of methods to control the hazard.

Fault Tree Analysis

To determine ways to reduce the risks, *fault tree analysis* (FTA) and *failure modes and effects analysis* (FMEA) are valuable tools (Stephenson 1991). FTA is a method that allows an investigator to go from the generic fault to the specific root causes for its occurrence. FTA is effective because it allows you to utilize qualitative analysis using deductive logic or quantitative analysis using reliability or failure data. By working your way backward through what has led to each level of failure, you can identify and then develop solutions to the most powerful root causes of a hazard.

Failure Modes Effect Analysis

FMEA is borrowed from the science of reliability. It is usually performed by reliability personnel to determine how the reliability of equipment affects the overall safety. The FMEA is generally an important part of the definition phase. When structured toward safety investigations, FMEA focuses on the single events that pose a state of hazard in the system. An FMEA identifies each subsystem of the system that may have failures and then determines what hazards may result from the failure. For example, a fire alarm system requires the tripping of a switch to activate it. What happens if that switch fails? At the risk of oversimplification, an FMEA can be used to identify the faults to investigate using FTA.

Once the hazards have been identified, there are a variety of techniques to develop acceptable solutions to address the hazards. The hazard reduction precedence should be followed as it is best, whenever possible, to remove the dependence on personal behavior from the hazard. In fact, an engineering solution often removes a barrier that leads to at-risk behavior. For example, it is better to design a machine guard that prevents someone from placing his or her hand in a pinch point than to depend on training employees not to place their hands near pinch points. Whichever method of problem solving is used to address hazards, it is important to follow some structured process as it is too easy to jump to wrong, or less effective, solutions when less rigorous procedures of analyses are not followed.

Operating and Support Hazard Analysis

This is very similar to the operating support hazard except that it is performed during the operating phase and the final stages of the construction phase. It is necessary to perform this analysis as the operating and support hazard may not identify all the hazards that may be generated by the actual operation of the system due primarily to interface synergism. Also, this analysis investigates issues surrounding support systems that may be technically outside of the original system. For example, a building on a major military complex once experienced a major flood due to the successful operation of its sprinkler system. Somehow the signal that was supposed to be received by an outside fire department failed, and due to a sensor within the building declaring a fire, the sprinkler system activated and remained discharging for an entire weekend until someone opened the building on Monday morning.

Fault Hazard Analysis

Fault hazard analyses (FHAs) are conducted as a subset of the SSHA, SHA, or operating and support hazard analysis. An FHA is a limited analysis in that it inductively investigates the effect of a single fault rather than the effects of multiple faults occurring at one time. It is similar to the FMEA. The final effect of a fault is traced through the subsystems and interfaces.

The following is a description of some of the statistical tools that may be used to determine actual probabilities of hazard exposure or system failure.

Statistical Tools

There are basically two ways of determining a probability—*a priori* and *a posteriori*. *A priori* (before the fact) probabilities are based on the inherent nature of the events. For example, before a coin is ever tossed, it is known that the probability of the coin landing with the "heads side up" on any given toss is 50 percent. Unfortunately, *a priori* probabilities do not often occur in occupational safety situations. An example of where an *a priori* probability may occur is a switch position. Disregarding human behaviors that may play a role, the probability that a three-position switch will randomly be in any given position is one in three (33.33 percent).

An *a posteriori* (after the fact) probability is based on past experience. This means that you must have some data on which to base the probability. An *a posteriori* probability is calculated by dividing the number of times something has occurred in the past by the exposures (times it could have occurred). Mathematically this appears as

P = occurrences / exposures

Since the laws of statistics allow you to expect the likelihood of something occurring in the future to be the same as has happened in the past, an a posteriori probability means that on any single future exposure you can predict the probability of a specific outcome. However, some issues must be considered. One is that there must be enough historical data on which to base the probability. Otherwise, the calculated probability may not be accurate because there have not been enough statistically significant occurrences. For example, using limited data to calculate the probability of having an automobile accident, one accident in ten trips would lead to a probability of .1 of an accident happening on any given future trip. However, if additional data were available, there may have really been two accidents in 100 trips, or a probability of .02. The space shuttle is an example of limited data. Prior to the accident, was it really logical to assume a 0.0 probability of an accident, and after the accident have there still been enough launches to predict realistically, based on *a posteriori* probability, a probability of an accident on any given launch? A general rule of thumb is to have at least thirty independent data points before even considering using *a posteriori* probabilities.

A second issue that is even more important from an occupational safety perspective is that you do not want to use historical data that is too old to be reliable. Processes or equipment may have changed that have had an impact on the previous experience. For example, automobiles are much safer today than sixty years ago and driving practices have changed, so it would not be valid to use all of the trips driven in the last sixty years as the pool of exposure to determine the probability of an accident. You should try to use as much historical data as are still valid.

Once you understand how to calculate the probability of an occurrence on any single, given trial, it is possible to determine the probability of a certain number of successes or failures occurring in the future dependent on the type of distribution of the exposure. The actual formulae used to do these calculations are beyond the scope of this chapter, but many of the more useful ones in occupational safety applications may be found in several of the references. However, one concept that needs to be understood is that the total probability of everything that can possibly happen is 1. Therefore, in any system the probability of success plus the probability of failure must equal 1. This means that if you know the probability of success or failure, you can compute the other by subtracting the known probability from 1 (Complimentary Law). Mathematically this appears as:

$$P_{(success)} + P_{(failure)} = 1$$

or

$$P_{(success)} = 1 - P_{(failure)}$$

or

$$P_{(failure)} = 1 - P(_{success)}$$

Since there are so many factors to consider on whether or not to trust an *a posteriori* probability, how confident can you be that the one you use is accurate? The solution is to generate a confidence limit when basing designs on desired probabilities of success.

Confidence Intervals

Based on an experienced failure rate, you can determine a more likely probability based on the degree of confidence you want. A 95 percent confidence interval means that there is less than a 5 percent chance that the true probability falls outside of the calculated interval. For example, if you had experienced 10 automobile accidents in 1,000 trips, would you want to use the *a posteriori* probability of .010 in predicting failure rates of a system? If you decide that you want to be 95 percent confident of your predicted probability, it would be better to develop a one-tailed, upper-bound confidence test. This would yield a probability of .017 (Walters 1995), so in designing the system we would want to expect a .017 probability of failure rather than the tested .010. This is a more conservative estimate of what to expect. The degree of confidence has an effect on the value used. A 90 percent test of the same parameter yields a probability of .015. Therefore, in most cases, safety predictions should use higher confidence intervals in order to err on the side of safety.

Event Systems

As stated previously, most systems consist of numerous subsystems. To determine the probability of success or failure of a system, it is often easier to determine the probabilities of success or failure of the subsystems and then to use one of various methods to generate a probability of success for the entire system. Event system analysis allows you to do this.

Systems are usually arranged in such a manner as to be considered series or parallel systems. However, problems arise when a system does not fall into a true series or parallel arrangement. Let us examine how to use event systems analysis to determine when systems should be designed in series and when they should be designed in parallel.

Series Systems. A series system is one in which all the subsystems are in a linear relationship. For example, a fire can be modeled as a series system consisting of subsystems of oxygen, fuel, ignition, and reaction (Figure 24-1). All four subsystems have to be present and working in order for a fire to occur—no fuel, no fire. Assume that it has been determined that the probabilities of success (occurrence) of the subsystems are .99, .3, .1, and .01, respectively. The probability of success of a series system is the product of the probabilities of success of all of the subsystems. Mathematically, this is

$$P_{(Fire)} = P_{(Oxygen)} \times P_{(Fuel)} \times P_{(Ignition)} \times P_{(Reaction)}$$

$$= .99 \times .3 \times .1 \times .01 = .000297$$

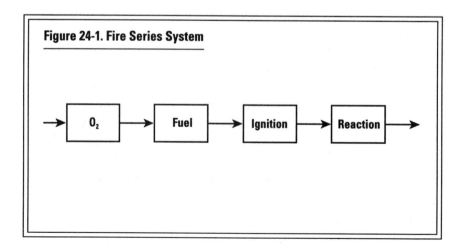

Figure 24-1. Fire Series System

The probability of failure of a series system is 1 minus the probability of success. In the previous example, the probability of failure (not having a fire) is 1 minus .000297, or .999703.

The success of the fire example is an undesired outcome—a fire. Therefore, it can be called an accident system. There are also series success systems. A success system is one in which the successful operation of the system is a desired outcome. For example, assume that there is a fire extinguishing system consisting of a water source, a water distribution system, and sprinkler heads. Once more, all three must work in order for the fire to be extinguished. The probabilities of the individual subsystems working are .98, .9, and .85, respectively. The probability of success of the systems then is

$$P_{(extinguish)} = .98 \times .9 \times .85 = .7497$$

and the probability of failure of the system is:

$$1 - P_{(extinguish)} = 1 - .7497 = .2503$$

Note that the probability of success of a series system is less than the individual probabilities of success of each of the subsystems. Therefore, you want to design accident systems in series because the more subsystems are in series, the greater the decrease in probability of the accident occurring. Conversely, you should refrain from designing success systems in series since the overall probability of success of the system is even less than the probability of success of any single subsystem. Success systems should be designed as parallel systems.

Parallel Systems. Parallel systems are those that allow a system to work if any of the subsystems works by itself. For example, the space shuttle has three auxiliary power units (APUs) on board arranged in such a way that any one of the three can supply sufficient power for the shuttle to operate safely (Figure 24-2). Just as in series systems, parallel systems can be either accident systems or success systems.

An example of an accident system in parallel is the numerous independent ways that an accident might occur. For example, let us use the possible ways that a person might fall off a ladder: the ladder might blow over, or it may have been improperly leaned against a building and slides, or the person may just reach too far and lose his or her balance. Consequently, there are numerous ways that the person could fall off the ladder. The probability of falling off the ladder could be determined by calculating all of the various ways that it could occur and then adding them together, but this would prove somewhat tedious. However, there is only one way that the person will not fall off the ladder: if all three of the possible ways to fall fail to occur. Using the Complimentary Law, it is easy to compute the probability of falling off the ladder by first determining the probability of not falling and subtracting that from 1. Mathematically, the success of any parallel system is

$$P_{(success)} = 1 - P_{(failure)}$$

Assume the probabilities of each of the ways of falling off the ladder are .001, .010, and .020, respectively. Using the Complimentary Law, the probabilities of not falling off the ladder for each are .999 (1 – .001), .990 (1 – .01), and .980 (1 – .02), respectively. Again, in order not to fall off the ladder, all three ways to fall must fail. Therefore, the probability of system failure is:

$$P_{(f)} = .999 \times .990 \times .980 = .969$$

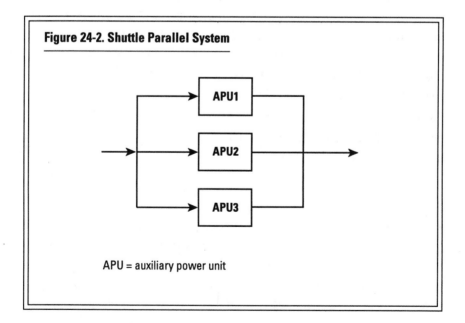

Figure 24-2. Shuttle Parallel System

APU1

APU2

APU3

APU = auxiliary power unit

Since the probability of success of the parallel system is 1 minus the probability of failure, the probability of success of the ladder accident system is

$$P_{(s)} = 1 - P_{(f)} = 1 - .969 = .031$$

Note that the probability of success of the system is higher than the probability of success of the individual subsystems. Accident systems should not be designed in parallel. If one of the ways the person could fall of the ladder could be eliminated, there would be a decrease in the overall probability of the person falling.

Success, or reliability, systems should be designed in parallel. Let us use the shuttle as an example. Assume the probabilities of success of the individual APUs are .900 (Figure 24-2). Since they are identical, we will assume identical probabilities of success. Since any one of the APUs will power the system, the only way for the system to fail is for all three to fail at once. The individual probability of failure is .100 (1 – .900) so the probability of system failure is

$$P_{(f)} = .100 \times .100 \times .100 = .00100$$

The probability of success is

$$P_{(s)} = 1 - P_{(f)} = 1 - .00100 = .999$$

Note that the probability of success of the system is considerably higher. One way to look at it is that if only one APU was on board, the probability of success of the shuttle would be 9 of 10 trips, or 900 of 1,000 trips. With the three APUs acting in parallel, the probability of success is 999 in 1,000 trips. Success (safety) systems should be designed in parallel.

Cut Sets

Sometimes a system cannot be designed as purely a series or parallel system. Figure 24-3 illustrates such an irregular system. Note that Block 2 leads to both Blocks 1 and 3. This means that there is no true series or parallel relationship among the subsystems. Therefore, it is impossible to use the formulae for series and parallel systems to determine success of the system.

To determine system success, you must first determine all the individual ways that the systems can fail. This is called determining the cut sets. By determining the minimum cut sets, it is then possible to calculate a probability of system success or failure. However, it is beyond the scope of this chapter to discuss this method, and the reader is referred to the references at the end of this chapter.

Fault Tree Analysis

FTA is a method that allows both quantitative and qualitative analysis of a system. The qualitative analysis allows for the determination of the root causes of a system failure. The analyst begins with the basic fault and works backward, identifying the causes of each level of failure. The analyst works through each level until he or she can no longer identify a cause for that level. At that point, it is determined that a basic fault has been reached.

Through the use of Boolean Logic, a system can also be quantitatively analyzed in order to determine the probability of a fault occurring. Figure 24-4 is an example of a fault tree.

In fault trees, the symbols have specific meanings. A rectangle indicates a fault requiring analysis. The "or" gate means that any of the faults going through it will by itself cause system failure (analogous to series systems). The "and" gate means that all the faults going through it must occur before there is system failure (analogous to parallel systems). The development of a fault tree continues until there is nothing but circles at the lowest level of every branch. A circle indicates a basic fault, which requires

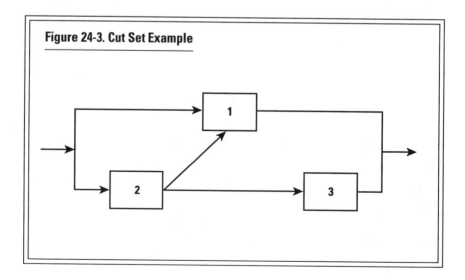

Figure 24-3. Cut Set Example

Figure 24-4. Sample Fault Trees

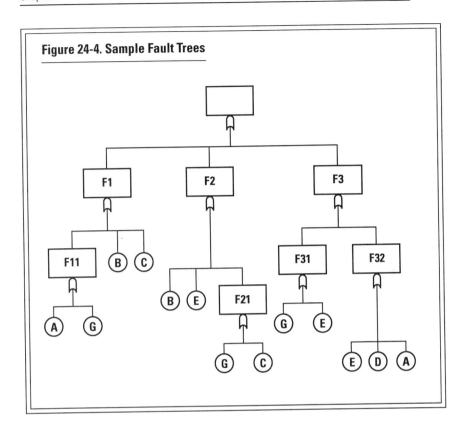

no further investigation. Safety professionals should be able to determine a fairly reliable probability for the occurrence of basic faults. Additional symbols may be used, but are not discussed here.

Quantitative analysis also allows the development of a Boolean-equivalent fault tree

$$P_{(A)} = .001 \quad P_{(B)} = .003 \quad P_{(C)} = .002 \quad P_{(D)} = .004 \quad P_{(E)} = .006 \quad P_{(G)} = .005$$

which simplifies the original tree and identifies the root causes of system failure. Using Boolean Logic, mathematically the fault tree in Figure 24-4 can be expressed as:

$$F = (A + G) + B + C + BE(G + C) + (G + E)ED \qquad A$$

which can be reduced to:

$$F = A + B + C + G$$

Using the given the probabilities of the basic faults in the sample fault tree, the probability of a fault occurring could then be calculated as:

$$F = .001 + .003 + .002 + .005 = .011$$

Figure 24-5 is the Boolean-equivalent fault tree of Figure 24-4. It is easy to see the value of this simplification as one can clearly see which areas need addressing in order to prevent the fault. For additional information on FTA, refer to the references at the end of the chapter.

CONCLUSION

Systems safety engineering is no longer a concept that is practiced by safety design engineers alone. The principles that allow for the proactive, safe design of equipment must be adopted by occupational safety professionals. When examined realistically, the principles are not difficult to adapt to that role.

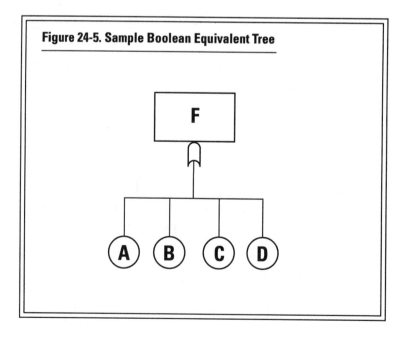

Figure 24-5. Sample Boolean Equivalent Tree

This chapter has noted some of the systems in place in the workplace and highlighted the general concepts such as the need to be proactive and yet perform different analyses at each stage of a system's life cycle. The failure to conduct these analyses may result in the introduction unexpected hazards. Each analysis serves a different purpose. The reasons for the various analyses and how to perform them were noted with suggestions on how to accommodate the basics to meet individual workplace needs.

The discussion of statistical tools points to the availability and importance of quantitative methods of analyses. The reader is encouraged to take advantage of the numerous references offered below.

REFERENCES AND SUGGESTED FURTHER READING

Brown, B. A. (February 1993). "Simplified System Safety," *Professional Safety*, p. 24.

Brown, B. and S. Hodges (August 1997). "A Behavioral Strategy for HAZMAT Safety, Cuts, Injuries, and Incidents," *Industrial Safety and Hygiene News*.

Esposito, P. (December 1993). "Applying Statistical Process Control to Safety," *Professional Safety*, p. 18.

Ferry, T. (1990). *Safety and Health Management Planning*, New York: Van Nostrand Reinhold.

Fitzgerald, R.E. (June 1997). "Call to Action: We Need a New Safety Engineering Discipline," *Professional Safety*, p. 41.

Goyal, R. K. (May 1996). "Cost-Effective Implementation of Process Safety Management in a Refinery," *Professional Safety*, p. 33.

Manuele, F. A. (July 1997). "Principles for the Practice of Safety," *Professional Safety*, p. 27.

Manzella, J. C. (May 1997). "Achieving Safety Performance Excellence through Total Quality Management," *Professional Safety*, p. 26.

Ragan, P. T. and B. Carder (June 1994). "Systems Theory," *Professional Safety*, p. 22.

Roland, H. E., and B. Moriarty (1990). *System Safety Engineering and Management, Second Edition*. New York: John Wiley & Sons, Inc.

Roughton, J. (August 1993). "Process Safety Management: An Implementation Overview," *Professional Safety*, p. 28.

Roughton, J. and D. S. Buchalter (January 1997). "OSHA's Process Safety Management Standard vs. EPA's Risk Management Plan: A Comparison of Requirements," *Professional Safety*, p. 36.

Rsiwadkar, A. V. (April 1995). "ISO 9000: A Global Standard for Quality," *Professional Safety*, p. 30.

Stephenson, J. (1991). *System Safety 2000*, New York: Van Nostrand Reinhold.

Sutton, I. S. (1992). *Process Reliability and Risk Management,* New York: Van Nostrand Reinhold.

U.S. Department of Defense *(1984). MIL-STD-882B 1984: System Safety Program Requirements.* Washington, DC: U. S. Government Printing Office.

Utley, P. T. (September 1994). "System Safety Hazard Analysis: A Tool for Determining Confined Space Entry Requirements," *Professional Safety*, p. 25.

Vincoli, J. W. (1993). *Basic Guide to System Safety.* New York: Van Nostrand Reinhold.

Walters, H. A. (1995). *Statistical Tools of Safety Management,* New York: Van Nostrand Reinhold.

Wang, J. and T. Ruxton (January 1997). "Design for Safety: U. K. Marine and Offshore Applications," *Professional Safety,* p. 24.

Weinstein, M. B. (July 1996). "Total Quality Approach to Safety Management," *Professional Safety,* p.18.

Chapter 25

Materials Handling

Richard Sesek and Donald Bloswick

INTRODUCTION

Materials handling takes place on some level in every business, ranging from the stock room of a small office to the plant floor of a large manufacturing facility. Although the size and scope of operations of these facilities differ dramatically, they may have similar hazards with similar solutions. In addition, material handling accidents can stop production in either facility. For example, a back injury to an office worker lifting reams of paper may have as devastating an impact on an office as does a serious forklift accident in a warehouse.

Whether the material is moved manually or with the assistance of mechanized equipment, materials handling is performed by virtually every worker in every facility. For many workers, material handling is their primary or only work responsibility. Even for workers who only incidentally perform materials handling tasks, these tasks may significantly impact the performance of their primary duties.

One method to increase the safety in materials handling is to eliminate it altogether. In other words, the fewer materials must be handled, the safer and more efficient the operation. In the initial design, or redesign of facilities, the objective is to eliminate unnecessary manual or mechanical handling of materials. Ideally, this is done when the facility layout and equipment are designed. The design should place the receiving area for raw materials as close as possible to the machinery that will use those materials. When the handling of materials cannot be eliminated altogether, an alternative is to use mechanical equipment in place of manual handling.

In advanced texts, the handling of materials by machine is often distinguished from manual handling. In this chapter, however, no such distinction is made and the term *materials handling* refers to both manual and mechanical means of handling.

CASE STUDY: IMPORTANCE OF MATERIAL HANDLING

Bill had stopped the conveyor on his production line many times, mostly to clear simple jam-ups of boxes. But today the motor was making noise. He noticed that the drive cover was loose and that some material appeared to be caught in the drive mechanism. Bill stopped the conveyor, removed the cover, and began to remove a piece of rag from the chain and sprocket drive. Suddenly, the conveyor restarted and Bill lost three fingers when his hand was drawn into the nip point.

Harold had been a forklift driver for many years and always checked trailers before driving into them from the loading dock, but today he and his coworkers were celebrating his impending retirement and he was eager to dig into the cake and see what going-away gifts his coworkers had brought him. There had never been any problems before, so Harold drove into the trailer to start the loading operation for the last trailer. The unchocked truck crept forward, dropping and wedging Harold's lift truck between the dock and trailer. Harold bruised some ribs and broke his arm when he fell from the lift.

When Don showed up for his shift on Thursday, the lifting device for the hoist in his work area was missing. Anxious to begin work, Don fashioned a lifting device using a lifting sling from a nearby workstation. The paper rolls weighed 850 lb. and the sling was rated for 1,000 lb., so Don assumed that it would be strong enough. Don rapped the sling around the rolls and tied a few knots to make it shorter. He tested his setup several times by tugging on the sling, and then he began to load rolls onto the printing press. After ten rolls had been loaded, the sling failed and dropped the rolls. Don's body and ego were both badly bruised when he turned to run and hit a post with his face and shoulder. Don was not seriously injured, but the machine required three days to repair fully.

SCOPE OF MATERIALS HANDLING

In the United States, the mishandling of materials is considered the single largest cause of accidents and injuries in the workplace (USDL 1996h).

Materials handling is estimated to account (at least in part) for 20–45 percent of all occupational accidents (NSC 1997c). This is not surprising when one considers the extent of materials handling operations. The sheer amount of materials handled is enormous. The average industry handles 50 tons of material for each ton of production (NSC 1997c). These figures and percentages indicate an area where major waste, loss of resources, and loss of productivity can occur and where management can effect major improvements in both productivity and profitability by reducing or eliminating accident-causing hazards.

A key to reducing materials handling accidents is to make the process as safe and efficient as possible. One of the most successful ways of preventing these accidents is through improved methods of operation, with increased emphasis on safety-conscious supervisory methods in materials handling. Management should ask the following questions [(adapted from NSC (1997c)]:

- Is the job designed to minimize manual handling?
- Are mechanical means in place to move materials?
- Are assistive devices utilized by operators (for example, hoists, lifts, hand trucks)?
- What are the inherent hazards of the materials moved?
- What are the hazards associated with the movement method?
- Are employees properly trained to safely handle the materials or use material handling equipment?
- Will protective clothing or other personal protective equipment (PPE) help prevent injuries?

Answering these questions will help the manager determine those areas where additional resources may need to be focused.

THE COST OF MATERIALS HANDLING ACCIDENTS

Handling materials manually is typically the slowest and least efficient method. More important, it is often the most hazardous method. Mechanization and automation are not new developments, but are being utilized with increasing frequency. Although mechanized methods typically result in fewer accidents and injuries, they create a new set of hazards that often have more severe consequences. For example, moving boxes by hand can result in a back injury, but moving them with a forklift truck can result in a fatality.

A great variety of manually operated accessories have been developed to assist workers with materials handling. These accessories include such simple items as special hand tools, bars, racks, dollies, and wheelbarrows. All of these call for careful storage, repair, and use. A piece of equipment or an accessory, regardless of its apparent simplicity, should not be put into use until its safe operation is fully understood by all users. A clear understanding of the hazards involved in both manual and mechanized methods can help the manager make better decisions regarding material handling options.

Back injuries are among the most common materials handling injuries and often result from lifting too heavy a load or from poor lifting technique. The Bureau of Labor Statistics reported that back injuries account for about 27 percent of nonfatal occupational injuries involving days away from work (USDL 1997).

In 1995, there were approximately 540,000 back injuries (USDL 1997) and the cost to society of low-back injuries is estimated to be between $50 and $100 billion per year (US DHHS 1996). Liberty Mutual Insurance (1997) reported in its 1996 Annual Report that 18 percent of all workers' compensation claims and 30 percent of costs are related to low-back pain. The total cost goes up considerably when combined with other types of materials handling injuries, such as overexertions, upper extremity cumulative trauma disorders, slips/trips/falls, and accidents involving machines. More than 50 percent of lost-time (involves days away from work) overexertion injuries incurred while lifting involve more than five days away from work and about 25 percent involve twenty or more missed days (USDL 1997).

The average workers' compensation claim for a low-back disorder is more than $8,000 (US DHHS 1996). If one considers the indirect costs of accidents such as hiring replacement workers, machine downtime, administrative costs relating to accident investigation, and reduced production, total accident costs climb quickly. The indirect costs of an accident are commonly estimated to be three to five times the direct or insured costs. Therefore, the true cost to a business for an average back injury may be closer to $25,000–$40,000 when such factors as lost production, overtime for replacements, and machine downtime are considered. If the company is operating at a 4 percent profit margin and incurs a back injury with a total cost of $25,000, the sales force must generate an additional $625,000 in sales to offset this injury ($625,000 x 0.04 = $25,000). Table 25-1 indicates how much revenue must be generated to pay for accident costs, given various profit margins.

It is important that managers realize that the incorporation of safety into all levels of a materials handling operation (design, training, management, medical, and so on) is cost-effective. Materials handling safety is important not only because it is "good for the worker" or "good public relations," but because it pays off financially in the long run.

It should also be noted that several areas of materials handling are covered directly by the Occupational Safety and Health Act (OSHAct). Therefore, action is not entirely voluntary. Some effort in the area of materials handling will be compulsory. For example, OSHA contains specific regulations regarding the use of industrial trucks, cranes, derricks, hoists, and other mechanical equipment. Management is also expected to provide safe equipment with which to handle materials and to ensure that employees are properly trained in their safe use.

MATERIALS FLOW

By analyzing the flow of materials, management can identify hazardous or potentially hazardous operations and locations. Material flow "bottlenecks"

Table 25-1. Sales Necessary to Offset the Cost of Accidents and Injuries at Different Profit Margins

ACCIDENT COSTS	COMPANY PROFIT MARGIN				
	2%	4%	6%	8%	10%
$10,000	$500,000	$250,000	$167,000	$125,000	$100,000
$20,000	$1,000,000	$500,000	$333,000	$250,000	$200,000
$50,000	$2,500,000	$1,250,000	$833,000	$625,000	$500,000
$75,000	$3,750,000	$1,875,000	$1,250,000	$938,000	$750,000
$100,000	$5,000,000	$2,500,000	$1,667,000	$1,250,000	$1,000,000
$500,000	$25,000,000	$12,500,000	$8,333,000	$6,250,000	$5,000,000
$1,000,000	$50,000,000	$25,000,000	$16,667,000	$12,250,000	$10,000,000

can be anticipated during the design phase and necessary steps can be taken to minimize the impact of these handling operations. Several techniques are available for determining potential hazards, including the job safety analysis.

To determine what materials handling processes can be eliminated, it is necessary to understand the materials-flow requirements. A thorough understanding of how materials must be processed, combined, and moved to produce a product is essential for eliminating unnecessary handling. In many cases, the process bottlenecks are also the biggest safety concerns. Eliminating these bottlenecks makes the operation both safer and more productive.

Eliminating unnecessary handling in the design phase is more cost-effective than trying to eliminate problems "after the fact." Also, some problems, which can be eliminated easily in the design phase, may be virtually impossible to control adequately once the facility is constructed. For example, machine locations within a plant are typically fixed and cannot be altered significantly after installation. Maximal optimization may not be achievable with alteration of paths and routes alone. Flow charts, flow diagrams, simulation, and similar techniques for displaying and analyzing information more graphically can be helpful in planning or revising material-flow patterns.

Material-flow patterns provide the information necessary to determine transport methods, routes, and aisleways so that the number of turns, blind corners, and crossing routes can be planned for or minimized. Considerations include the locations of warning signs or parabolic mirrors for increased visibility; physical barriers between pedestrians and equipment; one-way traffic zones; and training for both equipment operators and pedestrians.

The following basic materials-flow principles [adapted from Groover (1987a)] should be considered and implemented to the extent feasible and practical:

- Simplify the handling operations so that only activities essential to the transport are conducted.
- Minimize the distance materials must be moved.
- Minimize the time materials are not moving, but waiting to move (that is, loaded on or off transport vehicle, waiting in a queue, and so forth).
- Use gravity (consider safety and potential for product damage).
- Move loads both ways (move materials with equipment both directions–avoid unloaded traveling).

- Mechanize manual operations.
- Integrate materials handling with other systems (that is, storage).
- Integrate material information (identity, destination, origin, and so forth coded or marked on process bins, containers, or the product itself).
- Minimize the need to reorient or reposition transported items.

The first goal of material-flow analysis should be to eliminate unnecessary handling. The next challenge is to mechanize as much of the necessary handling as possible. Priority should be given to those strenuous tasks that present the largest ergonomic risk to workers. When feasible, the mechanization should involve full automation or should be controlled by the operator, without requiring direct manual handling.

When analyzing materials flow, expert assistance is particularly valuable to increase efficiency, maximize profits, and control hazards. The materials flow expert, however, will not always be a safety and health expert, so all planning should be carried out in conjunction with a safety specialist who can help design hazards out of the operation. It may also be appropriate to bring industrial engineering personnel into the review process to assist with line balancing and other production issues.

ERGONOMICS

Sometimes manual handling and lifting of materials is necessary, and ergonomic principles must be utilized to ensure that the manual handling is performed as safely as possible. The consideration of ergonomics in the design of manual material handling tasks can result in reduced physical stress and injury costs, as discussed previously.

There are two concepts that must be understood to appreciate the impact of the task on the cause and prevention of musculoskeletal injuries in industry: *moment and compressive force.* A moment is defined as the property necessary to cause or resist rotation of the body. This can be thought of as the effect of a force acting over a distance, or force x distance:

MOMENT = (Weight of Object) x (Distance from Center of
                                       Object Weight to Fulcrum)

A system is balanced, or in static equilibrium, when the force x distance (moment) on one side of the fulcrum equals the force x distance (moment) on the other side. In the case of manual material handling tasks, the moment about the low back is caused by the weight of the load, the distance that the load is held out from the body, the weight of the torso, and the torso flexion angle. These moments must be resisted by an equal and opposite moment caused by the muscles in the back.

Figure 25-1 represents a situation in which a 160-lb person holds a 50-lb weight out in front of the body with the torso flexed approximately halfway (45°). It is assumed that the upper body weighs approximately 50 percent of the total body weight. The clockwise moment caused by the load and body weight is 50 lb. x 24 in. (= 1,200 in.-lb.) caused by the load plus 80 lb. x 12 in. (= 960 in.-lb) caused by the upper body weight, for a total of 2,160 in.- lb. To balance this, the back muscle must generate the same 2,160 in.-lb. moment with a 2 in. moment arm (the approximate distance from the back muscles to the spine). This requires a force of 2,160/2 or 1,080 lb. The fulcrum (low back) must accept all the forces, for a total compressive force of 1,210 lb. (80 + 50 + 1,080 = 1,210 lb.). NIOSH (1983) has established that back compressive forces less than 770 lb. (action limit) "can be tolerated by most young, healthy workers" and that back compressive forces in excess of 1,430 lb. (maximum permissible limit) "are not tolerable in most workers."

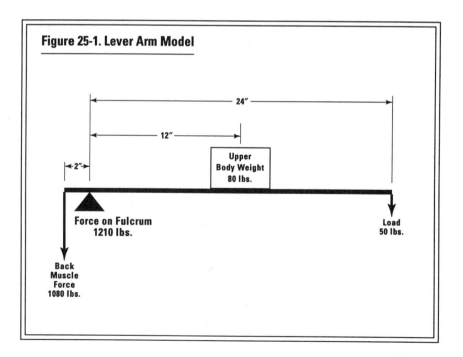

Figure 25-1. Lever Arm Model

Moments generated by loads in the hands can also cause stresses about the shoulder and elbow.

Manual material handling analysis can be used, not just to determine the stresses in a particular job, but to propose and justify the initial job design or job redesigns. For example, a biomechanical model analysis (University of Michigan 1989) was run to establish the relationship between load and lift posture. This analysis was conducted to establish acceptable load locations, in terms of the vertical distance from the floor and horizontal distance out from the ankles, for loads of different weights. The back compressive force was determined for a 50th percentile male (height = 70 in., weight = 166 lb.) for loads of 10, 20, 30, 40 and 50 lb. The results are noted in Table 25-2.

Fifty-pound loads lifted from the floor (or 4-in. pallet height) generate back compressive forces in excess of 770 lb. regardless of horizontal location. Loads in excess of 30 lb. lifted from the floor (or 4-in. pallet height) must be held close (12 in.) to the body. A vertical distance of approximately 36 in. appears to be acceptable for all loads when the horizontal distance is 20 inches or less. Note that these calculations do not recognize the fatigue caused by repetitive lifting or the additional stress resulting from torso twisting and/or constrained working postures. In addition, the back compressive forces for males of larger stature would be greater than those just noted.

In general, workplace factors that have been shown to contribute to musculoskeletal injuries during load handling include the following:

- Lifting heavy objects
- Moving and lifting bulky objects
- Lifting objects from the floor
- Lifting objects frequently
- Static bent postures

General lifting guidelines to reduce musculoskeletal stresses during manual material handling include the following:

- Keep the load close to the body.
- Use the most comfortable posture.
- Lift slowly and evenly (do not jerk).
- Do not twist the back.

Table 25-2. Back Compressive Force as a Function of Weight and Posture for a 50th Percentile Male

Vertical (in. from floor)	Load (lb.)	Horizontal Location (in. from center of ankles)		
		12	20	28
68	10	184	226	327
	20	236	307	416
	30	287	383	499
	40	335	456	577
	50	383	524	649
52	10	151	215	300
	20	206	296	399
	30	258	372	491
	40	309	445	576
	50	359	514	655
36	10	139	399	485
	20	194	486	590
	30	248	569	689
	40	300	648	**778**
	50	351	723	**867**
20	10	433	568	642
	20	503	668	**773**
	30	572	766	**902**
	40	640	**862**	**1029**
	50	706	**956**	**1154**
4	10	591	649	679
	20	674	763	**817**
	30	757	**878**	**956**
	40	**840**	992	**1094**
	50	**924**	**1106**	**1232**

Note: Numbers in bold indicate when last action limit of 770 lb. Of back compressive force is exceeded (NIOSH 1983).

- Securely grip the load.
- Use a lifting aid or get help.

Answers to the following questions regarding general job design can help reduce musculoskeletal stresses during manual material handling:

- Are strenuous lifts near waist level?
- Are machines difficult to load?
- Are material handling aids available and are the aids *usable*?
- Are containers satisfactory (size, handles, and so on)?
- Are walking/working surfaces clean and nonslip?
- Is task frequency acceptable?

Back belts are often issued in the hope that they will reduce back injury rates among employees who must lift or otherwise manually handle materials. One study has indicated a positive effect of back-belt use (Krause et al. 1996), but most studies fail to show consistently any clear biomechanical advantage to using back belts (Perkins and Bloswick 1995). There may be times when a back belt could be advantageous, particularly when prescribed by a physician, but at this time, there is no sufficient proof to warrant supplying them to all workers prophylactically.

EQUIPMENT

The size, weight, and shape of the material to be moved should dictate what type of equipment should be used to move it safely. The variety of handling devices can seem endless. Although most have specific requirements for their safe operation and use, some basic rules apply to each broad class or type of equipment. Nearly all materials handling equipment requires specialized training for safe operation, sometimes even extended courses and certification. Some representative types of equipment, including industrial trucks, cranes, hoisting equipment, conveyors, and automated material handling systems, are presented next. These brief overviews are intended only to alert the senior manager to some of the critical safety considerations involved in materials handling equipment and some of the basic requirements for their safe use.

Manual Hand Trucks

One of the most common devices used to assist in material handling is the hand truck or dolly. Hand trucks can be purchased or fabricated for objects of various sizes and shapes, but should be used only for the purpose for which they were designed. A hand truck designed for boxes may not be appropriate for lifting and moving drums. Common two-wheeled hand trucks appear easy to handle but can present problems for the unskilled user. The following guidelines can be helpful for hand truck users (NSC 1997c):

- Keep the load's center of gravity as low as possible.
- Place the load such that the weight is carried by the axle and not the handles. The truck should "carry" the load with the operator only balancing and pushing.
- Stabilize the load to prevent shifting, slipping, or tipping. For bulky or awkward loads, strap, chain, or otherwise secure the load to the truck.
- Load the truck to a height that allows a clear view when traveling.
- Avoid walking backward with a hand truck.
- When moving up or down an incline, keep the truck downslope from you.
- Move the truck at a safe speed that can always be quickly and safely stopped.
- Use hand trucks only for the purpose for which they were designed.

Four-wheeled trucks and carts require a few additional precautions. Unlike a two-wheeled truck, the four-wheeled truck can more easily be overloaded with the materials obstructing the operator's vision. Materials should be stacked so that they do not obstruct vision and should be evenly balanced to prevent tipping. Whenever possible, push rather than pull. This puts less stress on the back and will prevent heels and calves from being struck by the truck or cart.

Hand truck operators should be made aware of the hazards in their particular work areas as well as some general hazards such as running the wheels off bridge plates, platforms, or ramps; colliding with obstructions, uneven work surfaces, or other trucks; and pinching the hands or arms between the truck and other objects, particularly doorways. Special care should be taken when negotiating areas congested with people, traveling through doorways, or around blind corners. When not in use, hand trucks should be stored where they do not obstruct traffic or present a tripping hazard, preferably in an area designated and dedicated for their storage.

Powered Industrial Trucks

The powered industrial truck or forklift is one of the most common pieces of material handling equipment. Its power and versatility make it one of the most useful tools for quickly moving large amounts of material. Because it is so commonplace, it is often not given respect in accordance with its potential to do harm. The power and weight of these vehicles is often underestimated since they are highly maneuverable and relatively easy to control. Every year forklifts are involved in approximately 75–100 fatal injuries and an estimated 13,000 workers' compensation claims (Suruda et al., in press).

Industrial trucks are powered by electric motors or internal combustion engines. The designations specified by OSHA (USDL 1996e) are D, DS, DY, E, ES, EE, EX, G, GS, LP, and LPS. The first letter(s) indicates the power source (diesel, electric, gas, liquid propane) and the additional letter, if there is one, indicates added control of ignition sources. For example, *D* indicates a diesel engine–powered truck, whereas *DS* indicates diesel units with additional safeguards to the exhaust, fuel, and electrical systems. The type of lift used in an area must meet or exceed the designation requirement for that area. For more detailed information on truck designations, refer to the National Fire Protection Association's Fire Safety Standard for Powered Industrial Trucks (NFPA 1996).

Fueling Trucks. Refueling of industrial trucks should be conducted using approved dispensing pumps and only in areas and locations designated for refueling. Refueling should only be performed by trained and designated personnel with the truck engine stopped and no open flames, smoking, or other ignition sources nearby. Emergency refueling away from the designated pumping area should be done only using approved safety cans (NFPA 1996).

Recharging Truck Batteries. Many industrial trucks are battery powered. The most commonly used batteries are lead and nickel-iron containing corrosive chemical solutions (electrolyte) that are either acidic or alkaline. Hydrogen and oxygen gases are given off during the charging process. In certain concentrations and under certain conditions, these gases can be explosive (NFPA 1996). The charging process must be conducted only in designated areas that are equipped with facilities for flushing and neutralizing spilled electrolyte, proper fire protection, protection of the charging apparatus from damage by trucks, and adequate ventilation to prevent the buildup of vapors. In addition, whenever there is the possibility that the eyes or body may be exposed to the corrosive electrolyte solutions, eye washes and showers must be provided for quick flushing of the eyes and drenching of the body (USDL 1996c). When acid is diluted for use as electrolyte, the acid should *always* be added to water—never water to acid. This ensures that any resulting splash is as dilute as possible. A hoist, conveyor, or other mechanical handling equipment must be provided for handling and changing batteries when they are removed from trucks (USDL 1996e).

Operator Training and Operations. OSHA permits only trained and authorized operators to operate industrial trucks. This training should include the following items:

- Inspecting trucks prior to use
- Procedures for correcting truck defects
- Function and proper use of truck controls
- General truck-loading practices, including loading and unloading highway trucks, trailers, and railroad cars
- Truck operations such as traveling speeds, cornering speeds, driving near pedestrians, the importance of adequate clearances and of looking in the direction of travel
- Determining whether the load is safe to handle
- Correct piling/stacking of materials in stock
- Precautions when leaving a truck unattended
- Working in hazardous environments or with hazardous materials
- Refueling or recharging operations
- Specific hazards of the truck operators' tasks

All those who operate a forklift, even if only occasionally and in limited situations, must be trained. New hires, even those with previous training, need at least a test of operating proficiency, instruction in the hazards of the particular job, and a review of the facility's specific truck maintenance program. Operators with extensive experience on one brand, model, or type of truck still need time to become familiar with the placement of controls, different attachments, and handling characteristics of new trucks. Many commercial training programs are available through safety consultants, forklift dealers, or organizations like the National Safety Council. All training, however, should include practical proficiency testing and should be tailored to the specific facility in which it will be used. Videos and training booklets are good supplements to a training program, but should not be used alone and cannot take the place of hands-on training.

Industrial Truck Hazards. The typical safety manager cannot be expected to be familiar with all aspects of industrial truck operation, but several basic safety rules should be familiar to all. These rules include the following:

- Employees should never be permitted to stand under the elevated portion of the truck whether loaded or unloaded.
- Riders should not be permitted on industrial trucks unless the truck is designed for it, has a safe place provided for riding, and riding of trucks is authorized.
- Unattended trucks should have the load-engaging means (forks or lifting attachment) fully lowered, the brakes set, and the power off. Wheels should be blocked if the truck is on an incline.
- High-lift trucks must be provided with overhead guards for protection against falling objects.
- Trucks should not be used for lifting personnel unless they are specially fitted and equipped for doing so. These trucks should have
  — A firmly secured safety platform
  — A means for shutting off power to the truck from the platform
  — Fall protection and protection from falling objects as necessitated by conditions

Trucks must abide by all traffic regulations including plant speed limits, keeping the truck under control at all times, following other trucks at a safe distance (approximately three truck lengths), and yielding the right-of-way to other vehicles in emergency situations. Other trucks should not be passed at intersections, blind spots, or other dangerous locations. Operators should slow and sound their horns where aisles cross and where vision is obstructed. The truck should be driven only at a speed that allows the vehicle to be brought to a stop in a safe manner. Operators should travel with the load trailing whenever the load obstructs vision. Loose surfaces should be avoided and turns and grades negotiated at low speed. Grades in excess of 10 percent should be traveled with the load upgrade. Only stable, safe loads should be handled. The rated capacity should never be exceeded and must be adjusted downward for odd shapes and sizes of loads.

Maintenance and Repair. Trucks should be inspected prior to placement into service each day. When trucks are used around the clock, they should be inspected prior to each shift. Trucks that are in need of repair, are defective, or are in any other way not in safe operating condition must be removed from service until they can be repaired. All repairs must be completed by authorized maintenance personnel. Modifications to industrial trucks must not alter the relative positions of the functional components and additional counterweighting may not be added without first consulting the manufacturer. For more information on industrial trucks and their safe operation refer to ANSI/ASME Standard B56.1 Safety Standard for Powered Industrial Trucks "Low Lift and High Lift Trucks" (see ASME 1993) and OSHA Standard 29 CFR 1910.178 "Powered Industrial Trucks" (see USDL 1996e).

Powered Hand Trucks

A powered hand truck is a self-propelled truck that is controlled by a walking operator or standing rider. These are often referred to as "pallet jacks" or "mules." Most of the principles that apply to rider-operated powered industrial trucks can also be applied to powered hand trucks. In fact, the OSHA standard for industrial trucks, 29 CFR 1910.178, specifically includes powered hand trucks (USDL 1996e).

There are two basic types of powered hand trucks: low-lift trucks, which only lift the load high enough to permit horizontal movement of the load; and high-lift or "stacker" trucks, used when it is necessary to stack, tier, or palletize objects. These trucks are typically powered by battery-operated electric motors. These batteries are smaller than those used by industrial trucks, and the hazards associated with the charging and handling of these storage batteries are often overlooked. Appropriate safety precautions should be taken based on the number and types of batteries being used. For example, eye washes and safety showers should be located near the charging stations or wherever batteries are serviced and maintained.

The principle hazards associated with powered hand trucks (NSC 1991a) are as follows:

1) Contact with the device's moving parts: wheels, lifting devices, and so on

2) Falling loads due to improper stacking or operation of the truck

3) Collisions with the following:
   - Fixed objects such as storage racks or doorways
   - Other trucks or vehicles
   - Other employees

4) Injuries associated with mounting and dismounting a device not intended for riders

Many of the requirements for the safe operation of powered hand trucks are similar to those for full-size industrial trucks. At blind corners, doorways, and aisle intersections, the truck should be fully stopped and the warning device sounded. For easy and quick access, the horn or other warning signal should be operated from the control handle. Operators should not ride on a truck unless it is specifically designed for a riding driver, and other employees should never be allowed to ride on the truck. Trucks should always be operated at normal walking speed to allow the truck to be safely brought to a stop under all operating conditions. The transported materials should never obstruct the driver's view. Also, like industrial truck operators, powered hand truck operators must be trained and authorized to operate trucks. Trucks should be inspected prior to operation with necessary corrections made by authorized maintenance personnel before the truck is placed back into service. Operators must also be trained in the proper procedures for changing and charging truck batteries. In addition, like their larger cousins, powered hand trucks should be modified only in accordance with and after consulting the manufacturer.

In addition to these safety procedures, several safety rules are specific to powered hand trucks (NSC 1991a). All trucks should be equipped so that when the handle is fully raised or lowered the brakes are applied. When released, the handle should return to the fully upright position and the brakes should activate. Hand guards can be installed around the controls to prevent the hands and/or controls from contacting obstacles in tight quarters. The operator should face the direction of travel and walk ahead and to the side of the truck, except when it must be driven down an incline or close to an obstruction. When entering tight or restricted spaces, the truck should be operated with the operator trailing the truck to avoid being caught between the truck and an obstruction. Powered hand truck operators should always yield the right-of-way to pedestrians.

Skids, pallets, or the truck platform itself should be loaded in a stable manner to minimize the possibility that the load could fall on the operator or others. Trucks should be plainly and clearly marked with their rated load capacities. Ideally, trucks should be labeled with limits based on the product that is moved (for example, five boxes high, three across, two wide). Operators should be trained to refuse to move damaged, incorrectly loaded, or otherwise unsafe loads.

The operator should wear PPE as necessary for the work environment where the trucks are used. For example, if the truck is brought into an area requiring safety glasses, the operator should also be wearing safety glasses. When not in service, trucks should be parked in designated areas.

Front-End Loaders

Just as the forklift truck is the jack-of-all-trades in the manufacturing environment, so is the front-end loader to the construction and mining industries. These multipurpose vehicles are most commonly involved in accidents while backing up, but the most serious accidents occur when unloaded machines are driven at high speed (NSC 1990c). When traveling empty, these vehicles are typically much less stable and, at high speeds, have a tendency to bounce and weave. Operators must be aware of this limitation. Care must also be exercised when going up or down slopes or when traversing uneven ground, particularly with the load carried high, where it can create a lever arm and topple the loader. As with forklift trucks, the bucket should not be left unattended in the raised position nor should it be used as a work platform. Prior to operation, the driver should walk completely around the loader to inspect it, check the gages, sound the horn, and test the steering and brakes as he or she first begins to move the loader (NSC 1990c). The rules of operation and handling that apply to industrial trucks should also be applied to front-end loaders.

Highway Trucks and Railroad Cars

Most facilities unload their raw materials from or load their finished goods onto highway trucks and railroad cars. The loading and unloading of these vehicles addressed here.

Loading and Unloading Highway Trucks. One of the most serious accidents that can occur while loading or unloading highway trucks or trailers using industrial trucks is movement or "creeping" of the tractor and trailer. If a trailer moves forward due to the weight of forklift and load, the forklift can fall between the loading dock and trailer. To prevent movement of the tractor and trailer, the truck should be left in gear, the tractor's brakes should be set, and wheel "chocks" or blocks should be used to secure the rear wheels of the trailer. "Dock locks" or other recognized positive means for preventing the trailer from moving can be used in lieu of wheel chocks. If the trailer is loaded or unloaded without the tractor, jacks should be placed behind the rear wheels to prevent lifting of the front of the trailer during entry and, if necessary, near the trailer support for extra stability.

Loading and Unloading Railroad Cars. Wheel stops or other recognized protection should be in place to prevent railroad cars from shifting or moving during loading and unloading with industrial trucks. Positive protection and appropriate warning placards (typically a blue flag) should be in place to prevent the cars from being moved purposefully by a switch engine or car puller during loading or unloading (NSC 1990b). When rail tracks must be crossed with industrial trucks, they should be crossed diagonally. Industrial trucks and other equipment should not be parked within 8 ft of rail tracks.

Dockboards and Bridge Plates

Dockboards and bridge plates are used to bridge the gap between the loading platform of a facility and highway trucks, trailers, or railroad box-cars. The most serious accidents occur when the dockboard or plate shifts, moves, or fails while being driven over with an industrial truck or other material handling equipment so that the truck falls into the gap between the loading dock and shipping vehicle. Damaged dockboards can result in a tripping hazard as can smooth, wet, or oily ones. Workers may attempt to lift plates in an unsafe manner and suffer strains and sprains, or drop them on their feet. Improper storage of plates can result in injuries if they fall onto workers or cause damage if struck by an industrial truck. This damage can result in serious injury during use of the plates.

Bridge plates can be made of wood, steel, or light metals (aluminum) and must be designed to carry all imposed loads. Steel plates are the most common because of their strength and low cost. Many companies, however, choose magnesium or aluminum plates, despite their higher costs, because their lighter weight facilitates manual placement and removal of plates. Wooden plates or "gangplanks" are sometimes used for larger spans (more than 4 ft) because they have greater rigidity than unsupported metal plates (NSC 1990b). Regardless of the materials used, dockboards and bridge plates must be regularly inspected to ensure that they remain in good shape. Damaged plates should be removed from service until they can be properly repaired. Irreparable plates should be destroyed to prevent further use.

Plate dimensions are dependent on the loading dock or platform and the vehicle being loaded or unloaded. Ideally, the plate should span most of the width of the car or trailer door, thereby providing the most working room and minimizing the chance that an industrial truck will run a wheel off the edge of the plate. The plate should overlap the trailer or car door and the loading platform by at least 8 in. (20 cm) on each side (NSC 1990b). Plates should be rigid enough to resist springing or deflecting under load. A safety factor of at least four times based on a concentrated, maximum live load at the center of the span is recommended for bridge plates (NSC 1990b).

Dockboards and bridge plates must be secured or otherwise anchored prior to driving over them. They should be driven over slowly to avoid jarring the load loose. The combined weight of the industrial truck and its load should not exceed the rated load capacity of the dockboard or bridge plate (USDL 1996a).

Plates that are traversed by pedestrians should be provided with a non-slip surface to prevent slips and falls. The surface itself can be slip resistant, be painted with abrasive paint or covered with abrasive material, or be otherwise roughened to provide additional traction. The plate can be left smooth where industrial trucks drive and roughened only on the walking surface (the middle of the plate).

Employees are frequently injured during the placement or removal of plates. When handled manually, plates should be lowered into place or slid into position. Employees should not "walk," "tumble," or "drop" plates into position, because this can result in foot or leg injuries when the plates bounce. To facilitate safe manual handling of plates, handles or handholds should be incorporated into the plate or special lifting devices should be used. Enough personnel must also be assigned to the task to prevent musculoskel-etal stress. Manual material handling can sometimes be eliminated by fitting the plate with chains or lifting rings to allow it to be lifted by a forklift truck. However, these lifting rings or attachments should not present a tripping haz-ard to pedestrians or an obstacle to trucks while the plate is in use.

Cranes and Lifting Devices

A crane is simply a machine for vertical lifting and lowering, and for horizontal moving of loads. They are not designed for transporting loads great distances and are safe only when used within fairly narrow param-eters. Cranes and lifting devices should not be used to slide or pull materi-als horizontally; items should be lifted, then moved. Nonvertical lifting can lead to "swinging" of the load, resulting in injuries or property damage. Also, since cranes and lifting devices are designed for vertical lifting, nonvertically loaded cranes can become damaged and fail at significantly lower loads than their rated capacities. Because of the potential for injury, property damage, and product loss, only persons thoroughly familiar with crane functions and operations should be involved with their use. OSHA standards (USDL 1996f; USDL 1996g) contain some general requirements for cranes, including the following:

- Rated loads (lifting capacities) must be clearly marked on each side of the crane and on each lifting device used with that crane. These lifting capacities must not be exceeded except under planned engineering lifts supervised by a qualified engineer.
- Loads must be lifted and handled safely.
- Only employees designated, qualified, and authorized may operate cranes.
- Cranes must be thoroughly tested and inspected, including a load test, prior to use and after alteration.
- A preventive maintenance program based on the manufacturer's recommendations must be established and followed.
- Cranes must be inspected daily before use and, based on the level of service and operating conditions, must be periodically inspected more thoroughly.
- Modifications to cranes must be thoroughly evaluated and the crane's capacity rerated by a qualified engineer or the equipment manufacturer.

Training. Qualifications for operators vary with the type of crane, but all limit certification to the specific type of equipment on which the operator is trained and has been examined (ASME 1996). A thorough crane operation training program should include the following items:

- Review of potential and actual crane accidents.
- Design limitations and methods for calculating lifting capacities under different operating conditions.
- Erection, testing, and dismantling of cranes that are disassembled for transport to remote work sites.
- Selection, inspection, and use of steel wire and fiber ropes; chains, hooks, and slings; and below-the-hook lifting devices and attachments.
- Functional operational testing and mechanical inspection for damage of cranes prior to use. Maintenance personnel must periodically inspect for mechanical wear, lubricate, and perform preventive maintenance to repair and replace worn or damaged components.
- Safe crane operation and load handling. Operators must be familiar with all controls and what to do in emergency situations; how to

use wind indicators and other safety devices; how to adjust capacity for various boom angles and crane configurations; and safe techniques for lifting, lowering, traveling, and positioning loads.

- All applicable regulations, particularly those pertaining to operator duties.
- Methods of attaching loads including use of hooks, slings, and other devices.

Training should be formally presented with practical demonstrations using the specific types of devices that the operator will be authorized to operate. Training must be conducted prior to beginning work, documented, and should be reviewed annually or whenever the operator's understanding appears to be in question.

Inspection. All cranes and hoists must be inspected daily by operators prior to use. For around-the-clock operations, cranes should be inspected prior to each shift. These daily inspections should include both visual and functional tests. Abnormalities should be promptly reported to qualified maintenance personnel. When necessary, hoists should be removed from service with controls locked and tagged to prevent use until repairs can be made.

Functional operational tests should include hoisting and lowering, rotating and moving the device through its range of motion (for example, bridge and trolley travel), testing brakes and limit switches, and other safety features. One important safety feature that must be tested is the upper limit switch, which stops power to the hoist motor and prevents the load block from being drawn up into the hoisting mechanism.

Visual inspection should include the following components:

- Wire ropes (stretched, broken, kinked, twisted, or corroded strands)
- Chains (excessive wear; stretched, twisted, or distorted links)
- Controls or control pendant (loose or damaged, exposed wiring, damaged strain relief)
- Synthetic slings (cut or frayed threads, chemical burns, thermal damage)
- Hooks and below-the-hook lifting devices (bent, twisted, or corroded components)

If there is any question as to the condition of a hoist or a sling (or other lifting device), notify authorized maintenance or other qualified personnel and do not use the hoist.

The following guidelines can assist with developing or improving existing crane safety procedures:

- Obtain the relevant ANSI, ASME, and OSHA standards for the cranes, hoists, and lifting devices used at the facility.
- Develop a policy with respect to overhead lifting devices. Include provisions for training of authorized operators and maintenance personnel, inspection, preventive maintenance scheduling, and testing of new or altered devices. These activities should be well documented to ensure that all necessary actions are being performed and to demonstrate compliance with applicable standards and regulations.
- Determine inspection requirements and frequencies for cranes and hoists based on their level of service, operating conditions, and manufacturer's recommendations.
- Determine the lifting capacities for all cranes, hoists, and associated lifting equipment and ensure that these rated loads are clearly marked on the equipment.
- Develop operator instructions and daily inspection forms as part of the program.
- Provide additional training to authorized maintenance personnel including, when necessary, outside training courses with crane manufacturers, suppliers, or consultants.
- Review the program at least annually to ensure that all elements are strictly enforced.

Hoisting Apparatus

Every time a load to be lifted is attached to a hoist or crane, the art of rigging is being practiced. In the past, this job was performed only by a handful of specially trained individuals who often would fabricate their own slings and lifting devices. This required a complete understanding of the material strengths of the sling components and of rigging principles. Hoisting devices have proliferated at facilities and at work sites. As a result, the typical operator today is often not well versed in the limitations of slings and lifting devices, nor are they qualified to repair or modify them. Employees who use these devices must be taught how to use them for each particular lift that they perform. When an object cannot be lifted in a way that employees know is safe, assistance from a knowledgeable and qualified person should be sought. When possible, lifts should be preplanned so that the most appropriate rigging arrangement is used. In assembly-type manufacturing environments, where conditions are more constant, specially sized slings or special lifting devices can be fabricated or the transported item itself can be fitted with hooks or other places to attach the lifting hook. Again, understanding material-flow requirements in the planning stage can help ensure that important details such as these are addressed before the hoisting is performed.

All slings and below-the-hook lifting devices must be inspected daily prior to use and more thoroughly inspected periodically by a qualified individual (USDL 1996g). Periodic inspections are a function of frequency of sling use, severity of service conditions, the nature of the lifts being made, and experience based on the service life of slings used in similar circumstances (USDL 1996g). All employees who use slings or other lifting devices must be trained how to inspect these items for wear, damage, or other defects. Defective slings should be removed from service immediately, until they can be repaired by qualified personnel. Irreparable slings should be destroyed to prevent further use. Damaged items removed from service can be used as effective training aids to demonstrate what constitutes excessive wear and damage.

One effective way to guarantee proper sling inspection, storage, and maintenance is to store all slings in equipment tool cribs under the supervision of an authorized sling inspector. Any required maintenance such as lubrication of metal mesh slings or repairs can then be made by the qualified inspector.

The most common types of slings are chain, wire rope, metal mesh, natural fiber rope, synthetic fiber rope, and synthetic web. When choosing the best sling for a job, consider the size, weight, shape, operating environment, and the characteristics of the material moved (for example, sharp edges, smooth surface). Operators and authorized maintenance personnel must be familiar with how to use and care for whichever type is selected.

There are four primary factors to consider when using slings to lift a load (USDL 1996h):

1) Size, weight, and center of gravity of the load

2) Number of legs and the angles they make with the horizontal

3) Rated capacity of the sling

4) Usage history of the sling (for example, mishandling and misuse)

These factors must be evaluated and considered while following the safe operating practice required by OSHA (USDL 1996g). These practices are as follows:

1) Never use damaged or defective slings.

2) Slings must not be shortened with knots, bolts, or other makeshift devices.

3) Slings should never be loaded beyond their rated capacity.

4) Slings must be padded or otherwise protected from loads with sharp edges.

5) Suspended loads must be kept clear of all obstructions and never be lifted over or near employees.

6) Shock loading is prohibited as is pulling the sling from under a load.

7) The following steps should help meet the safe operating requirements outlined by OSHA (USDL 1996h):

   • When attaching a sling, inspect the item being lifted to verify that it is not secured or fastened to the floor or another item.

   • Slowly increase tension to avoid shock loading. Always accelerate and decelerate the load slowly.

   • Stop the load after it has been raised only a few inches and check the balance of the load.

   • Keep all personnel clear while the load is handled and never allow anyone to work under or near a suspended load. Never raise the load more than necessary.

   • Never leave an unattended suspended load. Loads should be lowered to the floor or positioned on a rack or other holding device when they are not being transported.

ASME Standard B30.20 (see ASME 1994) provides more detailed guidance on the rating, inspection, and modification of lifting devices.

Conveyors

A conveyor is defined by ANSI/ASME B.20.1 (see ASME 1993), as "a horizontal, inclined, or vertical device for moving or transporting bulk material, packages, or objects, in a path predetermined by the design of the device, and having points of loading and discharge, fixed or selective." Conveyors can be broken into two broad classes: unit-load and bulk handling systems. Unit-load conveyors are designed to carry discrete items such as boxes or finished products, whereas bulk conveyors carry continuous materials such as powders, ores, or other loose items. Consideration should be given to the specific type of conveyors used when developing a safety program.

Hazards associated with conveyors (NSC 1991b) include the following:

- Pinch points between moving conveyor components or items moving on the conveyor and stationary portions of the conveyor
- Pinch points caused by items "jamming-up" against one another on the conveyor
- Objects falling from the conveyor
- Falling components of the conveyor itself
- Workers falling from conveyors they attempt to cross or ride
- Bumping into conveyor components
- Accidental starting during maintenance or trouble-shooting
- Contact with power transmission components such as belts, chains, gears, or pulleys

Additional hazards associated with bulk material conveyors (NSC 1990a) can include the following:

- Fires from friction, overheating, or overload of electric motors
- Health hazards associated with irritating dusts
- Explosions from buildup of combustible dusts
- Increased potential for falling materials

The most serious injuries typically occur when an individual becomes entrapped in a pinch point or in-running nip point. Usually employees are pulled in when their hands, clothing, or tools are caught in a pinch point (NSC 1990a). Employees working with or near conveyors should wear close-fitting clothing that cannot easily become caught in the moving components of the conveyor. In areas where there is the potential for objects falling from conveyors, safety shoes should be required. Some bulk material handling environments may also necessitate the use of hard hats, goggles, respirators, or other PPE.

The most severe conveyor accidents occur during maintenance or servicing when the conveyor is accidentally restarted. Typically, the employee conducting the repairs is in a vulnerable position in close proximity with the power transmission apparatus and moving parts from which the guards have been removed. For this reason, strict adherence to the OSHA lockout/tagout standard is critical (USDL 1996b). Lockout/tagout ensures that all energy sources have been completely shut off, that residual energy has been dissipated, and that all energy sources are physically locked in the "OFF" position to prevent inadvertent operation. This is particularly important for conveyor systems since they often occupy a large area that cannot be continuously observed by maintenance personnel, and operators who are accustomed to continuous operation of the conveyors may attempt to restart them.

For both safety and convenience, many conveyors can be started and stopped from multiple locations. All controls should therefore be clearly marked and protected by recessing them or otherwise guarding them from accidental operation by employees or conveyed materials (NSC 1991b). Whenever there are two or more stop buttons on a conveyor, the conveyor should be wired to require a manual reset before the conveyor can be restarted after activation of a stop button. During restart, workers should be instructed to stand clear.

Start and stop controls should be clearly marked, remain free from obstructions, and be located so that as much as possible of the conveyor is visible. The locations and operations of all controls should be explained to employees working with or near the conveyor. A strong wire or cord can run the length of a conveyor so that a pull of the cord anywhere can activate the stop device and halt the conveyor. When multiple conveyors are used in conjunction with one another, it is necessary to mechanically and/or electrically interlock them.

All conveyors should be periodically inspected to ensure their continued safe operation with particular emphasis on safety features such as emergency stops, braking devices, and guard rails. Walking surfaces and passageways around, over, or under conveyors must comply with the ANSI/ASME B.20.1 conveyor standard (ASME 1993) and the NFPA 101 Life Safety Code (NFPA 1997). Hinged sections, to allow employees to pass through a conveyor line, should be fitted with interlocks to stop the conveyor while the section is up; made of lightweight materials; counterbalanced, or otherwise mechanically assisted to simplify access through the conveyor line; and fitted with a catch mechanism to hold them in place while they are open (NSC 1991b).

Screw conveyors are a special type of bulk material conveyor that move material with a continuous spiral screw mounted within a stationary, typically U-shaped, trough. An unenclosed screw conveyor presents a serious entrapment hazard. When accidents occur, they are generally very serious and include amputations and death (NSC 1997b). Covers and gratings must be securely fastened prior to operation of the conveyor. Feed openings, where materials enter the conveyor, should be fitted gratings. When unguarded housing or feed openings are functionally necessary, the conveyor should be otherwise guarded such as by fencing the area to prevent access to the screw mechanism. Employees should *never* walk on covers, gratings, or guards.

Manual roller conveyors rely on employee muscle or the force of gravity to move items. As a result, workers sometimes fail to exercise necessary precaution when working around roller conveyors. Employees can be injured if they attempt to climb up on roller conveyors to release stuck or jammed packages or try to cross them, and should be prohibited from doing so. Steel or wood plates can be installed between rollers to prevent an employee's limbs from being caught between rollers. The principal hazards associated with roller conveyors are loads falling off the conveyor and loads that "run away" (NSC 1997b). Guard railing on each side of the conveyor can help prevent materials from running off the edge of the conveyor. Guard rails are particularly advisable at corners and turns or where conveyors are elevated, and are absolutely necessary whenever conveyors are over areas where employees are located. To control the descent of heavy loads, retarders, brakes, or sections of powered conveyors can be installed.

Automated Guided Vehicles

An automated guided vehicle (AGV) is a material transport vehicle that carries a load from one location to another without operator intervention. AGVs are typically controlled by one of three methods: on-board control panel, remote call stations along the vehicle path, or central computer control. Central computer control requires continuous communication between the AGV and the computer regarding the AGVs whereabouts and status (Groover 1987a).

Examples of AGVs include driverless trains, automated pallet trucks, and unit- load carriers. Driverless trains are basically towing vehicles that pull one or more "cars" that are loaded either manually or automatically. Automated pallet trucks are typically loaded by an operator and then placed on the AGV path and the destination programmed manually. Unit-load carriers are typically the most automated and are usually loaded and unloaded by other machines (Groover 1987a). A "unit load" is a standard-sized container, bin, or pallet.

AGVs travel over prearranged routes with movement controlled by electromagnetic wires buried in the floor, optical guidance, infrared, inertial guidance (gyroscope), position-referencing beacons, or computer programming (NSC 1997a). They have great flexibility and are capable of delivery to various workstations.

AGVs must be equipped with a means for stopping if someone or something is in the path. This is usually achieved via a lightweight, flexible bumper that shuts off power and applies the brakes when contacted (NSC 1997d). The bumper must flex enough to allow the truck to stop completely before the truck contacts the pedestrian or obstacle. AGV bumpers should not need hardware or software logic or signal conditioning in order to operate. Also, AGVs in automatic mode must stop immediately when they lose guidance (NSC 1997a). Most vehicles are programmed to require manual reset before resuming motion (Groover 1987a).

AGV safety devices include monitors for guidance and velocity, guards, dead man controls, turn signals, and detection sensors (infrared optics and ultrasonics) to anticipate and avoid collisions (NSC 1997a). Blinking or rotating lights and/or warning bells can alert workers to the presence of AGVs in their work area, and turn signals can alert pedestrians to which way an AGV will be turning.

AGVs must have clearly marked and unobstructed aisles for operation. Employees must not ride these trucks. In addition, trucks should never be loaded or unloaded while in motion (NSC 1997d).

A thorough preventive maintenance program is important for AGVs and should include provisions for monitoring floor conditions, repainting of control lines and aisles, and periodic testing of AGV safety features such as the bumpers (NSC 1997a).

Automated Storage/Retrieval Systems

An automated storage/retrieval system (AS/RS) is defined by the Materials Handling Institute as "a combination of equipment and controls which handles, stores, and retrieves materials with precision, accuracy, and speed under a defined degree of automation" (Groover 1987b). AS/RSs utilize warehouse space efficiently and reduce labor costs, but because the equipment and automation costs are quite expensive, these systems require thorough economic analysis prior to implementation.

Typically AS/RSs comprise a series of storage aisles with one or more storage/retrieval (S/R) machines, usually one per aisle, used to deliver materials. The S/R machine is sometimes referred to as the "crane." Some systems are used in conjunction with human operators and are called "man-on-board" or "manaboard" systems, but this arrangement is typically restricted to small "order picking" jobs in which one or more small items are retrieved from bins (Groover 1987b). The S/R machine shuttles between pickup and deposit stations and the storage locations.

Fully automated systems typically utilize unit loads (standard-sized containers, bins, or pallets). The system must be designed for maximum storage (the largest item or amount to be stored) and maximum retrieval rates. This means that the system will be very "overdesigned" when utilization is low and will have no allowance for rush orders or breakdowns when utilization is very high (Groover 1987b). Automatic controls can be superseded or supplemented by manual controls under emergency conditions or man-on-board operation.

Carousel storage systems are types of AS/RSs that consist of bins or baskets fastened to carriers that are connected together and revolve around a long, oval track system. The purpose is to position the bins at the end of a rack similar to a dry cleaner's overhead rack. These are typically mechanized rather than automated, and loading and unloading of bins is generally manual (Groover 1987b).

Some persons predict that in the factory of the future, all manufacturing, material handling, assembly, and inspection will be done by computer-controlled machinery and equipment (Kalpakjian 1995). The role of humans will be primarily supervisory—maintaining and upgrading machines and equipment, shipping and receiving supplies and finished products, providing security, upgrading the monitoring of computer equipment and programs (Kalpakjian 1995). More information on AS/RS machines can be found in the ASME B30.13 Standard "Storage/Retrieval Machines and Associated Equipment" (ASME 1993).

STORAGE

The storage of raw materials and finished products is as important to the safety and efficiency of an operation as the handling and transportation of those goods. The storage of materials may interfere with plant operations, impede emergency exit, or present a hazard itself. For example, sharp or protruding edges, flammable or combustible material, unstable or shifting material, and material stored in passageways all present hazards. In addition, materials should not block access to machinery controls or safety equipment such as fire extinguishers, eye-wash stations, fire alarms, or first aid kits. If it is important that the first material placed into storage be the first removed and used (first-in/first-out method), then the storage method should allow for this without requiring any additional handling.

Storage space at most manufacturing facilities is a precious commodity and cannot be wasted or inefficiently used. Inventory, whether raw materials, work in progress, or finished products, always costs money. The just-in-time (JIT) production method seeks to eliminate inventory of materials altogether. In a JIT system, raw materials and supplies are received just in time to be processed. Subassemblies are produced from these supplies just in time for incorporation into finished products and these finished products are delivered just in time to customers. The reduction in storage space, material handling, and manpower are the most obvious cost savings associated with JIT manufacturing. Quality and efficiency are also increased because production problems become immediately evident, and scrap is reduced because bad components are not assembled and stockpiled (Kalpakjian 1995).

Whether a JIT or more traditional manufacturing method is utilized, some storage of materials or finished products will be necessary. Whether storage is temporary or long term, it should be orderly, secure, and tidy in order to conserve space and minimize hazards. When designating storage areas, considerations should be given to the materials being stored, their proximity to other materials and processes, the storage methods to be used (racks, pallets, bins, tanks, and so forth), and paths to exits. Management should recognize that good housekeeping is critical to safe storage. Rubbish, trash, and other waste should be disposed of at regular intervals to prevent fire and tripping hazards.

Adequate clearance must be provided from sprinklers to ensure proper function. Materials stacked too closely to sprinklers will block the flow of water and limit effectiveness. Typically, a minimum of at least 18 in. of clearance is needed. Depending on the class, quantity, and height of rack storage, in-rack sprinklers may also be necessary (NFPA 1995c). Exits and paths to exits must be maintained free of obstructions and other hazards. In addition, paths to exits should be clearly marked and visible from all locations in the storage area, and signs indicating escape routes should not be blocked by stored materials.

Storage racks should be securely bolted or fastened to the floor and walls to prevent tipping. These fastenings should be inspected periodically, particularly in areas where they may be damaged by forklifts. Where materials are stacked, provisions should be made to ensure secure, stable piles; for example, bags or sacks should be interlocked to stabilize the load. Markings may be provided to indicate the maximum height to which materials can be stacked to prevent the floor or rack load limit from being exceeded and to maintain proper clearance from sprinklers. Employees should be forbidden from climbing on storage racks to retrieve or store items.

Outdoor storage should be avoided when possible, but is necessary in many industries. Outdoor storage may be acceptable if the materials meet the following requirements (NFPA 1995b):

- They present a low fire hazard.
- The potential loss of the materials is low enough that building a storage facility cannot be justified.
- The severe fire hazard makes building a structure impractical due to the high potential for loss of the structure.
- The large volume and bulk make it impractical to construct a storage facility to protect the materials.

Providing automatic fire protection, such as sprinklers, is not always practical for outdoor storage. When automatic fire protection cannot be provided, the following control measures should be emphasized (NFPA 1995b):

- Control of potential ignition sources such as transformers, yard equipment, refuse burners, power lines, and vandals
- Elimination of adverse factors such as trash accumulation, weeds, and brush
- Provision of favorable physical conditions such as limited pile size, low storage heights, wide aisles, and use of fire-retardant covers or tarps
- Rapid and effective application of manual firefighting methods facilitated by the provision of alarms and strategically located hydrants and hoses

Sufficient safe clearance must be provided for aisles and passageways in material storage areas. Where mechanical materials handling equipment is used, additional space is needed, including the separate designation of pedestrian and vehicle paths. Permanent aisles should be appropriately marked. Ideally, pedestrian paths should be separated from vehicular traffic by physical barriers such as chains or guard rails.

Employees should travel through doorways provided for pedestrian use, not bay or dock doors used by forklifts and other handling equipment. These doorways are especially dangerous since equipment drivers may not be able to stop quickly enough to avoid pedestrians. For example, a pedestrian may step off the pedestrian path to walk through the open bay door to avoid opening the pedestrian door. A quick move by a pedestrian into the vehicular aisle may not allow enough time for the equipment operator to stop. Also, the visibility of both the pedestrian and the equipment operator may be impaired by changes in lighting or by flexible doors (such as overlapping plastic slats). The use of pedestrian doors should be enforced, and vehicle operators should slow, stop, and sound their horns as they enter doorways, intersections, or other limited-visibility areas.

Aisles and passageways must always be kept clear of obstructions and tripping hazards. Materials in excess of what is needed for the immediate operations should not be stored in work areas or paths to and from work areas.

Housekeeping

Strict housekeeping must be enforced to ensure that storage areas remain free from accumulation of materials that constitute trip, fire, explosion, pest harborage, or other hazards (USDL 1996d). Poor housekeeping is often indicative of poor safety practices and should provide a warning to management that the workplace is not only operating inefficiently, but possibly in an unsafe manner as well. Good housekeeping will help make unsafe conditions more obvious and provide an atmosphere more conducive to safe behavior.

Hazardous Materials

Hazardous materials are materials whose properties present health or safety hazards to workers, the facility, or the environment. Flammable, explosive, corrosive, extremely high- or low-temperature, toxic, and carcinogenic materials all present hazards beyond those associated with the handling of the materials themselves. The consequences of a material handling accident are also more severe, since they could result in a release and subsequent exposure of workers to the hazardous materials.

Special consideration must be given to materials handling operations that involve hazardous materials. Precautions may include preventive measures that include safe handling, spill prevention, and the use of PPE and clothing; emergency showers and eye-wash stations; and respiratory equipment. Preventive measures should also include the substitution of less toxic or corrosive materials, isolation of the hazardous process by the use of enclosures, and provision of adequate exhaust ventilation.

Managers should be familiar with safe work practices and current regulations for the hazardous materials under their supervision. When relevant, managers must also be familiar with the regulatory details governing the shipping and receiving of hazardous materials and wastes. The assistance of staff or consultants with special expertise may be necessary to achieve full compliance.

The environment in which the hazardous materials are stored, transported, and processed often must be controlled. Some locations where flammable materials are stored, handled, or used may require the use of explosion-proof, intrinsically safe, or otherwise protected equipment depending on the classification of the area (NFPA 1995a). Classification depends on the properties of the flammable vapors, liquids, gases, combustible dusts, or fibers that may be present and the likelihood that a flammable or combustible concentration or quantity is present (NFPA 1995a). The equipment used to move the materials must also meet or exceed these hazard classification requirements.

CONCLUSION

There is no doubt that the handling and storage of materials is essential to all industries. Estimates of material handling costs run as high as two-thirds of the total manufacturing cost, varying with the type and quantity of production and degree of automation (Groover 1987a). Efficient materials handling ensures that materials are where they are needed when they are needed. The high costs associated with materials handling accidents, including injuries, facility damage, equipment downtime, and product loss, make it critical for the manager to understand and control the hazards that can lead to accidents.

The best way to minimize these hazards is to study thoroughly the material-flow requirements for a facility and determine what material handling operations are absolutely necessary. Follow the materials from the beginning of the process to final shipment from the facility. Unnecessary handling should be eliminated and the remaining essential tasks analyzed to determine the most safe and efficient methods possible. Mechanical rather than manual means of handling should be used, whenever possible. Employees should be aware of the hazards associated with materials handling, whether manual or mechanically assisted, and know how to minimize these hazards. The storage of materials is also important and should not create fire hazards, impede emergency egress from the facility, cause tripping hazards or obstructions to paths, endanger employees or equipment with unstable stacking or piling, or encourage the harboring of rats or other pests.

To be effective, management must take responsibility for initiating and enforcing a strong safety plan committed to the safe handling of materials within a facility. The most effective materials handling program will have management commitment and employee involvement. Front-line supervisory staff must be convinced of the importance of safe material handling and storage and be held accountable for enforcing employee compliance with safe work practices. Employees must be given training in the proper selection, inspection, and use of material handling equipment and techniques. Regular inspections of the work environment, equipment, and work practices as well as initial training and periodic retraining of employees will help ensure that the work environment remains as safe as possible.

REFERENCES

American Society of Mechanical Engineers (1993). *Safety Standard for Conveyors and Related Equipment.* (ASME B20.1-1993). New York: ASME.

———. (1994). *Below-the-Hook Lifting Devices: Safety Standard for Cableways, Cranes, Derricks, Hoists, Hooks, Jacks, and Slings* (ASME B30.20-1993). New York: ASME.

———. (1996). *Overhead and Gantry Cranes: Top Running Bridge, Single or Multiple Girder, Top Running Trolley Hoist* (ASME B30.2-1996). New York: ASME.

Groover, M. P. (1987a). "Automated Material Handling." *Automation, Production, Systems, and Computer-Integrated Manufacturing.* Englewood Cliffs, NJ: Prentice-Hall, pp. 361–403.

———. (1987b). "Automated Storage Systems." *Automation, Production, Systems, and Computer-Integrated Manufacturing.* Englewood Cliffs, NJ: Prentice-Hall, pp.404–430.

Kalpakjian, S. (1995). "Computer-Integrated Manufacturing Systems." *Manufacturing Engineering and Technology, 3rd ed.* Reading, MA: Addison-Wesley, pp. 1171–1215.

Krause, J. F., K. A. Brown, D. L. McArthur, C. Peek-Asa, L. Samaniego, C. Kraus, and L. Zhou. (1996). "Reduction of Acute Low Back Injuries by Use of Back Supports" *International Journal of Occupational and Environmental Medicine* 2:264–273.

Liberty Mutual. (1997). *From Research to Reality: Liberty Mutual Research Center for Safety and Health Annual Report 1996.* Hopkinton, MA: Liberty Mutual.

National Fire Protection Association (1995a). *National Electric Code 1996 Edition.* (NFPA 70). Quincy, MA: NFPA.

———. (1995b). *Standard for General Storage* (NFPA 231). Quincy, MA: NFPA.

———. (1995c). *Standard for Rack Storage of Materials* (NFPA 231C). Quincy, MA: NFPA.

———. (1996). *Fire Safety Standard for Powered Industrial Trucks Including Type Designations, Areas of Use, Conversions, Maintenance, and Operation* (NFPA 505). Quincy, MA: NFPA.

———. (1997). *Life Safety Code* (NFPA 101). Quincy, MA: NFPA.

National Institute for Occupational Safety and Health (1983). *Work Practices Guide for Manual Lifting* (DHHS NIOSH Publication No. 81-122). Washington, DC: U.S. Government Printing Office.

National Safety Council (1990a). *Belt Conveyors for Bulk Materials* (Data Sheet 1-569 Reaf. 90). Itasca, IL: NSC.

———. (1990b). *Dock Plates and Gangplanks* (Data Sheet 1-318 Reaf. 90). Itasca, IL: NSC.

———. (1990c). *Front-End Loaders* (Data Sheet I-589 Rev. 90). Itasca, IL: NSC.

———. (1991a). *Powered Hand Trucks* (Data Sheet I-317 Rev. 91). Itasca, IL: NSC.

———. (1991b). *Roller Conveyors* (Data Sheet I-528 Rev. 91). Itasca, IL: NSC.

———. (NSC) (1997a). "Automated Lines, Systems, and Processes." In G. R. Krieger, and J. F. Montgomery, eds. *Accident Prevention Manual for Business and Industry: Engineering and Technology, 11th ed.* Itasca, IL: NSC, pp. 732–753.

———. (NSC) (1997b). "Hoisting and Conveying Equipment." In G. R. Krieger, and J. F. Montgomery, eds. *Accident Prevention Manual for Business and Industry: Engineering and Technology, 11th ed.* Itasca, IL: NSC, pp. 413–473.

———. (1997c). "Materials Handling and Storage." In G. R. Krieger, and J. F. Montgomery, eds. *Accident Prevention Manual for Business and Industry: Engineering and Technology, 11th ed.* Itasca, IL: NSC, pp. 375–412.

———. (1997d). "Powered Industrial Trucks." In G. R. Krieger, and J. F. Montgomery, eds. *Accident Prevention Manual for Business and Industry: Engineering and Technology, 11th ed.* Itasca, IL: NSC, pp. 506–528.

Perkins, M. S., and D. S. Bloswick (1995). "The Use of Back Belts to Increase Intraabdominal Pressure as a Means of Preventing Low Back Injuries: A Survey of the Literature." *International Journal of Occupational and Environmental Health* 1:326–335.

Suruda, A., D.S. Bloswick, M. Egger, H. Wing, and D. Lillquist (June 1997). "Fatal Lift-Truck-Related Injuries from Tipover and Rollover," *Material Handling Engineering,* pp. 71-75.

University of Michigan. (1989). *Center for Ergonomics Two Dimensional Static Strength Prediction Program Version 4.21* [computer software]. Ann Arbor: University of Michigan, College of Engineering.

U.S. Department of Health and Human Services (1996). *National Occupational Research Agenda* (DHHS NIOSH Publication No. 96-115). Washington, DC: U.S. Government Printing Office.

U.S. Department of Labor (1996a). OSHA Regulations (Standards) Part 1910 Subpart D: Walking-Working Surfaces, Section 30: Other Working Surfaces. [CD-ROM OSHA A 96-4].

———. (1996b). OSHA Regulations (Standards) Part 1910 Subpart J: General Environmental Controls, Section 147: The Control of Hazardous Energy (lockout/tagout). [CD-ROM OSHA A 96-4].

———. (1996c). OSHA Regulations (Standards) Part 1910 Subpart K: Medical and First Aid, Section 151: Medical Services and First Aid. [CD-ROM OSHA A 96-4].

———. (1996d). OSHA Regulations (Standards) Part 1910 Subpart N: Materials Handling and Storage, Section 176: Handling Material-General. [CD-ROM OSHA A 96-4].

———. (1996e). OSHA Regulations (Standards) Part 1910 Subpart N: Materials Handling and Storage, Section 178: Powered Industrial Trucks. [CD-ROM OSHA A 96-4].

———. (1996f). OSHA Regulations (Standards) Part 1910 Subpart N: Materials Handling and Storage, Section 179: Overhead and Gantry Cranes. [CD-ROM OSHA A 96-4].

———. (1996g). OSHA Regulations (Standards) Part 1910 Subpart N: Materials Handling and Storage, Section 184: Slings. [CD-ROM OSHA A 96-4].

———. (1996h). *Sling Safety* (OSHA Publication 3072). (rev. ed.). Washington, DC: U.S. Government Printing Office.

———. (1997). *Bureau of Labor Statistics News* (USDL Publication 97-188). Washington, DC.: U. S. Department of Labor.

Chapter 26

Computers and the Safety Professional

Mark D. Hansen

INTRODUCTION

Safety and health compliance is strongly regulated by the government. Most companies that have to deal with these regulations are forced to grapple with volumes of OSHA regulations. This includes understanding the regulations as well as having to refer to various parts, sometimes on a daily basis. Being compliant with these regulations is further complicated by numerous reports, many of which overlap and are due to the Government at different times during the year.

The problem is that safety and health professionals and managers spend an inordinate amount of time digging through regulations and filing reports rather than focusing on the safety and health job. By automating this process, the safety and health professional can spend more time out in the field implementing safety and health programs and less time ensuring that compliance documentation is filed on time, and can eliminate wasting time manually digging through regulations. Examples of automation include having Material Safety Data Sheets (MSDSs) available on computer at sites requiring employee review; and managing the document flow through the review process, integrating engineering drawings and other documents, and managing and tracking all the training requirements required for OSHA compliance with the use of computers.

With the prices of powerful personal computers (PCs) and Macintosh computers falling below $2,000, most safety and health professionals have access to more computing power than they have ever had before. Ironically, with all this computing power available, rarely are these machines fully utilized. Most of the time, we use computers for routine tasks such as word processing, spreadsheets, or databases.

CASE STUDY:
SPEEDING UP THE SAFETY PROFESSIONAL'S JOB

It is Monday morning and safety manager Jerry Smith of XYZ Chemicals is drinking his first cup of coffee. The phone rings. It's the plant manager and he wants a summary of all the OSHA recordables over the past five years involving acrylonitrile in the ethylene imine process. The reason? The customer is conducting an audit and he needs the information by 9 *AM*. The plant manager asks, "Do you have this information at your fingertips?" And, of course you respond by saying "yes," whether you have the information at your fingertips or not. You think to yourself, *so* much for finishing your coffee. But wait. You just databased your OSHA recordables and near misses by process, chemicals, Worker's Compensation Costs, Date, time of day, and so on. You call up the database and search with the appropriate parameters. Instantly you get three incidents that were recordables and two that were near misses. You compose an e-mail with the results, attach the incident reports, and send it all to the plant manager. You then call the plant manager and tell him the results and that you sent him an e-mail with your response. The plant manager breathes a sigh of relief. Meanwhile, you can finish your coffee before you make rounds.

In the past, you would have had to pull the incident reports over the last five years and manually search for the incidents and you may or may not have found all of them. Now, you not only look like a hero but your results are accurate and survive scrutiny. This is what computers were designed to do—maintain large amounts of information and present it to users in short order.

CHOOSING THE APPROPRIATE
PERSONAL COMPUTER

In choosing a portable/personal computer today, you must play a game of give-and-take in the area of attributes and functionality. The more options you choose, the better and faster for desktops, and the bigger and heavier the portable. For example, if you want a 3½-in. drive and a compact disk-read only memory, (CD-ROM) they will add considerable weight to your computer. Some of the attributes include weight, keyboard size, screen, pointing device, processor, memory, hard drive, speed, expandability, and disk drives.

The best place to start when choosing options is usually the guts of the computer: the processor, memory, and hard drive. How fast do you want the computer to run? How much random access memory (RAM) do you think you will need? How much storage space do you think you will need for documents? The ultimate questions would be, What are you going to do with your computer? Will you be writing documents? Will you be modifying computer-aided design (CAD) drawings? Will you be making high-tech graphic marketing demonstrations? Depending on what you will be doing, your needs may require more or less computing power.

The Processor

Looking at today's marketplace, you probably cannot even find a 386 processor. Besides, today's software requirements are increasing so quickly that a 486 processor is almost marginal for most applications. If you are going to invest in a portable/personal computer, you probably want it to keep pace with technology, so you might as well buy a Pentium, a 586 or a 686.

However, if you are planning to use graphics, multimedia, spreadsheets, or CAD drawings, you will probably need the computing power of a 586. If you need more performance, you might as well get the 120- to 180-MHz (or more) processor clock speeds. The 133-MHz processor clock speed is already becoming extinct, and the 120-MHz clock speed just does not seem to be fast enough for the "got to have it right now" generation. For those who may not know what clock speed means, clock speed is the speed at which the computer takes the software from memory and displays it on the screen.

One way to boost performance is the single-sheet paper scanner built into a keyboard, the other a CD-ROM drive that can write huge amounts of data to a cartridge-based compact disc. Another is the writeable CD-ROM drive, which holds 560 megabytes (MB) or more if you compress the data.

Extended-Data Output

The extended-data output (EDO) memory chip can boost performance by 10–15 percent. The hard drive—often the performance bottleneck—takes a mere three seconds to start up WordPerfect 6.1 for Windows.

Random Access Memory

RAM is the location where programs are stored when accessed from memory. For example, when you call up WordPerfect 6.1a, it is executed and stored in RAM while it is being used. Because software programs are growing to use more active RAM, such as many Windows-based programs, I believe in loading up the computer with the maximum, or 32–40 MB of RAM. I always look ahead and try to think what will happen in the next five years: I can see the Windows 2000 version requiring the maximum. Consequently, if you do not have the maximum, you will not be able to run the program you want to run. Also, think of running your favorite CD-ROM application: it could take fifteen minutes to load, and your battery might lasts only two hours!

Hard Drives

I suggested that the minimum hard drive be 1.2 gigabytes (GB), or even 2.1 GB because I do not believe in disk doublers. Disk doublers are used to double your hard drive. Some use compression algorithms to achieve this, and they seemingly double your hard drive. As a result, you can store more software on your hard drive and save some space and money. However, when disk doublers are used, I have seen far too often that you either sacrifice performance or risk catastrophic failures that cause the computer to lock up. This usually happens when it is most critical—when someone wants an answer right away, or during that meeting with the company president who is about to say yes to your safety budget. Memory is not cheap; it costs anywhere from 50 cents to $1 per Megabyte. But the extra expense is worthwhile because it gives you peace of mind.

For portable computers, one strategy is to buy multiple removable hard drives. That is, you use your portable/personal computer at home with the 1 GB hard drive. When it is time to hit the road, you use a tape backup and copy the needed applications and databases, to a 340 or 500 MB hard drive. You then remove the 1 GB hard drive and replace it with your 500 MB hard drive and you have road to go. If a catastrophic error occurs on the road, you at least have the data backed up at home. There is one catch: not all portable computers offer removable hard drives. For example, all IBM portables except for subnotebooks come with removable hard drives: however, none of Toshiba's notebooks have this capability.

The Screen

EGA monitors are history, and VGA monitors are on the brink of extinction. Especially with the arrival of Windows 95, Super VGA (SVGA) is in. Monitors come in all shapes and sizes, with resolutions from 800 x 600, 1,024 x 768, 1,024 x 1,024, 1,280 x 1,024 and 14-in. to 21 inch diagonal measurements. Monitor basics include refresh rate, pixel density, and scale to gray. Refresh rate is how many times the screen is redrawn. If the screen is redrawn seventy times per second, it has a refresh rate of 70 Hz. Pixel density can be broken down into input and output density. Most input is in the 200–400 dots per inch (dpi) range. On the output side, the video controller and monitor determine how reliable the document will be. A standard VGA controller with a 640 x 480 resolution is equivalent to 60 dpi. The next step up is SVGA, which has a resolution of 1,024 x 768, or 90 dpi. For high resolution, the next step is SVGA 1,280 x 1,024. This is the minimum tolerable resolution on a 17-in. monitor. For very high resolution, you will need a controller that produces 1,600 x 1,280 or 120 dpi on a 21-in. monitor. To show a complete page of a document on a screen, controllers often shrink the image and reduce the resolution. This is called *scale to gray.* This process makes the image worse because clusters of pixels are converted into a single gray pixel. Scale to gray determines the best shade of gray for this new pixel (from either four or sixteen combinations). The controller card selects the best shade to represent the lost (decimated) pixels.

I can remember when everyone was waiting to buy a portable computer for the color screen. When these screens first came out, the weight of the portable computers went up and the screens were not very clear and crisp. Again, the choice depends on what you are planning to do with your computer. If you are like me, working outside of the office, a black and white screen is sufficient. If you are making presentations using multimedia and you want them to be "sexy," color is probably a better choice. I will offer one alternative to color: usually, when you are marketing a remote customer for a consulting opportunity, he or she probably has a color monitor that you can hook up to your portable computer. However, that is not always the case and it takes more planning on your part to ensure that all required hardware is available.

Screen size should also be considered. As portable computers get smaller, so do the screens. As a result, extended use can wear on your eyes.

For prolonged use, I recommend the larger screens, such as the 9.4-in. or 10.4-in. screens.

The Battery for Portables

Just as you get into your application, while traveling on an airplane, the battery light flashes. There is no place to plug in to recharge your battery and you must shut down. The result? You end up wasting your time reading some airline Hollywood stars magazine. Battery life has been advertised as ranging from two to five hours. Since batteries of any kind usually last less than the advertised specification, go for the longest life. As a rule, you may want to cut the advertised battery life in half to arrive at a more accurate measure of how long the battery will last. Also, unless you run the battery all the way down, the battery develops memories that shorten the specified life over time. I recommend the lithium-ion battery over the nickel-cadmium battery. The lithium-ion battery costs more, but provides a longer life than the nickel-cadmium. For travel purposes, I also recommend a backup battery, so that you can simply replace the battery if necessary.

The Floppy Drive

I believe floppy drives will be here to stay even if there are removable hard drives. People will always want to transport software on the fly. For portables, when you return from your trip, you can copy all of the files to your home hard drive, rather than using a tape backup, which could take several hours. If you are looking at a portable computer without an internal floppy drive, because of weight or expense, you will probably end up buying an external floppy drive. An external floppy drive is bulkier and heavier than an internal drive, especially when you add cabling (which you also must keep track of in transit).

Expandability for Portables

The easiest way to expand the capabilities of your portable computer is to add a personal computer modular interface adapter (PCMIA) card. The PCMIA card allows you to hook up to fax/modems, add space to you hard drive, add CD-ROMs, capture video, or connect to a local area network (LAN). At a minimum, I recommend one PCMIA slot. For more versatility two are preferred.

Weight for Portables

Regarding weight for portables, the question is, *How much will you use your portable?* Notebooks are a little large and heavy (6–9 lb.) whereas subnotebooks are smaller and weigh less (5 lb. or so). However, if you are going to be typing a lot, you will be cramped, as the keyboard is condensed, making the keys smaller and closer together. Therefore, if you want smaller and lighter, there is a trade-off for usability.

The Pointing Device

There are several alternatives for pointing devices: a desk-mounted trackball or clip-on trackball; an internal trackball; an eraser-looking device in the center of the keyboard; a finger pad; and a regular mouse hooked to the portable computer. The clip-on trackball is okay, but it can cause fatigue in your right hand after prolonged use. Furthermore, it is very difficult for left-hand users. The internal trackball is centered below the keyboard, which conforms for left- and right-handedness. But, it is still awkward in those "in-flight" conditions. The eraser-looking device that sits in the middle of the keyboard looks promising, but I have not used it enough to be able to make a recommendation. Another alternative is just now coming out on the market: the finger pad. It uses your finger as the pointing device, with three buttons much like the three-button version of the mouse. The finger pad seems to be meeting with some success. Only time will tell. The final alternative is to bring along your regular mouse. No matter which pointing device you have for "in-flight" use, bring along you regular mouse for the "in-hotel" use. That way you are providing the most comfort wherever possible.

THE PURCHASE

Given all the information provided herein, now all you have to do is go out and buy the personal/portable computer that suits your needs. Once you have specced out your needs, there are several ways to go. One is to buy from a local dealer, with local support in case you have problems. Another is to buy from a mail-order house. This will save you a few dollars, and if you do not mind waiting a few weeks for the manufacturer to "burn-in" your portable computer (leave it on for several days for infant mortality— if it is going to fail it will usually fail during this time period), it is a safe gamble. However, some computers are sold only through retail stores. If you do buy by mail order, use a credit card to protect yourself against defects or loss. You may wish to try to buy a computer through the classified advertisements in your local newspaper. Be careful that the computer has everything you want and that you know the right questions to ask; otherwise you may get caught holding a lemon. Some questions you might consider asking are as follows:

- What is the brand name of the computer?
- What kind of processor does it have?
- What is the processor clock speed?
- How much RAM does it have?
- What is the upper RAM limit?
- How big is the hard drive?
- Have there been any problems?
- Has it had any warranty or other work done on it?
- Can it be upgraded?
- Why are you selling it?
- How old is the computer?
- Are you selling any peripherals with it?

Buying a personal computer is a major investment. Think about why you need a portable computer. Buy what will suit your current and foreseeable needs. Take the time to shop around for a bargain that has all the attributes you will need. Some of the entry-level, step-up are given in Table 26–1. and laptop systems, as of March 1997 that can be purchased retail are given in Table 26-1:

Some costs for mail-order systems may be slightly lower than retail, but you will need to know specifically what you want to purchase in order to get the appropriate the computer for your needs. Some mail-order systems are given in Table 26-2:

There is now a new option in buying a computer: buying a used computer. This is a new and thriving market. In 1995, U.S. consumers purchased 2.4 million used computers, up 50 percent from 1994. Of these computers, 62 percent were sold for less than $500. It is a relatively small market, however, when you consider that about 10.5 million computers

Table 26-1. Some of the Portable Computers Available, March 1997

ENTRY-LEVEL SYSTEMS	STEP-UP SYSTEMS	LAPTOPS
AST: Advantage 486/66, 8-MB RAM, 540-MB HD, 4X CD-ROM, 14.4-KBPS modem, 14-in. color monitor ($1,000)	Compaq: Presario 3000 Pentium/166, 24-MB RAM, 2-GB HD, 6X CD-ROM, 33.6-KBPS modem, 12-in. color LCD monitor, JBL speakers ($3,500)	Compaq: Presario 1000 Pentium/120, 16-MB RAM, 810-MB HD, 4X CD-ROM, 33.6-KBPS modem, 7 lb. ($2,500)
Compaq: Presario 4400 486/133, 8-MB RAM, 1.6-GB HD, 8X CD-ROM, 33.6-KBPS modem, 14-in. monitor ($2,000)	SONY: Video Audio Integrated Operation Pentium/166, 16-MB RAM, 2.1-GB HD, 8X CD-ROM, 28.8-KBPS modem, no monitor, speakers ($2,200)	IBM: Think Pad 560 Pentium/100, 8-MB RAM, 810-MB HD, speakers, no floppy, 4.1 lb. ($2,700)
NEC: Ready Series 9620 486/133, 16-MB RAM, 1.6-GB fHD, 8X CD-ROM, 33.6-KBPS modem, 14-in. monitor ($1,800)	IBM: Aptiva C-66 Pentium/166, 16-MB RAM, 2-GB HD, 8X CD-ROM, 28.8-KBPS modem, no monitor, speakers ($2,400)	Texas Instruments: Travelmate 6050 Pentium/150, 16-MB RAM, 1.3-GB HD, 28.8-KBPS modem, 6 lb. ($5,500)
Apple: Performa 6400 486/180, 16-MB RAM, 1.6-GB HD, 8X CD-ROM, 28.8-KBPS modem, 14-in. monitor ($2,400)	Hewlett-Packard: Pavillion 7285P Pentium/200, 32-MB RAM, 3.1-GB HD, 8X CD-ROM, 28.8-KBPS modem, no monitor, speakers ($3,000)	NEC: Versa 6030X Pentium/133, 16-MB RAM, 1.4-GB HD, 6X CD-ROM, 28.8-KBPS modem, 7 lb. ($6,500)
		Acer: AcerNote Nuovo Pentium/133, 16-MB RAM, 1.2-GB HD, 28.8-KBPS modem, 7 lb. ($4,000)
		Canon: NoteJet III Pentium/120, 8-MB RAM, 810-MB HD, 9.7 lb. ($5,800)
		Samsung: SENS 810 Pentium/120, 8-MB RAM, 810-MB HD, 4X CD-ROM, 7.4 lb. ($2,800)

were sold in 1995. Most disappear into a network of brokers. Considered as a best buy is the remanufactured PackardBell, a 486 that sells for $899.99, which is $400–$500 less than if purchased new. Although used computers do not have any guarantees, they do come with the manufacturer's warranty.

Table 26-2. Some of the Available Mail-Order Portable Computers Desktops

Desktops

Acer: AcerPower, Pentium/133 MHz, 16-MB RAM, 1.6-GB HD	($1,795)
NEC Pentium/133 MHz, 16-MB RAM, 1.2-GB HD	($1,659)
AST: Advantage, Pentium/100 MHz, 16-MB RAM, 1.2-GB HD	($1,449)

Laptops

Toshiba: Satellite 110CT, Pentium/100 MHz, 8-MB RAM, 772-MB HD, 6.9 lb.	($2,499)
NEC Versa 2000, Intel/DX4/75 MHz, 350-MB HD	($1,435)
AST: Ascentia J Series Pentium/133 MHz, 8-MB RAM, 800-MB HD, 5.8 lb.	($1,919)
Texas Instruments; Extensa 570CD Series, Pentium/100 MHz, 8-MB RAM, 810-MB HD, 6X CD-ROM, 6.4 lb.	($2,775)

The manufacturer will even let you test your software on their computer to see how it works. Compaq has a used computer store in which it sells computers that are hundreds of dollars less than new computers. However, most used computers are going overseas to Third World countries. For example, a Third World country may purchase a lot consisting of 42,000 computers for resell.

THE UPGRADE

As cost-efficient consumers, we upgrade and upgrade our computers until we wonder when we should buy new, or a new used computer. There is a strategy in determining when to buy newer technology. When the upgrade is within a few hundred dollars of buying new, it is probably better to buy new. For example, let us say that you find a laptop for $500. It has a 486 processor, a clock speed of 66 MHz, and a 120 MB hard drive. Considering that laptops are selling for $1,500–$5,000, this is a good deal. However, if the same laptop is selling for $1,500, it is probably not such a good deal. Furthermore, if you will need to add RAM or a modem, the cost will increase about $200–$500, making the total cost $1,000. You will need to evaluate whether it is worth buying used or new given current availability of technology. As of June 1996, used computer prices were as follows:

Dell 386/33: 4-MB RAM, 120-MB HD, 14-in. color monitor   ($350)
IBM 486/66: 8-MB RAM, 420-MB HD, 14-in. color monitor   ($825)
Toshiba 1960 CS: 486/66, 8-MB RAM, 420-MB HD,
    14-in. color monitor   ($1,000)
Compaq P100 486/66: 16-MB RAM, 1-GB HD,
    15-in. color monitor   ($1,685)
Apple Powerbook: 4-MB RAM, 320-MB HD   ($1,725)

Compaq and AST offer discounts at their refurbishment stores at 85 percent and 75 percent of retail, respectively. The bad news is that you have to visit their store in the local area—Houston and Fort Worth. So if you are planning your trip across the country, plan a long layover and stop in and pick up a laptop, at a fraction of the retail cost!

Some savvy must be exercised in purchasing used computers. Some things to look out for include "shaved" central processing units (CPUs), packaging, Federal Communications Commission (FCC) certification, and the prepayment requirement. "Shaved" CPUs are when dealers shave off the top of the chip and relabel it with a higher number. Others may even camouflage the CPU by hiding it under a heat sink so that you cannot easily see it. Dealers may even install dead chips in cache slots so that they look like legitimate installations, and then change the ROM BIOS so you will think a cache exists. Packaging is also important. This trick entails reboxing an older PC (for example, 486/33) and selling it as a newer PC (for example, 586/120). If the packaging is fresh but there is no computer-generated label with a serial number, chances are the dealer got fresh boxes from the manufacturer. You will want factory-sealed boxes with preprinted serial numbers that match those on the product. Also check the software. Without documentation you may not have the real McCoy. FCC certification is important, too. The FCC refuses to certify separate components. For example, the FCC issues one number for the motherboard, case, and power supply. If the FCC certification numbers on the PC are different or missing, the PC has been repackaged. The problem is that sometimes dealers try to sell PCs that have not received FCC certification as a whole, or they use low-cost power supplies that lack UL listing. Also beware of prepaying because the dealer could disappear overnight. Deep discount warranties are another clue that the dealer does not intend to be around when you come calling with a problem.

COMPUTER ARCHITECTURES

What is computer architecture? Computer architecture is a combination of hardware and software that, when integrated, works together to perform a set of specific functions. This architecture is usually depicted in block diagrams that look like functionality diagrams. Depending on the size and function of the application, architectures may vary from single-user systems to enterprise-wide solutions.

Single-User Systems

Single-user systems (Figure 26-1) usually consist of a CPU that, at a minimum, is a 386 on the low end up to a 586 Pentium on the high end, with clock speeds ranging from 16 to 100 MHz. The monitor is usually an EGA, at a minimum, up to a VGA, and sometimes it is a graphics monitor with a 21-in. diagonal screen for CAD capabilities. RAM can range anywhere from 1 MB up to 16 MB depending on the computing needs. Hard drives can range from 20 MB to 1 GB, also depending on the computing needs.

Printers that can be used with single-user systems vary from dot matrix to near letter quality, and from letter quality to laser printers. Scanners can be integrated from hand scanners to full-page scanners. Scanners usually have optical character recognition (OCR) capabilities to read in data and translate to ASCII characters. Some may even have optical storage such as CD-ROM to store and retrieve large amounts of information. A variation on the CD-ROM is the Write Once Read Many (WORM) drive, which enables users to generate, manipulate, exchange, and store information once and retrieve in a read-only fashion. This is helpful in total quality management and OSHA Product Safety Management (PSM) compliance environments that require time-dated sequential data on information used and transferred in and through company-wide systems.

Defining Client/Server Computing

Client/server computing is a form of distributed processing in which an application is split in a way that allows a front end (the client) to request services of a back end (the server). Client/server definitions are given below.

Figure 26-1. Single-User Computer Architecture

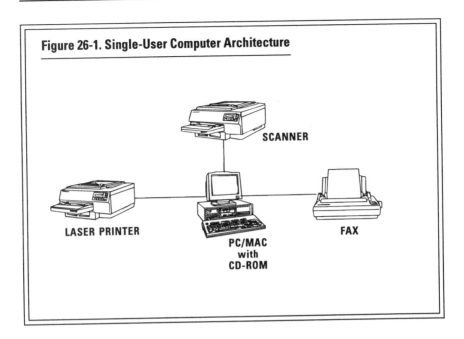

Typically, the front ends reside in the end-user desktop systems (PC, Macs, UNIX, workstations, and so forth). Back ends can reside in any network server, ranging from a local database server to the largest mainframe. The front-end and back-end portions of the application reside on different processing systems, usually separated by a network. Neither the application's front-end nor the back-end are complete applications themselves. Rather, they complement each other to form a complete application.

Client. A single-user workstation that provides presentation services and the appropriate computing, connectivity, and database services relevant to the business need.

Server. One or more multiuser processors with shared memory that provide computing, connectivity, and database services and interfaces relevant to the business need.

Client/Server Computing. An environment that appropriately allocates application processing between the client and the server. The environment typically is heterogeneous, with the client and the server communicating through a well-defined set of standard application program interfaces and remote procedure calls.

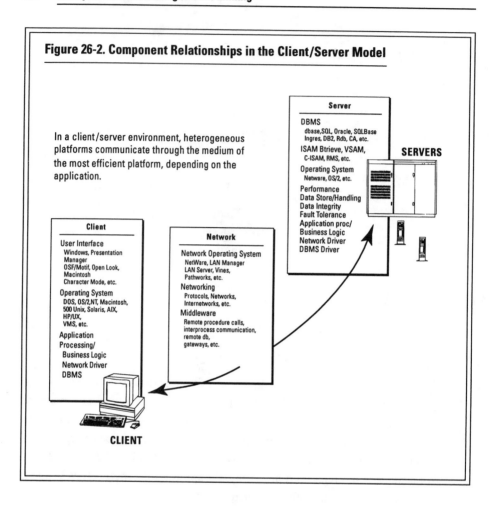

Figure 26-2. Component Relationships in the Client/Server Model

In a client/server environment, heterogeneous platforms communicate through the medium of the most efficient platform, depending on the application.

Client/server database systems are constructed such that the database runs on a database server while the database users interact with their own desktop systems (client), which are responsible for handling the user interface, including windowing and data presentation. Figure 26-2 illustrates the component relationships in the client/server model.

Simple Network/Unix Systems

With a very basic introduction to client/server technology, Figure 26-3 illustrates a potential application for simple and UNIX networks. These architectures are usually implemented at the low end for one to five concurrent ($25,000–$50,000) users to a high end of ten to fifty concurrent users ($100,000–$500,000), or department-level applications that generate, manipulate, exchange, and store common information. Some of the low-end and all of the high-end applications usually use fourth-generation program applications software to provide ease of use. Fourth generation languages allow users to reconfigure and manipulate the system without reprogramming the software. In the past, this kind of reconfiguration required programmers to recode the software.

Enterprise-Wide Solutions

Enterprise-wide solutions are generally solutions that allow users across the company to communicate, generate, manipulate, exchange, and store common information that may be in several different geographic locations and have up to 5,000 users. These architectures are usually implemented at

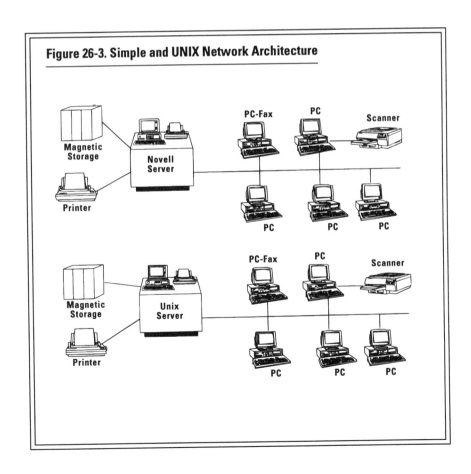

Figure 26-3. Simple and UNIX Network Architecture

the low end for 50 to 100 or more concurrent users ($100,000–$500,000) to a high end of 100 to 5,000 concurrent users ($500,000–$1,000,000 or more). All the low- and high-end applications of the enterprise-wide systems use fourth-generation program applications software to provide ease of use. The primary driving difference in the cost of each of these implementations is the number of concurrent users.

Many of the attributes of enterprise-wide solutions are similar to department-level solutions, except that system requirements provide larger amounts of information to be transferred and exchanged between locations and a larger number of concurrent users. Enterprise-wide solutions use a gateway to provide connectivity to an existing host system. The advantage of such an implementation is that if an existing system is currently being used, it is not taxed or slowed by this type of implementation.

Enterprise-wide solutions also use a gateway to provide connectivity to an existing host system. In this case, the existing system is a Unix mainframe. Since many computing systems today are currently implemented on mainframe platforms, this example illustrates how upgrades can take place without taxing or slowing the existing computing system. The architectures can also be implemented as stand-alone systems if existing systems do not exist, or if companies desire to replace their current computing system, if it is outdated.

STATE-OF-THE-ART TECHNOLOGIES

New state-of-the-art technologies are now available that can be implemented on single-user systems all the way up to enterprise-wide applications. These technologies typically use stand-alone PCs either as a single user or as the front end on a client/server network. These new technologies are "document imaging" of forms (for example, MSDSs), business process automation using "workflow" software, and computer output to laser disk (COLD).

Document Imaging

Document imaging is the core of automating the safety and health process for OSHA, EPA, and ISO compliance. Document imaging is central to the progression toward the "paperless" office. The "paper" from other organizations, such as documents and engineering drawings, can be scanned in and the paper literally thrown away. The reason the paper is thrown away is fourfold. First, the electronic copy is now legally considered to be the original. Second, the electronic copy is often better than the original (especially when dealing with MSDSs). Third, one of the main reasons for using document imaging is to eliminate the need for rooms full of filing cabinets containing engineering drawings and documents. Fourth, document imaging can be used to capture the signatures on all types of documents (for example, engineering drawings, OSHA-related testing documents).

There are two distinct methods to scan in information. One is to capture the physical image (or picture) of the document using a document scanner. This image is usually stored in a .TIFF format. The information on this image is usually information that cannot be changed and that can be displayed or printed when necessary.

The second method is to scan in documents of which the text is converted to ASCII data that can be shared in a database. The conversion process is accomplished using OCR software. This is important when dealing with data (for example, MSDSs) that need to be stored, retrieved, and manipulated from a database—for example, searching for the boiling point of a particular chemical to determine storage location at a plant site, or the determine compatibility between chemicals.

If the original document is of maximum quality, there are two methods to significantly improve the quality and readability. One is to use an intelligent character recognizer (ICR). This software recognizes and matches each character based on the image scanned-in. Another type of software program also accomplishes this and, additionally, fills in the image character with a shade of gray so that it is more readable. This is especially important for managing documents from multiple vendors such as MSDSs.

Document imaging can also provide virtually instantaneous access to engineering drawings and documents. This is particularly important when complying with OSHA 1919.119, Process Safety Management, which stipulates that:

> prior to the introduction of highly hazardous chemicals to a process the PSSR must confirm that construction and equipment are in accordance with design specifications.

Studies have shown that a document imaging system allows drawings to be updated six times faster than manual methods.

> the written program provides employees and their representatives access to process hazard analysis and all other information developed as required by the Process Safety Management standards.

Because documents are stored electronically they can be retrieved instantaneously and are less susceptible to fire and other catastrophes.

written process safety information will be compiled before conducting a process hazard analysis (PHA).

Document imaging enhances a company's ability to conform to this requirement by having all documents available electronically, by date of preparation, revision date, etc.

Document imaging can be used to reduce the paper in the process both reliably and accurately, as well as to provide easy access for manipulation. Document imaging provides the mechanism for progressing to electronic data management, which is the key to survival in the 1990s, from both a standpoint of cost and speed of access.

Business Process Automation Using Workflow

Business process automation is the mechanism for integrating document imaging and workflow technology to make computers even more useful in the 1990s.

Business Process Automation. Business process automation refers to automating manual processes using computer technology. For example, the physical filing of documents in a filing cabinet is automated in a database of some sort, or the manual review of a document is automated so that the document is routed electronically and sequentially to each person in the process. Business process automation includes integrating all electronic data in a compatible form. This involves converting different types of electronic images (for example, .TIF, .GIF, .WPG) and word- processing data (for example, WordPerfect, MultiMate, Word, ASCII) to a common form. However, merely automating a manual process is usually not sufficient. By using workflow, the manual process can be routed electronically and sequentially, rather than manually.

Workflow. Workflow refers to the electronic routing of documents and engineering drawings through a computer network. Workflow allows users the ability to comment, approve, or disapprove, and then send the information to the next person in the review process, or back for editing. Workflow can be used to route information, documents, engineering drawings, and so on, from local to enterprise-wide distributions.

Let us say that a new MSDS for a product you use in your manufacturing process arrives from a vendor. The first thing you might do is assess the impact in the quality assurance. If information is missing, the quality assurance individual may call the vendor for more information; if not, the MSDS is electronically forwarded to S&H (the Safety & Health Department), OPS (Operations) and Tech_Serv (Technical Services). The S&H Department may evaluate the MSDS for impact on how the product is to be used in the manufacturing process and the ramifications of use, waste, and disposal. Operations may or may not use the product but may need the "paperwork" for reference purposes.

Technical Services would use it to update all documentation for all operations and maintenance procedures. Next, it would be forwarded to Cost Eval (Cost Evaluation) to determine the cost impact of the new MSDS on the company. If the new MSDS has a cost impact, it would be forwarded to AFE (Authority For Expenditure) to make disbursements accordingly. If it requires more review from comments, it would be re-rerouted through the repair function and iterate until it is acceptable to proceed. At this point, if no action is decided for the new MSDS, it would be forwarded to the Reject Archive function and archived. On the other hand, if action is required, and once everything is completed, it would be forwarded on to initiate the change and archived in the Reject Archive function.

Benefits of Document Imaging and Workflow. Document imaging and workflow allow companies to work smarter rather than harder. This includes reducing the number of steps in a process, reducing the amount of time to do a task, doing the task with fewer persons, reducing the storage space to store all kinds of paper, and so on. Reducing the amount of time to do a task is accomplished by having all the appropriate information to complete a task electronically attached and immediately available. For example, engineering drawings stored electronically can be quickly accessed for review prior to the safety and health professional and/or manager signing off for changes or modifications. This is in stark contrast to the time-consuming, manual process of searching for engineering drawings in filing cabinets. Using this method also precludes the need to enter the same data into the system several times. Once data are entered, they never need to be reentered. It follows that reducing the number of steps also provides an opportunity to reduce the amount of time to do a particular task. If this is done properly, manpower can often be reduced. In this day and age of "right sizing" to maintain a competitive advantage, document imaging and workflow allow a company to reduce manpower without sacrificing the quality of work performed.

The individuals responsible for assembling and maintaining workflow maps are not involved in the process. By removing them from the process, the integrity of the process is maintained. This is crucial because if the person responsible for maintaining the workflow map is also in the process different activities could be circumvented by modifying the workflow map. However, if the person responsible for the workflow map is not in the process, no activity can be circumvented by anyone else in the process. This is important for OSHA, EPA, and ISO compliance. When a problem is being worked or an assembly is being accomplished, it is repeatable. It is repeatable because it is not a paper process that can circumvent any activity or organization (for example, safety and quality). The electronic mechanism inherent in workflow also lends itself to ease of auditing as everything done leaves a time/date audit trail. This is important not only for auditing but also for obtaining the latest copy of documents, engineering drawings, and so forth.

These technologies and processes are most successfully implemented in a computing environment that is flexible, open to various computing platforms and allows growth from small applications. Consequently, many companies are moving to a client/server architecture.

Computer Output to Laser Disk

It is estimated that U.S. companies spend nearly $20 billion per year spooling from 2 to 50 million pages of computer-generated reports per month to a printer or computer output microfilm (COM) equipment for distributing or archiving. Sorting through these vast amounts of growing paper files and fiche for OSHA, EPA, and ISO 9000 compliance is becoming more tedious, expensive, and time-consuming every day. Meanwhile, companies are looking to improve their bottom line and conserve valuable office space by harnessing this vast information store. To do this, users and, in particular, safety and health professionals need a quicker, more efficient method to access their mainframe data. COLD technology represents a promising solution.

Ideally, users generally prefer on-line viewing capabilities to the mainframe. Information managers, however, contend that they cannot afford the consumption of disk resources and the impact on mainframe processing as the volume of inquiries increases. The result—management information systems (MISs)—restrict the volume of available report data, as well as access to those reports, on the mainframe to the most current data only from selected reports. The reports are output on a daily, weekly, or monthly basis. The remainder of the reports have traditionally been output to printers and stored in paper files and on magnetic tape for subsequent access. As the paper files grow, access to them becomes laborious and expensive. Access from magnetic tape archives requires remounting the tape and taking up valuable mainframe resources that are better spent on current operations. To remedy the disadvantages of these traditional methods, MIS managers naturally looked to COM.

The technique of photographing microimages of computer-generated reports onto easily handled cards has become a multi-billion dollar-per-year business. There are few MIS groups today that do not use microfiche to some extent. When compared to paper output, microfiche output lowers the cost of storage by a factor of 10 or 20. This is in addition to the cost savings from reduced floor space and cabinet requirements of paper storage. Microfiche also provides users with automatic, in-line viewing via viewers and printers using automatic bar coding and readable indexes—a much more efficient method than searching through reams of paper.

COM technology, however, has its drawbacks. Once data are dumped from computer storage onto microfiche, they are no longer in data format. The data are transformed into a photographic picture that must be redigitized before it can be brought back into the computer system so that it can be distributed throughout the network and displayed. As an image, the data cannot be displayed on mainframe terminals; image-capable terminals are required.

Security is another problem with COM technology. A report or other confidential data is only as secure on microfiche as is the envelope in which it is delivered or shipped through the mail. In addition, the use of chemical developing agents poses ongoing costs and disposal problems. And, most of all, although more efficient than paper searches, retrieval time can be tedious and slow.

Technology Perspective. The feasibility of optical disk technology as a storage and retrieval medium for document images has become widely accepted. The advantages of this technology, namely, its faster retrieval rates over microfilm, its ability to compress images, its WORM security feature, all provide solutions to the disadvantages encountered with COM storage. Why not use optical disk for storing data and images?

A special software program that will compress and store computer-generated reports as data to optical disk is required. This software, combined with a special processor and stand-alone optical media, makes up what is commonly called a COLD subsystem.

The cost of storing data reports on microfiche as compared to optical media is reported to be much less, by a factor of about 10 to 1. Increases in user productivity have been attributed to the ability to retrieve pages in ten to fifteen seconds, as compared to the much longer time it takes to sort manually through fiche pages on a shared viewer. Windows-based retrieval software has enabled users to gain the benefits from COLD with very little up-front training. For some, it is just another Windows application. And, because the reports are stored as data, not images, there is a great deal of flexibility in developing index criteria, including the ability to cross-reference reports based on data in the report as well as automated indexing by downloading data from a mainframe.

Many COLD vendors estimate that companies switching from paper or microfilm to COLD can reduce their current distribution and archival costs by as much as 45–90 percent. In addition, vendors report that users can typically achieve a return on their investment within twelve to eighteen months after implementation.

COLD Subsystem Characteristics. One of the reasons COLD has become so economical as a replacement for COM is that COLD subsystems are available for use on stand-alone PCs, servers on LANs, and microcomputers that interface with mainframes, or they can be directly channel-attached to a mainframe. Data stored through these systems can also be accessed through remote workstations via communications or a modem. PC-based systems typically cost anywhere from $30,000 to $100,000, whereas mainframe-based systems cost upward of $150,000.

A typical COLD system consists of a jukebox or optical disk interface program; optical media plus an onboard processor for offloading compression, indexing storage, and retrieval functions from the mainframe; a magnetic disk for storing the index and directories for retrieving reports; and the user interface for report retrieval. Many COLD systems include global text retrieval and keyword searching capabilities; forms overlay for merging an image of a form with the data (thus re-creating an accurate representation of the original form); facsimile transmission; and the capability to export the data into other applications such as databases or spreadsheets. Figure 26-10 is an example architecture.

Since most of the COLD software runs on the subsystem processor, very little needs to be done to the existing mainframe software and hardware. From the mainframe standpoint, the administrator must configure two additional devices (to the mainframe) that the optical system emulates: a printer and a tape controller. The mainframe sees the optical disk as a printer; by printing one copy to an optical disk, any number of people can access it.

The administrator must also install the optical disk interface subsystem program in the subsystem library. Finally, the indexes for retrieving the reports must be identified so that the data is read onto optical disk, and the index file can be built.

Emerging Technologies

The advent of the writable CD-ROM technology may have a dramatic impact on the use of COM. Writable compact disks represent an ideal and low-cost alternative to microfiche for report distribution since they can store thousands of pages of ASCII text and typically conform to industry standards, such as the ISO 9660 standard for file and volume structure, allowing them to be read on any CD-ROM drive attached to a PC. In addition, writable CD-ROM technology eliminates the need to have CD pressed at an outside service bureau, thus saving companies thousands of dollars. Large organizations that need to distribute computer-generated report data to remote locations (field offices) will benefit from this technology. The typical cost for writable CD is less than $25 for each CD that can store 550 MB of data, or approximately 240,000 pages of ASCII text. One CD can also hold the equivalent of 444 fiche or 120,000 pages.

Relationship to Imaging

Many document imaging vendors allow data processing reports to be entered directly onto their imaging systems via COLD software. As imaging applications extend into the enterprise, the ability to handle multiple data types becomes imperative. The question is not whether to use COM, COLD, or imaging; rather, it is which method (or combination of methods) will meet an organization's requirements.

One concern is retrieval time. It takes much longer to retrieve an image of a report than it does an encoded page. However, not all documents can be stored as data. Some must be preserved as images because they contain signatures, handwriting, or other noncodable information. Others, even some data reports, will be accessed so infrequently that it is not cost-effective to keep them on optical disk. These documents will be better handled by an imaging system for distribution and archiving.

Implementing a COLD system can reduce operating costs, and paper distribution, storage, retrieval, and archive costs (including office space). On the other hand, implementing a COLD system can increase user productivity, report access, distribution, storage, and retrieval time. COLD systems can be easily integrated into current computing architectures. The low cost of CD, which store large amounts of information, help make COLD technology affordable in the 1990s. The relatively short period of time for a return on investment (twelve to eighteen months) makes COLD a desirable mechanism for achieving office automation.

PCs AND LOOKING TO THE FUTURE

What Does All this Technology Mean to Safety Professionals?

I see several benefits to safety professionals. One is virtual reality. Companies are now working on virtual reality training for forklifts. Just think, if we could train people on fall protection, confined space entry, or hot work using the computer, substantial savings in time would be realized and true hazards of the environment communicated. Consequently, safety professionals would have more time to "do safety." I also see the implementation of voice recognition systems. These systems could virtually eliminate carpal tunnel syndrome (CTS) related to office workstations as technology and the price of the systems permit. Just think, each CTS costs about $33,000 per person. Surely, these systems are less costly than that.

Computers, Computers Everywhere

As the price of computers continues to drop like that of calculators, computers are popping up everywhere. They first showed up at the airport in kiosks that allowed you to generate a map and directions of how to get to your hotel. Now, even airlines are getting into the act, putting them in their first-class cabins. I am sure that once they are proven successful, they will soon come to coach. Now we will have less wasted time in transit (or did you look forward to sleeping on the plane?). Hotels are also beginning to provide computers, ergonomic office furniture, and fax lines in their hotel rooms (for a price, of course). Have you ever heard of hoteling at work? Many offices are implementing a hotel concept. Several large companies have implemented this concept, but it is helpful especially for consultant companies. You have a common office where you can come in and plug in your laptop to a docking station, and the office is yours. Wow! What a concept!

Keeping Up with the Best Computer Will Be Difficult

With the newly offered 150- and 166-MHz Pentium chips, suddenly the 133-MHz chips that were top of the line are now midrange systems. Compaq has come up with two nifty hardware add-ons for its latest line of Presario desktop computers. These devices are not installed across the entire line, but only in certain machines. One is the single-sheet paper scanner built into a keyboard, the other a CD-ROM drive that can write huge amounts of data to a cartridge-based compact disc. Another is a writable CD-ROM drive that holds 560 MB more if you compress the data: extended data output (EDO). This kind of memory chip can boost performance by 10–15 percent. The hard drive—often the performance bottleneck—takes a mere three seconds to start up WordPerfect 6.1 for Windows.

More Bad News about Computers

I hate to bring some bad news to all this neat futuristic technology, but in looking at the computing power available today, I cannot help but look to the future and speculate on what will be available tomorrow. Trying to keep your computer upgraded to the latest technology will be like changing the tires on a moving car. No matter what you buy now, it will most likely be out of date in five years or less. Computer technology has at least redoubled itself every five years. Intel's cofounder Gordon Moore has espoused Moore's Law, which states, "Technology doubles every eighteen months." It does not take an engineer with an HP calculator to figure out that this is exponential growth in computing power every five years. So, in five years, the Pentium will be obsolete, having been replaced by the Spentium and subsequently the Repentium (the Dimtel will already be marketed as the next quantum leap in technology); 1 GB will be required to run your word processing software; CD-ROMs will barely be hanging in there as the transportable media of choice; modems will be superfast; no one will remember Windows 3.1 or Disk Operating System; subnotebooks will be considered too fat and heavy (notebooks will be on their way to the Smithsonian along with the desktop computer); and voice interface will replace keyboards and perhaps even pointing devices. However, we must all muddle along with the computers we have at the time. But, there is still some good news about the future. Technology is continuing to advance at a rapid pace. Some of the technology advances follow.

How about a Tabletop Particle Accelerator?

The University of Michigan's Center for Ultrafast Optical Science has developed a tabletop particle accelerator. Although considerably less powerful than Stanford's 3.7-km linear accelerator, these small devices, nonetheless, fire short, powerful bursts of electrons that can take snapshots of phenomena that occur in femtoseconds. This speed can be used in biology, chemistry, and physics to study virtually instantaneous reactions. These systems could supplant larger and costlier radio-frequency linear accelerators.

Fiber Optics

Because of Bob Lucky, formerly chief scientist of Bell Labs, fiber optics is coming to a home computer near you . . . but not until about 2010. I am sure that this will come to pass, but don't hold off buying your next PC because of this development.

Man as Modem

Neil Gershenfeld of the MIT Media Lab indicated that since the human body can conduct electricity, it could also be used for transmitting data. The lab is working on this concept right now. Imagine shaking hands and transmitting your business card or downloading data to a computer by touching it. Someday you might.

Brother Can You Paradigm?

The technology of each previous decade seems to be transforming that of the next decade. For example, in the 1970s, the invention of the microprocessor gave us the PC that transformed the 1980s. In 1980 the laser was discovered, affecting everything from laser surgery to telecommunications. In the 1990s it will be smart sensors that will show up in everything from toys to transportation. This technology will allow digital devices to communicate with each other in unique ways. Imagine a video camera with fuzzy logic sensors that tell it to start taping when your child is kicking the ball on the soccer field. This technology appears to be the next rage; however, there is no guarantee that they will make us smarter.

CD-ROMs in the Smithsonian?

In fact, CD-ROMs are already in the antiquation mode. Los Alamos National Laboratories in Los Alamos, New Mexico, has recently invented the High Density-ROM (HD-ROM). The HD-ROM uses a unique ion beam to inscribe information on pins of stainless steel, iridium, or other materials. The HD-ROM holds about *180 times* more information than a comparably sized CD-ROM and costs 0.5 percent that of CD-ROM costs. This type of technology growth is merely the beginning of such technology leaps that we are about to see in the next five years.

The Information Superhighway

The Information Superhighway is upon us and here to stay. The toll to use that highway seems to be getting cheaper, but the speed limit is going higher. What is true today as a limitation may not and probably will not be true tomorrow. What does this mean for safety professionals? Computer literacy is a must. Computer proficiency is a necessary basic requirement. To keep abreast of the computer technology will take a considerable effort, perhaps almost as much as keeping up with changes with OSHA regulations. Companies, customers, even safety professionals will demand computer proficiency in addition to proficiency in safety, just to maintain the status quo. The days of the computer-phobic are over. So, get on the On-Ramp of the Information Superhighway or get in line at the local employment office.

CONCLUSION

Staying compliant with OSHA in the 1990s certainly poses a challenge for safety and health professionals and managers. Automating the process to reduce time, steps, and manpower—while not only maintaining compliance but achieving superior performance—will help ensure company survivability in the 1990s. Document imaging, business process automation, reengineering, and COLD implemented on client/server architectures brings state-of-the-art technology to the desktop of the S&H professional. Instead of spending an inordinate amount of time wrestling with regulations, documents, engineering drawings, and filing reports, S&H professionals can now focus on the job for which they were hired.

SUGGESTED FURTHER READING

Addamo, F. (May/June 1993). "Imaging: A Critical Component of Process Safety Management (PSM)." *Document Imaging & Windows Imaging*, pp. 21–28.

Baum, D. (April 1993). "Client/Server Development Tools for Windows." *Computerworld*, pp. 73–75.

Blotzer, M. J. (May 1995). "Welcome to the World-Wide Web." *Occupational Hazards* 57(5):89–92.

Brauer, R. L. (1994). *Directory of Computer Resources.* Des Plaines, IL: American Society of Safety Engineers.

Dvorak, J. C. (August 1996). "Web Wake-up Call." *PC Computing*, p. 59.

Gangemi, R. A. (June 1996). "Internet: Web Sites We Love: CEOs Pick Their Favorites", *Inc. Managing Technology*, p. 134.

Hafner, K., and N. Croal (March 1996). "Getting up to Speed: The Old Fashioned Cable May be Your Ticket to the Fast Lane on the Infohighway." *Newsweek*, 46–47.

Krasowska, F. (October 1995). "Health and Safety on the Internet." *Occupational Health & Safety*, pp. 100–108.

Levy, S. (1996). "How to Cast a Wider Net: There Is a Lot of Trash on the WWW. But the Really Good Sites Point to the Future Glory." *Newsweek*, 4 March, p. 48.

Lyons, D. (May 1995). "Internet Means Business for Internetworking-Savvy VARs." *Varbusiness*, pp. 55–56.

O'Lone, E. J. (June 1993). Datapro, Document Imaging Systems, Management Issues." *Client/Server Computing*, 1–8.

Rigdon, J. (1996). Cyber-Trash Bogging Down 'Net': Success of Web May be Worst Enemy." *Houston Chronicle*, 4 February, 4D.

Shegala, K., and M. Richardson (October 1993) "Computer Output to Laser Disk (COLD)." *Datapro, Document Imaging Systems*, pp. 5060–5080.

Somerson, P. (August 1996). "Web Coma." *PC Computing*, p. 57.

Stephans, R. A., and T. W. Warner (1993). *System Safety Analysis Handbook, A Source Book for Safety Practitioners.* Albuquerque, NM: System Safety Society.

Stickle, R. W. (Spring 1997). "Double-Click, The Internet Part 1." *American Society of Safety Engineers, Engineering Division Newsletter, Division News*, pp. 15-17.

Stickle, R. W. (Fall 1997). "Double-Click, the Internet Part 2." American Society of Safety Engineers, Engineering Division Newsletter, Division News, pp. 2-3.

U.S Department of Labor (1993). 29 Code of Federal Regulations, 1910, General Industry. Washington, DC: Occupational Safety and Health Administration.

Chapter 27

The Internet and the Safety Professional

Mark D. Hansen

INTRODUCTION

The Internet is a complex interconnection of large database computers. Nationally, the Internet gets its main traffic-handling support through the National Science Foundation (NSF) system. This system allows scientists at universities and other academic institutions to access and share information. With the ever-increasing use of the system for nonacademic purposes, shunt routes and subnetworks are evolving to keep information moving rapidly and efficiently. The "Net" makes many EPA, OSHA, NIOSH, NSF, university, and institutional databases accessible from a personal computer (PC). The network, originally developed by the Department of Defense, began with just a handful of computers. Now, with the right hardware and software, anyone can access the Internet's vast store of information and its more than 2 million users. The range of topics and information available is constantly expanding.

USING THE INTERNET

To use the Internet, your computer should have either a 386 or a 486 processor. The faster the better. Most applications require Windows and a high-speed modem (14 kilobytes per second). There are three ways to access the Internet: gateways, remote modem, and direct Internet access. Gateways allows networks to "talk" to each other, but users of the network are limited in their ability to access fully all the tools of the Internet. In this group, users are limited to what they can access on the Internet by what a service provider allows them to access. This group consists of Prodigy, CompuServe, and America Online, among others. Table 27-1 gives a ranking of several online services.

The second group is remote modem access, which is done through a dial-up terminal connection. Users access the "host" computer using a modem, and their computer becomes a terminal on that mainframe. Users type commands on their computer, but the host computer executes the commands.

The third group is direct Internet access. This is the highest and most expensive level of connectivity for Internet access. At this level, users are directly connected to the Internet using high-speed telephone lines. Users are connected to the Internet (on-line) twenty-four hours a day, seven days a week. This group is the best method of connectivity to the Internet if there is a mainframe at your site with hundreds of users, but not too advantageous for single users. As technology progresses, there is a new twist to the third group: "on-demand direct connectivity." Since users rarely spend twenty-four hours a day on the Internet, some sites allow users to connect to the Internet using a high-speed modem and point-to-point protocol or serial line Internet protocol connection. Table 27-2 gives the services and cost breakdowns of providers.

Services Provided on the Internet

The most used service on the Internet is electronic mail, or e-mail. E-mail addresses are actually domains, and are based on Internet protocol addresses. Each server connected to the Internet has an address. Examples of domain addresses are as follows:

edu	U.S. Educational sites	mil	U.S. military sites
com	U. S. Commercial sites	org	U.S. organizations
gov	U.S. government sites	ca	Canada
net	Network administrative organizations		

Table 27-1. Ranking Computer Online Services

Rank	Service	E-mail	Internet Service	Other Services	Cost
1	Prodigy	Excellent	Excellent	Good	Excellent
2	CompuServe	Excellent	Good	Excellent	Excellent
3	America Online	Good	Good	Good	Poor
4	eWorld	Good	Excellent	Fair	Poor

Table 27-2. Costs of Online Services Based on Hours Used

Service	Cost per Month ($)				
	5h	10h	20h	30h	40h
America Online	$ 9.95	$24.70	$ 54.20	$83.70	$113.20
CompuServe	9.95	24.70	24.95	44.45	63.95
Prodigy	9.95	24.70	30.00	30.00	59.50
CRIS	30.00	30.00	30.00	30.00	30.00
Internet MCI	9.95	22.45	47.45	72.45	97.45
Netcom	19.95	19.95	19.95	19.95	19.95
PSI	29.00	29.00	29.00	30.50	45.50
Microsoft Network	9.95	22.45	19.95	42.45	62.45

How Much the Internet Is Being Used to Conduct Business and Exchange Information

In 1994, the number of Internet servers had grown by more than 500 percent. In the third quarter of 1994 alone, the number of commercial hosts had grown by 36 percent. The number of ".com" sites is now greater than the number of ".edu" (educational) sites, an indication that the Internet, which was originally geared toward scientists and academicians, now means business.

In 1997 two driving forces were behind the success of the Internet: 1) the World-Wide Web (WWW, W3, the Web), a framework for distributing multimedia documents on the Internet; and 2) the Mosaic browser interface. The key to the Web is the uniform resource locator (URL), which provides a standard way to specify the location and access of files on the Internet. For example, http//www.osha.gov is the URL for the OSHA server in Washington, D.C.

Spyglass Inc., of Naperville, Illinois, holds a master license for the popular Mosaic browser. Officials at the company say that Spyglass has licensed 13 million copies of Mosaic, and that shipments are running at 60,000 copies per month. So, the Internet is reaching critical mass; in other words, there are enough people on-line to make it worth the effort and expense required to develop a presence on the Internet. For now the biggest use of the Internet among businesses is sending electronic mail and moving files. But, increasingly, companies are using the Internet as a means to provide customers with support and information. And, next, they will be using it to do business. Indeed, $60 billion worth of goods and services will be purchased through the Internet in 1997, which is forecast to grow to $250 billion in 1998. This information gives credence to the statement that the Internet is here to stay and not just a passing fad.

What Are People Using the Internet For?

People are using the Internet for a myriad of reasons, including to

- Browse (79%)
- Gather information (77%)
- Be entertained (65%)
- Collaborate with others (54%)
- Work (51%)
- Access customer/vendor support (50%)
- Research competition (46%)
- Communicate internally (44%)
- Publish information (33%)
- Shop (14%)
- Sell (13%)

The following is a profile of Internet users:

1) Average household income of user ($69,000)
2) Primary activity on the Web:
   - Browsing (82%)
   - Entertainment (57%)
   - Work-related (51%)
3) Average age of user (35 years, male and female)
4) Population using the Internet
   - Women (30%)
   - Men (70%)
   - U. S. residents (76%)
   - California residents (10%)

- Windows users (52%)
- Mac users (26%)
- Once-per-day browsers (77%)

Safety-Related URLs Addresses

The following are safety-related URL addresses:

- Brookhaven National Laboratory, Safety & Health Protection Division— http://sun10.sep.bnl.gov/seproot.html
- ASSE
  San Jose Chapter home web page—http://www.best.com.80/~assegsjc; National home page—http://www.asse.org; E-mail address on CompuServe—73244,562; Engineering division E-mail—oshex@dreamscape.com or rick.stickle@nasa.gsfc.gov
- Canadian Centre for Occupational Health & Safety—http://www.rpi.ccohs.ca
- Environmental Health Center/National Safety Council—http://envirolink.org
- Federal and State Regulations—http://www.gate.net/~gwarbis/solutions
- FedWorld—http://www.fedworld.gov
- Firenet—http://life.anu.edu.au/firenet/firenet.html
- HazDat (Hazardous Substance Release/Health Effects Database)—http://atsdr1.atsdr.cdc.gov.8080/atsdrhome.html
- IAPPS (International Association of Personal Protection Specialists)—http://www.mps.ohio-state.edu/cgi/cgi-bin/hpp/laps_home.html
- Index of OS&H Resources on the Internet—http://turva.me.tut.fi/~tuusital/oshalinks.html
- List of Listerves Pages—http://www.clark.net/pub/listserv/listserv.html
- Martindale's Health Science Guide '95—http://sss-sci-lib.lib.uci.edu/~martindale/HSGuide.html
- National University of Singapore BioMed Web Server (natural toxins database)— http://biomed.nus.sg
- OSHA—http://www.osha.gov
- NIOSH—http://www.cdc.gov/niosh/homepage.html

- MSHA—gopher.msha.gov
- OSHA Salt Lake Technical Center (Federal Register and OSHA regulations)—http://www.osha-slc.gov
- Periodic Table—http://www.cchem.berkeley.edu/Table/index.html
- RISKWeb—http://riskweb.bus.utexas.edu./riskweb.html
- Safety Online—http://www.safetyonline.net
- Typing Injury Archive WWW Site—http://alumni.caltech.edu/~dank/typing-archive.html
- University of Illinois (Urbana) Division of Environmental Health Safety— http://romulus.ehs.uiuc.edu./DEHS/dehs.html
- U.S. Centers for Disease Control—http://www.cdc.gov
- U.S. Department of Energy - Office of Environment, Safety & Health— http://130.20.92.130:8001/esh/home.html
- U.S. Department of Health and Human Services—http://www.os.dhhs.gov
- U.S. EPA—http://www.epa.gov
- U.S. EPA Center for Exposure Assessment Modeling— http://ftp.epa.gov/epa_ceam/wwwhtmlceam_home.html
- U.S. FEMA—http://www.fema.gov
- U.S. Food and Drug Administration—http://vm.cfscan.fda.goc/index.html
- U.S. National Library of Medicine—http://www.nih.gov/
- World Health Organization—http://www.who.ch

Listserves

Listserves are mailing lists maintained by a LISTSERV program of a group of persons who share similar interests. Anyone can subscribe to a LISTSERV mailing list by sending a SUBSCRIBE command to the LISTSERV address. Any e-mail letter sent to the list's address is copied and mass mailed to the e-mail box of every person subscribing to the list. Everyone on the list can then reply to that letter, and the dialogue begins.

Safety and Health Mailing Lists

To subscribe to any of the lists in Table 27-3, send this message to the listserv:

        subscribe [name of list] [your real name]

Table 27-3. Directory of Safety and Health Mailing Lists

Name of List	Listserv Address	General Subject Area
BIOSAFETY	listserv@mitvma.mit.edu	Safe handling of biohazards
CHEMED-L	listserv@uwf.cc.uwf.edu	Chemical safety
CMTS-L	listserv@cornell.edu	Chemical management and tracking
DISPATCH	majordomo@tcomeng.com	Police, fire, EMS
EMERG-L	listserv@vm.marist.edu	Emergency services
FIRENET	listserv@life.ani.edu.au	Firefighting and emergency medical services
FIRST-AID	listserv@first-aid-request @rabble.uow.edu.au	Casual and professional users of first aid
HEALHTRE	listserv@ukcc.uky.edu	Health care reform
HELPNET	listserv@vm1.nodak.edu	Network emergency response planning
LEPC	listproc@moose.uvm.edu	Hazardous materials emergency response
RISKNET	Listproc@mcfeeley.cc. utexas.edu	Safety and health
SYSTEM-SAFETY	listserv@listserv.gsfc.nasa.gov	System safety

Table 27-4. Additional Safety and Health Mailing Lists

Name of List	Listserv Address	General Subject Area
COM-L	ccm-request@eja.nes.hscsys. edu	Critical care
DMATNEWS	listserv@medicam.norden1.com	Disaster medical assistance teams
EMC-PSTC	majordomo@ieee.org	Product safety
EMED-L	emed-l@itsa.ucsf.edu	For health care professionals
HAZMATMED	listserv@medicomm.norden1.com	Hazardous materials
SSAVETT_NCEMSF	majordomo@indiana.edu	National Collegiate EMS Foundations

To subscribe to any of the lists in Table 27-4, send this message to the listserv:

subscribe [name of list] [your e-mail address]

On the Lighter Side: Favorite Web Sites of Some CEOs

Now that you have seen some safety-related Web information, what about information for other people? What about CEOs? What are some of their favorite sites and why? Table 27-5 is a sample of some CEO's favorite Web sites and why they like them. Perhaps these sites can be useful to safety professionals, too.

A SUCCESS STORY OF THE INTERNET

It's 3 a.m. and Madeline Shea's 22-year-old son Michael isn't feeling well. Shea figures it's not something that merits awakening their doctor in the middle of the night. However, she would like to ask him about the problem while it's fresh on her mind. So she goes to her computer, accesses the Internet, and queries physician George Bergus via electronic mail. In the morning, she shows her husband a printout of her message, and when they arrive home that evening, they rescue Bergus's reassuring response.

An assistant professor of family practice at the University of Iowa, Bergus is one of a growing number of doctors who use e-mail for renewing prescriptions, making referrals, handling questions about minor ailments

Table 27-5. Sample of Web Sites Preferred by CEOs

CEO/Company	Favorite Site	Address	Why the Site is Favored
Cal Lai/Lai/ Venuti & Lai	Egghead	www.egghead.com	Ranks articles in order of importance.
Richard Gorgens/ Microsystems Software	Infoseeks	www.infoseek.com	Its matches are better than Yahoo's. Its engine is more selective.
Steven Papermaster	CNN Interactive	www.cnn.com	It is well formatted in BSG terms of easy access for relevant topics.
Brad Freeburg/ Lantronix	DEC Search Engine	www.altavista or www.dec.com www.digital.com	It is extremely easy to use
James Ackles/ LBS Capital Management	Microsoft	www.microsoft.com	It gives valuable information for free.
Peter Zandan	Suck	www.suck.com	It casts a jaundiced eye on all types of hype. It is a humorous reality check.

and symptoms, and communicating test results. By giving physicians a chance to think through their responses to patients' questions, e-mail can enhance medical care, says Bergus, who checks for messages up to three times a day. Bergus states, "It forces me to think before I write."

NEWLY EVOLVING INTERNET LANGUAGE

The Internet is giving rise to a new language. This language is called *Geekspeak*. Geekspeak is Internet jargon that is founded in acronyms and shortened speech. Some examples are as follows:

Pulling glass: Laying down fiber optic cable.

Tweak freak: A computer techie obsessed with finding the root of all tech problems, regardless of the relevance. A tweak freak might spend hours trying to track down something that could instantly be fixed by reinstating the software.

Twitch game: A computer or arcade game that is all hand-eye and little brain. Similar to thumb candy.

User eye-d: A face-to-face meeting with someone you've gotten to know over the net. My user eye-d with Robin was not what I expected.

Alpha geek: The most knowledgeable, technically proficient person in an office or workgroup. "Ask Larry; he's the alpha geek around here."

Barney page: Web page designed to capitalize on a current trend (such as Barney bashing).

Betamaxed: When a technology is overtaken by an inferior, but better-marketed technology.

Bio-break: Techie euphemism for using the toilet.

Cobweb site: A WWW page that has not been updated for a long time.

Synthespian: Synthetic actor. Used in 3D animation to describe sophisticated human forms that can be imported into a virtual world.

Geekspeak: Internet jargon founded in acronyms and shortened speech.

:):   Just kidding (also called a neticon).

BRB: Be right back.

LOL: Laughing out loud.

FAQ: Frequently asked questions.

As time goes on, this evolving language will reveal much about the multimillion user population on the Internet. Only time will tell if it is predominantly good or bad.

THE INTERNET PANACEA OR PLAGUE

A case study of a session in the life of an Internet user may reveal some serious posing problems that it must consider solving if the Internet is to grow as expected. In a typical session, it took ninety minutes for an experienced PC user to find the reference material he was looking for. When he finally found it, the information was a year old and had been deleted. He wandered into an electronic chat room where twenty-two persons were carrying on four different conversations in a tangled gabble.

This story is hardly unusual. Despite the hype, legions of computer users are finding the Web less of an electronic wonder and more of an electronic bore. They complain that access is slow, that Web sites are often difficult to find and contact, and that the content of many offerings is frivolous or hopelessly outdated.

It is not that the Web is a flop. Rather, groaning under 7.5 million daily users, and increasingly flooded with cyber-trash, the Web's success has become its worst enemy.

So, although the Web may indeed become a great thing in the near future, many now find it analogous to the PC in 1982—slow, clunky, and oversold. Spending an evening on the WWW is much like sitting down to a dinner of Cheetos. Two hours later your fingers are yellow and you're no longer hungry, but you haven't been nourished. Netscape stock has soared, more than sixty times its initial offering price. Many companies are using the Net to roll out mounds of electronic data or to advertise products. And speedier modems, better graphics capabilities, and better Web-navigating software may one day deliver the Web to heights equaling its hype. Indeed, Netscape estimates that as these improvements come on-line, many of the shortcomings that drive users crazy will evaporate by the end of 1998. There are tens of thousands of Web users who stay on-line for hours and consider it a cyber-miracle. But for scads of eager computer users, shoppers among them, who have been seduced by Web hype, their experience has left them in a cyber-funk. FIND/SVP, a New York-based market research firm estimates that about 9.5 million Internet users have visited Web sites. But, only 2 percent have found anything worth buying from Web-offered shopping services.

To navigate the Web's complex weave, most users have to type a string of ten to dozens of characters, starting with "http/www" for hypertext transfer protocol WWW. Next comes a word or two punctuated by periods, slash marks, ampersands, and other symbols that most persons have difficulty finding on their keyboards. One slip of the pinkie sidetracks the whole trip.

Http Doesn't Mean a Thing to Me

I would be just as happy if "Http" were just an "H." Users are supposed to key in one word and pull up dozens, if not hundreds of Web sites. They must sift through too many sites that have no meaning., some even misusing keywords to attract you to their site.

The Web Coma

You all know that zoned-out comatose daze you sometimes feel after pigging out at Thanksgiving, or fighting to stay awake during some endless human resources department sensitivity sermon, or slogging through an incomprehensible tax form. These are all trance-inducing events, but nothing like a *Web coma*. I have identified the following varieties.

Click Candy

The Web is the ultimate in attention deficit disorder, with some "Let's make a Deal" thrown in. Hypertext beckons from every corner of your screen. What's behind link number one? Go there now! No time to read the current page. Too hip, gotta go! Click! Gone!

Content? Who cares! It doesn't matter if each page is good or bad. If it is bad, you just warp right out of there! If it is good, you trace down every branch. Let the coma begin. It seems that everyone searches the Web with the thought that enlightenment is just one mouse click away. Cooler stuff is just a click away. You are hopelessly swept down the clickstream.

Search Me

Then you really need to do some serious research. So you fire up your favorite search engine. It spits out 90,000 hits. You dink with the parameters and it spits out 85,000 hits. You dink with it a little more and get no matches. Argh! You switch to your other favorite search engine and it spits out similar results. You get sucked in until you hear the cleaning lady come through the office, and when you finally look up again, the automatic timer has turned off the air conditioner.

Interglut

Internet is the coma generated by frenzied marketers desperate to catch on to the Web wave. Every product is being hooked to the Net whether it really makes sense or not. Like Tripp Lite advertising, it's UPS to be Internet ready (like cable ready). This surely does not add value to your search. You are getting heavy.

A FEW SOLUTIONS TO INTERNET PROBLEMS

As more and more people hook up to the WWW from home, frustration over limitations on "bandwidth" or line capacity is mounting. Most people use 14.4K-baud modems; some use 28.8K-baud hooked to their phone lines. That's fine for sending and retrieving plain text such as e-mail. But doing the same with graphics and sound at that speed is agonizingly slow. Full-motion video is impossible. Many exasperated Web surfers now log on after hours from their offices, which often have high-speed Internet connections.

A wire many people already have in their homes—the lowly coaxial cable—could change all that. In a handful of trials around the United States, cable companies are piping data into homes at unprecedented speeds. Because cable carries data through the wide channels used for television, it has the potential for moving data at 30 million bits per second. That's thousands of times faster than twisted copper phone wire and hundreds of times faster than phone companies' Integrated Services Digital Network (ISDN). If the cable companies' pricing models are to be believed, costs will drop too. At around $45 per month for unlimited access, cable connections will be cheaper than high-speed phone links. And the cable companies plan to compete directly with Internet providers like America Online, CompuServe, and Netcom by including an Internet connection in their service. Forester Research, Inc., predicts that by the end of the decade nearly 7 million households, using a device called a cable modem, will surf the Web using the same wires that pipe in HBO and the Home Shopping Network.

Superfast transmission speed is the good news. The bad news is that the monopolistic cable companies, many of them newcomers to the on-line universe, are bringing it. The ethos of the Internet is one of choice. If your cable provider is also your Internet access provider, the cable company is likely to want to control what you get, much as it does with cable programming. If you use phone lines to get to the Internet, you may be stuck with Telco. But, through them you can still choose your Internet provider.

Having control of the pipe is handy. Time Warner, which is testing its Linerunner cable-modem service with two hundred residents in Elmira, New York, conspicuously promotes the company's books, magazines, and other products. The Linerunner menu nudges users toward Pathfinder, Time Inc.'s own Web site. In fact, customers cannot get to the Web unless they go through Pathfinder first. Officials at Time Warner Cable defend the practice by pointing out that Pathfinder is one of the most popular sites on the Web.

@Home, formed by TCI-Communications, built its own network to connect to the Internet. They are building alliances with multimedia forms, CD-ROM companies, schools, and so forth. The move to cable has been slower than the cable companies had hoped. Only 5 percent of the cable systems in the United States can handle two-way data transmissions. An additional 10–15 percent were built to handle two-way traffic, but still aren't. That means that most cable networks allow only "downstream" transmission into the home: there is no way for the user to send commands and other messages back to the Internet. Cable companies are upgrading their wiring with fiber-optic cable to allow for high-capacity "upstream" traffic, but the transition is slow and expensive. As a rule, even when a system is capable of two-way traffic, it sends data downstream much faster than upstream.

Worldwide standards for cable modems are still being determined, and many of the companies building them still haven't gone into mass production. Until that happens cable modems will be expensive (about $400), and

Table 27-6. Comparison of Copper Wire, ISDN, and Coaxial Cable

	Copper Wire Phone Line	ISDN Phone Line	Coaxial Cable
Speed	14.4 Kbytes	128 Kbytes	10,000 Kbytes
Time to download a 1-MB file[a]	9.7 min.	66 s	0.8 s
Average monthly cost	Same as telephone	$40 plus usage	$45 flat rate[b]
Usage	98% now use dial-up modems	6.7 million homes by year 3000[b]	6.9 million homes by the year 2000[b]

[a] Without data compression
[b] Source of Estimates: Forrester Research, Inc.: NYNEX.

customers will likely lease them, much as they do the set-top device for cable TV service. It has been estimated that there will be 1 million customers by the end of 1996.

Internet Meltdown

With the increase in information transmission rates, the Internet performance should improve, right? Experts say, *wrong*. The increasing speed through the broadband and wireless only hastens the Internet's meltdown. As more and more people connect at faster and faster speeds, the Internet will get slower, not faster. Even though more conduits are cropping up, they are doing so without serious flow-control software in place. The World Wide Web is growing so fast that it may become the World Wide Fib.

Millions of Tourists on the Internet

The cool thing to talk about on the Net these days is the sudden *uncoolness* of the Net. It's the old Yogi Berra joke gone cyber: nobody goes online anymore because its too damn crowded. Worse, the newcomers are craven *arrivistes:* their inexperience is the virtual equivalent of Bermuda shorts and cameras around their necks. Well, tough luck, trendsetters—The Net, and particularly that multimedia, publishing-based portion of it called the WWW, is quickly coming of age as our next mass medium. So far everybody's been complaining about problems with ease of use and speed. But the more profound questions deal with content: Can you really draw a throng of people by offering bunches of bits? What do consumers really want from Web sites? Is there money in it?

CONCLUSION

Staying compliant with OSHA certainly poses a challenge for safety professionals and managers. Automating the process to reduce time, steps, and manpower—while not only maintaining compliance but achieving superior performance—will help ensure company survivability in the 1990s. Instead of spending an inordinate amount of time wrestling with regulations, documents, and engineering drawings, and filing reports, safety professionals can now focus on the job for which they were hired.

Using the Internet has become a mania for many users today. One way to determine whether a product is useful is if it begins to be used for many purposes, some for which it was not originally designed. For example, the

Internet was originally designed for universities and other academic institutions to share and exchange information. It has now grown to individuals and businesses. Surfing the "Net" can be informative and fun, but it can also be hazardous to your data. If you surf the Net, you may want to consider some form of antivirus program to protect your computer. Most safety professionals have become fairly dependent on computers to do their jobs. Locking up your computer and having to reboot is one thing; corrupting or losing all of your data is quite another. Surf's up! Be careful and watch out for the undertow.

SUGGESTED FURTHER READING

Addamo, F. (May/June 1993). "Imaging: A Critical Component of Process Safety Management (PSM)." *Document Imaging & Windows Imaging,* pp. 21–28.

Baum, D. (April 1993). "Client/Server Development Tools for Windows." *Computerworld,* pp. 73–75.

Blotzer, M. J. (May 1995). "Welcome to the World-Wide Web." *Occupational Hazards,* pp. 89–92.

Brauer, R. L. (1994). *Directory of Computer Resources,* Des Plaines, IL: American Society of Safety Engineers.

Hafner, K., and N. Croal (1996). "Getting up to Speed: The Old Fashioned Cable May Be Your Ticket to the Fast Lane on the Infohighway." *Newsweek,* 4 March, 46–47.

Krasowska, F. (October 1995). "Health and Safety on the Internet." *Occupational Health & Safety* 64(10):100–108.

Levy, S. (1996). "How to Cast a Wider Net. There Is a Lot of Trash on the WWW. But the Really Good Sites Point to the Future Glory." *Newsweek,* 4 March, p. 48.

Lyons, D. (May 1995). "Internet Means Business for Internetworking-Savvy VARs." *Varbusiness,* pp. 55–56.

O'Lone, E. J. (June 1993*).* "Datapro, Document Imaging Systems, Management Issues" *Client/Server Computing,* pp. 1–8.

Rigdon, J. (1996). "Cyber-Trash Bogging Down 'Net' Success of Web May Be Worst Enemy." *Houston Chronicle,* 4 February. 4D.

Somerson, P. (August 1996). "Web Coma." *PC Computing,* p. 57.

Stephans, R. A., and T. W. Warner (1993). *System Safety Analysis Handbook, A Source Book for Safety Practitioners.*

Stickle, R. W. "Double-Click, The Internet Part 1." American Society of Safety Engineers, Engineering Division Newsletter, *Division News,* pp. 15–17.

Stickle, R. W. "Double-Click, The Internet Part 2." American Society of Safety Engineers, Engineering Division Newsletter, *Division News,* pp. 2–3.

U.S. Department of Labor. (1993). 29 Code of Federal Regulations, 1910: General Industry. Washington, DC: Occupational Safety and Health Administration.

Chapter 28

Maintenance and Security

Richard W. Lack

INTRODUCTION

Any business or organization depends on sound maintenance and security management systems. It is also a fact that these two essential elements of management planning are often either taken for granted or not emphasized sufficiently in the total management strategy. The results of this neglect will be gradual deterioration of equipment and facilities and, ultimately, an expensive failure, breakdown, or other loss-producing event.

At a large refinery, the discharge pipe elbow from a high-pressure charge pump in the digestion unit suddenly burst, spraying hot caustic liquor over a large area and causing a plant shutdown for repair. Some dozen unit plant operators sustained caustic burns, fortunately none too severe. Investigation revealed that the pipe wall thickness had been worn down beyond safe limits due to erosion/corrosion. The maintenance engineering department had equipment to measure pipe wall thickness but the inspection program frequency had apparently fallen behind schedule. The case history that is presented subsequently illustrates what can happen when security management systems either are not in place or are not systematically applied.

The fundamental purpose of maintenance and security management systems is the protection of assets. Sound application of these systems will reduce the unit's exposure to losses that can result from the threat of intentional or unintentional acts.

In the case of security, inadequate maintenance or security measures could expose the unit to liability lawsuits on the basis of the forseeability of the criminal act.

This chapter approaches this subject from four broad perspectives:

1) Organization
2) Risk assessment
3) Security management systems
4) Maintenance management systems

As in all other aspects of management planning, strong management leadership and support are critical factors for the success of any enterprise. The overall unit goals and strategic plans must include sufficient funding for appropriate staff, equipment, and training.

Finally, I must emphasize that this chapter addresses an extremely broad and complex subject area; hence, the contents present only an overview. A comprehensive listing of reference sources is provided at the end of this chapter for those readers desiring more in-depth information on any topic discussed herein.

CASE STUDY: SECURITY AFFECTS LOSS

The management at a chemical process plant became aware of security concerns when it was informed of several unexplained losses:

1) Seventy-two pairs of gloves were drawn from stores by one supervisor.

2) Lead foil and a platinum crucible were discovered missing from the plant laboratory.

3) A contractor reported the loss of an electric welding machine.

4) Investigation by plant loss-prevention personnel uncovered numerous problems with the plant's security systems and procedures, including a general lack of security awareness by the plant employees. Examples were as follows:

   • Too many people had keys to the plant back gates.

   • The word around was that "you can walk out the gate with anything."

   • Supervisors were sending in requisitions to stores for miscellaneous small tools indicating the reason for this need as "theft" or "stolen."

- The union reported to management cases in which firearms and other dangerous weapons were being brought onto the plant and being used to "threaten people's lives."

Plant management, reacting to these concerns, requested help from the corporate security department. In response to this request, two corporate security professionals conducted an in-depth survey and audit of the plant facilities and procedures. Highlights of their findings and recommendations for improvement were as follows:

1) Paychecks were not signed for—some were collected by spouses.

2) No records were kept of lost badges.

3) No temporary passes were issued for visitors to the secured plant areas.

4) No procedure was established for response to bomb threats.

5) No controls were enforced on delivery truck drivers delivering materials to the plant.

6) Employees bringing personal tools on to the plant were not held accountable.

7) Security lighting in the employee car park and at the railroad and back gate was inadequate.

8) Building keys kept at the plant security office were not signed for when used by maintenance and operations employees.

All these and other problems were, of course, promptly addressed. But the fact that they existed points out obvious failures in the plant's maintenance and security systems.

THE ORGANIZATION

The management organization appropriate for the effective management of a unit's security and maintenance management systems will obviously vary widely. Factors to consider are the unit size, the type of business or operation, and the existing management culture or climate.

On the security management side, human resource, legal, and risk management functions should be involved in the process. For maintenance management, engineering, purchasing, and warehousing will be in close interaction. There are so many variables that a "boiler plate" organization simply cannot exist.

As an example, Figure 28-1 illustrates the management organization at a chemical plant. The plant had 1,500 employees and was a unit in a large corporate organization.

Organizational Relationships

Based on my experience, some important organizational relationships are as follows. First, there must be very close control and coordination between engineering, maintenance, and the various loss-prevention functions. At one organization, the maintenance manager, principal engineer, and director of construction support supervised their own departments. True, they did report to the same individual, but the control over facility and equipment modifications and new construction was inadequate. If any department needed to make an addition or change to its facility or equipment, it simply sent a work request to scheduling and control, which scheduled the appropriate maintenance crafts to carry out the work. There was no review for compliance with appropriate standards and no systematic updating of engineering drawings.

At this facility, and others I know of, complete buildings, lighting systems, roadways, and so forth technically did not exist based on the drawings in the engineering office! By contrast, at the plant whose organization is illustrated in Figure 28-1, corporate policy required that maintenance report to the chief engineer and that no changes or additions to plant buildings, facilities, or equipment were permitted unless authorized and supervised by engineering. In addition, each project incorporated a checklist that involved review, where necessary, by the safety, security, fire protection, and environmental departments.

Second, there must be a good relationship among purchasing and warehousing and maintenance and security and, indeed, among all the loss-prevention functions. Without specific controls and guidelines, potentially hazardous materials and/or equipment may find their way into your storeroom, process unit, or maintenance shops.

Finally, human resources and security supported by legal is a vital organizational relationship in the areas of employment and employee relations. New employee screening, control of alcohol and drug abuse, prevention of

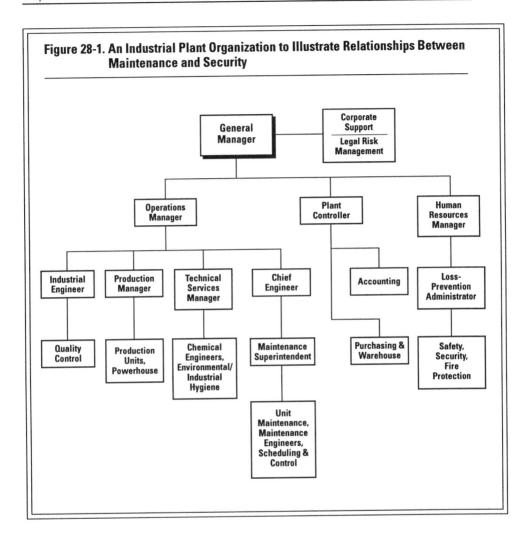

Figure 28-1. An Industrial Plant Organization to Illustrate Relationships Between Maintenance and Security

fraud and theft, and prevention of violence in the workplace are key areas requiring close coordination between these groups.

RISK ASSESSMENT: DEFINITIONS

The American Society of Safety Engineers' *Dictionary of Terms Used in the Safety Profession,* 3rd ed., defines risk and risk assessment as follows:

Risk—A measure of both the probability and the consequence of all hazards of an activity or condition. A subjective evaluation of relative failure potential. In insurance, a person or thing insured.

Risk assessment—The amount or degree of potential danger perceived by a given individual when determining a course of action to accomplish a given task.

The Canadian Society of Safety Engineering's *Canadian Dictionary of Safety Terms* defines risk assessment as follows:

Risk assessment—a) information, hazard analysis method and control program evaluation which permit estimation of risk. (b) (similar to the ASSE definition).

RISK ASSESSMENT TECHNIQUES

Process Safety Management Standard

The Federal OSHA Standard "Process Safety Management of Highly Hazardous Chemicals" contains requirements for the management of hazards associated with processes using highly hazardous chemicals. This standard requires employers to use one or more of the following methods to determine and evaluate the hazards of the process being analyzed:

- What-if
- Checklist
- What-if/checklist
- Hazard and operability study
- Failure mode and effects analysis
- Fault-tree analysis
- An appropriate equivalent methodology

Risk Evaluation and Control

A simple system widely used in industry, and most likely originally adopted from systems used by the military, utilizes the following criteria to assess the risk value of hazards:

Severity of injury/damage potential—high (6 points), moderate (3 points), low (1 point)

Frequency of employee exposure—high (3 points), moderate (2 points), low (1 point)

Probability of accident—high (3 points), moderate (2 points), low (1 point)

Violation of safety codes standards, rules, and so forth—yes (1 point), no (0 points)

Having established the risk factor on a "points" basis, the priority for control may be obtained from Table 28-1.

Table 28-1: Priority Corrective

Risk Value	Rating	Action
10 – 13 points	E	Immediate (today)
to 10 points	A	One week
6 – 8 points	B	One month
4 – 6 points	C	Three months
4 points	D	Six months, subject to further review

Security Risk Assessment

No security management plan can be effective unless it is based on a clear understanding of the actual risks it is designed to control. The following method is commonly utilized in the security profession. Only the framework is described here. [For more in-depth information, readers are referred to the *Protection of Assets Manual* (Walsh and Healy 1996).]

The system is termed *security vulnerability analysis.* Security vulnerability is defined as the probability of occurrence of a contingent loss event. The process involves identifying the risks, determining the loss probability, and criticality, and then selecting appropriate countermeasures to neutralize the risks.

Developing a loss event profile involves identifying the kinds of threats of risks affecting the assets to be safeguarded. These events are not speculative and will produce an actual loss measurable in dollars.

Determining a loss event probability requires assessment of the probability of those threats becoming an actual loss event. Probability being the likelihood of a loss occurring, the events can be grouped as follows:

- Virtually certain
- Highly probable
- Moderately probable
- Improbable
- Probability unknown

Assessing criticality of loss involves assessing the impact or effect on the assets or enterprise if the loss occurs. The criteria for this process are as follows:

- Fatal impact—one that spells the end of the business
- Very serious—major reinvestment will be needed to get back into business
- Moderately serious
- Relatively unimportant
- Seriousness unknown

THE RISK ASSESSMENT PROCESS

The foregoing is a very simplified description of some well-known risk assessment techniques. The selection of a particular assessment technique is less important than ensuring that such a process is actually undertaken and that the selected countermeasures are put into action.

The assessment process should be conducted by a designated team utilizing in-house expertise when this is available, or with outsource assistance when appropriate. In case of security risk assessment, top management, risk management, human resources, and legal counsel should be involved along with security, safety, fire protection, and other specialized professionals as needed.

Typical Security Risks

- The following are typical security risks that may affect your organization:
- Natural catastrophe (earthquake, tornado, flood)
- Industrial disaster (explosion, fire, major accident)

- Sabotage and malicious destruction (terrorism, arson, bomb threats, labor violence, civil disturbance)
- Criminal acts (theft, forgery, fraud, industrial espionage, hijacking)
- Conflicts of interest (kickbacks, employees with own business)
- Other risks (drugs and alcohol in the workplace, workplace violence, violations of security regulations)

Neutralizing Security Risks

There is an old saying in the security profession that goes as follows:

Target + opportunity = vulnerability

Target   + opportunity + motive = vulnerability actualized!
(particularly an "attractive"
target such as a small high-
value item)

Just as in other fields of loss prevention, the emphasis in security management should be to design a secure and safe place of work. After all feasible physical security measures have been installed, the employer should implement specific security systems and procedures necessary for the protection of personnel and property. The basic principle of physical security is the *layering approach.*

Outer Layer—First Line of Defense. This generally defines the perimeter of the premises and creates a physical and psychological deterrent to those who might innocently enter. This defense line will only delay intrusion and channel personnel and vehicles through entrances provided. Controls at this line are fencing, walls, lights, alarm systems, and security officer patrols.

Secondary Layer—Second Line of Defense. Building exteriors are the next layer. Vulnerable areas in this layer would be doors, windows, tunnels, manholes, sewers, roof, ventilating shafts, outside fire escapes, and building walls.

Tertiary Layer —Third Line of Defense. Offices, vaults, safes, secure storage areas, and research areas are designed to protect against surreptitious entry.

Security Alarm Sensors and Closed-Circuit TV

It is beyond the scope of this chapter to describe the vast and growing array of security devices that are now available. These devices can be a most effective form of defense, but their use must be approached as an engineering project, not one to be accomplished by the in-house maintenance force. Security and fire professionals should be involved along with qualified engineers of the project outsourced to a qualified consulting firm to design and install the system.

Providing adequate lighting around buildings, parking lots, and so forth is a well- recognized crime deterrent. Figure 28-2 is an example of a large corporate organization's standard for protective lighting.

Adoption of specific security systems and procedures will vary with the nature and size of the business or operation and its location. The following is a list of key security procedures that are almost universal in their application:

- Security loss investigation and report
- Control of visitors
- Identification badge control system
- Property pass control system

**Figure 28-2. Example Security Standard for Protective Lighting
Illuminated Parking Areas and Exterior Entrances**

Policy

Company parking areas and exterior entrances must be adequately illuminated. Parking areas must be free of obstructions that could provide cover for an intruder or attacker.

General Provisions

Any area where employees or authorized visitors park a vehicle and walk to the location must be illuminated to an intensity equal to or greater than one foot candle. This is roughly equal to the amount of light an average individual would require to read subheadings in a newspaper held waist high.

Any exterior door used for employee entrance and exit must have a minimum five-foot candle illumination.

Pedestrian doors accessible to the public shall not be obstructed with bushes, shrubs, or other materials that could hide an attacker.

Exterior lights should be programmed by automatic timers to account for seasonal charges. These lights should be covered with lens protectors and should be operational during nondaylight hours.

- Protection of proprietary information
- Control of vehicles entering and leaving premises
- Emergency procedures—for example, bomb threats, threats and violence control, HAZMAT, fire

SECURITY MANAGEMENT SYSTEMS

Naturally, a great variety of security management systems have been developed by industries and organizations worldwide. I have seen many excellent programs, and the following is one that I have found works well and is consistent in its approach with safety and fire protection management systems.

Seven-Part Security Management System

The seven-part security system consists of the following key elements:

1) *Manual.* Contains all the policy, procedures, and guidelines necessary for administration of the program.
2) *Coordinators and committees.* Assigned individuals and/or committees responsible for assisting line management with the administration of the "unit's" security management systems.
3) *Training and awareness.* Programs designed to promote employee security awareness and specialized security training provided for employees.
4) *Surveys and audits.* Systematic inspections to identify potential security risks and to assess the effectiveness of existing security protective systems and procedures.
5) *Risk control.* A procedure to ensure that when security problems are identified, they are resolved in a timely fashion.
6) *Risk assessment.* An ongoing program to review regularly the list of identified security risks and the status of control measures adopted.
7) *Incident investigation.* A system for investigating and reporting all security losses and incidents in order to determine the source and causes and establish measures to prevent future exposure to these risks.

Security Manual

The foundation of a sound security management system is the security manual. Table 28-2 illustrates a sample classification index for a unit

Table 28-2. Example Security Manual Contents and Classification Index

Tab No.	Section	Policy-Procedure Classif. Assigned Range	No.	Subject	Date Issued Revised Reviewed
1	Administration	110		Organization	
		120		Manuals	
		130		Law enforcement agencies	
		135		Professional associations	
		140		Security during construction	
		145		Purchasing and material control	
		160		Security loss/incident investigation reports	
		170		Security reports	
		180		Employee relations	
		185		Employee background investigations	
2	Physical standards	210		Facility design	
		215		Perimeter barriers	
		217		Exterior doors and windows	
		220		Protective lighting	
		225		Safes, vaults, and secure storage areas	
		230		Parking and traffic control	
		231		Shipping/receiving areas	
		240		Communications systems	
		245		Lock and key control	
		250		Access-control system	
		251		Intrusion detection systems	
		255		CCTV surveillance	
		260		Computer and EDP equipment	
		265		Special equipment	
		270		Firearms	
3	Protection services programs	310		Risk assessment	
		320		Control of visitors	
		322		Control of vehicles entering premises	
		324		Control of employees entering and leaving premises	
		325		Traffic control	
		328		Control of material and equipment leaving premises	
		329		Protection of proprietary information	
		335		Handling suspicious mail and/or packages	
		340		Safeguarding cash funds and blank checks	
		350		Security surveys and audits	
		360		Security committee	
		365		Security coordinators	
		370		Security department	
		380		Risk assessment and review	
		390		Security awareness	
		391		Employee security orientation	
		395		Security training	
4	Emergency procedures	400		Crisis management plan	
		410		Threats and violence control plan	
		420		Bomb threat procedure	
		430		Earthquake emergency plan	
		440		Severe weather emergency plan	
		450		Major accident/fire emergency plan	
		460		Hazardous material emergency plan	

security manual. The manual should be widely distributed throughout the "unit's" organization. All members of management, supervision, and team leaders should be included in the distribution. Development of the manual contents should be the responsibility of top management with input from in-house security professionals or outsourced to consultants. All procedures and guidelines should be regularly reviewed and audited to ensure that they are current and being implemented effectively.

Sources of Help

Readers are referred to the references at the end of this chapter for additional resources and more in-depth information. In addition, personnel assigned security management responsibilities should consider becoming a member of related professional associations. The largest association in the workplace security field is the American Society for Industrial Security (ASIS). The ASIS has chapters in practically every state in the United States and in a number of other countries. The ASIS can be contacted at American Society for Industrial Security, P.O. Box 1409, Alexandria, Virginia 22313; tel. (703) 522-5800.

Security practitioners are also encouraged to obtain Certified Protection Professional designation. Details on how to enroll for this certification process are available from the ASIS.

MAINTENANCE MANAGEMENT SYSTEMS

Anyone who has worked in a process plant knows the critical value of the maintenance function. The quality of the maintenance leadership is considered by many to be more important toward an overall plant's success than that of the production and process units.

A strong, well-organized maintenance department will ensure that the best quality materials and equipment are used for repairs and maintenance, and that appropriate preventive maintenance procedures are systematically carried out on all process systems and plant machinery and equipment.

Primary Elements of a Maintenance Management System

The *Maintenance Managers Guide,* used by the U.S. Department of Energy, specifies eighteen areas of consideration:

1) Program policies

2) Work order system

3) Materials control system

4) Equipment records system

5) Preventive maintenance program

6) Mandated preventive maintenance

7) On-call maintenance program

8) Job planning and procedures

9) Work sampling

10) Maintenance scheduling

11) Backlog control

12) Performance measurement and improvement

13) Control versus actual performance

14) Prioritization system

15) Engineered time standards

16) Cost identification and controls

17) Repair versus replacement criteria

18) Personnel training

Most of these elements need no explanation. Commentary on a selected few follow.

Work Order System. Work order or service request systems are common in industry. What is less common is the effective merging of this system with security and safety systems for the control of physical hazards and security risks.

The work order format should have space to indicate that the work is a safety/security category, and the completion date requested should be based on the requester's risk assessment (see the risk priority rating system described previously).

At one organization, the scheduling and control section developed a computerized record system for all safety/security work orders. This enabled all concerned to "track" the status of all these projects and to follow up as necessary on any that may have overrun their risk priority completion dates.

Preventive Maintenance Program. The efficient organization of preventive maintenance is a complex and skilled activity requiring systematic record keeping, inspection procedures, job analysis, and work programming and planning. The security and safety functions will need to contribute to these programs based on regulatory requirements, other standards, or accident/incident experience.

Mandated Preventive Maintenance. Mandated of preventive maintenance calls for close coordination between the security/safety functions and scheduling and control. Operations and maintenance teams are usually too busy with their daily tasks to be relied on to systematically remember to perform all necessary Preventive Maintenance (PM) checks and to maintain all the proper records. This is where the value of an efficient scheduling and control unit really comes into play. Aided by a computer, it can ensure that these vital inspections and tests are carried out and that proper records are maintained and available for review by regulatory agencies.

Scheduling and control should obtain input not only from security and safety, but also fire protection, environmental, and motor vehicle maintenance (for Department of Transportation requirements). Other departments will be involved as applicable to the "unit's" nature of business (for example, aircraft operations and maintenance for FAA requirements, food processing operations for FDA requirements).

Without reference to a particular regulation or standard, the following are examples of equipment or systems for which mandated preventive maintenance is required:

- Elevators and escalators
- Hoists
- Cranes and lifting tackle
- Industrial trucks
- Steam boilers
- Air receivers
- Ventilation systems
- Fire alarm systems
- Fire suppression systems
- Intrusion alarm systems
- Radiation sources
- Fall protection systems and equipment
- Breathing apparatus

Job Planning and Procedures. Performing a maintenance task calls for adequate engineering and technical backup before, during, and after the job is completed. Maintenance tasks should be properly sequenced to ensure that adequate parts, materials, and equipment have been ordered or are on hand before work begins.

For complex or major repetitive jobs, maintenance should develop guideline procedures for reference use by crews concerned. At a major process plant in Texas, where I once worked, the maintenance manager had established a maintenance manual that contained a number of what were called standard maintenance practices (SMPs). In the safety-related area, some SMP titles the manager had developed were

- Scaffold Control and Design
- Inspection of Mobile Cranes
- Inspection of Power-driven Bridge Cranes and Hoists
- Inspection of Slings

Repair versus Replacement Criteria. Criteria for replacing machinery, equipment, or systems should be established with input from the following functions:

1) *Accounting*—to provide depreciation and write-off figures, as well as cost factors.

2) *Engineering*—to be aware of new developments in equipment or process design, as well as safety engineering factors.

3) *Production* particularly in cases in which the decision may be influenced by expected or scheduled production activity.

4) *Quality control*—to be able to furnish relevant data on waste, rework, substantial work, and so on.

5) *Maintenance*—who may have the most accurate information on when a piece of equipment is no longer repairable or operable.

6) *Security/safety/fire protection*—who can provide information on existing or pending regulatory requirements and other recommendations regarding the safe operation of the equipment or process.

At a large municipal organization, some of its employees had experienced back problems, especially those who spent much of their working day behind the wheel. The maintenance shop's past practice in purchasing replacement vehicles was to order the standard cab layout. The extra cost to order ergonomic seats with lumbar support for the driver was more than justified. In the interim, pending vehicle replacement, special fitted lumbar supports were also made available.

Building Maintenance

So far, the discussion in this chapter has addressed maintenance as it pertains to equipment and processes. Obviously, buildings require maintenance to prevent hazardous conditions from developing. The structural parts of buildings that require routine attention are as follows:

- Foundation, footings, and column bases
- Structural members
- Walls
- Ceilings and floors
- Roofs, gutters, and roof-mounted items
- Tanks, tank towers, and stacks
- Loading platforms
- Basements
- Vaults, manholes, and tunnels

Inspection of buildings should be conducted on a scheduled basis by qualified persons from the in-house engineering staff or outsourced to qualified consulting firms. Professional associations and insurance companies can also provide helpful guidelines and/or inspection services.

Lighting Systems

Problems with lighting systems can seriously affect the safety and security of employees and visitors. The appropriate maintenance section (such as electric shop) must maintain adequate replacement materials, and service calls for lighting problems must be handled on a priority basis. Engineering with input from security and safety should establish criteria to ensure that lighting meets applicable regulations and standards.

Instrument Systems: Process, Safety, Security, and Fire Protection

It goes without saying that problems with any instrument system affecting the security and safety of personnel requires top-priority response by maintenance. Since this is a very specialized field, it is wise to select a number of technicians from the electric/instrument shop and have them receive special training on the maintenance of security and safety instrument systems. If this expertise is not available in-house, it should be outsourced to a qualified firm that can provide on-call service.

Accident Prevention

Some of the worst accidents I have personally had to investigate have involved maintenance employees. The pressure is always on maintenance to avoid the shutdown of production processes or utilities. Safe procedures such as high-voltage switching, lockout/tagout, confined space entry, and excavation must be rigidly implemented in order to protect personnel and property. Some other nearly universal key maintenance-related safety procedures are as follows:

- Identification of piping systems
- Safe opening of pipe flanges
- Installing pipe blinds
- Use of slings
- Erection and use of scaffolds
- Use of ladders
- Fall protection
- Use of aerial lifts
- Handling and storage of compressed gas cylinders
- Welding and cutting containers that have held combustibles
- Welding and burning permit system
- Working with asbestos/lead-containing materials

This is by no means a complete list of common maintenance tasks involving severe potential hazards. Maintenance personnel are expected to encounter additional hazards, which must be dealt with as their employer's procedures require. Job hazard analysis procedures should also be followed when performing all maintenance tasks.

Equipment Identification

Maintenance, in coordination with engineering and production, must ensure that all systems and equipment are properly labeled as to unit number, description, equipment, or system controlled and, in the case of piping systems and tanks, identification of the contents.

At an alumina refinery in Australia, two mechanics were assigned to work on a defective process pump. The production supervisor, who was standing below, pointed up into the unit and indicated the location of the pump, on a platform above a process vessel. The mechanics fetched their tools and then went up into the unit. They started to dismantle an adjacent identical pump that was under pressure. As they opened a pipe flange attached to this pump, hot caustic liquor suddenly burst out and sprayed both workers, causing severe chemical burn injuries. After this accident experience, the plant management revised their existing lockout/tagout procedures to include an additional orange equipment identification tag. This tag was attached to the actual equipment to be repaired or serviced, rather than the control devices isolating the equipment. This procedure avoids the possibility of confusion between identical pieces of equipment located side by side or in the same immediate area.

CONCLUSION

Security and safety considerations in the maintenance program are often too little appreciated but have tremendous importance. They can be compared to the essential maintenance activities on a space shuttle, or an aircraft, where proper functioning can be a life-or-death matter.

This chapter has discussed the most essential areas of security and safety interactions with maintenance. Again, the purpose of this chapter is to provide an overview for those with management responsibilities in this area. In addition, it is intended to point out how closely related these functions are and that security and safety management systems are only as effective as the maintenance management systems that support them.

REFERENCES AND SUGGESTED FURTHER READING

Abercromie, S. A., Ed. (1988). *The Dictionary of Terms Used in the Safety Profession, Third Edition.* Des Plaines, IL: American Society of Safety Engineers.

Basic Guidelines for Security Investigations. (1981). Arlington, VA: American Society for Industrial Security.

Bird, F. E. (1988). *Management Guide to Loss Control.* Loganville, GA: Institute Publishing.

Bird, F. E., and G. L. German (1996). *Practical Loss Control Leadership.* Loganville, GA: Det Norske Veritas.

Combating Workplace Violence: Guidelines for Employers and Law Enforcement. (1995). Alexandria, VA: International Association of Chiefs of Police.

D'Addario, F. J. (1989*). Loss Prevention through Crime Analysis.* National Crime Prevention Institute. Stoneham, MA: Butterworth.

D'Addario, F. J. (1996*). The Manager's Violence Survival Guide.* Chapel Hill, NC: Crime Prevention Association.

Ellis, J. N. (1993). *Introduction to Fall Protection,* Des Plaines, IL: American Society of Safety Engineers.

Fay, J. (1993). *Encyclopedia of Security Management.* Newton, MA: Butterworth-Heinemann.

Fennelly, L. J. (1997). *Effective Physical Security, 2nd Ed.* Newton, MA: Butterworth-Heinemann.

Grimaldi, J. V., and R. H. Simonds (1993). *Safety Management, 5th Ed.* Boston, MA: Irwin. Des Plaines, IL: American Society of Safety Engineers.

Grose, V. L. (1987). *Managing Risk/Systematic Loss Prevention for Executives.* Englewood Cliffs, NJ: Prentice-Hall.

Grund, E. V. (1995*). Lockout/Tagout.* Itasca, IL: National Safety Council.

Hopf, P. S. Ed. (1979). *Handbook of Building Security Planning and Design.* New York, McGraw-Hill.

Johnson, W. G. (1980*). MORT Safety Assurance Systems.* New York: Marcel Dekker.

King, R. W. (1979). *Industrial Hazards and Safety Handbook.* London, UK: Newens-Butterworth.

Kletz, T. A. (1990). *Critical Aspects of Safety and Loss Prevention.* London, UK: Butterworth.

Kletz, T. A. (1983). *HAZOP & HAZAN Notes on the Identification and Assessment of Hazards.* Rugby, UK: Institution of Chemical Engineers.

Lack, R. W. ed. (1996). *Essentials of Safety and Health Management.* Boca Raton, FL: CRC/Lewis.

Lack, R. W. ed. (1997). *Security Management.* Itasca, IL: National Safety Council.

Larr Enterprises Ram Consulting (1987). *Canadian Dictionary of Safety Terms.* Kleinburg, Ontario, Canada: Canadian Society of Safety Engineering.

McGoey, C. E. (1990). *Security Adequate . . . or Not?* Oakland, CA: Aegis Books.

Nertney, R. J. (1978). *Safety Considerations in Evaluation of Maintenance Programs.* Idaho Falls: System Safety Development Center.

Oliver, E., J. Wilson, and T. Slater (1988). *Practical Security in Commerce & Industry.* Aldershot, UK: Gower.

Post, R. S., and D. A. Schachtsiek (1986). *Security Managers Desk Reference.* Newton, MA: Butterworth-Heinemann.

Process Safety Management of Highly Hazardous Chemicals. (February 1992). Title 29 Code of Federal Regulations Part 1910.119. FR 57-6356. Washington, DC: OSHA Publications Office.

Purpura, P. (1991). *Security and Loss Prevention: An Introduction.* Newton, MA: Butterworth-Heinemann.

Rekus, J. F. (1994). *Complete Confined Spaces Handbook.* Boca Raton, FL: CRC Lewis.

Risser, R. (1993). *Stay Out of Court: The Managers Guide to Preventing Employee Lawsuits.* Englewood Cliffs, NJ: Prentice-Hall.

Roper, C. A. (1997). *Physical Security and the Inspection Process.* Newton, MA: Butterworth-Heinemann.

San Luis, E., L. A. Tyska, and L. J. Fennelly (1994). *Office and Office Building Security, 2nd Edition.* Newton, MA: Butterworth-Heinemann.

Seldner, B. J., and J. P. Cothrel (1994). *Environmental Decision Making for Engineering and Business Managers.* New York: McGraw-Hill.

Sennewald, C. A. (1985). *Effective Security Management.* Newton, MA: Butterworth-Heinemann.

Suprenant, B. A., and K. D. Basham (1985). *Excavation Safety.* Addison, IL: Aberdeen Group.

Swartz, G. (1997). *Forklift Safety.* Rockville, MD: Government Institutes, Inc.

Walsh, J., and R. J. Healy (1981). *Principles of Security Management.* Long Beach, CA: Professional Publications.

Walsh, T. J., and R. J. Healy, Eds. (1996). *Protection of Assets Manual.* Santa Monica, CA: Merritt.

Chapter 29

Environmental Issues

Jack E. Daugherty

INTRODUCTION

Government regulations can make reasonable persons say unreasonable things and make gentle persons curse. Perhaps this is because regulations are written in a legal tone and frequently refers back to previous paragraphs, or to paragraphs as yet unread, or even to other documents. As more and more safety professionals assume responsibility for environmental compliance, they need to develop a basic understanding of environmental laws and regulations.

This chapter briefly examines the evolution of environmental regulations in the United States and then looks at some major issues currently facing many compliance managers and planners. These laws are the Resource Conservation and Recovery Act (RCRA), the Comprehensive Environmental Response, Compensation, and Liability Act (CERCLA), and Title III of the Superfund Amendments and Reauthorization Act (SARA III). The latter law is sometimes known as the Emergency Planning and Community Right-to-Know Act (EPCRA). In addition, the OSHA standard on Hazard Communication (HAZCOM or Employee Right-to-Know) and the Toxic Substances Control Act (TSCA) are discussed.

Finally, this chapter considers how a safety manager might sort out the issues covered by these laws and regulations and looks at planning for full compliance. Some environmental issues not covered here, due to lack of space, are the Safe Drinking Water Act, the Clean Water Act, the Clean Air Act, the Federal Insecticide, Fungicide, and Rodenticide Act, the Oil Pollution Prevention Act, and Pollution Prevention Act, among others.

CASE STUDY:
CONSEQUENCES OF DODGING ENVIRONMENTAL LAWS

As an introduction to the necessity for environmental laws, consider the following case study. In 1973, two former employees of Allied Chemical Company, Virgil A. Hundtofte and William P. Moore, formed a company they named Life Sciences Products and won contracts with both Allied and Hooker Chemical to produce the pesticide Kepone. Life Sciences set up operations in a former gasoline filling station in Hopewell, Virginia, not far from a large Allied operation. The facility was served by a public sewer line. The city's sewage treatment plant discharged into the James River, a commercial fishing stream that empties into the Chesapeake Bay, which is a major fishing and shellfishing ground.

Two weeks into operation, recurrent surges of smelly, frothy, white wastewater were seen in the sewer line and at the sewage treatment plant. Two months into operation, the Hopewell sewage treatment facility experienced a dieback of sewage-digesting bacteria. Soon afterward, Kepone turned up in fish, shellfish, waterfowl, and other aquatic organisms taken from the James River and the Chesapeake Bay, as far as 60 miles downstream. Kepone was also found in sediment samples. The Hopewell public works department disposed of its digester sludge, which was contaminated with Kepone, along the James River floodplain.

Air samples taken one block from the Life Sciences plant assayed at 400,00 parts per million (ppm)—or 40 percent Kepone. The response of the EPA and the Virginia State Water Control Board was to raise the discharge limits of Kepone.

Then, a Life Sciences employee complained to OSHA about exposure to Kepone fumes and dust during production. OSHA responded but issued no citation due to insufficient evidence. When an employee was diagnosed with Kepone poisoning, the Virginia Health Department launched an investigation of its own. Blood samples were taken from employees. Levels of Kepone in blood samples ranged from 2 to 72 ppm. The historical documentation of Kepone in human blood was 5 ppm. Thirty-one employees were hospitalized. The Virginia Health Department closed Life Sciences and OSHA joined it in an inspection of the facility.

The health inspectors found several inches of Kepone dust covering the floors of the plant. Kepone dust filled the workplace. Employees reported that they had complained to their supervisors, but that they were met with a lack of concern. The inspectors learned that neighborhood traffic sometimes halted because of low visibility caused by particulate matter discharged from the plant's vents. A nearby ice distribution plant complained that ice was being contaminated and that its employees were suffering from eye and skin irritation because of vapors emitted from Life Sciences. Overall, numerous health and safety regulations were cited as being violated in the operation of the plant. A $16,500 penalty was imposed.

Shortly after the OSHA/Health Department visit, the EPA and the Virginia State Water Control Board made a joint inspection of the Life Sciences plant. Life Sciences was routinely discharging up to 600 parts per billion (ppb) of Kepone into the sewer, exceeding the 100 ppb limit. Kepone was also illegally discharged directly into the James River by both Life Sciences and Allied Chemical.

The ultimate toll on public and employee health was that seventy persons were made ill, of which twenty were permanently hospitalized with untreatable brain and liver damage. Slurred speech, loss of memory, and eye twitching were widespread among the victims, and sterility was especially noted among the plant workers.

The Life Sciences Products company was fined $3.8 million, but the company was worth only $32. Life Sciences was unable to pay its penalties or the cost of cleaning up the damage it had done. Virgil Hundtofte received five years of probation and a fine of $25,000 for misdemeanor conspiracy in exchange for cooperating with the government in preparing criminal cases against the other principles. William Moore was found guilty on a charge of criminal conspiracy and 153 counts of illegal discharge. He received a five-year suspended sentence and was fined $25,000.

Allied Chemical was fined $13.24 million for 940 misdemeanor conspiracy charges, which it did not contest. Allied Chemical settled with the court by paying $8 million into the Virginia Environmental Endowment. Much of the settlement was recouped as a tax deduction because the company claimed that a charitable contribution had been made to the Endowment.

The city of Hopewell was fined $10,000 for three counts of willfully, negligently, and unlawfully allowing Life Sciences Products to discharge illegally into the Hopewell sewage treatment plant. Hopewell claimed immunity and appealed, but the appellate court disagreed with the city.

Here is a situation in which OSHA and EPA regulations plus state health department and water control board rules were violated. The primary difference between this case and many others is that many of these violations were not due to ignorance or confusion concerning regulations, but to apparently blatant attempts to dodge them. At least the courts thought so.

BACKGROUND

Manufacturing plants, especially small- to medium-sized ones, typically assign environmental responsibilities to an overworked individual who is talented in his or her field, but who is a complete novice in environmental management. More often than not, this novice is appointed the plant safety manager. Typically, such safety-environmental managers work under tremendous stress—one might even say under duress. Particularly stressing are the complex environmental regulations about which the safety manager is ignorant.

Codified OSHA standards are found in chapter 29 of the Code of Federal Regulations (29 CFR), and the environmental regulations are found at 40 CFR. However, when transporting wastes, one must also be familiar with 49 CFR [Department of Transportation regulations]. A general familiarity with all these rules is quite feasible. The technical and legal issues covered by the rules do have a logical basis, though not always apparent to those unfamiliar with the development of the rules. Fortunately, compliance management is no monolith and can be broken into basic building blocks, which are more readily digested.

RCRA/HAZARDOUS WASTE MANAGEMENT AND DISPOSAL

Until 1976, when the U.S. Congress passed RCRA, no differentiation was made between hazardous and other kinds of wastes generated by industry and commerce. Before RCRA, acceptable disposal techniques included discharging aqueous and other liquid wastes into a stream or river; emitting gas, vapor, or particulate waste as air plumes; and burying containerized or even bulk wastes as landfill debris. This was state-of-the art waste management until 1976 (the year of both Love Canal and RCRA) and was still common practice until 1980 (CERCLA).

The primary objective of RCRA is to provide a comprehensive means of managing hazardous waste without harm to public health or the environment. Because Congress intended to protect the environment through every phase of the waste stream's life, the concept of cradle-to-grave management was derived. To implement the cradle-to-grave idea, EPA wrote a set of regulations that covered hazardous waste generators; transporters; and treatment, storage, and disposal facility (TSDF) operators.

Hazardous waste is a threat to the quality of the environment that will not go away. Not all wastes are imminently hazardous, but natural processes have unlimited time to undo humanity's safeguards for isolating waste materials. Unfortunately, respect for RCRA is weakened because EPA overregulates hazardous waste in some cases and underregulates it in others. Some hazardous waste facilities still operate under interim permits without meeting financial assurance requirements. Yet, bankruptcy of hazardous waste management facilities leads to far more instances of ongoing public health risk than do specific incidences of noncompliance with regulatory minutiae.

Novices are typically confused by the term "hazardous waste." Some wastes, hazardous by edict, do not seem to be that hazardous in the environment. Other wastes, defined as nonhazardous, may be riskier than first thought. For example, EPA has resisted regulating used oil as hazardous waste although evidence of a public health risk continues to mount. Used oil is heavily trafficked on public roads daily. On the other hand, debris and other material that contact a listed hazardous waste are hazardous wastes by definition. However, such waste often has no real hazardous characteristic of its own. Consequently, precious hazardous waste landfill space is taken up by debris that could safely be stored in nonhazardous landfills.

The term "hazardous waste" covers materials that range from synthetic organic chemicals to heavy metals, inorganic sludges, and dilute aqueous waste streams. Hazardous wastes may be solid, liquid, or gas. They may be pure materials that have no further use; complex mixtures, residues, or effluents from manufacturing operations; or discarded products, contaminated containers, soil, or debris.

Keep in mind that solid waste and hazardous waste are regulatory, not scientific, terms. Hazardous waste is a subset of the set of all solid wastes. Solid wastes designate those wastes that have been disposed in landfills, piles, and surface impoundments from time immemorial: trash, debris, containers, bulk liquids, bulk solids, and powders.

RCRA defines solid waste as

- Garbage, refuse, or sludge from waste treatment plants, water supply plants, and air pollution control facilities

- Discarded solid, liquid, semisolid, or containerized gaseous material from industrial, commercial, mining, agricultural, and community activities

RCRA defines waste that is exempt from regulation as solid waste as

- Solid or dissolved material from domestic sewage and irrigation return flows
- Industrial discharges from point sources subject to permits under Section 402 of the Clean Water Act
- Source, special, or by-product material as defined by the Atomic Energy Act of 1954

What these definitions mean is that any waste produced by an industrial facility is a RCRA-regulated solid waste. The only exceptions are domestic sewage, Section 402 waste, and nuclear waste regulated by the Atomic Energy Act.

Purpose and Scope of Identification Rule

RCRA regulations found at 40 CFR 261 Subpart A are titled Identification and Listing of Hazardous Waste—General. Section 261.1 identifies those solid wastes that are subject to regulation as hazardous wastes under other rules and are therefore subject to the notification requirements of RCRA. Subpart A not only defines the terms *solid waste* and *hazardous waste,* but also identifies those wastes that are excluded from regulation and establishes special management requirements for hazardous waste produced by conditionally exempt small-quantity generators and hazardous waste that is recycled. Subpart B of 40 CFR 261 establishes criteria used by the EPA to identify characteristics of hazardous waste and to list particular hazardous wastes. Subpart C identifies characteristics of hazardous waste, and Subpart D lists particular hazardous wastes.

Solid waste is a term that applies only to wastes, that also are hazardous for purposes of regulation. The term does not apply to such materials as nonhazardous scrap, paper, textiles, or rubber that are otherwise not hazardous wastes and that are recycled. A material that is neither a regulated solid waste nor a hazardous waste is still a solid waste and a hazardous waste if EPA, by virtue of its authority to inspect facilities and gather information about hazardous waste, says it is a hazardous waste and/or solid waste. RCRA (the law) defines hazardous waste as any solid waste or combination of solid wastes that, because of quantity, concentration, or physical, chemical, or infectious characteristics, may 1) cause or significantly contribute to an increase in mortality or an increase in serious irreversible or incapacitating reversible illness, or 2) pose a substantial present or potential hazard to human health or the environment when improperly treated, stored, transported, disposed of, or otherwise managed. EPA may also regulate an otherwise uncovered material as hazardous waste, if it brings suit to restrain an imminent and substantial endangerment to health or the environment and the statutory elements of the term are established.

Before a material can be either a solid or hazardous waste, it must first be a waste. A material is a waste if it has been declared a waste or has waste-like characteristics or is inherently a waste.

A spent material is one that has been used, but as a result of contamination can no longer serve the purpose for which it was produced, without additional processing. Sludge has the same meaning. A byproduct is not a primary product and is not solely or separately produced by the production process. Process residues such as slags or distillation column bottoms are examples of byproducts. The term *byproduct* does not include a coproduct that is produced for general use in the same form that it is produced by the process.

A material is reclaimed if it is processed to recover a usable product, or if it is regenerated. Recovery of lead values from spent batteries and regeneration of spent solvents are two examples of reclamation. A material is used or reused if it may be used or reused as an ingredient or an intermediate in an industrial process to make a product. For example, distillation bottoms from one process may be used as feedstock in another process. However, a material will not satisfy this condition if distinct components of the material are recovered as separate end products. When metals are recovered from metal-containing secondary materials, the secondary materials are not used or reused.

A material may be used or reused in a particular function or application as an effective substitute for a commercial product. The example EPA gives is the reuse of spent pickle liquor as a phosphorous precipitant and sludge conditioner in wastewater treatment. Scrap metal is bits and pieces of metal parts, bars, turnings, rods, sheets, wire, or metal pieces that are combined together with bolts or soldering, which, when worn out or superfluous, can be recycled. Radiators, scrap automobiles, and railroad box cars are cited by EPA as examples of the latter.

A material is recycled if it is used, reused, or reclaimed. A material is accumulated speculatively if it is accumulated before being recycled. A material is not accumulated speculatively if it is potentially recyclable and has a feasible means of being recycled. During any calendar year, the amount of material that is recycled, or transferred to a different site for recycling, must equal at least 75 percent by weight, or volume, of the amount of that material accumulated at the beginning of the year in order to have been accumulated under speculation. In calculating the turnover percentage, the 75 percent requirement must be applied to each material of the same type that is recycled in the same way. Equivalent recycling recovers the same usable material from the waste material or recovers materials that are used in the same way. Materials accumulating in units that would be exempt from regulation cannot be included in the calculation. Materials that are already defined as solid wastes also may not be included in the calculation. Materials are no longer accumulated speculatively once they are removed from accumulation for recycling.

Materials are solid wastes when they are applied to or placed on the land in a manner that constitutes disposal. If the materials are used to produce products that are applied to or placed on the land or are otherwise contained in products that are applied to or placed on the land, they are also solid wastes. In these cases, the product itself remains a solid waste. Under this rule, the following materials are considered solid wastes:

- Spent materials
- Sludges listed at 40 CFR 261.21 or 261.32
- Sludges exhibiting a characteristic of hazardous waste
- Byproducts listed at 40 CFR 261.31 or 261.32
- Byproducts exhibiting a characteristic of hazardous waste
- (Discarded) commercial chemical products listed at 40 CFR 261.33
- Scrap metal

Commercial chemical products listed in Section 261.33 are not solid wastes when discarded if they are applied to the land and that is their ordinary manner of use.

Materials are solid wastes when they are burned to recover energy or used to produce a fuel or are otherwise contained in fuels. In these cases, the fuel itself remains a solid waste. The following materials are solid wastes under this rule:

- Spent materials
- Sludges listed at 40 CFR 261.21 or 261.32
- Sludges exhibiting a characteristic of hazardous waste
- Byproducts listed at 40 CFR 261.31 or 261.32
- Byproducts exhibiting a characteristic of hazardous waste
- [Discarded] commercial chemical products listed at 40 CFR 261.33
- Scrap Metal

However, commercial chemical products listed in Section 261.33 are not solid wastes if they are themselves fuels.

Some materials are solid wastes when reclaimed, as the following:

- Spent materials
- Sludges listed at 40 CFR 261.31 or 261.32
- Byproducts listed at 40 CFR 261.31 or 261.32
- Scrap metal

Certain materials are solid wastes when accumulated speculatively:

- Spent materials
- Sludges listed at 40 CFR 261.31 or 261.32
- Byproducts listed at 40 CFR 261.31 or 261.32
- Byproducts exhibiting a characteristic of hazardous waste
- Scrap metal

Some materials are solid wastes when they are recycled in any manner due to their inherently waste-like characteristic. What are inherently waste-like materials? Such materials belong to the set of solid waste because they are not useful in any manner or are too dangerous to be recycled in any manner. These materials are solid wastes even when recycled. Materials identified by Hazardous Waste Numbers F020, F021, F022, F023, F026, and F028 are inherently waste-like. These materials are certain wastes from the production or manufacturing use of trichlorophenol, tetrachlorophenol, pentachlorophenol, tetrachlorobenzene, pentachlorobenzene, hexachlorobenzene, intermediates used to produce pesticide derivatives of the phenolics, or discarded formulations of these materials. The significance is that although the materials themselves are hazardous, dioxin species typically contaminant these materials. Dioxin may be the most hazardous of all anthropogenic materials. F021 material (pentachlorophenol and its intermediates) is exempt when used as an ingredient to make a product at the site of generation.

Another category of inherently waste-like materials are the secondary materials that are fed to a halogen acid furnace and that subsequently exhibit a characteristic of a hazardous waste or are listed as a hazardous waste. They, too, are solid waste even when recycled. Brominated material is not covered by this rule if the bromine concentration of the material is greater than 45 percent. The material must also contain less than a total of 1 percent of toxic organic compounds listed in Appendix VIII of 40 CFR 261. Finally, the material must be processed continually on-site in the halogen acid furnace via direct conveyance called *hard piping*.

Fortunately, not all materials are solid waste when recycled. Materials are not solid wastes when they are used or reused as ingredients in an industrial process to make a product, provided the materials are being recycled and not reclaimed. Materials that are used or reused as effective substitutes for commercial products or returned to the original process from which they are generated, without first being reclaimed or land disposed, are not solid waste. To be exempt, the material must be returned as a substitute for feedstock materials. When the original process, to which the material is returned, is a secondary process, the materials must be managed such that there is no placement on the land in order to retain the exemption.

Other recycled materials are solid wastes, even if the recycling involves use, reuse, or return to the original process. Materials used in a manner constituting disposal or used to produce products that are applied to the land fit this category. Also, materials burned for energy recovery, used to produce a fuel, or contained in fuels are included. Materials accumulated speculatively and those F-wastes discussed previously as inherently waste-like are also included.

The significance of this discussion is that solid wastes are hazardous wastes when they are listed by EPA as a hazardous waste or exhibit a hazardous characteristic. If the waste is not a solid waste, it cannot be a hazardous waste. The criteria developed to determine if a solid waste is hazardous include ignitability (I), corrosivity (C), reactivity (R), and toxicity (T). Additionally, some materials are listed as hazardous wastes due to their origin of source, such as wastewater treatment sludges, plating materials, and residue from specific industrial processes.

The burden of RCRA compliance falls chiefly on either the generator or the TSDF, representing the cradle (generation) and the grave (ultimate fate) of every hazardous waste stream. Determination of whether a solid waste is hazardous, maintenance of hazardous waste on-site, tracking the off-site life by means of a manifest, and ensuring that the TSDF is permitted by EPA are the general obligations of the generator. Waste management in perpetuity, unless the hazardous characteristics are neutralized or the material is destroyed such that the residue has no hazardous characteristics, is the general obligation of the permitted TSDF. However, if the TSDF becomes an abandoned site or the site of uncontrolled releases, the generator may be named a potentially responsible party (see CERCLA next) in spite of having an untainted compliance record.

CERCLA/HAZARDOUS SUBSTANCE RELEASES

The U.S. law popularly called Superfund is perhaps the most misunderstood legislative act that has ever been written. Superfund was enacted by Congress in 1980, four years after the law that required environmentally friendly waste management (RCRA). Superfund, or the Comprehensive Environmental Response, Compensation, and Liability Act, was enacted to break the deadlock that city, township, county, state, and federal governments had gotten themselves into at Love Canal, New York, and Woburn, Massachusetts.

Essentially, CERCLA sets aside faultfinding until after the cleanup is effected. To expeditiously mitigate either an impending environmental disaster or one in progress, EPA may place the liability for costs on potentially responsible parties (PRPs). If these parties fail to participate for any reason, EPA may use the Hazardous Substances Response Trust Fund (the "super" fund) to conduct the work and recover the costs from the PRPs after the fact, regardless of who is to blame.

It sounds like a simple plan that ought to work, but it does not. Why? Because being named a PRP is not about being guilty, necessarily. Until Superfund, in English Common Law countries, such as the United States, the guiltless did not pay for a crime or civil wrong.

Being named a PRP is not about guilt, but proximity. Are you near the incident due to generation, transportation, brokerage, financial control of operations, site management, or any type of leverage over the site management? If so, you may be a PRP. The cause of this precedent-setting law was that the American people watched the evening news for months to see government officials from Washington to Niagara Falls wring their hands while nothing got done. The sight of paralyzed bureaucrats was offset by angry, frustrated citizens whose lives were being affected by neighborhoods destroyed by some bubbly, gooey chemical stuff no one could pronounce, much less understand. The national outrage at the situation precipitated CERCLA.

Unfortunately, even with CERCLA, nothing gets done. As you might guess, the introduction of an entirely new concept into U.S. law—where the innocent, nearly innocent, slightly guilty, and guilty are placed together in the position of financing a remedial action—gave U.S. trial lawyers an opportunity for endless litigation before the fact. The schedule of events was not supposed to happen that way. The cleanup was to start 120 days after PRP notification and to be completed in another 120 days. Then the lawsuits were to have been filed—after the fact. Unfortunately, most Superfund sites identified in the 1980s are still on the National Priorities List (NPL) awaiting remedial action due to unresolved legal challenges.

Funding for CERCLA was to have expired in 1985. Congress examined the law, made major changes, and extended the time of its coverage. The changes were so great that essentially a new law was passed. We know it as the Superfund Amendments and Reauthorization Act (SARA) of 1986. Title III of SARA has its own title: EPCRA, or SARA III, which is discussed subsequently. For now, suffice it to say, that the main provisions of EPCRA are the right-to-know provisions for the community and legislated emergency response to accidental chemical releases. Despite the fact that we call it by a different name, EPCRA has its basis in CERCLA.

Diverse Supervision by EPA

Most aspects of CERCLA are implemented by EPA, so you will not find extensive regulations giving the minutiae of Superfund. Other federal agencies also have defined roles under CERCLA. For instance, the Department of Defense (DOD) and Department of Energy (DOE) have roles when land under their control are involved in the release. Chemical weaponry left stockpiled at a Colorado military base after the United States signed an agreement not to produce or use such weapons is an example of DOD Superfund involvement. Many air force and naval bases contain large stretches of land contaminated by fuel, industrial cleaning and paint stripping solvents, and metal finishing compounds. Military entities may supervise their own remediation projects, but states may participate as they have a stake in any land and aquifer contamination within their boundaries. Otherwise, states play a major role in supervising cleanup activities under the guidance and overall supervision of EPA.

Litigation is not the only thing to slow the progress of eliminating NPL sites. Broader events within EPA have also slowed progress of both RCRA corrective actions and CERCLA cleanups. The hazardous waste programs have been criticized by virtually all interested parties—certainly by all affected groups. Citizens, industry, environmental organizations, states, and Congress have all complained that EPA moves too slowly. The agency is accused of being ineffective in protecting public health and the environment despite complex, costly, and sporadic action.

Liability Established by EPA

As previously mentioned, guilt is the least examined factor in assigning cleanup responsibility. Implementation of CERCLA actions has proved to be a complex, multifaceted process. Congress essentially gave EPA the authority to address hazardous substance emergencies; established a trust fund to provide fast response and cleanup activities; created a taxing mechanism to generate revenue for the trust fund; and established broad liability for all who generate, transport, store, treat, and dispose of hazardous wastes.

Priorities for Cleanup and Removal

Removal actions are projects that deal with emergencies such as Love Canal. Under the hazardous substances removal program, well over a thousand actions have been approved. Although EPA has spent far more money than first envisioned by Congress, the removal of hazardous substances in the environment is barely started.

Tens of thousands of remedial action sites have been identified. Remedial actions are long-term, permanent cleanups. More than twelve hundred have been placed on the EPA's National Priorities List, but the agency wants to add more.

Hazardous Substance Response Trust Fund

Section 211 of CERCLA establishes the taxing structure that generates the Superfund. In general, the fund acquires money from three sources. One source is a tax levied on crude oil received at any U.S. refinery and for all petroleum products entering the United States for consumption, use, or warehousing. Another tax supplying Superfund is levied on forty-two listed chemicals when sold or used by a manufacturer, producer, or importer thereof. The third source is general tax funds that may be budgeted to the EPA for CERCLA remediation when the other sources fall short of meeting current needs.

Sections 231 and 232 of CERCLA call for a postclosure tax and trust fund. Hazardous wastes that are shipped to a TSDF are taxed to provide funds for a trust to ensure that postclosure care expenses can be met.

RCRA *versus* CERCLA

Despite different scopes, purposes, and methods, RCRA and CERCLA are related. One distinct difference between RCRA and CERCLA is the noticeable lack of regulatory activity associated with CERCLA. Only the National Contingency Plan and certain requirements for liability generated additions to the Code of Federal Regulations. On the other hand, RCRA spawned extensive regulations by EPA.

Proactive *versus* Reactive Waste Management

The second major difference between RCRA and CERCLA lies in their functions. RCRA is a *proactive* approach to solid waste management. It addresses present and future activities aimed at preventing uncontrolled, unpermitted contact with the environment; that is, RCRA prevents future problems potentially caused by present actions or inaction. CERCLA remedies existing or pending releases; that is, CERCLA addresses current problems caused by past actions. Both laws, however, establish trust funds for postclosure activities for disposal sites. Both laws also specifically preclude the compensation of persons whose health is impaired by exposure to released hazardous substances.

OSHAct with Regard to RCRA/CERCLA

OSHAct was enacted to provide a safe, healthy workplace for all working Americans. Specific health and safety standards have been promulgated in the twenty-ninth chapter of the Code of Federal Regulations (29 CFR). Some states have also promulgated their own safety programs.

Protecting employees who engage in RCRA or CERCLA response activities requires knowledge of the types of hazards to which each worker is exposed. Response supervisors must also know the protective measures appropriate to specific hazards and applicable OSHA standards. However, OSHA standards provide minimal protection. Therefore, each response team should have access to a professional health and safety specialist who is familiar with good industrial hygiene and safety practices. This professional also ought to have experience in the unique problems of hazardous waste management.

Safety Systems Require Continual Improvement

Generic site safety plans are typically prepared for remedial actions and site cleanups. This makes the plan applicable to response activities at many hazardous substance release sites. Given the time constraints of remedial sites, though, safety systems should be reviewed and continually improved. The site safety plan, particularly, should be updated for each site. Specific materials present and physical characteristics of each site vary, but though the generic types of hazards may be very similar from site to site.

Remedial Site System Safety

A systems safety approach to remedial action is recommended. Malasky (1982) defines a system as a composite of personnel, tools, equipment, facilities, environment, and software. A system may have any level of complexity.

A hazardous substance release site may be considered a system. Any component, such as soil, groundwater, or liner, may be considered a system also. During the remedial action project, the entire Superfund or RCRA correction action site, including all remedial actions, is considered a system. System safety provides an optimum degree of safety established within the constraints of operational effectiveness, time, and cost that is achievable through all phases of the system lifecycle. The systems safety approach presupposes an understanding of, and accounting for, all interfacing parties.

Stakeholders

The first step in the systems safety approach is to identify interfacing parties, called *stakeholders*. The following are potential stakeholders for a Superfund site:

1)   Hazardous waste/substance site owner/operator
2)   Hazardous waste transporters to/from site
3)   Hazardous waste generators shipping to site

4) Population in nearby areas or surrounding site
5) Future population in area
6) Shareholders of stakeholder companies
7) Federal, state, county, and municipal agencies
8) Employees of stakeholder companies and their families
9) Employees of regulatory agencies (inspectors, compliance officers)
10) Consumers of products manufactured by generators
11) Legislators
12) Professionals engaged in hazardous waste management and removal or remediation.

Some stakeholders are named PRPs by law; others are stakeholders because they administer and enforce the law. The livelihood of nearby families may hinge on the project. The concern of shareholders and consumers is financial, unless they also fall into one of the other stakeholder groups. Financial interests are proportional to the cost in evaluating the feasibility of a given remedial measure and the cost of implementing the measure.

System Safety Process

The mechanism for achieving a workable system safety framework is a systematic process that specifically designs each site. The site safety plan is tailored to the unique needs of the site and is structured in such a way as to include the entire lifecycle of the system.

First, the process addresses the entire life cycle of the project in order to provide for the complete integration of all functions. Second, adequate training is conducted before the first operations task is fulfilled. And third, the integration of functions is accomplished in a cost-effective manner, within the constraints of technological feasibility. The process also maintains a level of safety that protects the general public and the environment while protecting the on-site workers.

Commonality of Requirements

The site safety plan must apply to all personnel and activities at the site for which it is written. The owner/operator must maintain full control of and responsibility for the management and implementation of the process in cooperation with a management committee representing all the major functions. In the absence of an owner/operator at a CERCLA site, the PRP committee should appoint someone to oversee the operations and maintain a carefully written log of daily events. In the absence of an owner/operator at a remedial action, the prime contractor, whose employees and subcontractors are performing the hazardous work, works in cooperation with the local EPA emergency response coordinator, but assumes control of the project and the system safety process.

The basic assumption of the site safety plan is complete and voluntary compliance with all applicable laws, regulations, and standards. Performance reviews and contract renewal evaluations should include an assessment of activities related to onsite safety and health.

In addition to RCRA, CERCLA, and OSHAct, other laws and regulations must be considered in designing a systems safety approach to the hazardous substance release problem. Regulations include the following:

- 29 CFR 1910: General Industry Standards
- 29 CFR 1926: Construction Standards
- 40 CFR 260-270: Hazardous Waste Regulations
- 49 CFR: Hazardous Materials Transportation

OSHA's General Industry Standards apply when operating processes at the remedial site, as if the cleanup activity were a little factory unto itself. However, the Construction Standards apply when conducting demolition of structures or infrastructure; conducting earthwork: constructing structures or infrastructures; or maintaining equipment and facilities by painting or covering. Besides these cited regulations, specific state, county, or municipal regulations may also apply, especially those that are more stringent than the federal regulations. Also state, county, or municipality may have transportation requirements that are more stringent than those of the U.S. DOT.

Responsibilities and Authorities

The system safety organization should have direct access to senior operational management. A system safety representative would ideally participate in the highest level of decisionmaking, so that decisions are based on solid information.

Each project should have a director of safety and health and a director of environmental engineering assigned to it. The size of the system safety group, including the occupational safety and health group and the environmental group, depends on the area covered by the site and variables known prior to planning for remedial action.

Specific Responsibilities. The primary responsibilities of the systems safety group are planning and implementation of the plan. First, the group must ensure proper plan development in order to meet the basic assumptions of the planning process. Second, the group must implement the plan soon after it is endorsed by top management.

The plan should have specific provisions for authority and accountability. Authority can be passed down the organization chart, but accountability should always remain with the one who delegated the authority. In other words, when a senior delegates a task to a junior, the authority to accomplish the task should also be delegated, but the accountability for the task remains with the senior. The senior may hold his or her subordinates accountable, as any supervisor holds one of his employees accountable, but to the overall organization, the senior is still accountable for the task. This may seem contrary to typical human resources management, but it is the only effective way to accomplish difficult and dangerous missions satisfactorily.

Some of the responsibilities of the systems safety group are as follows:

1) Develop training requirements, including, but not limited to, occupational safety and health, hazard awareness, toxicology, performance of emergency procedures, personal protective equipment, decontamination, accident investigation, recordkeeping, and supervisory training.

2) Review engineering plans, providing system safety analysis and advising upgrades for the system.

3) Complete and report hazard analyses as appropriate.

4) Develop occupational, public health, and environmental safety processes.

5) Provide input to management decision making.

6) Provide adequate safety and health staff to meet the needs of the site.

7) Interface with other departments to ensure cooperation between contractors in achieving system safety objectives.

8) Develop and implement system safety milestones, and evaluate the status of the process at each milestone.

9) Develop and implement emergency action plans; accident investigation procedures; environmental and personal biological exposure monitoring procedures; recordkeeping requirements; databases; and site closure and post-closure procedures.

10) Acquire pertinent operating licenses.

11) Review the process, as necessary, and recommend site changes and safety process revisions.

12) Conduct system safety audits of all integrated functions to ensure compliance with, and ongoing improvement of, safety processes and activities.

13) Verify that company, contractor, and visiting personnel are aware of, and unerringly comply with, regulations and policies.

14) Require contractors to submit documentation of system safety and also require that liaison personnel be provided by the owner/ operator or managing contractor.

15) Develop system safety processes that address specific needs at given milestones.

Process Criteria. A number of system safety process criteria need to be defined in the site safety plan. These criteria include hazard level categories, system safety precedents, and analytical techniques. The hazard level categories are defined in Table 29-1.

Table 29-1. Hazard Level Categories

Classification	Level of Safety	Hazard Result
Class IV	Safe	Does not result in injury and/or damage to equipment.
Class III	Marginal	Results in release of hazard substance to the soil, which is detected and corrected before public exposure occurs.
Class II	Critical	Results in release and subsequent contamination of soil and ground or surface water. Subsequent public health exposures may or may not result in illness, injury, or death. Release requires specific remedial measures to correct the problem.
Class I	Catastrophic	Results in release with subsequent contamination and public health effects. Remedial action may not mitigate future exposures. A significant present and future risk of illness, death, and/or genetic damage exists.

Management must enable all departments with authority to prevent or control Class I or II hazards. However, management is ultimately responsible for prevention or control of these hazards. Successful prevention and control of hazards presumes technological feasibility. When the elimination of a hazard is infeasible, engineering controls such as isolation, containment, or limitation must be implemented to minimize both the severity and probability of occurrence. Administrative controls may also be required. PPE should always be the last line of defense.

After Class I and II hazards are addressed, attention must be turned to Class III and IV hazards. The following is a list of control priorities for Class III and Class IV hazards:

- Engineer systems to meet overall preventive needs
- Engineer controls to prevent the release of hazardous substances from becoming a public health exposure
- Implement safe work practices, administrative controls, and training to protect a public health exposure
- Provide PPE to protect workers from released materials

The impact of a specific remedial technology on the environment, in addition to the impact of the hazardous substance release, must be evaluated before work begins. All technically feasible steps to mitigate further degradation of the environment must be taken. An environmental impact statement (EIS) may be required at the end of the foregoing evaluation. An EIS is strictly required when one of the PRPs is a federal entity, and some states require private concerns to prepare EISs. When the EPA manages a project, an EIS is required.

Another responsibility of the owner/operator is to prepare a preliminary hazard analysis (PHA), such as a fault tree analysis, that determines various means by which a release with either an occupational or public health exposure can occur. The PHA should also define the means of control available. A quantitative analysis and sensitivity study of the PHA will reveal potential exposures at the site.

Four Phases of the Task

System safety tasks are accomplished in four distinct phases: 1) concept formulation, 2) development and production, 3) operations, and 4) disposal.

Concept Formulation. The first phase is defined by three major tasks. A review is made of all environmental studies, with particular attention to the hydrogeological, geological, and meteorological factors. During the investigation for environmental impact, potential remedial technologies should also be determined. Next, management must be informed about the system safety process and its requirements. Coordinating and liaison activities of management are initiated. Finally, management must be made aware of the identity of all input groups, each of which must be assigned its own responsibilities within the overall project.

Development and Production. In the second phase, a number of tasks must be completed by the systems safety group, including the following:

1) Review of purchasing documents and approved of materials and equipment to be purchased.

2) Validation of quality control testing and documentation procedures, as well as support procedures.

3) Development and implementation of site control and decontamination procedures.

4) Development of operations manuals and procedures for the remedial action selected from among the alternatives.

5) Conduction of inspections required by specific regulations and establishment of appropriate recommendations for changes in items 1 through 4

Operations. Four tasks are required in the operations phase:

1) Complete preplacement and periodical physical examinations, job training, and safety training

2) Implementation of emergency procedures; and determination of the need for and scheduling of refresher training in these areas

3) Validation of installation, inspection, and maintenance procedures, as well as proper operation of monitoring and warning systems

4) Review and approval of engineering changes as proposed and implemented

5) Development and use of accredited protocols for sampling, analysis, and documentation of materials produced by or left as residue of the remedial action

Disposal. The final part of the systems safety process is the disposal phase. A closure and postclosure plan is developed and implemented after approval of the appropriate regulatory enforcement authority. The closure plan describes how the site will be decontaminated and closed down at the end of the remediation. If not closed clean, a postclosure plan describes how monitoring and maintenance of the site will be carried out over a specified period of time, usually thirty years. The postclosure plan also establishes a postclosure fund or otherwise transfers risk for future accidents or unanticipated releases of hazardous substances from the site.

Documentation. Although documentation is not a distinct phase of work, the final step in system safety is documentation. Documentation starts at the beginning, but some amount of documentation will necessarily have to wait until all the other work is completed. Procedures developed during the planning, development, and active periods of the project are collected with supporting documentation that demonstrate compliance with OSHA and EPA regulations. Operating, security, monitoring, and inspection logs; analytical results; and other data from the four systems safety phases are assembled and archived for future reference. Do not assign a destruction date to the files, but retain these records permanently.

Incident Response. Environmental incidents involve a release or threat of a release of hazardous substances. Such an incident presumes an imminent and substantial danger to public health or to the environment. Special problems are encountered in each incident. Response personnel have to evaluate these problems and determine an effective course of action to mitigate the danger while not creating additional risk.

An incident, by nature, represents a set of circumstances that is potentially hostile to humans. Substances that are flammable, explosive, corrosive, radioactive, toxic, or biologically active may affect the response team, the general public, or the environment. Workers may slip, trip, or fall; be struck by objects; be caught between objects; or be subject to danger from water, electricity, or heavy equipment. Injury and illness may occur due to physical stress or weather conditions. Conditions created by the incident can vary widely, causing a broad range of unexpected risks to the response team, public health and welfare, or the environment.

Although each incident is unique, common requirements do exist. For example, incident responses typically require PPE for the health and safety of the responders either from unknown hazards or from unknown concentrations of known hazards present.

Response Team Protection. Chemicals represent can be inhaled, ingested, or absorbed, or be destructive to the skin. Chemicals may be present in the air by eddy diffusion or droplets of liquids can become airborne by mechanical means to splash on the skin and eyes. They can be accidentally ingested by relaxing the level of protective clothing too soon or not practicing good industrial hygiene, such as washing the hands before redonning protective equipment to reenter the hot zone. Most chemicals present a minimal hazard when used as intended, but almost any chemical is potentially dangerous if uncontrolled. Toxic chemicals may cause no apparent illness or may cause death or serious medical problems immediately. On the skin, you may observe no demonstrable effects, but the fatty tissue may be dissolving away, such as when a highly alkaline substance absorbs through the skin. Certain substances absorb through the skin or lungs readily and enter the bloodstream, which flows to all parts of the body, leading to possible systemic poisoning.

Creating an Effective Site Safety Process

To reduce the risk to workers who respond to hazardous substances incidents, an effective site safety process should be implemented. The following elements should be included:

1) Safe work practices
2) Engineered safeguards
3) Medical surveillance
4) Environmental and personal monitoring
5) Personal protective equipment
6) Education and training
7) Standard operating procedures (SOPs) in which safety is integral

While most of these elements are self-explanatory, SOPs are worth discussion.

Standard Operating Procedures. As part of an integrated process, SOPs provide instructions on how to accomplish specific tasks in a safe manner. Sometimes, however, safe procedures are independent of task, and such written procedures are more commonly called safety procedures, standard safety procedures, or standard operating safety procedures.

The applicability of a safety procedure, as opposed to a task-specific SOP, is determined by the overall situation at the site. A safety procedure usually requires some on-the-scene modifications to match prevailing conditions at the site. PPE, for example, is an initial consideration at any site, but the exact need and the type of equipment to be used requires a case-by-case evaluation. Another example of a safety procedure is the first entry made at a site, which requires an assessment of the conditions prevailing in order to determine the exact approach.

Instructions for writing safety procedures and SOPs are gleaned from several sources. Management guidelines, policies, plans, programs, procedures, and other documents that refer to administrative, technical, or safety matters provide uniform instructions for accomplishing specific tasks or related groups of tasks.

State and federal agencies also provide guidance as well as specific requirements. Site control and entry are especially covered by government guidance documents, which illustrate the technical factors to be considered in developing instructions and procedures. For any given incident, the government-recommended procedures should be adapted to the present situation while other safety procedures are being developed.

Basic Site Safety

Basic safety principles are listed as follows:

- In contaminated areas, prohibit eating, drinking, chewing of gum or tobacco, smoking, or hand-to-mouth transfer of materials.
- Require thorough washing and rinsing of hands and face upon leaving work area or before eating, drinking, or other hand-to-mouth transfer of materials.
- Immediately after decontamination and removal of garments, require a thorough washing and rinsing of the entire body.
- Prohibit excessive facial hair.
- Discourage contact with potentially contaminated surfaces.
- Because medicines and alcohol can potentiate the effects of toxic chemicals, do not allow workers to enter the hot zone for a minimum of twenty-four hours after consuming alcohol or taking medicine.

The prohibition on excessive facial hair is made on the presumption that it interferes with respirator fit. Otherwise, frequent fit testing would be required. Instruct workers to avoid puddles, mud, and discolored surfaces that may be contaminated. Instruct them not to kneel on the ground or to lean on, sit on, or place equipment on drums, containers, vehicles, or the ground in the hot zone. Strongly discourage recreational use of alcohol for the duration of the response. For long response actions, provide three-day recreational weekends with an alcohol abstinence beginning twenty-four hours before returning to work.

Site Safety Protocol. Remedial sites are divided into three zones for safety. The contaminated area is called the *exclusion zone* or sometimes the *hot zone.* Only workers who have been thoroughly trained in all aspects of remediation operations and safety are allowed to enter this zone. Immediately upwind of the exclusion zone is the *decontamination zone.* At each end of the decontamination zone is an access control point with log books that record who entered or exited, when, and why. Between the two access control points, running the length of the decontamination zone, is the decontamination corridor. Often the corridor is a trailer, for privacy and modesty. This is where workers suit up on the way into the exclusion zone, or decontaminate and strip on the way out. The following is a typical decontamination sequence:

1) In the exclusion zone, drop segregated equipment.
2) Wash outer gloves and boot covers.
3) Rinse outer gloves and boot covers.
4) Remove tape.
5) Remove boot covers.
6) Remove outer gloves.
7) Crossing the hot line into the decontamination zone, wash suit and safety boots.
8) Rinse suit and safety boots.
9) Remove safety boots.
10) Remove fully encapsulated suit and hard hat.
11) Remove self-contained breathing apparatus backpack.
12) Wash inner gloves.
13) Rinse inner gloves.
14) Remove face piece.
15) Remove inner gloves.
16) Remove inner clothing.

17) Take a shower.

18) Redress with clean clothes (those removed prior to entering the exclusion zone).

19) Crossing the contamination control line, enter the support zone.

Some decontamination control procedures place the decontamination control line between steps 16 and 17, with part of the decontamination area in the support zone. If using this scheme, be sure to paint a decontamination control line on the floor to indicate in which direction progressively increasing contamination lies. The advantage of the listed method is that once you exit the decontamination area into the support zone, you are in uncontaminated surroundings. Conversely, once you enter the decontamination area, you are potentially exposed to contamination. The disadvantage is that your support operations work clothes are inside the decontamination trailer with potential contamination. Either way, a decontamination station assistant should be assigned to work in the decontamination area, keeping everything cleaned up, and moving other workers along a progressively more or less contaminated path depending on whether they are entering or leaving the exclusion zone.

The third zone of a remedial site is the *support zone,* where a command post, administrative offices, cafeteria or canteen, restrooms, and other supporting services are found.

On sites where flammable materials are present, the introduction of sparks or other ignition sources must be strictly forbidden. Also, prohibit exclusion zone operations unless backup personnel are in the support zone and assistance in the decontamination zone is available.

Before entry to the exclusion zone, all personnel must know where to find the emergency telephone numbers for the nearest medical facility, ambulance service, fire department, police department, and poison control center. Normally, this is accomplished by posting a large sign near a telephone set with an external line.

Initially and periodically, a technician in a fully encapsulated suit and SCBA needs to determine atmospheric concentrations in the exclusion zone. The periodicity of monitoring depends on the situation and site conditions. The following hierarchy of air monitoring is recommended:

1) *Monitor for flammability.* The atmospheric concentration of flammable vapors must be less than 10 percent of the lower explosive limit (LEL).

2) *Monitor for oxygen deficiency.* The atmospheric concentration of oxygen must be greater than or equal to 19.5 percent.

3) *Monitor for atmospheres that may be immediately dangerous to life and health (IDLH).* Only SCBAs may be worn under these conditions.

4) *Monitor for record purposes.* Monitor the concentration of any contaminant present that could have long-term health implications.

Years ago an instrument called a flame safety lantern, which had an open flame, was used to check for oxygen deficiency, making it imperative to check for flammable vapors before checking for oxygen concentration. With modern handheld, direct-reading instruments, the order of steps 1 and 2 is not critical, but both should be accomplished before checking for IDLH.

Entrance and exit to the exclusion zone must be planned with emergency escape routes in mind. Warning signals for site evacuation must also be established. Personnel should practice unfamiliar operations in the support zone before attempting them in the exclusion zone.

Personnel in the exclusion zone must use the buddy system when respiratory protection is required. At a minimum, a third person, fully equipped as a safety backup, is required during initial entries. During continual operations, exclusion zone workers may act as safety backup for each other. Support zone personnel provide emergency assistance and rescue.

PPE is issued as follows:

1) General:
   - Hard hat
   - Coveralls
   - Safety boots
   - Work gloves
2) Site-specific:
   - Fully encapsulating chemical impervious suits
   - Monitoring instruments and dosimeters
   - Respiratory protective devices

The following are EPA-designated levels of personal protection for remedial sites:

- Level A: Self-contained respiratory protection (positive-pressure demand SCBA), fully encapsulated suit, and percutaneous agent protection in an IDLH environment.

- Level B: Respiratory protection (positive pressure demand) and skin contact protection in an IDLH environment.
- Level C: Air-purifying cartridge respiratory protection and skin contact protection in a less than IDLH environment.
- Level D: Standard safety protection when no exposure to a toxic substance is anticipated. A five-minute escape pack and an air-purifying respirator are carried by personnel as a precautionary measure.

Communications are maintained with workers in the exclusion zone until they cross the hot line into decontamination, and even then if no assistant is present. Radios are the communications equipment of choice these days, though other means have been used such as telephones and hand signals. Emergency communications need to be prearranged in case of radio failure, the need for evacuation, or another contingency arises.

Visual contact must be maintained between the work pairs in the exclusion zone and the safety personnel. Entry team members should remain close together and work should be planned accordingly. Consequently, they can assist each other more readily in case of emergency. Wind socks or other indicators should be positioned at several highly visible places on the site so that personnel can escape upwind.

Minimize the number of personnel and equipment that enter the exclusion zone, thereby minimizing decontamination requirements. You will appreciate this practice if you ever have to decontaminate an eighteen-wheeler, an airplane, or a bulldozer. Keep as much of the operation in the support zone as possible. Those things that need to be used in the exclusion zone should have a storage shed or layout yard just on the hot side of the hotline when not in use. Decontaminate these items at the end of the job, unless they become so contaminated as to present a special hazard to workers in the meantime. Have entry, exit, and decontamination procedures planned and approved before going into the exclusion zone for the first time. Also, practice outfitting on the way in and undressing and decontaminating on the way out before ever crossing the hotline.

Medical Program. To safeguard the health of response personnel, a medical program is developed and maintained. Two essential elements of the program are routine health and emergency treatment. A pre-employment

physical examination establishes a person's baseline state of health and fitness for strenuous work while wearing PPE. Arrange for special medical examinations in case of exposures to toxic agents. The following are some suggestions for preventive medical surveillance:

1) Initial examination
   - Personal/family medical history
   - Work history, including industrial hygiene reports
   - Standard occupational physical examination
   - Visual acuity, distant vision, and color perception
   - Audiogram
   - Pulmonary function screening: forced vital capacity and forced expiratory volume in one second
   - Posterior-anterior chest X-ray
   - Electrocardiogram for males over age 35 and females over 40
   - Routine urinalysis, including occult blood
   - Complete blood count
   - SMAC-23 blood profile to include gamma glutamyl transpeptidase

2) Periodic examinations: At least annually. All initial tests are repeated except the chest x-ray, which is repeated every third year unless other tests indicate otherwise.

3) Follow-up Examinations: Unscheduled examinations that assess problems uncovered by other examinations or by known or suspected exposures or by symptoms of exposure.

SARA III/EMERGENCY PLANNING AND COMMUNITY RIGHT-TO-KNOW

SARA Title III of 1986, also known as EPCRA, establishes requirements and guidelines for federal, state, and local governments, as well as businesses and industry, to protect the public health and welfare and the environment by gathering and sharing information, which is intended to be used for emergency planning and action. Section 313 of SARA III also provides for gathering information from industry that is to be made available to the public for discerning the environmental health of its communities. This act builds on older laws that made ineffective attempts to accomplish the same goal by other means. Rather than undoing sections of other laws, which would have been a complex and probably unsuccessful task, SARA III builds on them and, in doing so, provides us with a tapestry of requirements and guidelines that weave in and out of other laws. For instance, SARA III dovetails into the OSHA standard on hazard communication called Employee Right-to-Know.

Nevertheless, SARA III is quite complex with detailed information requirements and mandatory detailed reports, all of which are available to the public. The law does provide some trade secret information or confidential business information to be kept from public view, but, nevertheless, all information must be submitted to the appropriate government offices. Some special terms used in association with SARA III requirements are as follows:

- CERCLA   Comprehensive Environmental Response, Compensation, and Liability Act of 1980
- CGL   Comprehensive general liability (insurance) policy
- EHS   Extremely hazardous substance
- FDA   Food and Drug Administration
- FIFRA   Federal Insecticide, Fungicide, and Rodenticide Act
- HazChem   Hazardous chemical, as defined by OSHA Hazard Communication Standard
- HS   Hazardous substance
- LEPC   Local Emergency Planning Committee
- MSDS   Material Safety Data Sheet
- NRC   National Response Center
- NRT   National Response Team
- RAT   Regional Assessment Team
- RQ   Reportable Quantity
- RRT   Regional Response Team
- SARA   Superfund Amendments and Reauthorization Act of 1986
- SERC   State Emergency Response Commission

- SIC   Standard Industrial Classification
- TPQ   Threshold Planning Quantity

Background of SARA

CERCLA provided authority for federal cleanup of hazardous substance release sites and required companies to report to the NRC the release of any hazardous substance to the environment above a certain quantity (RQ). CERCLA lists 717 hazardous substances that could, upon release, pollute the environment and cause public harm. CERCLA was set to expire in 1985 when a terrible chemical release accident in Bhopal, India, aroused the public. As Congress debated what, if anything, needed to be legislated regarding chemical safety, the U.S. chemical industry steadily lobbied its overall excellent safety record. Although the record had some notable blemishes, such as chemical disasters in Flixborough, England, and Seveso, Italy, the industry was making headway among legislators when the same accident that happened in Bhopal occurred in Institute, West Virginia. In spite of the fact that the plant crew in Institute were quickly responsive and prevented deaths and injuries such as were suffered in Bhopal, public pressure forced the legislators to take decisive and effective action.

Examining case histories of chemical releases in which the public was affected, or could have been, it becomes evident that the release does its damage before emergency personnel arrive. Early warning and emergency response planning are needed for public and environmental protection. Hence, Title III was added to SARA as it was passed in 1986.

Whereas CERCLA aims to cleanup abandoned hazardous substances that present a threat to the public or environment, Title III intends to coordinate emergency planning. Such planning is required at all levels from federal to local government. Planning is also required by the potential release sources: industries. Therefore, Title III built on the EPA's Chemical Emergency Preparedness Program (CEPP), which was already operational. The public is also given access to information about the chemical substances in use at and released from any given facility.

Release Notification

Notification requirements under CERCLA and SARA III stand independent of each other. If a notification is required under CERCLA, but not

SARA III, it must be made appropriately; if required under SARA III, but not under CERCLA, notify. Civil and criminal penalties are built into the law similar to other environmental laws, and these penalties can be imposed for failure to comply with notification requirements.

Emergency Planning and Notification. The first important subtitle of SARA III sets the framework for emergency planning in Sections 301, 302, and 303 and for emergency notification in Section 304. This section of the law develops state and local government emergency response and preparedness capabilities. The law demands better coordination and planning, especially within municipalities. Regulated facilities that have EHSs, as defined under the Act, on their premises, in quantities greater than the TPQ, are required to cooperate with state and local planning officials (SERC and LEPC) in preparing comprehensive emergency plans (see Sections 302 and 303).

Section 304 also requires facilities to report accidental releases of EHSs and CERCLA hazardous substances to state and local response officials (SERC and LEPC). This notification is required if a reportable quantity leaves the property or is evaporated or enters the ground. CERCLA requires the notification if the RQ occurs during a twenty-four hour period. SARA notifications are made immediately to the LEPC, the SERC, and the NRC. The initial report may be made by telephone, by radio, or in person. A follow-up written report is due to the LEPC and SERC. Section 304 emergency reporting does not have trade secret protection. CERCLA reporting is made directly to the NRC.

The second SARA III subtitle establishes the community right-to-know requirements in Sections 311 and 312 and an emission inventory, or toxic chemical release reporting, in Section 313. Section 311 requires regulated facilities to submit MSDSs to local and state officials. Section 312 requires an annual inventory of hazardous chemicals, which is submitted to local and state governments (fire department, LEPC, and SERC) for public use and emergency planning. The forms for these inventories are called Tier I or Tier II forms because the second form requires an additional level, or tier, of information. Tier I has fallen into disuse, so expect your local government to require the inventory report on Tier II forms. The Tier II form includes, for each chemical reported, the name as it appears on an MSDS, total amount on the premises, storage and uses processes, and trade secret claims, if any. The forms must be made available to the public (a sanitized version is used where trade secret claims are made) on demand during normal working hours.

Under Section 313, businesses are required to submit annual reports on the amounts of chemicals their facilities release into the environment, either routinely (whether by permit or not) or accidentally. Congress wanted the federal government and the public to know about releases of toxic chemicals into the environment, how much, and by whom. Releases to air, water, and land are reported.

The final subtitle, Sections 321–330, gives trade secret protection, provides for citizen suits, and requires businesses to make information available to the public.

HAZCOM/EMPLOYEE RIGHT-TO-KNOW

In 1987 OSHA expanded the scope of the Hazard Communication Standard to cover all employees exposed to hazardous chemicals. OSHA intended to ensure that such workers would have the right to know the identities and hazards of chemicals in their workplace. The assumption is that knowledgeable workers will modify their behavior to handle hazardous chemicals more carefully. A second assumption is that employers will use the added data to design and use better programs and controls to prevent exposure to the chemicals. OSHA presumed that the hazard communication program would become an ongoing effort and that long-term efforts would go beyond mere compliance with the standard in terms of improving the quality of work life. Table 29-2 summarizes the Hazard Communication Standard.

OSHA continues to issue record citations for noncompliance with HAZCOM. The trouble is that the information given on MSDSs is frequently useless. Employees cannot understand MSDSs and employers cannot use MSDSs. Companies that prepare MSDSs use broad, generic statements (presumably to protect themselves from liability) that provide no useful information to those who need hard data. Yet, OSHA compliance officers still issue citations to hapless users of chemicals. HAZCOM is better than nothing but it has no punch where it counts: with the manufacturers, who must start writing MSDSs that mean something.

Table 29-2. Hazard Communication Standard

INITIAL TASKS

Written Program
  Develop a program addressing who, what, when, where, and how.

Chemical List
  Develop a chemical list.

MSDSs
  Acquire MSDSs.
  Make them available to employees.

Labels
  Label in-house containers.
  Do not remove labels as received.

Training
  Train employees to read and understand MSDSs and to recognize hazards of chemicals.

Nonroutine Uses
  Identify nonroutine uses.

Contractors
  Notify contractors of chemicals in their work area.

UPDATING TASKS

Written Program
  Review annually.
  Disseminate tasks and delegate.

Chemical List
  Update list.
  Review uses.
  Substitute lesser hazard chemicals.
  Minimize inventories.
  Decrease number of chemicals in use.

MSDSs
  Update MSDS files.
  Review completeness and accuracy.
  Notify maker of inaccuracies.
  Acquire MSDSs before purchase.
  Acquire MSDSs for all hazards.

Labels
  Update in-house labels.
  Replace worn/faded/torn labels.
  Review labels for accuracy.
  Notify manufacturers with inaccurate labels.
  Label byproduct hazards.

Training
  Train transfers and new employees.
  Update training for chemicals used.

Nonroutine Uses
  Expand chemical lists.
  Review operations for contaminating by products.
  Integrate with emergency drills.

Contractors
  Require contractors to provide a list of chemicals and MSDS used on their jobs.
  Restrict contractor chemical storage to specific location on property.

For instance, Product A arrives on the shipping dock with an MSDS. You look it over. The first thing you notice is that under the ingredients section is the statement: "This product contains no hazardous chemicals as defined by OSHA." Then you notice that the emergency treatment and health effects sections are two pages long. You also see that some unidentified component has an OSHA PEL assigned. Another has no PEL, but the American Conference of Governmental Industrial Hygienists (ACGIH) has established an 8-hour threshold limit value (TLV-TWA). The physical and chemical information is for the mixture, so you have no idea what you are dealing with here. You decide to use this material under a local ventilation hood just to be on the safe side. How much airflow do you need? So you examine the section on protective measures. The manufacturer tells you to "use adequate ventilation." How will you get rid of any residue? The manufacturer says to "dispose in accordance with all applicable federal, state, and local laws and regulations."

Far too many MSDSs are no more useful than the hypothetical one just presented. A few are well written and useful because they go well beyond the OSHA HAZCOM standard and give you some really useful information, such as exact evaporation rate instead of "slower than ether." Or they list the aquatic toxicity so that you can determine whether to treat the material as a wastewater. Or they tell you what hazardous waste codes might apply to this material in its used condition. One or two companies list odor thresholds. A few report the findings of toxicological studies on the components.

Inventory and Use Control

A collection of MSDSs can aid a facility in the control of inventory and usage. Although not directly the intent of OSHA, this control fits into the ultimate aim of the standard by reducing potential for exposure. Continual review of MSDSs identifies chemicals that are no longer needed so that they can be disposed of properly. Using MSDSs to track inventory allows smaller size plants to avoid the process safety standard by controlling inventory of regulated chemicals below the threshold level. This works well for avoiding SARA III 313 reports too. Despite the overall deficiency of useful information, most MSDSs supply enough information for inventory control of tightly regulated chemicals. They can also be used to track where and how the chemicals are being used in order to stop poor work practices. Even a typical poorly written MSDS has sufficient information to make comparisons between alternative products. Do not take the information too literally, though, because one or both MSDSs being compared may have gross inaccuracies, understatements, or overstatements. Nevertheless, many MSDSs can be used for pollution and exposure prevention screening purposes.

TOXIC SUBSTANCES CONTROL ACT

In 1971, the President's Council on Environmental Quality developed a legislative proposal for coping with the increasing problems of toxic substances. In response to growing public concern as evidenced by the phenomenon called *chemophobia,* Congress held endless hearings and debates over the next five years. On October 11, 1976, President Gerald Ford signed the Toxic Substances Control Act (TSCA) into law. TSCA became effective January 1, 1977, and granted EPA broad authority to anticipate and address chemical risks and to regulate the chemical industry.

The purpose of TSCA is to protect public health and the environment from chemical substances with significant risk. EPA is authorized to obtain information from industry on the production, use, distribution, health effects, environmental effects, and other matters concerning chemical substances and mixtures. After giving consideration to cost, risks, and benefits of a substance, EPA may regulate the substance's manufacture, processing, distribution into commerce, use, and disposal. Pesticides, tobacco and its products, nuclear material, firearms and ammunition, food, food additives, drugs and medicines, and cosmetics are exempt from TSCA because they are regulated under other laws.

The primary goal of TSCA is to regulate chemicals in commerce, without regard to their specific use or area of application. Thus, EPA controls a wide range of chemical hazards, particularly at their source, before they present a danger to the environment.

Some of the major provisions of TSCA are as follows:

- *Notification.* Manufacturers of new or existing chemical substances must notify EPA ninety days ahead of new commercial production or application.

- *Unreasonable risks.* The manufacture of new chemical substances can be restricted or delayed when information is inadequate to evaluate the health or environmental defects involved, if EPA decides the substance may present an unreasonable risk.

- *Significant new use rules (SNUR).* EPA may adopt rules governing the manufacture, processing, distribution, labeling, or disposal of a harmful chemical substance or mixture.

- *Imminent hazard.* EPA can obtain an injunction to protect the public and environment from a chemical substance or mixture that presents an imminent hazard.

- *Recordkeeping.* Reports and records are required on chemical substances and mixtures produced for commerce, and to provide safety, health, and environmental data on their products.

- *Testing.* Manufacturers and processors of potentially harmful chemical substances and mixtures are required to test and evaluate their effects on health and the environment.

- *Information.* Manufacturers or processors must immediately notify EPA of any substance or mixture that may cause or contribute to health or environmental risks.

- *Enforcement.* EPA is authorized to seize chemical substances or mixtures that may have been manufactured or processed in violation of TSCA. Production facilities, warehouses, and tank farms may be locked down by EPA while courts consider the evidence.

Some activities that are specifically prohibited by TSCA are as follows:

- Failure or refusal to comply with any rule or order issued under TSCA.
- Commercial use of any chemical substances or mixtures that user knows were manufactured, processed, or distributed in violation of TSCA.
- Failure or refusal to establish or maintain records required by TSCA.
- Failure or refusal to submit reports, notices, or other information requested by EPA per TSCA.
- Failure or refusal to permit access to or copying of records required by TSCA.
- Failure or refusal to permit entry or inspection as authorized by TSCA.

TSCA provides civil and criminal penalties, which apply to both corporations and individual employees. Any person who commits one of the previously listed prohibited activities faces a civil penalty of up to $25,000 for each violation. Each day that such a violation continues constitutes a separate violation. Any person who knowingly or willfully commits one of the prohibited activities faces a criminal fine of up to $25,000 for each day of violation and/or imprisonment for up to one year. This is in addition to any civil penalties levied. TSCA is extremely complex and the penalty for noncompliance is severe. Any company that manufactures, processes, or distributes chemical substances or mixtures as defined by TSCA would be prudent to seek both legal and technical expertise to interpret how the law and its most recent regulations impact business operations with the TSCA provisions listed previously.

SORTING OUT COMPLIANCE

Though confusing at first, compliance with myriad regulations becomes easier as you understand that any regulation is a set of instructions on how to obey a law. Laws and regulations are written by persons who speak the same language we do, and although it may seem at times that someone has tried purposely to confuse the rest of us, that is not the case. Ambiguous language is an attempt to cover too much ground. By being too specific, a law becomes hard to uphold.

CONCLUSION

To get a grip on full compliance, you have to itemize events and tasks that are necessary. Many subscription services do this for you, but anyone can sit down and draw up a compliance checklist, given time and motivation. Then it becomes a matter of sorting out the action items by date due and, where no dates is specifically given, prioritized by degree of risk. Next, look at the document you are developing and consider the need for funds, management approval, data, and other needs and rearrange the order of items accordingly. Rewrite the items in the form of objectives and place milestone dates on them. You now have a compliance plan. Stick to it.

REFERENCES

Blackman, W. C., Jr. (1996). *Basic Hazardous Waste Management. 2nd Edition.* Boca Raton, FL: CRC Press/Lewis Publishers.

Cockerham, L. G. and B. S. Shane (1994). *Basic Environmental Toxicology.* Boca Raton, FL: CRC Press/Lewis Publishers.

Daugherty, J. E. (1996). *Industrial Environmental Management: A Practical Handbook.* Rockville, MD: Government Institutes.

EPA 560/4-88-001 (February 1988). *The Emergency Planning and Community Right-to-Know Act: Section 313 Release Reporting Requirements.* Office of Pesticides and Toxic Substances.

Koren, H. and M. Bisesi (1996). *Handbook of Environmental Health and Safety: Volume I, Principles and Practices. 3rd Edition.* Boca Raton, FL: CRC Press/Lewis Publishers.

Malasky, S.W. (1982). *System Safety, Technology and Application.* New York: Garland STPM Press.

Sullivan, T. F. P., Ed. (1993). *Environmental Law Handbook, 12th Edition.* Rockville, MD: Government Institutes, Inc.

Wentz, C. A. (1989). *Hazardous Waste Management.* New York: McGraw-Hill Publishing Company.

Off-the-Job Safety

John Myre

INTRODUCTION

It is a few years in the future, and a new drug-resistant virus suddenly appears in the United States. It strikes indiscriminately. Newborns and senior citizens are felled. It takes a particularly heavy toll on teenagers and young adults. Over 80,000 die each year, and millions more are disabled—some of them permanently. A person will leave home in the morning, and later in the day a loved one will receive the terrible information that he or she is dead or seriously ill.

The disease quickly becomes front-page news. There is no cure, but preventive measures are found and publicized. As these measures are developed, organizations create elaborate plans to inform their employees.

Still the disease rages. It is usually contracted as the result of a failure by the individual to follow the proper preventive measures. Worse, many times a person contracts the disease as a result of failures by others to follow the preventive steps.

Unfortunately, there is a similar cause of deaths and injuries in our midst today. It is called off-the-job accidents, but the reaction is very different as compared to our fictional virus.

When the news media report on off-the-job accidents, it is usually in the context of counting the number of deaths during a holiday period, or highlighting a spectacular incident. Most organizations ignore the subject since they do not have to report the results to the government or their shareholders. Most individuals assume that these accidents only happen to the other guy, never to them or their family.

Only a few organizations try to raise the level of off-the-job safety awareness, such as the National Safety Council, some consumer groups and publications, and organizations that focus on one particular type of accident, such as Mothers Against Drunk Driving or the U.S. Coast Guard.

In the face of this indifference to off-the-job accidents, why should your company be concerned what employees and their families do away from the work site? The purpose of this chapter is to demonstrate the many benefits your company will receive from a comprehensive off-the-job safety program; to provide an outline to use in organizing the program; and to encourage the development of a personal safety program by each employee and family member.

WHY EVERYONE SHOULD BE INTERESTED

Every organization profits from a comprehensive off-the-job safety program. The evidence is compelling: the safety professional, the organization itself, and the employees all benefit.

Benefits to the Safety Professional

The job descriptions of safety professionals emphasize on-the-job safety. Reducing workers' compensation claims and costs is the number one priority. Most safety professionals feel that they are so busy preventing workers' compensation cases that they have no time for off-the-job safety, even if they wished to. That is a short-sighted impression.

At the work site, virtually every organization has sound safety practices in effect. Safety procedures applicable to each job are well known through investigations of previous accidents, company and industry safety practices developed over the years, and government regulations. The problem is getting employees to follow these procedures. That is where off-the-job safety programs can make a significant difference.

Many organizations approach on-the-job safety as a "switch" to be turned on when the employee comes to work. In fact, safety is a 24-hour attitude, not a "switch" to be turned on when employees arrive. The same safety rules will apply whether you are driving a company car to a meeting or taking the kids to a birthday party.

By developing an effective off-the-job safety program, on-the-job safety results will improve. Employees have fewer on-the-job accidents when safety becomes part of their value system and lifestyle. An off-the-job safety program will help promote safety as a "value."

Another benefit to the safety professional is a reduction in workers' compensation claims that are associated with off-the-job injuries. Some workers' compensation claims may begin as a minor off-the-job injury. These claims can be minimized by preventing off-the-job accidents from occurring in the first place.

Also, employees work more safely when they are not distracted by the concerns associated with injuries to family members.

Financial Benefits to the Organization

Each year, many U.S. organizations pay more than $400 per employee to cover health care costs and related expenses resulting from off-the-job accidents to employees and their families, most of which could have been prevented. To estimate your organization's costs, multiply the number of employees by $400. If the organization is a for-profit company, divide that amount by the pre-tax profit margin to determine the revenue required to pay for off-the-job accidents. For example, a company with one thousand employees, and a 10 percent pre-tax profit margin, needs $4 million in revenue just to pay for off-the-job accidents.

Other Benefits. In addition to the reduction in medical and other costs, other significant benefits to the organization are realized.

1) Sixty percent of the accidents that keep employees off the job occur away from work. A successful off-the-job safety program results in on-the-job productivity increases.

2) In today's shifting job climate, employees often feel undervalued and expendable. When morale suffers, an organization's product can suffer, too. An off-the-job safety program demonstrates that a company cares about the well-being of its employees and their families.

3) An off-the-job safety program complements an organization's wellness program, and can be viewed as an extension of its wellness program.

Benefits to the Employee

By providing practical safety tips and promoting safety awareness, off-the-job safety programs prevent deaths and injuries to employees and their families. Only when organizations prosper can they pay competitive wages and benefits. The reduction in costs and the increase in productivity will increase profits.

OFF-THE-JOB ACCIDENTS

Let's take a look at the human toll taken by off-the-job accidents. The lifetime odds of being killed in an off-the-job accident are approximately 1-in-50. As difficult as that may be to believe, the facts support the statement. Rounding the numbers for the sake of clarity, here is how the number is derived:

Approximately 4 million persons are born each year in the United States. In 1995, approximately 88,000 persons died from unintentional injuries off the job. The number of persons killed in each age bracket is relatively constant every year, for example, the number of one-year-olds killed is about the same, as with two-year-olds, and 55-year-olds. During a person's lifetime, more than 85,000 of his or her peers will be killed in some type of off-the-job accident. Dividing 4 million by 85,000 gives us the 1-in-50 approximation.

Table 30-1 gives recent off-the-job statistics from the National Safety Council, which are reason enough to have a 24-hour safety attitude.

Here are some more reasons to have a 24-hour safety attitude:

- The lifetime odds of being killed in a motor-vehicle accident are about 1-in-100.
- Accidents are the leading cause of death for persons aged 1 to 38.
- More than 15 million person suffer temporary or permanent disabling injuries from off-the-job accidents each year.
- Accidents rob Americans of more years of life before they reach age 65 than any other cause of death, including cancer, heart disease, homicide, and AIDS. (Table 30-2)

In view of the alarming statistics of the human toll taken by off-the-job accidents, why has off-the-job safety been largely ignored?

Is an "Accident" Really an Accident?

A basic reason that employers and society do not pay more attention to off-the-job safety is that the word *accident* is used incorrectly. The dictionary defines accident as "an unexpected and undesirable event, something that occurs unexpectedly or unintentionally, fortune or chance." There is

Table 30-1. The Human Costs of Accidents

CAUSE OF DEATH	NUMBER OF DEATHS
Motor-vehicle	41,500
Home	
Poisonings	8,500
Falls	7,300
Fires and burns	3,600
Suffocation	1,900
Firearms	800
Drowning	800
All other[a]	3,500
Total	**26,400**
Public	
Falls	4,600
Drowning	2,700
Air transport	800
Water transport	800
Firearms	500
Railroad	500
Fires and burns	300
All other[b]	9,900
Total	**20,100**

[a] Most important types are electric current; hot substances, corrosive liquids, and steam; and explosive materials.

[b] Most important types are medical and surgical complications and mishaps; suffocation by ingestion; poisoning by solids and liquids; excessive heat or cold; and other transport.

no quarrel with the undesirable reference, but the belief that accidents are unexpected or the result of fortune or chance is misleading.

For example, is an accident unexpected when someone using a ladder reaches out too far, instead of taking the time to reposition the ladder, and then falls? Does an accident occur by fortune or chance when a person consistently tailgates and then slams into the driver ahead of him or her in a moment of inattention? Is it fate when a boater drinks too much and then

Table 30-2. Summary of Years of Potential Life Lost Due to Premature Deaths

CAUSE OF DEATH	YEARS LOST
Accidents [a]	2,006,000
Cancer	1,861,000
Heart disease	1,423,000
AIDS	954,000
Homicide and legal intervention	869,000
Suicide	689,000
Stroke	245,000
Liver disease	222,000
Pneumonia and influenza	174,000
Diabetes	160,000

[a] Includes over 5,000 on-the-job accidents.

collides with another boat on a lake at night? The obvious answer is *No!* Most off-the-job accidents can be better described as failures. They are failures on our part and failures on the part of others. Stating that someone was killed or injured in an accident tends to exonerate the person responsible.

The National Safety Council has recognized that usage of the word "accident" is inappropriate, and has removed the term "accidental" from its mission statement. The statement formerly read "The mission of the National Safety Council is to educate and influence society to adopt safety and health policies, practices and procedures that prevent and mitigate human and economic losses arising from *accidental causes* and adverse occupational and environmental health exposures." It now reads "The mission of the National Safety Council is to educate and influence society to adopt safety, health and environmental policies, practices and procedures that prevent and mitigate human suffering and economic losses arising from *preventable causes.*" The Council said that it will seek better ways to describe such incidents and encourage persons to think of risk causes. The word accident is inaccurate in safety usage because it implies bad luck, chance, or fate.

Lack of Awareness of Costs and Potential Savings

Most employers are simply unaware of what their human and financial costs are. In fact, off-the-job accident costs will exceed on-the-job accident costs in many organizations. Some of this lack of awareness is due to the fact that off-the-job medical costs do not receive much attention in business publications. When employers focus on medical costs, the primary emphasis has been on finding less expensive medical treatment. That's fine as far as it goes, but it is more cost effective to eliminate the causes of the expenditures in the first place.

This lack of knowledge is easily remedied by making a cost study. After the costs are determined, for-profit companies should divide the total costs by their pre-tax profit margin to determine the revenue needed to pay for off-the-job accidents. This amount will be a real eye-opener for employers.

Table 30-3 is a summary of a recent cost study conducted by a corporation with more than 60,000 employees. More than 90 percent of the costs represented medical costs.

The actual costs were even higher. These figures in Table 30-3 do not include long-term disability payments, training costs, some recurring medical expenses, lost sales, and lost productivity. For example, a serious car accident might require lifetime back therapy for an employee. Those costs will probably not be charged to the accident medical categories in future years.

Table 30-3. Off-the-Job Accident Cost Study

CATEGORY	# OF CASES	LOST DAYS	WAGES PAID	MEDICAL EXPENSES	BENEFITS	CLAIMS ADM. COSTS	TOTAL
Employees	1,715	12,158	$972,640	$6,881,694	$291,792	$203,952	$8,350,078
Dependents	3,128			12,550,860		371,989	12,922,849
Total	4,843	12,158	972,640	19,432,554	291,792	575,941	21,272,927

Figure 30-1 presents an example of a cost study worksheet. Any claims administrator can provide the medical expenses by summarizing costs for the universal accident medical codes 800-999.

Examples of Successful Programs

Organizations with strong off-the-job safety programs cut their expenses significantly. Following are some examples:

Figure 30-1. An Example of a Cost-Study Worksheet

Summary of Off-the-Job Accident Costs

Medical expenses for employees and dependents . . . .$ _____

Wages paid to injured employees . . . . _____

Benefits paid to injured employees . . . . _____

Claims administration costs . . . . _____

Costs to train replacement workers . . . . _____

Reduced productivity . . . . _____

Lost sales . . . . _____

Other . . . . _____

TOTAL COSTS . . . .$ _____

Types of Injury and Poisoning [1]

800-829—Fractures
830-839—Dislocations
840-848—Sprains and strains of joints and adjacent muscles
850-854—Intracranial injury, excluding those with skull fracture
860-869—Internal injury of chest, abdomen, and pelvis
870-897—Open wound
900-904—Injury to blood vessels
905-909—Late effect of injuries, poisonings, toxic effects, and other external causes
910-919—Superficial injury
920-924—Contusion with intact skin surface
925-929—Crushing injury
930-939—Effect of foreign body entering through orifice
940-949—Burns
950-957—Injury to nerves and spinal cord
958-959—Certain traumatic complications and unspecified injuries
960-979—Poisoning by drugs or medicinal and biological substances
980-989—Toxic effects of substances chiefly nonmedicinal as to source
990-995—Other and unspecified effects of external causes
996-999—Complications of surgical and medical care not elsewhere classified

[1] Accidents are classified on the basis of the Ninth Revision of the International Classification of Diseases (ICD).

The off-the-job accident rate of Du Pont's employees is less than one-half the rate of the general public. Du Pont has an aggressive off-the-job safety program that includes employee meetings and safety articles in its company publications. Frequently during safety meetings, employees devote half the meeting time to off-the-job safety topics. Du Pont also tracks the off-the-job accident rates of its employees by plants. And safety is one of the factors that goes into determining the performance evaluation of managers.

Several years ago, Armco launched a program called "Safety for the Family." The results were striking. Reports showed that after four years, the number of Armco employees killed in off-the-job accidents was down from sixteen to only one. The total number of off-the-job injuries was down 47 percent in five years. Days lost from work because of off-the-job injuries dropped 36 percent.

A mining company in a small community used several safety program tools to prevent injuries to employees and their families over a vacation period. Not one employee was injured or involved in a serious accident during the vacation. The previous year twenty-three employees had serious accidents, with two fatalities.

Not My Responsibility

Historically no one claims responsibility for off-the-job safety. The focus of safety professionals is on-the-job safety, and wellness experts emphasize nutrition and exercise. In addition, many employers do not feel that it is their place to tell people how to live their private lives, or that they can affect how people lead their lives. However, this philosophy of avoidance is at odds with the concept behind wellness programs.

Why do employers think it is good business to influence a person's eating habits and exercise patterns, but then ignore trying to influence his or her safety habits? Wellness and off-the-job safety programs have the same goals, for example, in both programs the safety professional is trying to influence people in a way that helps the employer and the employees. In this regard, the emphasis on wellness programs and the lack of emphasis toward off-the-job safety is short-sighted. Eating a piece of chocolate cake once in a while will not harm most individuals, but virtually everyone who falls down a flight of stairs will require medical treatment.

Avoiding Accidents Is Merely Common Sense

Some employers feel that avoiding off-the-job accidents merely requires common sense. That is no more true than saying avoiding on-the-job accidents merely requires common sense. Simply the routine acts of driving a car, cooking, or cutting the grass require knowledge and a safety-conscious attitude. Just one mistake can cause a costly and tragic accident. For example, almost any employer with a vehicle fleet will have an extensive driver safety program. And yet this same organization will provide no defensive driving material to the rest of its workforce and the employees' dependents. It doesn't matter where an automobile accident occurs, the medical and other expenses will still impact the bottom line.

GETTING SUPPORT FOR OFF-THE-JOB SAFETY IN YOUR ORGANIZATION

Selling Off-the-Job Safety

The ideal off-the-job safety program would start with full top management support. Even if you are not able to obtain this support, the benefits obtained from a local or division program make the effort to implement a program worthwhile.

To gain the support of your boss or the company president, you will probably need to sell them on the value of an off-the-job safety program, using some of the material and arguments previously discussed. To sell off-the-job safety, you must be personally committed and passionate about the topic, and willing to mount a sales campaign.

The key in selling safety is to concentrate on the benefits to the person you are trying to sell. A simple way to remember this is to use the call letters of everyone's favorite radio station—WIFM. The letters stand for: What's In it For Me.

Some of the key steps required to gain the support of people in the organizational hierarchy are as follows:

1) Determine or estimate the human costs. Statistics on fatalities should be available from the human resources organization. Injury statistics will be more difficult to obtain, but averages are available from the National Safety Council. Applying those averages to your employee and dependent base will help illustrate the magnitude of the problem.

2) Determine off-the-job accident costs and the revenue needed to pay for the costs. Figure 30-1 is a model for the cost study.

3) Develop a budget for the program and compare it to the expected savings. Estimating savings is difficult, but a sound basic program

should be able to reduce medical costs for the 800–999 classification codes by at least 10–15 percent, and probably more over a period of time.

4) Sell other organizations on the benefits to them of an off-the-job program. Get the human resources personnel involved. They stand to gain considerably from the reduction in medical costs. Also, ask the sales organization to help you package your presentation in the most attractive format.

5) Sell the program up the line. The greater your sales success, the more impact you will have. It might be necessary to demonstrate success at the local level before you are able to convince others of the benefits of a more encompassing program.

6) Track the savings and costs. Publicize the savings to obtain continued funding for the program through the years. Off-the-job safety is a permanent commitment.

PLANNING YOUR PROGRAM

Teamwork often makes the difference between success and failure. An off-the-job safety program provides opportunities for forming and using several types of teams made up of managers and employees. These teams organize and conduct the overall program, develop special promotion activities, solve specific problems, and present recommendations to management. The types of teams used include steering committees, project teams, and safety circles. The teams provide for the involvement of individuals at all levels within the organization, as well as for meaningful study and analysis of off-the-job safety problems.

Assuming that you are able to obtain top management support, the subsequently presented guidelines should be followed. Modifications to the plan may be required depending on the scope of your program, but attention to the key steps outlined is necessary to create local or company-wide programs.

The Steering Committee

The steering committee provides management control of the program. It performs the following:

1) Identifies the work to be done in the program.

2) Authorizes the time, people, space, and money needed to implement the program.

3) Evaluates the program's progress toward goals and objectives.

4) Guides and coaches the efforts of the people who contribute to the program.

The steering committee meets on policy and decisions. It should be chaired by a senior level manager. It is usually made up of key functional managers, plus employee representatives. The committee's activities should be exclusively focused on the off-the-job safety program.

The steering committee's initial jobs are to identify loss exposures and develop effective project teams. It is important that team members be well versed in both safety fundamentals and the concepts of off-the-job safety programs. Therefore, training of some team members may be required. After any training, the committee directs compilation of the preliminary loss exposure inventory. It then sets the program's objectives and standards, and begins to recruit individuals for project teams.

After the program is organized, the committee concentrates on review and compliance. It hears recommendations presented by project teams and approves goals and budgets to meet program objectives. It also coordinates program activities with other functions of the safety and health program. Finally, the committee may audit managers' compliance with program standards. All these functions require high levels of authority to commit people and resources. Consequently, it is vital that senior managers not delegate committee responsibilities to subordinates.

Loss Exposure Identification

An effective safety and health program starts with a determination of loss exposures. To do otherwise is to manage by blind guessing. Table 30-4 is a summary of possible off-the-job safety topics. To get an idea of what employees consider important, the steering committee can give a copy of the list to employees and ask them to circle topics on which they would like more information.

Handout surveys normally have a very poor response, so the purpose of the survey must be explained when the forms are distributed. This can be done at a safety meeting or at a similar gathering of employees. A convenient return method can increase the rate of response. Collection boxes or return-addressed forms or envelopes can be provided if personal collection by the supervisor is not desirable. E-mail is another possibility.

Table 30-4. Possible Off-the-Job Safety Topics

Driving Topics

Backing up	Long distances and	Railroad crossings	Small cars
Bad weather	fatigue	Recreational	Speed
Car seats	Maintenance	vehicles	Teen drivers
City driving	Mature drivers	Rental cars	Trucks on the
Defensive driving	Nighttime driving	Road emergencies	highway
Diverted attention	Occupant	Rural roads	Wildlife collisions
Expressway	protection	Rush hour	Winter driving
Fall driving	Prescription drugs	Shopping for a	
		safer car	

Year-Round Topics

Alcohol	Choking	Hearing protection	Persons with
Appliances	Crafts	Hobbies	disabilities
Babysitting	Disaster planning	Home maintenance	Pets
Back injury	Electrical	Home safety	Poisoning
Bathroom	Eye protection	checklist	Prescription errors
Bowling	Falls	Home safety	Running
Child care	Firearms	devices	Shift work
Children	Fitness	Household	Shopping
Infant	Flammable liquids	chemicals	Sleeping
2-5 years	Furniture	Kitchen	Street crime
6-12 years	refinishing	Ladders	Toys
13-17 years	Garage	Older adults	Travel
Childproofing your	Hand tools	Parties (hosting)	Workshop
home	Head injuries	Pedestrians	

Spring and Summer Topics

ATVs	Heat stress	Pesticides and	Swimming
Barbecues	Helmets	insecticides	Diving
Bees	Hiking	Playgrounds	Pools
Bicycling	Inline skates and	Skin cancer	Vacations
Boating	skateboards	Sports (adults and	Water safety
Camping	Horseback riding	kids)	Water skiing
Canoeing	Insects	Spring cleaning	Wildlife
Fireworks	Kites	Storms	Yard and garden
Fishing	Lawn mowers	Summer activities	equipment
Food poisoning	Painting	Summer driving	
Gardening		trips	

Fall and Winter Topics

Burns	Fireplaces	Hunting	Snowboards
Carbon monoxide	Heating systems	Firearms	Snow clearing
Chain saws	Prevention	Other risks	Snowmobiles
Fall activities	Wood burning	Hypothermia	Soccer
Falls on ice and	stoves	Latch-key children	Winter activities
snow	Football	School	
Fire	Halloween	Skiing	
Detection and	Holidays	Sledding/skating	
response			

Selecting Off-the-Job Safety Topics

Since motor-vehicle deaths account for almost 50 percent of off-the-job fatalities, and numerous injuries, any safety program must begin with regular emphasis on various driving topics. However, the topics selected will vary by location. For example, an urban area will probably wish to cover highway and rush hour driving, whereas a rural area may emphasize the special hazards faced by drivers on rural roads and at railroad crossings.

Beyond driving topics, the interests of the employees will dictate the topics selected. Some employee groups may want information on outdoor activities, whereas others may want more information on children's activities.

A key point to consider is that virtually all the topics listed in Table 30-4 should be covered over a 4 to 5-year period, or at least made available to employees. For example, few employees will have trampolines in their backyards, so that topic may not be identified frequently on the employee response form. However, at some point in their lives, many children may be exposed to a trampoline in someone else's yard. In one case, a young man visiting a friend on a spring night jumped off a trampoline and landed awkwardly. He is now a quadriplegic, with medical expenses exceeding $1 million. If the family had received some periodic safety information on the dangers of trampolines, the tragedy might have been avoided.

Guidelines for Effective Programs

Once the general exposures are known, planning begins on the best way to control risks. Some risks are adequately controlled with knowledge of safe practices, so the program calls for education. The control of other risks requires skill or training, and the program calls for ways to get supervised practice, such as through a recreation club or sports team. Sometimes the hazard can be controlled with personal protective equipment (PPE), and the program requires sources of equipment, proper fitting, and instruction on equipment use.

No two organizations are alike. However, there are some universal guidelines for effective off-the-job safety programs conducted by organizations. These guidelines are as follows:

1) Overall direction should be given by an experienced professional.

2) A well-organized program encourages individual participation and

produces the best results. The major activities need to be planned and budgeted so that the necessary personnel and resources will be available.

3) Employee participation in activities should be spontaneous and voluntary. Each program activity should include an invitation for employees to participate in planning, preparing, presenting, and following through. In the beginning, selected employees may have to be asked to help, but a quality program motivates continuing participation.

4) The program should be family oriented. The family is the key group for sharing ideas and changing behavior. Spouses influence each other's actions. Parents influence children and vice versa. Studies show consistently better results when the message gets into the home and when activities include families.

IMPLEMENTING YOUR PROGRAM

Project Teams

One effective way of developing comprehensive safety topic promotions is the use of project teams. Superficial programs have little, if any, effect. Comprehensive promotion programs require a good deal of work—too much for one or two persons. The team approach is useful and gets managers and employees involved in the program. In most cases, the core of the team should consist of employees drawn from various departments. Each of these persons can add specific knowledge of both problems and methods. When combined, their talents are most effective.

A team should be organized for each topic to be covered. Many of the exposures are seasonal, and the remainder can be scheduled according to the degree of risk. Six topics a year is reasonable for most organizations. Team memberships should also be rotated to give all an opportunity to participate, to bring fresh ideas to the program, and to allow rest periods between projects for the team leaders. However, a person may be a team member on one project, however, and a team leader on another project four or six months later.

Other Topics. As noted previously, many other topics deserve presentation to employees. These topics do not require individual project teams and campaigns. However, a separate project team should be assigned to research these topics and develop material for the employees. Information is available from newsletters and services that focus on off-the-job safety issues. Some of these publications allow you to copy the material. A list of resources is supplied at the end of this chapter.

Conducting a Campaign

The agenda for a typical theme campaign includes the following:

- Appointing team members
- Training the team
- Preliminarily discussing the topic
- Surveying loss exposures by reviewing the inventory and talking with employees
- Outlining theme campaign activities
- Assigning responsibilities
- Projecting research
- Developing alternatives
- Briefing management
- Preparing final materials
- Conducting the campaign
- Critiquing program results

The following is a summary of possible approaches to off-the-job topics:

- Contests—essays and drawings (use outside judges)
- Defensive driving courses normally offered to employees driving company vehicles
- Displays
- Incentive awards for a group
- On-line database of topics that employees can access
- Outside speakers (for example, talks during lunch hour)
- Perfect attendance awards
- Personal protective equipment (suggest standards and where it can be obtained, or consider offering equipment at cost)
- Posters
- Publications—external
- Publications—internal (include employee stories in which protective gear or following safe practices prevented or mitigated injury)

- Recreational programs (use qualified outside instructors)
- Reference booklets
- Safety calendars
- Safety fairs, family nights, or picnics (use community groups and local merchants)
- Safety meetings
- Vacation/holiday programs
- Video library (for example, defensive driving tapes employees can show to the family)
- Videotape, film or slide shows

Steps in a typical promotional campaign might include the following:

1) Conduct a review to verify the loss exposures and potential risks in order to confirm that these are properly assessed in the original inventory.

2) Write motivational messages for the senior manager and department heads to announce the theme and stimulate employees' interest.

3) Develop or obtain educational material to improve people's knowledge regarding the theme and their personal exposures.

4) Develop guides and visual aids for supervisors to use in safety talks with their employee groups.

5) Develop lesson plans for special educational programs to teach people how to identify related hazards or to teach standard safety practices.

6) Make up displays, posters, and signs to reinforce the information presented in the planned educational activities.

7) Develop guidelines for employees to use in their personal safety programs. These can include safety features to look for when buying tools or materials, self-inspections for compliance with safety standards, and safe practices to learn or to teach to family members.

8) Make up employee and family contests that can be used to stimulate education regarding, and performance of, safety activities. For example, some organizations call employee homes after information is disseminated. If a family member can answer some relatively easy questions that indicate he or she read the material, he or she receives or is eligible for a prize.

9) Outline key points for briefings to senior management on the content and conduct of the theme campaign.

Three essentials of the successful project theme campaign are a definite beginning, adequate preparation time, and a definite end. The campaign should have a kickoff date to announce it to employees through various media, such as newsletters and posters. The scope of the campaign and the activities planned should be introduced and preliminary materials distributed. Depending on how much time team members can devote to the campaign during their work, the preparation must start two to three months ahead of the kickoff date.

Finally, the campaign needs to build to a climax and have a distinctive closing date. It should not be allowed to die of old age and lack of interest. Employees will continue the preventive efforts in their personal safety programs, and supervisors will continue to review and reinforce the key points, but the campaign as such should end, and all the special notices, posters, and displays should come down. If appropriate, the theme can be repeated at a later time, with a slightly different approach and renewed emphasis.

Safety Circles

Some continuing off-the-job safety problems can be addressed by safety circles. The circle is a small group of individuals with a common interest or problem. This interest may be a specific sport or hobby, children's safety, a community activity, or a technique such as first aid. The circle does not follow the normal structure of the organization. People belong because of their common interest. Although there is a leader to move the discussions along, there is no authority figure—hence the term "circle." All members are equal. Circles may be organized within departments, or across departmental lines.

In addition, safety circles can periodically become the nucleus of a project team when a decision has been made to emphasize the circle's activity as part of the overall program. A circle facilitator is designated to lead the program. This person is the liaison between the circles and the steering committee.

Organizing safety circles must not be a hurried process. Attempts to put a large number of circles into operation at the same time will probably guarantee failure. One or two circles should be started, and their success used as the catalyst for others. Fewer circles will let the facilitator divide his or her time in order to give each circle meaningful help. A suggested course of action should include:

1) Selecting critical loss exposures. The steering committee selects a few exposures that have the highest potential for off-the-job accidents.

2) Selecting circle leaders who are best qualified or most experienced in the activity associated with the critical exposures.

3) Training circle leaders in basic problem-solving techniques, group dynamics, personal and group communications techniques, accident causes and controls, and safety circle procedure.

4) Recruiting circle members by identifying employees who are interested in the topic, are good team workers, and have a good reputation for job safety.

5) Training team members. This is a learn-by-doing approach. After an initial orientation in the circle procedure, the leader prompts the selection of a problem to work on. The first problems should be simple ones so that the circle can concentrate more on learning the method.

6) Conducting the circle activity. The leader asks members to volunteer for researching various topics, for bringing in resource people for questioning, and for analyzing the data gathered.

7) Keeping progress data. Periodically, the leader reports on each circle's activities to the steering committee and presents the problems identified and the extent of research conducted.

Personal Safety Programs

Each employee's personal life is different. To make an off-the-job safety program truly effective, it must be tailored to the individual. Family size, ages of family members, type of home and furnishings, domestic activities, recreational interests, and personal travel all differ.

The approach needs to be one of getting employees interested in starting personal safety programs and then providing the tools they need. These tools include education in safety and health fundamentals, information on how to organize personal safety programs, assistance with the more difficult problems, and resources for ideas and materials.

In addition to providing periodic safety information through company and outside publications, a library of information on off-the-job safety topics should be available for employees and families. One way of providing this information is through an on-line facility.

To facilitate development of personal safety programs, you may wish to provide binders with a table of contents for broad categories, and perhaps some initial articles on general safety topics. To add further emphasis, the binder could be presented in a special safety meeting to review basic safety concepts and off-the-job hazards.

Encourage employees to file additional information on the hazards faced by their families, and review it periodically. For example, a family that boats should review their boating safety file before each season.

Key Elements. The parts of a personal safety program are called "elements." The elements are as follows:

1) *Leadership:* Within the family there must be a safety leader. One person must inspire and challenge the others to act safely. The leader sets a safety policy and encourages the setting of standards for safe conditions and practices. The leader prompts the others and ensures that they are adequately equipped and educated. The leader also sets the example.

2) *Training and education:* Lack of knowledge or skill is a basic cause of many accidents. Training activities ensure that everyone acquires the knowledge needed for safety in all activities. Sometimes this is achieved by formally teaching safe practices; at other times, it is achieved through self-study.

3) *Engineering controls:* There are safety standards for homes, vehicles, and public areas. These standards are the result of accident investigations. Someone must research the standards and verify that they are met.

4) *Purchasing controls:* Many tools and materials used in homes, hobbies, or other forms of recreation have hazardous properties. Become familiar with them, tell family members about the hazards that exist, and buy the items that are least hazardous.

5) *Personal protective equipment:* Some hazards can be controlled through proper PPE. This equipment can range from clothing to protect against poisonous plants, to face shields or goggles to protect against chemicals or flying objects. The person involved must study the specific hazard and obtain suitable protection.

6) *Emergency preparedness:* Natural disasters and technological accidents can affect personal safety; the effects vary from one situation to the next. As a family, consider potential disasters, make emergency plans, and hold emergency drills.

7) *Care of the injured:* First aid can prevent complications from injuries. Suitable first aid kits need to be obtained and individuals trained in first-aid techniques.

8) *Inspections:* Make periodic examinations of facilities, equipment, materials, and practices to ensure that they continue to meet safety standards.

9) *Family meetings:* People need to be reminded about key aspects of safety and the prevention of accidents. Hold family meetings during the year to discuss safety in various off-the-job activities. To ensure their involvement, let the children assume some of the leadership.

CONCLUSION

Employees and families must have a 24-hour safety attitude and know safe practices before they can be expected to take actions that will avoid accidents. Also, they must be periodically reminded of the wide range of safety hazards they face in everyday life. Without these reminders, we all tend to become complacent.

Safety awareness can achieved only through a comprehensive off-the-job safety program. Of course, no organization can provide enough information to an employee to guarantee a risk-free life. Even the most experienced and well-educated person will not anticipate every hazard, but he or she will discover the most critical safety problems.

Once employees have a true understanding of safety fundamentals, they can start to develop their personal safety programs. As a result, they often find that they need and want more information.

You are probably skeptical as to how much of an impact you can have on the behavior of employees and their dependents away from work. We need only look at heart disease, cancer, and AIDS to see how behavior can be changed through comprehensive education programs. We cannot change everyone's behavior, but we can change enough people to have a significant impact on the death and overall accident rates.

What then can you do? First, determine off-the-job accident costs. For-profit companies should divide that amount by the pre-tax margin to obtain the amount of revenue required to pay for accidents. This information, and knowledge of the benefits derived from an off-the-job safety program, will help obtain the management support that is vital to the success of any safety program. Second, develop an information program for and with employees and families. Put a structure in place that uses the talents of all groups in the organization. Third, regularly furnish employees with information on a wide range of topics, and encourage employees to review the material with their families.

It is time to turn the spotlight on this long neglected part of our health care crisis. Off-the-job accident prevention is truly a win-win program, an opportunity for employers to do the right thing—and make money doing it.

ADDITIONAL RESOURCES

AAA Foundation For Traffic Safety
1440 New York Ave., N.W.
Washington, DC 20005
 (202-638-5944)
 Videos, pamphlets

American Academy of Pediatrics
AAP Publications
PO Box 927
Elk Grove Village, IL 60009
 (800-433-9016)
 TIPP-The Injury Prevention Program

Canadian Safety Council
1020 Thomas Spratt Place
Ottawa, Ontario K1G 5L5
 (613-739-1535)
 Living Safety magazine (quarterly)

National Safe Kids Campaign
Children's National Medical Center
111 Michigan Avenue, NW
Washington, DC 20010
 (612-295-4135)
 Resource catalog and list

National Safety Council
1121 Spring Lake Drive
Itasca, IL 60143
 (630-285-1121)
 Family Safety & Health magazine
 (quarterly)
 The Off-The-Job Safety Program Manual
 (programs, topics, resources)
 Accident Facts booklet (statistics,
 resources)
 Safety, Health and Environmental
 Resources catalog (booklets, videos)

Parlay International
P.O. Box 8817
Emeryville, CA 94662
 (800-457-2752)
 Reproducible pages

Safety Times
1265 Rogue River Ct.
Chesterfield, MO 63017
 (800-952-1363)
 Newsletter (bimonthly) and Reproducible
 Articles

REFERENCES

American Heritage Dictionary, Second College Edition (1982, 1985).

National Safety Council (1996). *Accident Facts*. Chicago, IL: National Safety Council.

National Safety Council (1984). *The Off-The-Job Safety Program Manual*. Chicago, IL: National Safety Council.

REFERENCES

Index

SARA III. *See* Superfund
  Amendments and Reauthorization
  Act
Second injury funds, workers'
  compensation insurance, 70
Security alarm sensors, 727-729, *728*
Security management systems, 729-
  731
  manual, 729-731, 730t
Security professional, responsibilities
  of, 25-26
Security risk assessment, 725-726
Self-insurance, 79
Sensors, security alarm, 727-729, *728*
Sensory impairments, persons with,
  114-119
Shareholders, responsibility of
  corporation to, 139
Significant new use rules,
  environmental hazards, 777
Single parents, in workplace, 111-112
Site safety process, environmental
  hazards, 763-764
Smell, loss of, persons with, 116-117
Social stress, occupational stress and,
  469-470
Special audits, 393
Speech impairment, persons with, 117
Sprinkler systems, for fire protection,
  501
Staffing, 285-294
  competencies, assessment of, 293-
    294
  development level, 291-292
  knowledge, assessment of, 291-
    292
  professional competencies, 289-
    291
Stages in resource protection
  technology, *299*
  organizational structure, 302-304
    active, 303
    adaptive, 303
    dynamic, 303-304, *305*

reactive, 303
strategy formulation, 299-302
Standpipe, hose systems, for fire
  protection, 502
Statement of mission, in resource
  protection technology, *307*
Statistical tools, in systems safety
  engineering, 623-625
Steering committee, off-the-job safety
  program, 791-792
Storage/retrieval systems, automated,
  for materials handling, 663-664
Strategic planning, overview, 8-9, *9,* 11
Stress, occupational, 461-473
  body response, 462-463
  contributing factors, 467-470
    domestic stresses, 469-470
    personal health habits, 467-469
    social stresses, 469-470
  ergonomics and, 441-475
  predisposing factors, 463
  preventive measures, 470-473
    early detection, 471-472
    motivation, 470-471
  women in workplace, 107-108
  work environment, 463-467
    environmental stress, 465-466
    job demands, 463-465
    morale, 466-467
Structural fit concept, in resource
  protection technology, *305*
Subrogation, workers' compensation,
  72-73
Substance abuse programs, reduction
  of workers' compensation costs
  through, 87
Substance hazards, design safety and,
  275
Substitution, in industrial hygiene
  management, 424-425
Superfund Amendments and
  Reauthorization Act, emergency
  planning, community right-to-
  know, 769-773

Government Institutes Mini-Catalog

ENVIRONMENTAL TITLES

PC #		Title	Pub Date	Price
629		ABCs of Environmental Regulation: Understanding the Fed Regs	1998	$49
627		ABCs of Environmental Science	1998	$39
585		Book of Lists for Regulated Hazardous Substances, 8th Edition	1997	$79
579		Brownfields Redevelopment	1998	$79
4088	◉	CFR Chemical Lists on CD ROM, 1997 Edition	1997	$125
4089	💾	Chemical Data for Workplace Sampling & Analysis, Single User Disk	1997	$125
512		Clean Water Handbook, 2nd Edition	1996	$89
581		EH&S Auditing Made Easy	1997	$79
587		E H & S CFR Training Requirements, 3rd Edition	1997	$89
4082	◉	EMMI-Envl Monitoring Methods Index for Windows-Network	1997	$537
4082	◉	EMMI-Envl Monitoring Methods Index for Windows-Single User	1997	$179
525		Environmental Audits, 7th Edition	1996	$79
548		Environmental Engineering and Science: An Introduction	1997	$79
643		Environmental Guide to the Internet, 4rd Edition	1998	$59
560		Environmental Law Handbook, 14th Edition	1997	$79
353		Environmental Regulatory Glossary, 6th Edition	1993	$79
625		Environmental Statutes, 1998 Edition	1998	$69
4098	◉	Environmental Statutes Book/CD-ROM, 1998 Edition	1997	$208
4994	💾	Environmental Statutes on Disk for Windows-Network	1997	$405
4994	💾	Environmental Statutes on Disk for Windows-Single User	1997	$139
570		Environmentalism at the Crossroads	1995	$39
536		ESAs Made Easy	1996	$59
515		Industrial Environmental Management: A Practical Approach	1996	$79
510		ISO 14000: Understanding Environmental Standards	1996	$69
551		ISO 14001: An Executive Repoert	1996	$55
588		International Environmental Auditing	1998	$149
518		Lead Regulation Handbook	1996	$79
478		Principles of EH&S Management	1995	$69
554		Property Rights: Understanding Government Takings	1997	$79
582		Recycling & Waste Mgmt Guide to the Internet	1997	$49
603		Superfund Manual, 6th Edition	1997	$115
566		TSCA Handbook, 3rd Edition	1997	$95
534		Wetland Mitigation: Mitigation Banking and Other Strategies	1997	$75

SAFETY and HEALTH TITLES

PC #	Title	Pub Date	Price
547	Construction Safety Handbook	1996	$79
553	Cumulative Trauma Disorders	1997	$59
559	Forklift Safety	1997	$65
539	Fundamentals of Occupational Safety & Health	1996	$49
612	HAZWOPER Incident Command	1998	$59
535	Making Sense of OSHA Compliance	1997	$59
589	Managing Fatigue in Transportation, *ATA Conference*	1997	$75
558	PPE Made Easy	1998	$79
598	Project Mgmt for E H & S Professionals	1997	$59
552	Safety & Health in Agriculture, Forestry and Fisheries	1997	$125
613	Safety & Health on the Internet, 2nd Edition	1998	$49
597	Safety Is A People Business	1997	$49
463	Safety Made Easy	1995	$49
590	Your Company Safety and Health Manual	1997	$79

Government Institutes

4 Research Place, Suite 200 • Rockville, MD 20850-3226
Tel. (301) 921-2323 • FAX (301) 921-0264
Email: giinfo@govinst.com • Internet: http://www.govinst.com

Please call our customer service department at (301) 921-2323 for a free publications catalog.

CFRs now available online.
Call (301) 921-2355 for info.

GOVERNMENT INSTITUTES ORDER FORM

4 Research Place, Suite 200 • Rockville, MD 20850-3226
Tel (301) 921-2323 • Fax (301) 921-0264
Internet: http://www.govinst.com • E-mail: giinfo@govinst.com

3 EASY WAYS TO ORDER

1. Phone: (301) 921-2323
Have your credit card ready when you call.

2. Fax: (301) 921-0264
Fax this completed order form with your company purchase order or credit card information.

3. Mail: **Government Institutes**
4 Research Place, Suite 200
Rockville, MD 20850-3226   USA
Mail this completed order form with a check, company purchase order, or credit card information.

PAYMENT OPTIONS

❑ **Check** (payable to Government Institutes in US dollars)

❑ **Purchase Order** (This order form must be attached to your company P.O. _Note_: All International orders must be prepaid.)

❑ **Credit Card**   ❑ VISA   ❑ MasterCard   ❑ American Express

Exp.___/____

Credit Card No. _____

Signature _____

(Government Institutes' Federal I.D.# is 52-0994196)

CUSTOMER INFORMATION

Ship To: (Please attach your purchase order)

Name: _____

GI Account # (7 digits on mailing label): _____

Company/Institution: _____

Address: _____
(Please supply street address for UPS shipping)
_____

City: _____ State/Province: _____

Zip/Postal Code: _____ Country: _____

Tel: (____) _____

Fax: (____) _____

Email Address: _____

Bill To: (if different from ship-to address)

Name: _____

Title/Position: _____

Company/Institution: _____

Address: _____
(Please supply street address for UPS shipping)
_____

City: _____ State/Province: _____

Zip/Postal Code: _____ Country: _____

Tel: (____) _____

Fax: (____) _____

Email Address: _____

Qty.	Product Code	Title	Price

❑ **New Edition No Obligation Standing Order Program**
Please enroll me in this program for the products I have ordered. Government Institutes will notify me of new editions by sending me an invoice. I understand that there is no obligation to purchase the product. This invoice is simply my reminder that a new edition has been released.

Subtotal _____
MD Residents add 5% Sales Tax _____
Shipping and Handling (see box below) _____
Total Payment Enclosed _____

Within U.S:	Outside U.S:
1-4 products: $6/product	Add $15 for each item (Airmail)
5 or more: $3/product	Add $10 for each item (Surface)

15 DAY MONEY-BACK GUARANTEE

If you're not completely satisfied with any product, return it undamaged within 15 days for a full and immediate refund on the price of the product.

SOURCE CODE: BP01